CHICAGO PUBLIC LIBRARY

W9-ABC-893

617
.P552
1989

Phillips, Roger,
1932-

Mushrooms of North
America.

$38.95

DATE			

© THE BAKER & TAYLOR CO.

Mushrooms of
North America

for Amy Foy-Phillips

Mushrooms of North America

by
Roger Phillips

assisted by Geoffrey Kibby & Nicky Foy
with editing help from Alick Henrici,
Richard L. Homola, Currie D. Marr, Rodham E. Tulloss
Layout Jill Bryan

Little, Brown and Company
Boston • Toronto • London

Acknowledgments

Literally hundreds of people have helped in this project on all levels: checking identifications, lending or collecting specimens, guiding me to their favorite spots, giving us bed and board, entertaining the children — entertaining me, come to that. I do hope all those that I have not been able to include will forgive me.

Joseph F. Amaratti, Ron and Cathy Annala, David Arora, Alan Bessette, Howard E. Bigelow, Gerald Bills, Bill Bokaitis, Pat Brannen, Karen and Bill Brusher, Bill Cibula, Marshall E. Deutsch, Liz Farwell, Ray and Bernice Fatto, Robert L. Gilbertson, Martina Gilliam-Davies, Martha Hacker, John Haines, Roy E. Halling, Herbert H. Harper, Kenneth A. Harrison, Alick Henrici, Oswald Hilber, Ursula Hoffmann, Alma Homola, David Hosford, Bob and Genia Hosh, Jacqui Hurst, David T. Jenkins, Kasha Jenkinson, Geraldine C. Kaye, Rick W. Kerrigan, Sue Kibby, Paul Kroeger, Jenny and Sandy Kronick, Pat and Jim Kronick, David L. Largent, Gary H. Lincoff, Mr. and Mrs. Harry Lubrecht, Dee Lusk, Buck McAdoo, Currie Marr, Andrew S. Methven, Susan and Van Metzler, Linda Meyer, Paul Meyer, Gerry and Chris Miller, Orson K. Miller, Jr., Meinhard Moser, Greg Muller, Lorelei Norvell, Clark Overbo, Bob and Barbara Peabody, David Pegler, Ronald H. Petersen, Scott A. Redhead, Derek Reid, Fred Rhodes, Jim Richards, Sam Ristich, Maggi Rogers, Clark T. Rogerson, Rhoda Roper, Mike and Sue Rose, Rama L. Sharber, Rolf Singer, Dorothy Smullen, Brian Spooner, Philippa Staniland, Geoff Stone, Walt Sturgeon, Bob and Alice Targan, Harry D. Thiers, Jim Trappe, J. C. and Sarah Elizabeth Trial, Ron and Jane Trial, Rodham Tulloss, Greta Turchick, Bruce and Janice Vansant, Nancy Smith Weber, Carl B. Wolfe, Greg Wright, Gene Yetter, Gail and Sophie Zawaki, Dan, Steve.

I am grateful to Geoffrey Kibby for the photographs of *Leucocoprinus birnbaumii* (p. 35), *Russula mutabilis* (p. 108), *Russula polyphylla* (p. 111), *Russula elegans* (p. 115), *Russula perlactea* (p. 123), *Russula cessans* (p. 131), *Cortinarius elegantoides* (p. 143), *Hebeloma sacchariolens* (p. 179), *Hebeloma edurum* (p. 180), *Agaricus silvaticus* (p. 190), *Craterellus cinereus* var. *multiplex* (p. 216), *Boletus projectellus* (p. 218), *Boletus pseudosensibilis* (p. 222), *Tylopilus ballouii* (p. 242), *Sarcodon joeides* (p. 277); to Greg Wright for the photographs of *Amanita calyptroderma* (p. 19), *Amanita magniverrucata* (p. 23), *Cortinarius cedretorum* (p. 147), *Psilocybe cubensis* (p. 200), *Boletus barrowsii* (p. 233), *Boletus satanas* (p. 237); and to Gene Yetter for the photograph of *Boletus stramineum* (p. 223).

I would also like to especially thank the following organizations: Boston Mycological Club, Long Island Mycological Club, New Jersey Mycological Association, North American Mycological Association, Northeast Mycological Foray Organization, Oregon Mycological Society, Spokane Mushroom Club, Texas Mycological Society, Triangle Area Mushroom Club, West Michigan Mycological Society, Wisconsin Mycological Society

Copyright © 1991 by Roger Phillips

All rights reserved. No part of this book may be reproduced in any form or by any electronic or mechanical means, including information storage and retrieval systems, without permission in writing from the publisher, except by a reviewer who may quote brief passages in a review.

First Edition

Library of Congress Cataloging-in-Publication Data

Phillips, Roger, 1932–
   Mushrooms of North America / by Roger Phillips; assisted by Geoffrey Kibby & Nicky Foy.
      p. cm.
   Includes bibliographical references.
   1. Mushrooms — North America — Identification. 2. Mushrooms — North America — Pictorial works. I. Kibby, Geoffrey. II. Foy, Nicky. III. Title.
QK617.P552 1991
589.2′22′097 — dc20   89-37050
                                                 CIP

HC: 10 9 8 7 6 5 4 3 2 1

PB: 10 9 8 7 6 5 4 3 2 1

*Published simultaneously in Canada
by Little, Brown & Company (Canada) Limited*
         Printed in Hong Kong

Many mushrooms are poisonous, some deadly poisonous. The author and the publisher have made every effort to ensure accuracy in this book, but in the end the responsibility for eating any mushroom must rest with the individual; for instance, there are people who are allergic to *all* species of mushrooms. If you do collect mushrooms to eat make sure that your identification checks out in every detail. Beginners should never eat wild mushrooms until they have had their own identifications checked by an expert in the field. Neither the publisher nor the author accept responsibility for any effects that may arise from eating any wild mushrooms.

Contents

Introduction

My book *Mushrooms and Other Fungi of Great Britain and Europe* took five years to complete, during which time more than 25,000 specimens passed through my hands. Two years later, I decided to tackle North American mushrooms. As I thought I knew it all, I assured the publishers that I could produce a book containing 1,000 illustrations in five years. On the face of it, it should have been a piece of cake. But I forgot one thing — the sheer size of America. On my map it is only ten inches from Maine to Arizona, no farther than from London to Scotland, which I drive in a day and still have a few hours to collect; but looking for mushrooms in America is like searching the whole of Europe and then throwing in the Near East for good measure.

In August 1983 I duly set off with my family, my wife, Nicky, and our daughter, Phoebe, then only seven months old. Phoebe quickly learned to crawl in the woods along a county road near Bangor, Maine, where we were taken by Dick Homola on our first American outing at the Eighth Annual Northeastern Mycological Foray. It was a culture shock. Things are very different across the water; in Britain our annual mycological society foray is much more of an academic affair, Latin rather than English being the official language. However, we all changed gear and quickly (I hope) adapted to the lively and entertaining style of a weekend on an American campus.

The study of mushrooms in North America is both ahead and behind in comparison with mycology in Europe. Many excellent monographs of genera have been produced, almost all of them with Alexander H. Smith as co-author, a truly herculean volume of work demonstrating the enormous energy he must have put into his lifetime's study. More modern American monographs are available than monographs of European genera, but in Europe, on the other hand, there are three exceedingly good books dealing with the larger subject of the Agaric Flora in its entirety: *Flore Analytique des Champignons Supérieurs* by R. Kühner and H. Romagnesi, *Keys to Agarics and Boleti* by Meinhard Moser, and *The New Check List of British Agarics and Boleti* by R. W. G. Dennis, P. D. Orton, and F. B. Hora. This difference reflects the different state of knowledge on the two continents.

In Europe if something is found that cannot be named it creates a great deal of excitement, and the collection will rapidly be passed on to a mycologist who is an authority on that group so that it can be considered for classification and publication. In America if you find a mushroom that fails to fit any of the known species, its advent will be noted as: "Oh no, not another new species!" Alexander Smith has said that at least one-third of North American species are as yet undescribed. After traveling all over North America, from swamps to deserts to the high Rockies, and seeing the diversity of habitat and climate that have to be dealt with, I would be amazed if only that proportion remained to be described. This makes mycology in North America a most exciting subject; there is so much important and original work to be done in pushing forward the boundaries of science. To take just one group, the underground agarics that fail to rise and expand above the soil: in Europe such things are almost unknown, yet in America there are dozens of species near the West Coast, a most fascinating area of study.

How I tackled the book

This book is a collection of mushrooms that I have found or been lent by other collectors. Although it may seem an obvious thing to say, this approach makes the book very different from most general works, which are written to a list of mushrooms based on their recorded frequency. What happens when you work in my way is that the book represents the actual incidence of mushrooms in the wild. This can be seen by looking at the very large number of cortinarius and russula illustrated; both genera are exceedingly common across North America, with seven hundred and three hundred species respectively, but in practice they are often underrecorded on forays because of the difficulty of naming them. Though I believe my approach to be more realistically accurate, it can have its drawbacks: when I arrived in England after a three-month collecting trip with a tremendous preponderance of collections of these two genera, I then had to spend the next nine months struggling to come up with names. And this was the case with all the common genera. But, and this a big but, I have had the most tremendous amount of help from American mycologists, and I want to mention in particular Dr. Currie Marr, who painstakingly went through all my ramarias, and Rod Tulloss, who did the same for my amanita collections.

What is a mushroom?

A mushroom is only the reproductive part (known as the fruit body) of the fungus organism, which develops to form and distribute the spores.

Fungi are a very large class of organisms and have a structure that can be compared to plants; but they lack chlorophyll and are thus unable to build the carbon compounds essential to life. Instead, in the same way that animals do, they draw their sustenance ready-made from living or dead plants, or even animals.

A fungus begins as minute, hairlike filaments called hyphae. The hyphae develop into a fine, cobweblike net that spreads through the material from which the fungus obtains its nutrition. This net is known as the mycelium. Mycelium is extremely fine and in most cases cannot be seen without the aid of a microscope. In other cases, the hyphae bind together to make a thicker mat (tomentum), which can readily be observed. To produce a fruiting body, two mycelia of the same species band together in the equivalent of a sexual stage. Then if the conditions of nutrition, humidity, temperature, and light are met, a fruit body will be formed.

The larger fungi are divided into two distinct groups:

    (i) The spore droppers, Basidiomycetes (pp. 14–300). In this group the spores are developed on the outside of a series of specialized, club-shaped cells (basidia), which form on the gills, spines, tubes, or other spore-bearing surfaces. As they mature, they fall from the basidia and are normally distributed by wind. Most of the fungi in this book are of this kind, including the gilled agarics, the boletes, the polypores, and the jelly fungi.

    (ii) The spore shooters, Ascomycetes, or "Ascos" (pp. 301–313). The spores in this group are formed within flask-shaped sacs (asci). When the spores have matured, they are shot out through the tips of the asci. The morels, cup fungi, and truffles are in this group.

My daughter Phoebe helping me with a great *boletus* collection

described the species (the species' author). For instance, "Fr." after a mushroom name means that Fries, a Swedish mycologist, first described it. An author's name in parentheses followed by another name means that the second author has redescribed the mushroom, usually placing it in a different or new genus.

Collecting

When making a collection it will facilitate identification if you note the salient characteristics of a species in the same order that they are mentioned in the text. For example:

Cap. Note the size, the shape, the colors, the texture. Is it smooth or sticky, fibrous or scaly?

Gills. Note the color, the shape, the attachment to the stem.

Stem. Note the height and width, the colors, the texture, the presence of a ring, volva, root, or basal bulb. (Make sure you get the whole stem out of the ground.)

Flesh. Note the color, the texture. Is it fibrous or crumbly? Does it exude milk? Check the smell and taste. (To ascertain taste, nibble a little bit and break it up on the tip of your tongue without swallowing it, spit out the remains and completely clear your mouth; if done carefully, even the most poisonous species can be tasted in this way, though in practice you will soon learn to recognize and so avoid tasting the main poisonous genera.)

Spores. Note the spore color. This can sometimes be observed beneath the cap on the grass or leaves. If this is not the case, make a spore print (see below).

Habitat. Note if it grows on wood, soil, or manure. Does it grow in grassland or woodland? Under or near what species of tree or plant? Is the soil chalky or acid? The range of the species indicates where specimens have been found, though they may exist elsewhere; in this book, "North America" refers to the area north of Mexico.

How to use this book

Mushrooms and fungi are a large and very diverse group of organisms, and although this book only deals with the larger forms, there is nevertheless a bewildering selection to choose from. When you have leafed through the pages and got a general feel for the diversity of species that are illustrated, the next step in making an identification is to refer to the pictures for beginners (pp. 10–11) and then the keys (p. 12).

Positively identifying a collection of mushrooms is a very tricky business, even for an expert, and if you are going to eat some of the specimens you collect you must be *absolutely positive* about your identification. This book will help in that you can check the illustrations against your ideas, but no book will ever be able to give you the experience you need to be certain. If you want to learn about mushrooms, the only sensible way is to go out collecting with experts and listen to what they have to tell you and question them about everything that you can think of; then use a book for comparison.

The name or abbreviation after the Latin name of each mushroom refers to the mycologist who originally

Collecting for the pot

Worldwide the eating of wild mushrooms is gaining in popularity, with fashionable restaurants now boasting of their wild mushroom specialties. This rise in popularity of mushrooms for the table brings with it an increase in the probability of poisoning cases. The first rule is never eat any mushroom that you are not certain is edible. The second rule is keep one mushroom from any collection you eat in the refrigerator uncooked so that if you should develop any nasty symptoms you can give the evidence to a hospital.

The photographs

The majority of the pictures were taken on a Nikon 35mm camera with a Macro lens and extension tubes as needed. The light source was a Japanese flash pack powered by 2,000 joules of strobe light. The aperture was f.16. The film stock was Kodak Ektachrome 64ASA.

Poisonous species

This is a short list of the main deadly poisonous species and genera. Anyone considering eating wild mushrooms should look up each reference and learn to recognize those mentioned.

The most dangerous species are members of the genus *Amanita*, the most common of which are *Amanita virosa* (p. 29) and its twin, *Amanita bisporigera* (p. 28), along with the European Death Cap, *Amanita phalloides* (p. 28), which seems to be on the increase in North America. *Amanita virosa* is the main cause of fatalities, but many other members of the genus are poisonous to some degree. The two main toxins involved are amanitin and phalloidin. Therefore I believe no species of the genus Amanita should be eaten.

Many members of the genus *Inocybe* (p. 181) are poisonous, some deadly, and the main poisonous constituent is muscarine. Eating any member of this genus is to be avoided.

Modern research has shown that many more species of the genus *Cortinarius* contain poison than was previously thought. All should be avoided, including *Cortinarius armillatus* (p. 159), which is generally considered edible!

Gyromitra esculenta (p. 302), which is commonly eaten in some areas of North America and Europe, is quite definitely a deadly poisonous species. It is said that drying or boiling and discarding the water dissipates the gyromitrin poisons. However, many deaths have occurred in Europe from this species, and I believe it should never be eaten. Other gyromitras also contain the same poison, and it is thought that helvellas, some species of peziza, and morels also contain it to some degree.

Some clitocybes, especially the small white ones, are also deadly poisonous. They may contain muscarine as do the inocybes.

Galerinas, especially *Galerina autumnalis* (p. 186), and some lepiotas contain the deadly amanitin poisons.

Entoloma sinuatum (p. 135), formerly known as *Entoloma lividum*, is known to be extremely poisonous. *Paxillus involutus* (p. 171) and *Omphalotus illudens* (p. 207) are poisonous. Many other species will cause extreme upsets through being very acrid. Some species of russula and lactarius will do this. *Coprinus atramentarius* (p. 205) will cause violent symptoms if alcohol is drunk at the same time.

Note: this is not a complete list, so reference to the text for any mushroom collected must also be made.

Poisoning symptoms

Symptoms of poisoning may show up shortly after eating a meal of mushrooms or they may not be noticed for as long as ten days. They may show up as one or a combination of the following: stomach pains, diarrhea, vomiting, sweating, chronic thirst, hallucinations, slowing of the heartbeat, liver failure, coma.

If you have any of these symptoms, or if you realize that you may have eaten poisonous mushrooms, seek medical assistance at once.

Chemicals

Several chemicals are useful in establishing the identification of certain mushrooms:

SV (sulpho-vanillin — a few crystals of vanilla dissolved in 2ml conc. sulphuric acid + 2ml distilled water to give a yellow solution): a drop placed on a russula stem discolors violet-purplish in most cases or carmine in other species.

$FeSO_4$ (ferrous sulfate — a solution or crystal): applied to russula stems may cause significant color change.

Melzer's reagent (1.5g iodine, 5g potassium iodide + 100g chloral hydrate dissolved in 100ml warm distilled water) —

amyloid reaction: positive if reagent added to a mass of spores gives a blue-black color change;
dextrinoid reaction: positive if the spore mass turns reddish brown with the reagent.

Ammonia (NH_4OH — a 50 percent aqueous solution): reaction varies.

Phenol (a 2 percent aqueous solution): reaction varies.

NaOH (sodium hydroxide) or KOH (potassium hydroxide), a 30 percent aqueous solution: reaction varies.

Death Cap *Amanita phalloides*

Edible species

I adore eating mushrooms. They are a subtle, exciting, and interesting addition to my diet, but I only eat well-known edible species. The golden rule for safety if you are unsure is *Don't*.

Here is a short list of those that I enjoy most, the majority of which are quite common:

Cep *Boletus edulis* (p. 233) and its varieties
Meadow Mushroom *Agaricus campestris* (p. 189)
Chanterelle *Cantharellus cibarius* (p. 214)
Black Trumpet *Craterellus fallax* (p. 213)
Oyster Mushroom *Pleurotus ostreatus* (p. 206)
Wood Blewit *Lepista nuda* (p. 134)
Yellow Morel *Morchella esculenta* (p. 301)
Black Morel *Morchella elata* (p. 301)
Sulphur Shelf or **Chicken of the Woods Mushroom**
 Laetiporus sulphureus (p. 260)
Shaggy Mane *Coprinus comatus* (p. 204)

Remember, most mushrooms must be well cooked before eating, and with blewits and morels it is essential. I have no space here to give recipes, but I have illustrated all my favorite dishes in my volume on cooking from nature, *Wild Food*, also published by Little, Brown.

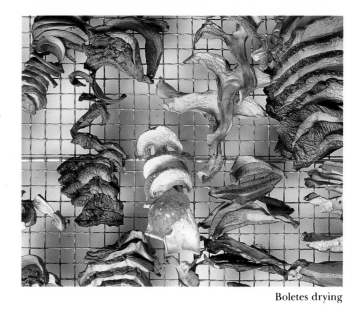

Boletes drying

Drying mushrooms

I have fixed a fine wire mesh 5cm (2in) above a radiator. This is an excellent arrangement, as it gives heat and a good flow of air, the two essentials for quick and effective drying. I have tried drying in a slow oven with the door open, but I find it difficult to control the temperature. It is also possible to buy electric food dryers that are most effective. Large housewares catalogues normally feature one or two models.

Spore prints

Making a spore print is an enormous help in identifying genera and species; the spore color is given for most species in the text. To make a spore print, take a fresh, mature cap and lay it on a clean piece of glass. Cover it by placing an upturned glass over it, which stops it drying out. In addition, it may be necessary to moisten the cap with a drop of water. Left overnight, or possibly longer, the cap should give you a good print. Scrape the spores together with a clean razor blade and you will be more easily able to observe the color. The spore droppers (Basidiomycetes) will make a print below the cap you put out. The spore shooters (Ascomycetes) will make a halo around the cap.

Dichotomous keys

A dichotomous key is designed to help sort genera and/or species into groups so that individual specimens can be identified. In this book I have provided only simple keys to the main genera. The system involves considering each pair of contrasting characteristics in order. Having decided which of the pair is appropriate, look at the end of the line, where you will be directed to the appropriate section of the book or told to continue with the next pair of characteristics. Carry on like this until you get a positive answer. See page 12.

Cep *Boletus edulis*

Beginners Key

On these two pages are shown some of the most common groups, or genera, of mushrooms and fungi that are illustrated. In identifying a mushroom, color of cap, stem, and gills should always be noted, as should the habitat; most mushrooms grow in association with trees, but they may be found on many varied types of soil or substrate — grassland, leaf litter, living trees, rotten wood, dung, or dead creatures. Start by learning to distinguish the most common and important genera:
Amanita, Russula, Lactarius, Cortinarius, and *Boletus.* More than half the collections you make will be a member of one of these genera.

MYCENA (p. 79) Small or tiny mushrooms, usually with a conical cap on a long, delicate stem. They are often found growing in troops or clumps. They grow on the ground, in litter, or on dead wood. A few exude a latex when damaged. Check for unusual smells. Also check to see if the gills have a dark edge. Some species have sticky stems or caps. The spores are white. The caps may be striate or partly striate.

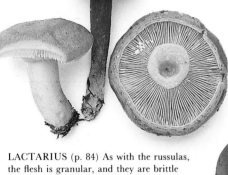

LACTARIUS (p. 84) As with the russulas, the flesh is granular, and they are brittle in the same way. They are distinct in that the damaged flesh or gills exude a colored latex, or milk, the color of which varies; it may be blue, white, yellow, orange, red, or colorless. Also note the color may change in a few minutes after it is exposed to the air. The taste of the milk is another important character. Take a tiny drop on your tongue and then wipe it away; the taste is sometimes very hot or acrid, sometimes mild. Note any special odor.

TRICHOLOMA (p. 41) This genus usually fruits late in the season, often into winter. They are robust and fairly large, fleshy mushrooms, not hygrophanous (that is, they do not change color when wet). The stem is fairly short, normally not longer than the cap is wide, mostly without a ring. The caps are usually gray, brown, or white and may have a sticky texture or may be dry and silky or scaly. Smell and taste are important; many have a distinctly mealy smell and taste. The spores are white.

RUSSULA (p. 108) One of the commonest genera, starting to appear quite early in the summer. They often have brightly colored caps: red, yellow, green, violet, and so on; the stems are normally as long as the cap is wide. The flesh is brittle rather than flexible because it is made up of rather granular cells (like a sugar lump). The gills are white, cream, or pale yellow, and the spore color is also white, cream, or yellow. Taste is an extremely important character, so take a quick nibble and then spit it out; it may take half a minute to decide if the flavor is hot, bitter, or mild. Note any special odor.

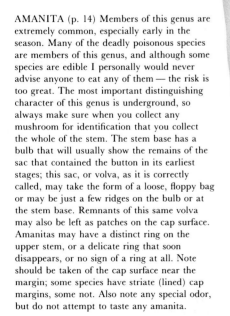

AMANITA (p. 14) Members of this genus are extremely common, especially early in the season. Many of the deadly poisonous species are members of this genus, and although some species are edible I personally would never advise anyone to eat any of them — the risk is too great. The most important distinguishing character of this genus is underground, so always make sure when you collect any mushroom for identification that you collect the whole of the stem. The stem base has a bulb that will usually show the remains of the sac that contained the button in its earliest stages; this sac, or volva, as it is correctly called, may take the form of a loose, floppy bag or may be just a few ridges on the bulb or at the stem base. Remnants of this same volva may also be left as patches on the cap surface. Amanitas may have a distinct ring on the upper stem, or a delicate ring that soon disappears, or no sign of a ring at all. Note should be taken of the cap surface near the margin; some species have striate (lined) cap margins, some not. Also note any special odor, but do not attempt to taste any amanita.

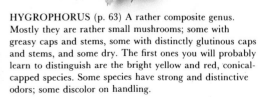

HYGROPHORUS (p. 63) A rather composite genus. Mostly they are rather small mushrooms; some with greasy caps and stems, some with distinctly glutinous caps and stems, and some dry. The first ones you will probably learn to distinguish are the bright yellow and red, conical-capped species. Some species have strong and distinctive odors; some discolor on handling.

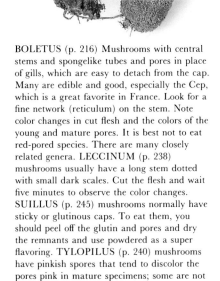

CORTINARIUS (p.139) This is the largest genus in North America and also worldwide. Their most important characteristic is that all have rusty-brown spores. Although this sounds like a difficult character to recognize in the field, in practice it is very helpful. As the spores mature, they will change the gill color to rusty brown, and this may be quite noticeable; also the spore deposit can often be seen on the stem or ground under the cap. Because it is so large, the genus is normally divided into six main subgenera (a seventh subgenus, *Cortinarius*, is very small): *Myxacium*, cap and stem sticky, often long-stemmed; *Phlegmacium*, cap only sticky, stem often short with an abrupt bulb; *Sericeocybe*, cap not sticky or hygrophanous, often silky or shiny, the stem typically swollen; *Leprocybe*, distinguished by the luminous reaction under ultraviolet light, generally resembling *Sericeocybe* but often with more acidic yellow, green, or olivaceous colors; *Telamonia*, cap changing color when wet (hygrophanous), size varying from small to large; note the color of the young gills, as when mature they will all be a similar rusty brown; *Dermocybe*, small mushrooms generally with long, slender stems and conical caps, often strongly colored reddish or greenish. I do not think it is safe to eat any species of cortinarius.

AGARICUS (p. 188) This is the genus from which the cultivated mushrooms have been bred. Some can be found in woods, but the best-known species are found in grassland or on roadsides. They can be split into two main sections: the ones that show signs of yellow when bruised and the ones that show pink in the flesh, especially when cut. They have a distinctive ring on the stem, and the mature gills are dark brownish or purplish black. Some species have a distinctive aniseed smell. Many are edible, but check carefully any that bruise yellow, as the yellow stainer can make you really ill.

RAMARIA (p. 293) mushrooms are the big, fleshy, much-branched groups of coral fungi. Some are said to be good to eat, but others are diuretic and best avoided. Note the color of the branch tips and observe any color change when bruised or cut.

BOLETUS (p. 216) Mushrooms with central stems and spongelike tubes and pores in place of gills, which are easy to detach from the cap. Many are edible and good, especially the Cep, which is a great favorite in France. Look for a fine network (reticulum) on the stem. Note color changes in cut flesh and the colors of the young and mature pores. It is best not to eat red-pored species. There are many closely related genera. LECCINUM (p. 238) mushrooms usually have a long stem dotted with small dark scales. Cut the flesh and wait five minutes to observe the color changes. SUILLUS (p. 245) mushrooms normally have sticky or glutinous caps. To eat them, you should peel off the glutin and pores and dry the remnants and use powdered as a super flavoring. TYLOPILUS (p. 240) mushrooms have pinkish spores that tend to discolor the pores pink in mature specimens; some are not edible as they are incredibly bitter.

HYDNUM (p. 273) and the closely related genera are distinguished by having spines in place of gills. Cut specimens in half and note color changes. Odor is also important. The common *Hydnum repandum* (p. 273) is an excellent edible.

CHANTERELLES (p. 213) are some of the best and most popular edible species. I prefer the Black Trumpet to anything else except morels. The gills are shallow ridges or almost nonexistent, and they run from the cap down the stem (decurrent). Beware the toxic *Omphalotus illudens* (Jack O'Lantern, p. 207), which grows in large clumps on wood and has proper gills rather than ridges.

POLYPORES (p. 252) are fungi with tubes and pores in place of gills. The tubes are not detachable from the cap as in boletus. There are a few with central stems that grow on the ground, but the majority of the genera in this group are brackets that grow on dead or living trees. Many species grow to an enormous size and some species will live for a few years growing a new pore surface every season. Most species are tough and woody and impossible to eat, but a few are prized edibles, such as Chicken of the Woods Mushroom (*Laetiporus sulphureus*, p. 260). Fresh young specimens are the ones to go for; they may weigh twenty pounds or more!

Generic Key

Simple keys to the agarics. The gilled mushrooms (agarics) with central stems that are illustrated have been divided into four separate keys to the main genera: (A) white-, cream-, and yellow-spored; (B) pink-spored; (C) ochre- to brown-spored; (D) purple-brown- to black-spored. The mushrooms that have either a stem that grows laterally from the cap or only a rudimentary stem start on p. 205; the spore colors in this group are mixed.

KEY A Mushrooms with gills, a central stem, and spores white, cream, or yellowish

1 With volva or hooplike remains on stem base: *Amanita* (p. 14)
1 Without volva   2
   2 Stem with ring
           (a) *Lepiota* (most) and allies   (p. 30)
           (b) Usually in clumps: *Armillariella mellea* (p. 37)
   2 Stem without ring   3
3 Exudes milk when damaged
           (a) Large to average types: *Lactarius*   (p. 84)
           (b) Small with narrow stems: *Mycena* (part) (p. 79)
3 Not exuding milk   4
   4 Cap and stem brittle, crumbly, average to large size:
                       Russula (p. 108)
   4 Not brittle or crumbly   5
5 Growing on the remains of other mushrooms
           (a) *Asterophora*   (p. 78)
           (b) *Collybia tuberosa*   (p. 62)
5 Not growing on mushroom remains   6
   6 Small (usually under 3cm [1¼in] across cap)
           (a) Gills decurrent: *Omphalina*   (p. 77)
           (b) Caps conical, stems tall and narrow: *Mycena* (p. 79)
           (c) Stems tough, cartilaginous: *Marasmius*   (p. 74)
           (d) *Lepiota* (part)   (p. 30)
   6 Larger (over 3cm [1¼in] across cap)   7
7 Gills decurrent
           (a) Gills thick, waxy: *Hygrophorus*   (p. 63)
           (b) Gills just shallow wrinkles: *Cantharellus* group   (p. 212)
           (c) Gills very blunt, forking: *Hygrophoropsis aurantiaca* (p. 72)
           (d) Gills thin, crowded: *Clitocybe* and allies   (p. 51)
7 Gills not decurrent   8
   8 With deep tap root: *Xerula* (p. 40)
   8 No tap root   9
9 Stems cartilaginous: *Collybia* and allies (p. 57)
9 Stems not cartilaginous   10
10 Gills thick, waxy: *Hygrophorus* (p. 63)
10 Gills not thick, waxy   11
11 Growing in clumps, stems fused at base
           (a) *Armillariella tabescens* (p. 37)
           (b) *Lyophyllum* (part)   (p. 48)
11 Not in clumps: *Tricholoma* and allies (p. 40)

KEY B Mushrooms with gills, a central stem, and spores pink

1 Gills decurrent: *Clitopilus* (p. 134)
1 Gills not decurrent   2
   2 Growing on wood: *Pluteus* (p. 138)
   2 Not growing on wood   3
3 Mature gills salmon pink
           (a) Gills free: *Pluteus*   (p. 138)
           (b) Gills attached: *Entoloma* and allies (p. 135)
3 Mature gills not pink: *Lepista* (p. 134)

KEY C Mushrooms with gills, a central stem, and spores ochre to rust-brown

1 Spores dull ochre to brown, never rust-brown   2
1 Spores rust   3
   2 Stem with a ring
           (a) *Agrocybe*   (p. 195)
           (b) Growing with sphagnum: *Pholiota myosotis* (p. 179)
   2 Stem without a ring
           (a) Cap fibrous, silky, or scaly, often conical: *Inocybe*   (p. 181)
           (b) Cap often sticky: *Hebeloma*   (p. 179)
3 Stem tall and slender, deeply rooting, small to large:
                     Phaeocollybia (p. 169)
3 Not deeply rooting   4
   4 Small to very small
           (a) *Conocybe*   (p. 185)
           (b) *Galerina* (part) (p. 186)
   4 Stems shorter, more robust   5
5 Cap margin inrolled, gill decurrent: *Paxillus involutus* (p. 171)
5 Cap margin not inrolled, gills not decurrent   6
   6 On wood, wood debris, or burned ground
           (a) *Galerina* (part) (p. 186)
           (b) *Gymnopilus*   (p. 172)
           (c) *Pholiota*   (p. 175)
   6 Not on wood, wood debris, or burned ground   7
7 Stem with ring: *Rozites caperata* (p. 168)
7 Stem without a ring, at most a few remnants:
                     Cortinarius (p. 139)

KEY D Mushrooms with gills, a central stem, and spores purple-brown to black

1 Cap turns to black slime (deliquesces): *Coprinus* (most)   (p. 204)
1 Not deliquescing   2
   2 Cap and stem very brittle/fragile: *Psathyrella* (p. 201)
   2 Not very brittle/fragile   3
3 Gills mottled, cap smooth: *Panaeolus* (p. 205)
3 Gills not mottled   4
   4 Gills very decurrent: *Gomphidius* and allies (p. 212)
   4 Not so   5
5 With a persistent ring, gills free   *Agaricus* (p. 189)
5 Without persistent ring but some with veil remnants on the stem, gills attached
           (a) Spores blackish: *Coprinus*   (p. 204)
           (b) Spores purple-brown: (p. 186)
           Hypholoma also *Stropharia* and allies   (p. 196)

A simple key to the main genera of boletes: the mushrooms with tubes and pores rather than gills (from p. 217). The polypores that have central stems and thus resemble the boletes immediately follow them (p. 253).

KEY E Mushrooms with tubes and pores and a central stem

1 Cap and stem with large floppy scales, gray or blackish-colored: *Strobilomyces* (p. 244)
1 Cap and stem without large floppy scales, variously colored   2
   2 Spore print pinkish or flesh-pink, mature pores may discolor pink   *Tylopilus* (p. 240)
   2 Spore print brownish or yellowish, not pinkish   3
3 Cap sticky, stem often glandular, spotted, sometimes with a distinct ring   *Suillus* (p. 245)
3 Not so   4
   4 Stem tall with distinct scales or dots, which darken in age: *Leccinum* (p. 238)
   4 Not so, stem sometimes with a network (reticulate)
           (a) *Boletus*   (p. 217)
           (b) *Gyroporus* (p. 236)

Glossary

adnate (of gills) connected to stem by whole depth of gill, e.g., *Stropharia aeruginosa* (p. 199)

adnexed (of gills) connected to stem by part of the depth of the gill

agaric general term for a fungus with gills

amyloid turning blue-black in iodine solutions such as Melzer's reagent (see Chemicals, p. 8)

appendiculate fringed with remains of the veil, e.g., *Psathyrella velutina* (p. 203)

appressed closely flattened onto surface

Ascomycetes one of the major groups of fungi, containing all those producing spores in asci which are liberated by pressure

ascospore reproductive cell of the Ascomycetes

ascus (plural **asci**) elongated cell in which ascospores are produced

basidia club-shaped cells on which spores are produced in Basidiomycetes

Basidiomycetes a major and very diverse group of fungi, including gill fungi, boletes, polypores, clavarias, jelly fungi, and Gasteromycetes, characterized by the presence of basidia

basidiospore reproductive cell of the Basidiomycetes

binding hyphae much-branched, thick-walled hyphae without dividing cell walls which bind other hyphae together

bulbous swollen into a bulb

caespitose joined in tufts, e.g., *Hypholoma fasiculare* (p. 187)

campanulate bell-shaped, e.g., *Conocybe lactea* (p. 185)

capillitium mass of sterile threadlike fibers among the spores in the Gasteromycetes which may aid spore dispersal

capitate with a round head

cartilaginous firm but flexible, as opposed to granular

cheilocystidia cystidia on the gill edge

chlamydospore a thick-walled, nondeciduous spore

chrysocystidia cystidia with granular contents which turn yellow in alkali solutions

clamp connection a hyphal outgrowth connecting the two adjoining cells resulting from a cell division, bypassing the dividing cell wall and apparently involved in the movement of nuclei

clavate clublike, e.g., *Clitocybe clavipes* (p. 52)

cortina (adjective **cortinate**) weblike covering running between stem and cap edge, enclosing the gills

cortinal zone faint remnant of cortina on stem

crescentic crescentlike in form

cuticle the surface tissue layer of the cap or stalk

cystidiole a sterile cell protruding beyond the spore-bearing surface

cystidium (plural **cystidia**) sterile cell, variable in shape, occurring between basidia in the spore-bearing surface, or in other parts of the fruit body

decurrent (of gills) running down the stem, e.g., *Pleurotus ostreatus* (p. 206)

decurrent tooth (of gills) where only the narrow end portion of the gill runs down the stem

dendroid treelike

dermatocystidia cystidia on the cap surface

dextrinoid turning reddish brown with iodine solutions such as Melzer's reagent (see Chemicals, p. 8)

dichotomously (branched) branching repeatedly in two

dimitic having two kinds of hyphae

eccentric (of stem) off-center, not centered in the cap

emarginate (of gills) see **sinuate**

equal (of stem) being of the same thickness over its entire length

fibril a small fiber

fibrillose covered with small fibers

filiform threadlike

fimbriate fringed

flexuose, flexuous undulating

floccose cottony, covered with cottony tufts

free (of gills) not connected to stem, e.g., *Amanita muscaria* (p. 14)

fugacious short-lived, fleeting

fusiform spindle-shaped, narrowing at both ends

fusoid somewhat spindle-shaped

Gasteromycetes a large, diverse group within the Basidiomycetes characterized by the basidiospores maturing within the fruit body; includes puffballs, earthstars, stinkhorns, and bird's-nest fungi

generative hyphae thin-walled, branched hyphae with dividing cell walls, giving rise to other types of hyphae, e.g., binding hyphae

germ pore a differentiated area in a spore wall which may give rise to a germination tube

glabrous smooth, hairless

glandular dots moist, sticky spots on surface of stem

gleba fleshy mycelial tissue that contains the spore-bearing cavities present in Gasteromycetes

gloeocystidia thin-walled cystidia with refractive, frequently granular contents

granulate covered with tiny particles

hyaline translucent or transparent, colorless

hygrophanous dark-colored and appearing water-soaked when wet, drying paler

hymeniform resembling a hymenium but lacking functional basidia

hymenium spore-bearing surface

hypha (plural **hyphae**) a single filament, the basic unit forming the fungus (adjective **hyphal**)

immarginate without a distinct edge

innate inseparable, bedded-in

iodoform a crystalline compound of iodine, used as an antiseptic, with a distinctive smell (iodine)

lageniform shaped like a narrow-necked flask

lanceolate elongate and tapering toward both ends

latex a milky, usually white juice exuded by the gills of lactarius species when cut or broken

marginate (bulb) having a well-defined edge, e.g., *Cortinarius calochrous* (p. 145)

monomitic having only one kind of hyphae

mycelium (plural **mycelia**) vegetative stage of a fungus, comprising a threadlike to feltlike mass

palmate having lobes radiating from a central point, like fingers on a hand

papillate having a small, nipplelike protuberance

paraphyses sterile hyphal filaments interspersed between the asci

partial veil see **veil**

pedicel a small stalk

pellicle a detachable skinlike cuticle

peridioles pea-shaped structures containing the spores

perithecia flask-shaped spore-producing chambers found in the Pyrenomycetes group of Ascomycetes

pleurocystidia cystidia on gill sides

pore (of polypore) the mouth of a tube

pruinose having a flourlike dusting

punctate minutely dotted or pitted

pyriform pear-shaped

recurved bent back

reflexed turned sharply back or up

resupinate lying flat on the substrate, with the spore-producing layer outward

reticulum a network of raised ridges found on surface of stem or spores of some mushrooms

rhizoid rootlike structure

rhizomorph cordlike structure comprising a mass of hyphae

ring remains of a partial veil, only present in some agarics (see **veil**)

ring zone faint mark where ring has been

saccate baglike

sclerotium (plural **sclerotia**) firm, rounded mass of hyphae, often giving rise to a fruit body

scurfy surface covered with tiny flakes or scales

sensu lato in the broad sense

septate divided by cell walls

septum (plural **septa**) a dividing cell wall

sessile without a stem

seta (plural **setae**) a stiff hair or bristle

sinuate (of gills) = **emarginate** notched just before joining the stem, e.g., *Hebeloma crustuliniforme* (p. 180)

sphaerocyst a globose cell

spore general term for the reproductive unit of a fungus, usually consisting of a single cell which may germinate to produce a hypha from which a new mycelium arises (see **ascospore, basidiospore**)

spore print deposit of spores falling from a cap placed gill or pores downward on a sheet of paper or glass

squamous, squamulose having small scales

squamule a small cap

stellate starlike

striate with fine lines

sub- (prefix) not quite, somewhat, e.g., subglobose, almost spherical

sulcate grooved

tomentum thick, matted covering of soft hairs (adjective **tomentose**)

trimitic having three kinds of hyphae

tuberculate with small wartlike nodules

tubes spore-producing layer in certain fungi, e.g., *Boletus edulis* (p. 233)

umbo a central hump on a cap like a shield boss

umbonate having an umbo

universal veil see **veil**

veil protective tissue enclosing the developing fruit body; **universal veil** encloses the whole developing fruit body, **partial veil** (of agarics and certain boletes) joins the edge of the cap to the stem, enclosing the developing spore-producing surface and in some genera later forming the ring or cortina (adjective **velar**)

ventricose inflated or swollen

vermiform wormlike

verrucose with small rounded warts

vesicle small bladderlike sac (adjective **vesicular**)

vesiculose formed of vesicles

vinaceous wine-colored

volva cuplike bag enclosing stem base in some agarics, the remains of the universal veil

μ (mu) a micron, $1\mu = 0.001$ millimeter (one thousandth of a millimeter)

Fly Agaric *Amanita muscaria* var. *muscaria* life-size

Basidiomycetes

White- and Cream-spored Agarics pp. 14–133.
See Key A, p. 12

AMANITA *Always make sure to collect the whole stem base so you can check the shape, color, and type of volva (cup at stem base). This volva is the remains of the universal veil that covers the whole mushroom in the early stages; remnants of it may also adhere to the cap in the form of patches. Some have rings on the stem, some not. The gills are free (not attached to the stem) or nearly free. Some amanitas have distinctive smells. I advise against eating any members of this genus, as there are many deadly poisonous species.*

Fly Agaric *Amanita muscaria* var. *muscaria* (L. ex Fr.) Pers. **Cap** 5–25cm (2–10in) across, convex to flatter, sometimes slightly wavy or depressed with a lined margin; blood red to orange-red, becoming lighter toward the margin; smooth, a bit sticky when moist, dotted with flaky patches of whitish volval remnants sometimes almost in concentric rings. **Gills** free to adnexed, crowded, broad; whitish. **Stem** 50–180 × 3–30mm (2–7 × ⅛–1¼in),

hollow to stuffed, sometimes tapering slightly toward the top; white to cream with cottony scales; the hanging, white, moderately fragile ring is at the top of the stem and is lined above and woolly below; the oval or ball-shaped basal bulb has a shallow rim on the upper portion which is the remnant of the volva, which also leaves rings of cottony warts on the lower stalk. **Flesh** white, but yellowish beneath cap cuticle. **Odor** faint, **Taste** pleasant. **Spores** broadly ellipsoid, nonamyloid, 9.4–13 × 6.3–8.7μ. Deposit white. **Habitat** singly or in small groups on the ground in mixed coniferous and deciduous forests. Uncommon in the East; fairly common in the Northwest. Found in eastern and western North America. **Season** July–October (November–February in California). **Deadly poisonous. Comment** In Europe and Russia this mushroom was eaten for its hallucinogenic properties, largely before the introduction of alcohol, but it contains considerable quantities of toxins and has definitely been known to cause death.

Amanita muscaria var. *formosa* (Pers. ex Fr.) Bertillon in De Chambre, **Cap** 4.5–16cm (1¾–6¼in) across, convex to flat with a slightly lined margin; pale yellow to orangy yellow, becoming lighter toward the margin; smooth, slightly sticky when moist, and dotted with flaky patches of whitish volval remnants. **Gills** free, crowded, broad to narrow; pale cream. **Stem** 40–150 × 7–30mm (1½–6 × ¼–1¼in), stuffed, tapering slightly toward the top; white to cream or pale yellow; finely hairy, cottony, or scaly; the creamy-white to yellowish ring near the top of the stem droops and collapses against the stalk, and often disappears. **Flesh** white, but yellowish beneath cap cuticle. **Spores** broadly ellipsoid, nonamyloid, 8.7–12.9 × 6.3–7.9μ. Deposit white. **Habitat** singly, in groups, or in fairy rings on the ground in mixed coniferous or deciduous forests. Common in the Northeast. Found widely distributed in North America. **Season** June–November (November–February in California). **Poisonous.**

Amanita muscaria var. *alba* Pk. **Cap** 4–21cm (1½–8¼in) across, convex to plane, sometimes with a slightly depressed disc and a faintly lined margin; white to silvery white; smooth, slightly sticky when moist, and dotted with small cottony patches or pointed warts of pale brown volval material which stick quite firmly to the cap and are often arranged in concentric rings. **Gills** free to adnexed, close, moderately broad; white to cream. **Stem** 50–140 × 7–20mm (2–5½ × ¼–¾in), stuffed, tapering slightly toward the top; white, turning yellowish when bruised; finely hairy or cottony above the ring, roughly hairy to scaly below; the pale cream or pale yellow drooping ring near the top of the stem soon collapses; the white oval to ball-shaped basal bulb has a rim of volval material at its top, with flaky, pale brown patches and rings on the lower stem. **Flesh** white, but yellowish beneath the cap cuticle. **Spores** broadly ellipsoid, nonamyloid, 7.9–14.1 × 6.3–9.4μ. Deposit white. **Habitat** singly or in small groups on the ground in mixed coniferous and deciduous forests. Occasional. Found scattered throughout northern North America. **Season** August–September. **Poisonous.**

Amanita pantherina var. *pantherinoides* (Murr.) Jenkins **Cap** 3–10cm (1¼–4in) across, convex to flat with a faintly lined margin; yellow-brown on disc, more honey yellow elsewhere; smooth, sticky when moist, dotted with small, whitish, cottony patches of volval material, and becoming minutely hairy toward the margin. **Gills** free to narrowly adnexed, close; white. **Stem** 40–110 × 5–11mm (1½–4½ × ¼–½in), stuffed becoming hollow, tapering slightly toward the top; white to pale cream; smooth with a large, white ring almost at the top of the stem; a white, oval-shaped basal bulb with the whitish, woolly volval material usually closely attached to the stem base with a slight, wavy, unattached margin. **Flesh** white. **Spores** subglobose to ellipsoid, nonamyloid, 9.1–11.2 × 6.3–7.7μ. Deposit white. **Habitat** singly on the ground in mixed coniferous and deciduous forests. Rare. Found in the Pacific Northwest. **Season** September–October. **Not edible** — possibly poisonous.

Amanita frostiana (Pk.) Sacc. **Cap** 2–8cm (¾–3in) across, convex becoming flat with a fairly distinctly lined margin; bright orange, slightly darker at the disc; smooth, sticky when moist, and dotted with yellow or cream cottony patches of volval material, becoming woolly toward the margin. **Gills** free, close; white. **Stem** 47–62 × 4–11mm (1¾–2¼ × ³⁄₁₆–½in),

(continued on page 17)

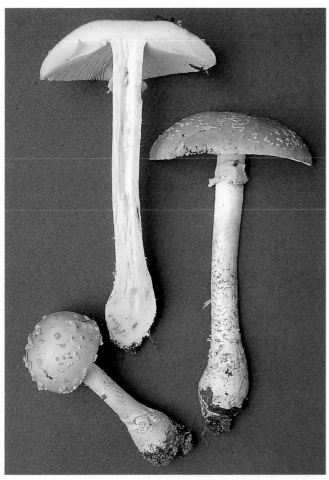

Amanita muscaria var. *formosa* ⅓ life-size

Amanita muscaria var. *alba* ⅓ life-size

Amanita pantherina var. *pantherinoides* ⅔ life-size

Amanita frostiana ⅔ life-size

Amanita pantherina var. *multisquamosa* ⅓ life-size

Amanita pantherina var. *velatipes* ⅓ life-size

Amanita albocreata ⅔ life-size

American Caesar's Mushroom *Amanita caesarea* ½ life-size

(continued from page 14)
stuffed, tapering slightly toward the top; white
to yellowish, slightly hairy, a yellowish,
drooping ring that sometimes falls off in age; a
white oval-shaped basal bulb with yellowish,
cottony patches of volval material on it and the
lower stem. **Flesh** off-white, yellowish. **Spores**
globose, nonamyloid, 7–10.2 × 7–10.2μ.
Deposit white. **Habitat** singly or in small
groups on the ground in mixed woods or under
conifers. Rare. Found in eastern North
America. **Season** August. **Not edible.**

Amanita pantherina var. *multisquamosa* (Pk.)
Jenkins **Cap** 3–11cm (1¼–4½in) across,
convex becoming flat, occasionally with a
depressed disc and a fairly distinctly lined
margin; whitish with a tan-colored disc; sticky
when moist, smooth, spotted with soft, white
patches or warts of volval remnants that are
more woolly at the margin. **Gills** free,
crowded, numerous; white. **Stem**
35–130 × 3–12mm (1¼–5 × ⅛–½in), usually
hollow, tapering slightly toward the top;
whitish; surface with minute patches or hairy
scales; basal bulb almost round, with a volva
like a distinct collar with a free margin at the
top of the bulb; a persistent, whitish, pendant
ring is often flaring with a thickened edge.
Flesh white. **Spores** subglobose to ellipsoid,
nonamyloid, 8.7–10.5 × 6–8μ. Deposit
white. **Habitat** growing on the ground in mixed
coniferous and deciduous woods. Occasional.
Found in eastern North America and possibly
the Pacific Northwest. **Season** August.
Poisonous.

Amanita pantherina var. *velatipes* (Atkinson)
Jenkins **Cap** 7–18cm (2¾–7in) across, convex
becoming flatter with a lined margin; creamy
to whitish yellowish, darker in the center;
smooth, sticky when moist, with virtually
concentric rings of thin, flattened, cottony
warts of whitish volval material. **Gills** free,
close; white. **Stem** 80–200 × 8–20mm
(3–8 × 5/16–¾in), hollow or stuffed and tapering
slightly toward the top; white; smooth or finely
hairy on the upper section, becoming strongly
hairy and scaly toward the base; the white ring,
which is above the middle of the stem, is often
inverted and has a thickish, irregular edge that
flares upward, then collapses and droops; the
white oval-shaped basal bulb often has membranous woolly limbs with whitish cottony
patches around its apex and the lower stem,
and the volva sheaths the base of the stem like
a stocking. **Flesh** thin. **Spores** broadly
ellipsoid, nonamyloid, 7.9–13.2 × 6.3–7.9μ. Deposit white. **Habitat** singly or in small groups
on the ground in mixed coniferous and deciduous woods. Frequent. Found in eastern North
America. **Season** August. **Poisonous.**

Amanita albocreata (Atkinson) Gilbert **Cap**
2.5–8cm (1–3in) across, convex becoming flat;
white to pale yellow or with a pale yellow disc;
sticky when moist, smooth, spotted with easily
removed flaky patches or warts of whitish
volval remnants. **Gills** free or narrowed,
adnexed, crowded, moderately broad; white.
Stem 50–150 × 5–13mm (2–6 × ¼–½in),
stuffed, with minute scaly patches, tapering
slightly toward the top; no ring; the almost
round basal bulb is white; the sheathlike volva
has a free margin and a few woolly flakes or
patches near the base. **Flesh** thin; white.
Spores subglobose to ellipsoid, nonamyloid,
7–9.4 × 6.3–8.7μ. Deposit white. **Habitat** on
the ground in coniferous and deciduous woods

Species close to *Amanita caesarea* (see Comment below) ½ life-size

and sometimes in open, grassy areas.
Infrequent. Found in northeastern North
America. **Season** June–August. **Not edible.**

American Caesar's Mushroom *Amanita
caesarea* (Scop. per Fr.) Grev. sensu Jenkins
Cap 6–22cm (2¼–8½in) across, convex
becoming flatter, usually with a slight umbo
and a margin with rather long striations;
orange to reddish orange on the disc, fading to
yellow-orange on the margin; smooth, slightly
sticky when moist, very occasionally with a few
white patches of volval remains. **Gills** free,
crowded, fairly broad; yellow. **Stem**
90–230 × 9–31mm (3½–9 × ⅜–1¼in), tapering
toward the slightly expanded apex, stuffed to
hollow; pale yellow, occasionally with darker
orange fibers or patches; smooth to hairy-scaly
with a prominent, pendant, membranous, pale
yellow to pale orange ring attached to the
upper stem; no basal bulb but a large, white,
deep volval sac. **Flesh** white to yellowish.

Odor pleasant. **Taste** pleasant. **Spores**
ellipsoid, nonamyloid, 7.8–9.7 × 5.4–6.7μ.
Deposit white. **Habitat** singly or in groups on
the ground in mixed coniferous and deciduous
woods. Rare but sometimes locally abundant.
Found in midwestern and eastern North
America. **Season** July–October. **Edible** but
with the usual cautions that apply to all
amanitas; be sure of your identification.
Comment There is much dispute about the
name that should be given to this mushroom,
but the mushroom itself is well known. In New
Jersey there is another amanita that
superficially resembles *Amanita caesarea*, but the
colors are very different. The center of the cap
is normally brown and the margin yellow
rather than the orange of *caesarea;* my
specimens also lack the orange tones on the
stem. This species has been recognized in
thepast but only mentioned as a color form of
caesarea. The spores support this position in
that they are the same size.

Amanita farinosa ⅔ life-size

Amanita spreta ⅓ life-size

Amanita fulva ⅔ life-size

Grisette *Amanita vaginata* ⅔ life-size

Amanita farinosa Schw. **Cap** 2.5–7cm (1–2¾in) across, broadly convex to flat with an upturned margin that is distinctly striate to plicate-striate; whitish gray but overlaid with a dense layer of mealy, brownish-gray, powdery volval material. **Gills** free, close, broad; white. **Stem** 30–65×3–9mm (1¼–2½×⅛–⅜in), tapering slightly toward the top; dirty white and smooth or with a white powder; no ring; smallish, white, oval-shaped basal bulb with a brown-gray band of volval remnant around its top. **Odor** strong, mink smell in old specimens. **Spores** ellipsoid, nonamyloid, 6.3–9.4× 4.5–7.9μ. Deposit white. **Habitat** singly or scattered on the ground under coniferous and deciduous trees; also in grassy wood edges. Infrequent. Found widely throughout North America. **Season** June–November. **Not edible** — avoid; dangerous.

Amanita spreta (Pk.) Sacc. **Cap** 5–11cm (2–4½in) across, convex becoming more flatly convex with a shortly striate margin, grayish brown; smooth, occasionally a few white, membranous patches of volval remnants. **Gills** free to just reaching the stem, moderately crowded, numerous; white. **Stem** 55–95×7–14mm (2¼–3¾×¼–½in), stuffed, tapering slightly toward the top; whitish; smooth to minutely hairy, moderately hairy toward the base; no basal bulb but membranous, white, cuplike volva at the base; small, white, membranous, persistent, drooping ring toward the top of the stem. **Flesh** white. **Spores** elongate to cylindrical, nonamyloid, 10.8–12.5×6.5–7.5μ. Deposit white. **Habitat** on the ground in mixed coniferous and deciduous woods. Quite common. Found in southeastern North America as far north as New Jersey. **Season** August. **Not edible.**

Amanita fulva (Schaeff.) per Pers. **Cap** 4–9cm (1½–3in) across, ovoid at first, expanding to almost flat with a low umbo and a distinctly grooved margin; orange-brown; slightly paler toward the margin; smooth, slightly sticky when moist then dry. **Gills** free, close, broad; white to creamy. **Stem** 70–150×5–12mm (2¾–6×¼–½in), slender, hollow, quite fragile, tapering toward the top; white tinged with orange-brown and very fine white hairs; no ring; no basal bulb, but base of stem encased in large baglike volva, white tinged with orange-brown. **Flesh** white. **Odor** not distinctive. **Taste** not distinctive. **Spores** globose, nonamyloid, 9.7–12.5×9.7–12.5μ. Deposit white. **Habitat** singly or in small groups on the ground in deciduous and coniferous woods. Fairly common. Found widely distributed throughout North America. **Season** July–September (January–March in California). **Edible** but I would avoid it as I would all amanitas, because there are so many deadly poisonous species.

Grisette *Amanita vaginata* (Bull. ex Fr.) Vitt. sensu lato **Cap** 5–9cm (2–3½in) across, oval then convex expanding to almost flat with an umbo and a distinctly lined margin; grayish brown, darker toward the disc, lighter toward margin; smooth and sticky when moist. **Gills** free, moderately close, moderately broad; whitish. **Stem** 65–120×4–14mm (2½–4¾×³⁄₁₆–½in), hollow, tapering slightly toward the top and often with a bloom; white flushed with cap color; no ring; no basal bulb, but base enclosed in a large, white, thin bag-like volva, often torn with age. **Flesh** white. **Odor** not distinctive. **Taste** not distinctive. **Spores** subglobose, nonamyloid, 9.1–11.2×8.7–10.5μ. Deposit white. **Habitat** singly or in groups on the ground or in grass in

open, mixed coniferous and deciduous woods. Common. Found widely distributed throughout North America. **Season** May–September (November–February in California). **Edible** but best avoided because of other deadly poisonous amanitas.

Amanita calyptrata Pk. **Cap** 7.5–26cm (3–10¼in) across, convex, then broadly convex to flatter with a margin distinctly lined with warts; color varies from whitish yellow to greenish to orange-brown and yellowish on the margin, with a thick, white volval patch on the disc; sticky when moist, smooth. **Gills** adnate to free, crowded, broad; white to pale yellowish. **Stem** 100–240×8–30mm (4–9½×⁵⁄₁₆–1¼in), appearing bulbous but no basal bulb; whitish to yellowish, darkening where handled; smooth to minutely hairy. **Veil** membranous partial veil forms a large but fragile skirtlike ring on the middle or upper stem; the large volva is thick, white, membranous, and saclike. **Flesh** white but yellowish next to cap. **Spores** ellipsoid to elongate, nonamyloid, 9.1–14.6× 5.9–7.9μ. Deposit white. **Habitat** singly, scattered, or in groups on the ground in mixed woods. Sometimes common with madrone and coast live oak. Found in the Pacific Northwest, south to central California. **Season** spring, September–November. **Edible. Comment** This photograph shows the pale yellow form that is normally collected in spring.

Amanita ceciliae (Berk. & Br.) Bas sensu lato **Cap** 5–12cm (2–4¾in) across, convex to flat with an upturned, deeply lined margin and a low umbo; brownish black to brownish gray, darker at the disc, paler toward the margin; smooth, slightly sticky when moist, with loose, charcoal-gray patches of volval remnants dotted around the cap. **Gills** free, close; white. **Stem** 50–160×7–15mm (2–6¼×¼–½in), hollow or lightly stuffed, tapering slightly toward the top; dingy white with flattened grayish hairs; no ring; no basal bulb, but loose, cottony, brownish or charcoal-colored patches of volval remnants dotted around stem base and lower stem. **Flesh** thin, soft, white. **Odor** faint or none. **Taste** slight. **Spores** globose, nonamyloid, 10.2–11.7×10.2–11.7μ. Deposit white. **Habitat** singly or scattered on the ground in mixed coniferous and deciduous forests and also in open woods. Infrequent. Found widely distributed throughout North America. **Season** July–October (November–March in California). **Not edible. Comment** Formerly known as *Amanita inaurata* Secr.

Amanita sinicoflava Tulloss **Cap** 2.5×6.5cm (1–2½in) across, broadly bell-shaped, then convex becoming flatter with a small, distinct umbo and downcurving lined margin; olive-tan to brownish olive, sometimes darker at the disk, occasionally paler at the margin; slightly sticky to dry. **Gills** free to narrowly adnate, close, broad; white or creamy, faintly tinged orange. **Stem** 68–135×6–12mm (2¾–5¼×¼–½in), hollow, tapering toward the top; whitish to graying, paler toward the top; hairy, becoming darker when handled, with faint longitudinal lines particularly near the base; no ring; no basal bulb, but remains of a whitish to gray submembranous sac, sometimes dotted with brown-red spots, collapsed around the base. **Flesh** white. **Odor** none. **Spores** subglobose or occasionally ellipsoid, nonamyloid, 9.1–12.2×8.4–11.5μ. Deposit white. **Habitat** singly or occasionally in small groups in sandy or loamy soil or in moss in mixed coniferous or deciduous woods. Infrequent. Found quite widely distributed in eastern North America. **Season** June–October. **Not edible.**

Amanita calyptrata ⅓ life-size

Amanita ceciliae ¼ life-size

Amanita sinicoflava ½ life-size

Turnip-foot Amanita *Amanita daucipes* ⅔ life-size

Amanita crocea ⅓ life size

Amanita longipes ⅓ life size

Turnip-foot Amanita *Amanita daucipes*(Montagne) Lloyd **Cap** 5–20cm (2–8in) across, convex to obtuse, then expanding, flattened; surface smooth, pale cream to pinkish, salmon pink to pale orange, covered with numerous fine and minute pointed warts joined into patches; margin with irregular fragments of partial veil left hanging. **Gills** free, crowded; pale cream. **Stem** 50–100 × 15–25mm (2–4 × ½.–1in), above ground, swelling abruptly below the soil level into a large turniplike tuberous base, often as long as the stem and 50–60mm (2–2¼in) across; upper edge of bulb with coarse, thick cracks and recurved flakes of volval material; color of upper stem pale pinkish cream, bulb salmon orange staining darker orange, red-brown, or vinaceous. **Veil** variable, remnants of the partial veil leaving a ring zone at apex of stem. **Flesh** thick, solid; white, not staining. **Odor** of old ham to sweet, nauseous. **Taste** not distinctive. **Spores** ellipsoid, 8.6–11.7 × 5.2–7.2μ. Deposit white. **Habitat** loose, sandy, or calcareous soils. Found from Ohio to Tennessee and New Jersey to South Carolina. **Edibility not known** — best avoided. **Comment** One of the most spectacular of the amanita species, with very distinct coloring.

Amanita crocea (Quél.) Singer **Cap** 3–13cm (1¼–5in) across, convex becoming flattened or turning up at the lined, occasionally splitting margin, with a broad umbo; pale yellow-orange to apricot, paler toward the margin; smooth, slightly sticky when moist; random patches of pale, membranous volval remnants on cap. **Gills** adnexed or free, crowded; cream. **Stem** 60–150 × 6–20mm (2¼–6 × ¼–¾in), stuffed to hollow, tapering toward the top; pale cream covered all over with silky or cottony

tufts, the same color as the cap; no ring; the nonbulbous base is encased in a thick, persistent, lobed volva that is white on the outside and flushed with the cap color on the interior surface. **Flesh** thin; white, to pale orange beneath the cap cuticle. **Odor** sweet. **Taste** sweet and nutty. **Spores** subglobose, nonamyloid, 10.3–12.9 × 10.3–12.9μ. Deposit white. **Habitat** singly on the ground in mixed coniferous and hardwood forests. Rare. Found from New York west to Colorado. **Season** August. **Edible** but I would avoid it as I would all amanitas, because there are so many deadly poisonous species.

Amanita longipes Bas ex Tulloss & Jenkins **Cap** 2.5–8cm (1–3in) across, subglobose then broadly convex; white, pale grayish brown, or buff at center, with a dense covering of soft floccose-pulverulent veil sometimes forming larger warts at center. **Gills** adnate, crowded, narrow, with traces of floccose veil on margins; whitish. **Stem** 25–140 × 5–20mm (1–5½ × ¼–¾in), tapered upward then flaring at extreme apex; base swollen, usually slightly rooting; surface with similar floccose veil material to cap, sometimes staining brick red at base. **Flesh** white to slightly grayish. **Odor** none or very slight, of disinfectant(?). **Taste** not distinctive. **Spores** cylindrical, amyloid, 8.4–17.5(20.3) × 4.2–7(7.7)μ. Deposit white. **Habitat** singly in sandy soil under oaks and pines. Found from Massachusetts to Alabama. **Season** August–September. **Edibility not known** — doubtful.

Amanita crenulata Pk. **Cap** 2.5-5.5cm (1–2¼in) across, convex then flat with a lined margin; grayish buff, tinging yellow with age; smooth; dotted with pallid cottony volval patches. **Gills**

Amanita crenulata ½ life-size

free or just reaching the stem, crowded; white. **Stem** 25–70 × 4–11mm (1–2¾ × ½–1in),. somewhat tapering upward with a large ovoid basal bulb; white; ring delicate becoming evanescent; woolly with volval remnants near the bulb. **Flesh** white staining slightly yellow. **Odor** slight. **Spores** globose to broadly ellipsoid, nonamyloid, 7.9–12.6 × 6.3–11.7μ. Deposit white. **Habitat** on the ground in mixed woods. Uncommon. Found in northeastern North American. **Poisonous.**

Amanita cokeri ⅓ life-size

Amanita cokeri (Gilbert & Kühner) Gilbert **Cap** 7–15cm (2¾–6in) across, flatly convex with veil remnants hanging from the margin; shiny white; smooth, slightly sticky when moist, and dotted with largish white to brownish warts of volval material which become flatter and more cottony toward the margin. **Gills** free or just attached, close, broad; white to pale cream. **Stem** 110–195 × 10–20mm (4½–7¾ × ½–¾in), solid or stuffed, tapering slightly toward the top; white; finely silky, then hairy becoming scaly toward the base; the white, membranous hanging ring near the top of the stem has a thickened edge (sometimes double) and is lined above and torn and woolly beneath; the large, spindle-shaped rooting basal bulb has white or pale brown pyramidical warts and recurved scales of volval material, usually arranged in concentric circles at the top of it. **Flesh** white. **Odor** slight. **Spores** ellipsoid, amyloid, 10.1–12.6(12.9) × 6.3–7.3(7.7)μ. Deposit white. **Habitat** singly or scattered on the ground in mixed coniferous or deciduous woods, especially oak and pine. Common. Found in southeastern North America. **Season** July–November. **Not edible. Comment** The spores of my specimen are rather small, and possibly further examination of this species is necessary.

Amanita atkinsoniana Coker **Cap** 6–13cm (2¼–5in) across, convex becoming flatter or concave with veil fragments hanging from the margin; white to cream and pale graying brown, lighter toward the margin; veil becomes yellow and slimy in age; small, reddish-brown warts from the volva become loose, cottony patches on the cap margin. **Gills** free, crowded, moderately broad; pale cream with a faint reddish stain. **Stem** 65–210 × 10–30mm (2½–8¼ × ½–1¼in), usually tapering slightly toward the top; whitish; smooth to finely hairy; a pale, fairly fragile ring persists for a time, then collapses against the stem; the turnip-shaped basal bulb is usually covered with volval remnants forming rings of reddish-brown warts on the bulb, sometimes extending slightly up the stem. **Flesh** white, occasionally

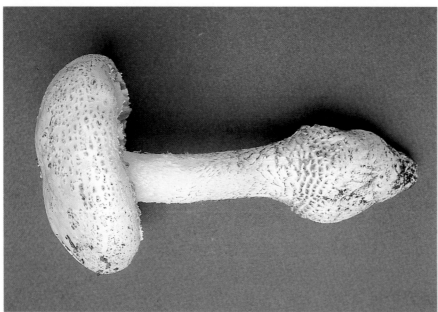

Amanita atkinsoniana ⅔ life-size

Amanita onusta ¾ life-size

staining yellowish or pinkish. **Odor** faintly of chloride of lime. **Spores** ellipsoid, amyloid, 9–12.9 × 5.3–7.9μ. Deposit white. **Habitat** singly or scattered on the ground in coniferous and mixed woods. Fairly common, particularly in the Southeast. Found widely distributed in eastern North America. **Season** August. **Not edible.**

Amanita onusta (Howe) Sacc. **Cap** 2.5–10.5cm (1–4¼in) across, convex to flatter or concave with a low, broad umbo and veil remnants hanging from the margin; whitish to pale gray; slightly sticky when moist, otherwise dry with dark brownish-gray warts of volval material which become woolly on the margin. **Gills** free to just adnexed, close; whitish to creamy yellow. **Stem** 35–155 × 6–15mm (1¼–6 × ¼–½in), solid, tapering slightly toward the top; gray or brownish gray toward the base, paler toward the top; finely hairy to woolly; a delicate, whitish to creamy-gray ring that usually falls away; deeply or sinuously radicating, slender basal bulb has brownish-gray warts and downward-curving scales of volval material on it and the lower stem. **Flesh** whitish to pale gray. **Odor** often of chloride of lime. **Spores** broadly ellipsoid, amyloid, 9–12 × 5.2–7μ. Deposit white. **Habitat** singly or in groups on the ground in mixed coniferous and deciduous forests. Locally quite common. Found in eastern North America. **Season** August–September. **Not edible.**

Amanita rhopalopus Bas f. *rhopalopus* **Cap** 4.5–18cm (1¾–7in) across, bell-shaped to convex becoming flatter with volval remnants hanging from the margin; white; dotted with white or pale brown cottony patches or warts of volval material. **Gills** free to just reaching stem, crowded; white to pale cream. **Stem** 60–195 × 8–25mm (2¼–7¾ × ⁵⁄₁₆–1in), solid, tapering slightly toward the top; white; smooth to minutely hairy; the delicate ring falls away as the cap expands; the spindle-shaped basal bulb is deeply rooting and sometimes has cottony patches or warts of volval material on it and the lower stem but these are also frequently absent. **Odor** strong. **Spores** ellipsoid, amyloid, 8.5–11 × 5.5–7μ. Deposit white. **Habitat** singly or in groups on the ground in mixed coniferous and deciduous forests. Rare. Found in northeastern North America. **Season** August. **Not edible.**

Amanita magniverrucata Thiers & Ammirati **Cap** 4–13cm (1½–5in) across, almost round becoming broadly convex then flat with a strong incurved margin that becomes flatter with age and is hung with white, cottony veil fragments; white, creamy to yellowish buff or darker when handled; dry or slightly sticky when moist, covered with large, pyramidical warts that become flattened in age. **Gills** adnexed to adnate, crowded; white or creamy, lightly powdered. **Stem** 70–115 × 10–25mm (2¾–4¾ × ½–1in), solid, tapering slightly upward, with a rooting basal bulb; white with brownish-yellowish stains; hairy to scaly below, smooth or lined above the ring; white, membranous partial veil forms fragile, skirtlike ring on the upper stem; volva leaves concentric rows of scales and warts at top of bulb and on lower stem, which sometimes disappear in age. **Flesh** thick, firm; white. **Odor** disagreeable in age. **Spores** ellipsoid to subglobose, smooth, amyloid, 8.1–12.7 × 5.5–8.3μ. Deposit white. **Habitat** singly or in groups on the ground under oak and pine. Frequent. Found in coastal forests of California. **Season** October–April. **Edibility not known** — avoid; possibly poisonous.

Amanita rhopalopus f. *rhopalopus* ¼ life-size

Amanita magniverrucata almost life-size

Amanita abrupta ½ life-size

Amanita silvicola ⅓ life-size

Amanita excelsa var. *alba* ⅓ life-size

Amanita flavorubescens ⅔ life-size

The Blusher *Amanita rubescens* ⅓ life-size

Amanita abrupta Pk. **Cap** 4–10cm (1½–4in) across, flatly convex to flat with a margin that is hung with small fragments; white; smooth, shiny, and dry; covered with small, white, conical warts that appear to be woolly at the margin. **Gills** free, crowded, narrow; white. **Stem** 65–125 × 5–15mm (2½–5 × ¼–½in), solid to stuffed, tapering toward the top; white; slightly hairy or smooth with a few warts of volval remnants; the subabrupt to abrupt basal bulb usually large; the white ring near the top is thin and drooping, usually with a thick edge, lined above and woolly below. **Flesh** moderately thick at center of cap, thin toward margin; white. **Spores** globose to ellipsoid, amyloid, 6.5–9.5 × 5.5–8.5μ. Deposit white. **Habitat** in mixed coniferous or deciduous woods. Common. Found in eastern North America. **Season** September–November. **Not edible** — avoid; dangerous.

Amanita silvicola Kauffman **Cap** 5–12cm (2–4¾in) across, convex to flat and sometimes wavy, margin often incurved and with hanging veil fragments; white; patches of dirty white, woolly veil remnants on the cap; a bit sticky. **Gills** free to just reaching the stem, crowded, narrow to broad; white. **Stem** 60–100 × 10–25mm (2¼–4 × ½–1in), solid, tapering slightly at the top; white with occasional brown stains; large, rarely rooting basal bulb with woolly veil remnants on the margin; evanescent white ring toward the top of the stem. **Flesh** quite thick; white. **Spores** ellipsoid, amyloid, 8.3–11.2 × 4.5–6.2μ. Deposit white. **Habitat** scattered to gregarious on the ground in coniferous or mixed woods, particularly under Douglas fir. Fairly common. Found in the Pacific Northwest and California. **Season** September–October (November–February in California). **Not edible** — avoid; dangerous.

Amanita excelsa var. *alba* Coker **Cap** 5.5–7.5 cm (2–2¾in) across, conical to convex to flat; white and slightly tan-colored on the disc; smooth, slightly sticky when moist; faintly shining and dotted with cottony white or pale tan patches of volval material that become brownish when dry. **Gills** just free, close; white. **Stem** 75–90 × 9–14mm (2¾–3½ × ⅜–½in) stuffed to solid, tapering slightly toward the top; white but bruises brownish, minutely hairy; the thin, white, delicate, membranous ring collapses near the top of the stem and turns brownish; the oval-shaped basal bulb has small brown or pinky-brown volval patches on it. **Flesh** white. **Odor** slight. **Taste** slight. **Spores** ellipsoid, amyloid, 6.3–7.5 × 4.2–5μ. Deposit white. **Habitat** singly on the ground in mixed coniferous and deciduous woods. Rare. Found in southeastern North America and Michigan. **Season** September. **Not edible.**

Amanita flavorubescens Atkinson **Cap** 4.5–11cm (1¾–4½in) across, convex becoming flatter with age, with a very faintly lined margin; golden yellow to brownish yellow; smooth, slightly sticky, with thick, loose, yellow patches of volval remnants dotted around the cap. **Gills** free, close, moderately broad; whitish. **Stem** 50–130 × 8–20mm (2–5 × ⁵⁄₁₆–¾in), tapering slightly toward the top; white to yellowish and minutely hairy, sometimes developing wine-red stains on the lower stem; the thin, yellowish ring sometimes hangs like a ragged skirt near the top of the stem; the oval-shaped basal bulb sometimes has yellowish cottony patches of volval remains on it. **Flesh** firm; white. **Spores** ellipsoid, amyloid, 9.7–10.2 × 6.3–7μ. Deposit white. **Habitat** singly or in small groups on the ground under hardwoods or mixed woods and in open urban areas. Occasional. Found in eastern North America. **Season** May–October. **Not edible** — possibly poisonous.

The Blusher *Amanita rubescens* Pers. **Cap** 5–15cm (2–6in) across, convex then flattened; rosy brown to flesh color, sometimes darker reddish brown; very variable, whitish, yellowish, and olive variants are known; usually covered with white or reddish cottony patches. **Gills** free, crowded; white, spotted reddish where injured. **Stem** 60–140 × 10–25mm (2¼–5½ × ½–1in), white, flushed with cap color; white above the striate, membranous ring, becoming reddish near the bulbous base, which may have scattered scaly remnants of veil. **Flesh** white, stained pinkish red where injured. **Odor** pleasant. **Taste** pleasant. **Spores** ovate, amyloid, 8–9 × 5–5.5μ. Deposit white. **Habitat** in mixed woodlands. Common. Found in eastern North America and California. **Season** July–October (February–April in California). **Edible** when cooked, but as with all amanitas extreme caution is recommended. You *must* be sure of its identity! I advise not eating any amanita as the safest policy.

Amanita franchetti ⅓ life-size

Yellow Patches *Amanita flavoconia* ½ life-size

Volvate Amanita *Amanita volvata* ⅓ life-size

Amanita citrina ⅓ life-size

Amanita franchetii (Boud.) Fayod syn. *Amanita aspera* (Fr.) Quél. **Cap** 4.5–10cm (1¾–4in) across, convex becoming flat; straw-colored to yellowy brown or grayish brown; smooth, sticky when wet, then becoming dry, dotted with yellowish patches of volval remains. **Gills** free or slightly adnexed, close, broad; whitish or tinged with yellow. **Stem** 60–140 × 10–20mm (2¼–5½ × ½–¾in), stuffed, tapering slightly toward the top; white; smooth or slightly woolly; white to pale yellow hanging ring on upper stem; ball-shaped basal bulb dotted with yellowish patches of volval remnants. **Flesh** white, but yellowish brown beneath cap cuticle and sometimes bruising reddish brown around insect holes at base. **Odor** faint, not distinctive. **Spores** ellipsoid, amyloid, 7.5–9.2 × 5.5–6.5μ. Deposit white. **Habitat** scattered on the ground under conifers and in mixed deciduous woods. Fairly common in the West, occasional in the East. Found in west and east North America. **Season** August–October (November–February in California). **Not edible.**

Yellow Patches *Amanita flavoconia* Atkinson **Cap** 3–7cm (1¼–2¾in) across, ovoid at first, then expanding to convex or flat with umbo; bright yellow to orange, with small bright yellow veil fragments loosely spread over surface; margin of cap without radial grooves. **Gills** free or slightly adnexed, crowded; white or with faint flush of yellow. **Stem** 50–100 × 5–15mm (2–4 × ¼–½in), white to yellow, with swollen basal bulb, covered on lower half with yellow floccose-crumbly veil fragments; with membranous white or yellow ring. **Flesh** white, unchanging. **Odor** slight, pleasant. **Spores** ovate-elliptic, smooth, amyloid, 7–8(9) × 4.5–5μ. Deposit white. **Habitat** in mixed woods. Very common. Found in most of eastern North America. **Season** July–October. **Edibility uncertain** — best avoided. **Comment** Most likely to be confused with the much rarer *Amanita frostiana* (page 15), which differs in its striate cap margin, nonamyloid, globose spores, and often marginate basal bulb of stem.

Volvate Amanita *Amanita volvata* (Pk.)Mar. **Cap** 5–10cm (2–4in) across, convex to flat, margin with radial striations; white bruising reddish brown; smooth, surface often with numerous patches of thin veil remnants. **Gills** free, broad, crowded; white. **Stem** 50–100 × 10–15mm (2–4 × ½in), white; surface floccose-shaggy to powdery, also discoloring; base emerging from a thick, membranous, fleshy, cuplike volva; no ring is left on the stem. **Flesh** white. **Odor** not distinctive. **Taste** not distinctive. **Spores** elliptic, smooth, amyloid, 9–11 × 6–7μ. Deposit white. **Habitat** in open, often sandy soils in mixed woods and roadsides. Common. Found in eastern North America. **Season** July–October. **Not edible.**

Amanita citrina (Schaeff.) per Pers. **Cap** 5–12cm (2–4¾in) across, convex then flat; pale lemon yellow to greenish yellow, usually covered with many patches of felty whitish veil; smooth, dry or viscid when wet. **Gills** free, broad, crowded;

white. **Stem** 60–120 × 10–15mm (2¼–4¾ × ½in), with a large basal bulb; white to discolored buff; upper edge of bulb grooved, gutterlike; apical ring well developed, yellowish, striate. **Flesh** firm; white. **Odor** of raw potatoes, very distinctive. **Taste** pleasant. **Spores** globose, smooth, amyloid, 7–10 × 7–10μ. Deposit white. **Habitat** in oak and pine woods. Occasional. Found in eastern North America. **Season** August–November. **Not edible. Comment** *Amanita citrina* var. *lavendula* Coker (as *A. mappa*) differs in its flush of lavender, in the universal veil, and sometimes in the streaks on the cap and is probably a distinct species in its own right.

Cleft-foot Amanita *Amanita brunnescens* Atkinson **Cap** 4–15cm (1½–6in) across, bluntly convex then flattened with central umbo; deep brown to olivaceous or grayish, paler near margin, with darker radiating fibers showing through surface; smooth, viscid when wet. In the var. *pallida* Krieger the cap is

completely white to pale cream or pale citrine; both types bruise reddish brown. **Gills** free, crowded, broad; white bruising reddish brown. **Stem** 50–150 × 10–20mm (2–6 × ½–¾in), swelling at base to a distinct sharp-edged bulb that is usually cleft with vertical splits at margin; pale dirty white to brownish with darker stains, particularly at the base; smooth to slightly roughened; apex with a flaring, pendant white ring that soon collapses against stem. **Flesh** firm; white bruising reddish brown. **Odor** not distinctive. **Taste** not distinctive. **Spores** globose, smooth, amyloid, 8–10 × 8–10μ. Deposit white. **Habitat** in deciduous woods, especially under oak. Abundant. Found widely distributed in much of eastern North America from Canada to Florida, west to Michigan and Texas. **Season** July–October. **Probably poisonous** and like all amanitas is to be avoided. **Comment** This should be carefully compared with the deadly *Amanita phalloides* (page 28).

Cleft-foot Amanita *Amanita brunnescens* ⅓ life-size

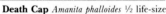

Death Cap *Amanita phalloides* ½ life-size

Amanita bisporigera ⅓ life-size

Death Cap *Amanita phalloides* (Fr.) Link in Willd. **Cap** 6–15cm (2¼–6in) across, convex then flattened; variable in color but usually greenish or yellowish with an olivaceous disc and paler margin; also, paler and almost white caps do occur occasionally; smooth, slightly sticky when wet, with faint, radiating fibers often giving it a streaked appearance; occasionally white patches of volval remnants can be seen on cap. **Gills** free, close, broad; white. **Stem** 60–140 × 10–20mm (2¼–5½ × ½–¾in), solid, sometimes becoming hollow, tapering slightly toward the top; white, sometimes flushed with cap color; smooth to slightly scaly; the ball-shaped basal bulb is encased in a large, white, lobed, saclike volva. **Veil** partial veil leaves skirtlike ring hanging near the top of the stem. **Flesh** firm, thicker on disc; white to pale yellowish green beneath cap cuticle. **Odor** sickly sweet becoming disagreeable. **Spores** broadly ellipsoid to subglobose, amyloid, 8–10.5 × 7–9µ. Deposit white. **Habitat** singly or in small groups on the ground in mixed coniferous and deciduous woods. Formerly rare but seemingly spreading and becoming frequent, especially in northern California. Found in eastern North America, the Pacific Northwest, and California. **Season** August–November (November–January in California). **Deadly poisonous. Comment** This is the most deadly fungus known, and despite years of detailed research into the toxins it contains, no antidote exists against their effects on the human body. Poisoning by *Amanita phalloides* is characterized by a delay of

between six and twenty-four hours from the time of ingestion to the onset of symptoms, during which time the cells of the liver and kidneys are attacked. However, if a gastroirritant has also been consumed — e.g., as the result of eating a mixed collection of mushrooms — gastric upset may occur without the characteristic delay, masking this vital diagnostic evidence. The next stage is one of prolonged and violent vomiting and diarrhea accompanied by severe abdominal pains, lasting for a day or more. Typically this is followed by an apparent recovery, when the victim may be released from the hospital or think his ordeal over. Yet within a few days death may result from kidney and liver failure. Although *Amanita phalloides* contains many poisonous compounds, it is believed that only the group known as amatoxins are responsible for human poisoning; the others (phallotoxins) are thought to be rendered harmless by being neutralized by other compounds or not being absorbed from the intestinal tract, by being present in very low concentrations, or by being so unstable as to be destroyed by cooking or digestive juices. The amatoxins, however, are fully active orally. The main constituent of this group is α-*amanitin*, which through its effect on nuclear RNA in liver cells causes the end of protein synthesis, leading to cell death. When filtered through the kidneys, it attacks the convoluted tubules and instead of entering the urine is reabsorbed into the bloodstream and recirculated, causing repeated liver and kidney damage. As with any hepatic disease,

treatment relies on the monitoring of blood chemistries, urine output, and so on, and the maintenance of fluid and electrolyte balance. In cases of amatoxin poisoning, mortality is 50 to 90 percent, and any chance of survival depends on early recognition.

Amanita phalloides var. *alba* (Vitt.) Gilbert. differs from the type in being entirely white throughout; like *Amanita phalloides*, it is also deadly poisonous.

Amanita bisporigera Atkinson **Cap** 3–10cm (1¼–4in) across, convex to flat or depressed; white, sometimes with a faint tinge of pale brown on the disc; smooth and slightly sticky when moist. **Gills** free to just reaching the stem, crowded, attenuate; white. **Stem** 60–140 × 7–18mm (2¼–5½ × ¼–¾in), solid, tapering slightly toward the top; white; often woolly or scaly; ball-shaped basal bulb; the white ring near the top of the stem is thin, delicate, and drooping or shredded in a mature specimen; the volva is a white membranous sac. **Flesh** white. **Spores** globose, amyloid, 7.8–9.6 × 7–9µ. Basidia mostly 2-spored. Deposit white. **Habitat** singly in mixed coniferous and deciduous forests. Fairly common. Found in eastern North America. **Season** June–September. **Deadly poisonous. Comment** This is extremely similar to *Amanita virosa* (below), which has mostly 4-spored basidia.

Destroying Angel *Amanita virosa* (Fr.) Bertillon in De Chambre **Cap** 5–12cm (2–4¾in) across, convex-conical at first, then

expanded with broad umbo; pure white; smooth, slightly viscid when moist. **Gills** free, crowded; white. **Stem** 90–120 × 10–15mm (3½–4¾ × ½in), usually swelling toward base; white with surface often disrupted into shaggy fibrils; base enclosed in a baglike, white, sheathing volva; apex of stem fragile, ring often torn or incomplete. **Flesh** firm; white. **Odor** sweet and sickly. **Spores** globose, amyloid, 8.5–10(11) × (7)7.5–9μ. Deposit white. **Habitat** in mixed woodlands. Common. Found in many parts of North America. **Season** June–November. **Deadly poisonous** — many deaths are caused by this fungus in North America. **Comment** Flesh turns instantly golden yellow with KOH, differentiating this species from the very similar *Amanita verna* (Bull. per Fr.) Roques (found in the Pacific Northwest), which has a smooth stem.

Symptoms of poisoning *Amanita virosa* and its relative *Amanita bisporigera* (above) both contain the deadly amatoxin poisons, and since they are so common in North America they have been responsible for many cases of severe poisoning and death. The first symptoms of poisoning are vomiting, persistent diarrhea, and severe stomach pains; the onset of symptoms normally occurs some eight to ten hours or as long as twenty-four hours after eating a meal containing these amanitas. After this there may be a period of apparent improvement before the second effect of the poisoning occurs; this is a deterioration in function of both the liver (hepatic failure) and the kidneys (renal failure). These will show up in the patient as yellowing or discoloration of the whites of the eyes and skin, as in hepatitis, and also in discoloration of the urine. Thus, it is crucial not to leave the patient untreated during the first stages. The sufferer should be taken immediately to the nearest hospital and the doctors informed that mushrooms were eaten during the past few days so that there is no possibility the doctors will misidentify the cause of the poisoning.

Amanita porphyria (A. & S. ex Fr.) Secr. **Cap** 3–10cm (1¼–4in) across, convex becoming flattened with an incurved margin: smooth, slightly sticky when wet; pale grayish brown with vinaceous flush; some woolly patches of ash-colored volval remnant on cap. **Gills** adnexed to free, close, moderately broad; white. **Stem** 50–120 × 6–20mm (2–4¾ × ¼–¾in), stuffed to hollow, tapering slightly toward the top; whitish with small gray fibers; thin, fragile, hanging gray ring halfway up, nearer the top; soft, large, rounded basal bulb encased in a short, fragile volva with a distinct collar. **Flesh** firm, whitish. **Odor** slight. **Taste** unpleasant. **Spores** globose, amyloid, 7.2–10.3 × 6.7–10.1μ. Deposit white. **Habitat** singly or scattered on the ground under conifers and in mixed woods. Infrequent. Found widely distributed in eastern North America and the Pacific Northwest. **Season** July–October. **Not edible.**

Destroying Angel *Amanita virosa* ½ life-size

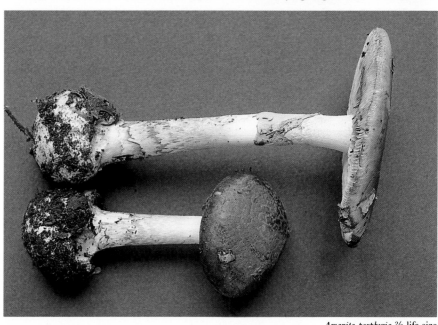

Amanita porphyria ⅔ life-size

Parasol Mushroom *Lepiota procera* ⅓ life-size

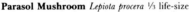

Lepiota rhacodes var. *hortensis* ⅓ life-size

LEPIOTA *Typically with a ring and free gills, the large species often have a detachable ring that can be moved up and down the stem. Smell is important and should be noted, as should color changes that can be seen on cutting or bruising. Cap scales are also a distinctive feature.*

Parasol Mushroom *Lepiota procera* (Scop. ex Fr.) S. F. Gray syn. *Macrolepiota procera* (Scop. ex Fr.) Singer **Cap** 7–20cm (2¾–8in) across, button, spherical or egg-shaped, expanding to flattened with a prominent umbo; pale buff or gray-brown; covered in darker brown shaggy scales. **Gills** free, close, broad; white, darkening somewhat in age, with woolly edges. **Stem** 130–400×8–15mm (5–16×⁵⁄₁₆–½in), very slender with a bulbous base; white with a gray-brown felty covering which becomes split into snakelike markings as the stem expands. **Veil** white, membranous partial veil leaves a large, double-edged movable ring on the upper stem. **Flesh** thin, soft; white. **Odor** slight, not distinctive. **Taste** sweet. **Spores** ovoid, with a germ pore, dextrinoid, 15–20×10–13μ. Deposit white. **Habitat** singly, widely scattered, or in groups in open woods, old pastures, and along trails. Sometimes quite common. Found in eastern North America, west to Michigan and Nebraska. **Season** June–October (November–December in Florida). **Edible** — excellent. **Comment** In Europe this lepiota can be found up to 50cm (20in) across, and all collections are much larger than those found in North America.

Lepiota rhacodes Pilát var. *hortensis* **Cap** 5–18cm (2–7in) across, subglobose expanding to convex, becoming flatter or slightly umbonate, with a fringed or shaggy margin; reddish brown at first, then breaking into coarse scales

as the cap expands, revealing the fibrous, whitish subcutaneous layer, the center remaining brown; dry, coarse, with scales, but smooth on the disc. **Gills** free, close, broad; white to cream, eventually dirty buff or brown-spotted. **Stem** 50–180×10–30mm (2–7×½–1¼in), stout with a large, subspherical bulb, which has a raised rim; white discoloring brownish below the ring; smooth, dry. **Veil** white or brownish membranous partial veil forms a large, thick, double-edged ring on the upper stalk. **Flesh** thick, fairly firm; white, discoloring orange or reddish when cut. **Odor** pleasant. **Taste** pleasant. **Spores** ellipsoid, with a large germ pore, smooth, dextrinoid, 6–13×5–9μ. Deposit white. **Habitat** singly or several or in fairy rings on rich soil in gardens, along roads, in open fields and woods. Common particularly in the West. Found widely distributed in North America. **Season** September–October (November–February in California). **Edible** with caution. **Comment** This well-known edible has been reported to cause gastric upset in some individuals.

Lepiota americana Pk. **Cap** 3–15cm (1¼–6in) across, oval becoming broadly convex, then flat with an umbo; background whitish at first, then reddening when mature and bruised; dry and smooth but quickly breaking into large, coarse, reddish-brown or dingy pinkish-buff scales. **Gills** free, close, broad; white staining pinky-buff. **Stem** 70–130×5–20mm (2¾–5×¼–¾in), often enlarged at or below the middle and tapering toward the base; white at first, staining or aging pinkish or reddish brown; smooth with appressed silky hairs. **Veil** membranous partial veil leaves a white double-

edged ring on the upper stalk which may disappear in age. **Flesh** thick, firm; white staining yellowish then reddish brown. **Spores** broadly ellipsoid, smooth, dextrinoid, 9–10.5×7–8.5μ. Deposit white. **Habitat** singly or in dense clusters in fields, waste places, and sawdust piles and around stumps. Found widely distributed in eastern North America, west to Michigan. **Season** June–October. **Edible** — good.

Lepiota clypeolaria (Bull. ex Fr.) Kummer **Cap** 2–8cm (¾–3in) across, broadly bell-shaped becoming flatter with a low, broad umbo; margin ragged from veil remnants and sometimes recurved in mature specimens; cinnamon brown on the disc, paler and more yellowish tawny near the edge; dry, velvety, quickly breaking up into scales. **Gills** free, close, narrow; white. **Stem** 40–120×3–8mm (1½–4¾×⅛–⁵⁄₁₆in), fragile; whitish or yellowish at the top, becoming more dingy brownish below in age; covered in soft, cottony, shaggy scales. **Veil** white, leaving a slight ring on upper stem. **Flesh** thin, white. **Odor** mild or slightly pungent. **Taste** not distinctive. **Spores** fusiform, smooth, dextrinoid, 12.4–18×4.4–5.5μ. Deposit white. **Habitat** singly, scattered, or in small groups in mixed woods, hardwoods, or sometimes swamps. Sometimes quite common. Found widely distributed in North America. **Season** July–November in the East, November–February in California. **Poisonous.**

Leucoagaricus naucinus (Fr.) Singer syn. *Lepiota naucina* (Fr.) Kummer **Cap** 5–10cm (2–4in) across, convex expanding to almost flattened; whitish becoming flushed flesh-color or pale ochre-cream; smooth, silky, or breaking up into scales in age. **Gills** free, close; white, becoming pale flesh-colored with age. **Stem** 50–150 × 5–15mm (2–6 × ¼–½in), with an enlarged base; white sometimes bruising in age; dry, smooth. **Veil** white, membranous partial veil forms a double-edged, sleevelike ring on the upper stem. **Flesh** thick; white in cap, browning in the stem. **Spores** broadly ellipsoid to ovoid, smooth, with an apical pore, dextrinoid, 7–9 × 5–6μ. Deposit white or very pale pink. **Habitat** singly, scattered, or in groups on lawns, pastures, waste ground, and sometimes in woods. Common and often abundant. Found widely distributed in North America. **Season** August–November, sometimes earlier in the West. **Edible** with caution. Although this is a well-known edible there have been several reports of poisoning. Furthermore, it might be confused with a deadly amanita. Best avoided.

Lepiota cortinarius Lange **Cap** 3–10cm (1¼–4in) across, convex becoming nearly flat with an irregularly raised margin; white background with a brown disc; covered in small, tawny, hairy scales arranged concentrically. **Gills** free, crowded, subdistant; white sometimes staining yellowish. **Stem** 30–90 × 7–20mm (1¼–3½ × ¼–¾in), enlarged into a small, flat bulb at the base; same color as cap; smooth at the top; hairy-scaly below, similar to cap. **Veil** cortinate, forms a weblike, evanescent ring. **Flesh** thick, firm; white. **Spores** oblong-ovoid, truncate at base, 7.5–10 × 3–4μ. Deposit white. **Habitat** on needle carpets under conifers or in mixed woods and cedar swamps. Quite common. Found in eastern North America west to the Great Lakes and rarely in the Pacific Northwest. **Season** September–October. **Not edibile** — probably poisonous.

Lepiota bucknallii (Berk. & Br.) Sacc. **Cap** 2–4.5cm (¾–1¾in) across, bell-shaped becoming broadly convex with an umbo; margin has some hanging veil remnants at first; whitish to pale lilac; mealy. **Gills** free; creamy, pale yellow. **Stem** 20–45 × 3–5mm (¾–1¾ × ⅛–¼in), hollow, sometimes tapering slightly at the base; pale lavender at the top becoming dark purple below, especially when bruised; somewhat scurfy. **Veil** transient. **Flesh** thin, whitish to lilac. **Odor** strong, of gas or tar. **Spores** elongate, ellipsoid, weakly dextrinoid, 7.5–10 × 3–3.5μ. Deposit white. **Habitat** on humus in mixed woods. Found widely distributed in the Pacific Northwest and Idaho. **Season** September–October. **Not edible** — possibly poisonous. **Comment** This mushroom was photographed in England.

Lepiota cepaestipes (Sow. ex Fr.) Pat. syn. *Leucocoprinus cepaestipes* (Sow. ex Fr.) Pat. **Cap** 2–8cm (¾–3in) across, oval becoming broadly bell-shaped to nearly flat with a distinct umbo; white to pale pinkish, more yellowish or darker in age; dry, powdery, mealy, becoming warty or scaly in age and margin clearly lined. **Gills** free, crowded, narrow; white. **Stem** 40–120 × 3–6mm (1½–4¾ × ⅛–¼in), sometimes swollen in places, with an enlarged base; white becoming tinged with flesh-pink, bruising straw yellow; smooth or with a slight bloom. **Veil** forms a large, thick, white, easily detachable ring on the upper part of the stem. **Flesh** thin; white, sometimes bruising yellowish. **Odor** mild. **Taste** mild. **Spores** broadly ellipsoid, with a germ pore, smooth, weakly dextrinoid, 8.5–9 × 6.5–7μ. Deposit white. **Habitat** in dense tufts on wood chips, sawdust, rich soil, and organic matter. Common in the East. Found widely distributed in North America. **Season** June–September. **Not edible** — possibly poisonous.

Lepiota americana ⅓ life-size

Lepiota clypeolaria ⅔ life-size

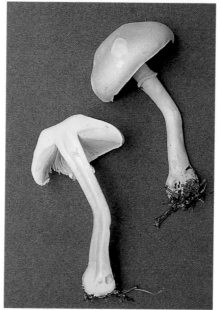

Leucoagaricus naucinus ½ life-size

Lepiota cortinarius ⅔ life-size

Lepiota bucknallii ⅔ life-size

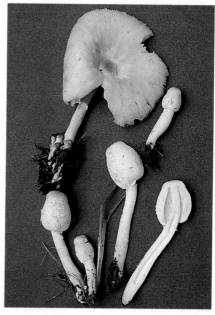

Lepiota cepaestipes ⅓ life-size

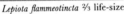
Lepiota flammeotincta ⅔ life-size

Lepiota rubrotincta ½ life-size

Lepiota flammeotincta Kauffman **Cap** 1–6cm (½–2¼in) across, broadly convex becoming flat or uplifted at the margin or sometimes broadly umbonate; whitish background with brown, reddish-brown, purplish-brown, or almost black hairs or scales; surface staining red when bruised, then slowly turning dark purplish brown; dry. **Gills** free, close; white edges, sometimes brownish. **Stem** 30–95 × 1–6mm (1¼–3¾ × ¹⁄₁₆–¼in), slender, slightly thicker toward the base; white above the ring; hairy like cap; quickly bruising scarlet like cap surface, then turning dark brown. **Veil** white, membranous partial veil forms a sleevelike ring on the middle or upper stalk. **Flesh** white staining pink, red, or orange when bruised, then fading. **Spores** ellipsoid, smooth, dextrinoid, 6–8.5 × 4–4.8μ. Deposit white. **Habitat** singly or in small groups in woods, under trees, along trails and paths. Frequent. Found along the Pacific Coast. **Season** September–November. **Edibility not known.**

Lepiota rubrotincta Pk. **Cap** 3–8cm (1¼–3in) across, ovoid with a flattened top, becoming convex then flat or broadly umbonate or with an uplifted margin; background whitish, center deep red to chestnut; surface dry, smooth and pinkish brown all over, then breaking into flat, minutely hairy scales varying in color from cinnamon buff to coral pink, reddish, or reddish orange, margin paler. **Gills** free, close, narrow; white. **Stem** 40–160 × 4-10mm (1½–6¼ × ³⁄₁₆–½in), stuffed becoming hollow, club-shaped at base, which often extends quite deeply into the soil; white, discoloring a little in age; smooth to milky. **Veil** white, membranous partial veil leaves a persistent hanging ring on the upper stem. **Flesh** thick; white. **Spores** ellipsoid, smooth, dextrinoid, 6–10 × 4–6μ. Deposit white. **Habitat** singly, scattered, or in small groups in the ground or in soil, compost, leaf litter, and humus in deciduous woods. Sometimes common. Found widely distributed throughout North America. **Season** July–November. **Not edible.**

Green-gilled Lepiota *Chlorophyllum molybdites* (Mayer ex Fr.) Mass. *Lepiota morganii* (Pk.) Sacc. **Cap** 5–30cm (2–12in) across, hemispherical to broadly convex becoming flatter; whitish underneath, covered with thin layers of pale pinkish-buff volval tissue which breaks up into many small scales and patches as the cap expands; dry, smooth or minutely hairy below, with scales curling upward in age. **Gills** free, close, broad; whitish slowly becoming dirty gray-green or darker. **Stem** 50–250 × 8–25mm (2–10 × ⁵⁄₁₆–1in), sometimes enlarging toward the base; whitish, slowly becoming dingy gray; smooth. **Veil** membranous, large, white, leaving double-edged, persistent pendant ring on the upper stalk. **Flesh** thick; white, discoloring dingy red when bruised. **Odor** faint and pungent or none. **Taste** mild or none. **Spores** ovoid or ellipsoid, smooth, thick-walled with small germ pore at tip, 8–13 × 6.5–8μ. Deposit green. No pleurocystidia. **Habitat** often forming fairy rings on grassy places such as lawns, meadows, and wasteland. Found widely distributed in North America but very common in the Gulf Coast area and Colorado. **Season** July–September. **Poisonous. Comment** Many people have reported this mushroom as edible, but it definitely contains toxins. These may be reduced by boiling, which may account for some people's eating it without symptoms of vomiting, cramps, and diarrhea.

Green-gilled Lepiota *Chlorophyllum molybdites* ½ life-size

Lepiota cristata life-size

Lepiota sistrata 1¾ times life-size

Lepiota cristata (Fr.) Kummer **Cap** 1–6cm
(½–2¼in) across, irregularly bell-shaped and
umbonate; center reddish tawny, background
white; dry, quickly breaking up into small
reddish-brown scales often arranged
concentrically. **Gills** free, close, narrow; white,
becoming brownish with age. **Stem** 20–80 ×
2–5mm (¾–3 × ³⁄₃₂–¼in), fragile, tender,
sometimes thicker below; white tinged flesh-
color, becoming more pinky-brown below;
smooth with a few faint hairs at the top. **Veil**
sometimes forms a thin, white, fragile ring on
the middle or upper stem, sometimes
disappears. **Flesh** thin, fragile; white. **Odor**
unpleasant, strongly fungusy or mild. **Taste**
pleasant. **Spores** bullet-shaped, smooth,
dextrinoid, 6–7.5 × 3-4.5μ. Deposit white to
pale buff. **Habitat** scattered or in groups on the
ground under shrubs and trees, especially
spruce, redwood, or cypress, and on lawns.
Common. Found widely distributed in North
America. **Season** June–October (November–
February in California). **Not edible** — possibly
poisonous. **Comment** In Europe the strong,
pungent smell of this mushroom is considered
a major diagnostic character; in North America
the smell is generally slight and mild, but
occasionally strong-smelling collections are
found.

Lepiota sistrata (Fr.) Quél. syn. *Lepiota seminuda*
(Lasch) Kummer. **Cap** 1–3cm (½–1¼in)
across, conico-convex with a distinct umbo;
white with a flesh-colored tinge; mealy. **Gills**
free, close; white flushed pale pinkish. **Stem**
15–25 × 1–2mm (½–1 × ¹⁄₁₆–³⁄₃₂in), thin,
fragile; white tinged pinkish toward the base;
finely mealy. **Veil** transient. **Flesh** thin; white
in cap, pinkish in lower stem. **Spores** ellipsoid,

smooth, 3–4 × 2–2.5μ. Deposit white. **Habitat**
singly, scattered, or in groups in humus under
conifers or hardwoods. Quite common. Found
mainly east of the Great Plains, but also occurs
in the Pacific Northwest and California.
Season July–October (November–February in
California). **Edibility not known.**

Leucocoprinus birnbaumii (Corde) Singer syn.
Lepiota lutea (Bolton) Quél. **Cap** 2.5–6cm
(1–2¼in) across, oval becoming broadly
conical or bell-shaped then umbonate and flat,
with the margin distinctly lined in mature
specimens; bright yellow to greenish yellow or
pale yellow, the disc sometimes darker or more
brownish; dry, powdery to mealy or minutely
scaly in age. **Gills** free, crowded; yellow or
pale yellow. **Stem** 30–100 × 1–5mm
(1¼–4 × ¹⁄₁₆–¼in), slender, but slightly
enlarged at the base; yellow; dry, smooth, or
powdery. **Veil** yellow partial veil forms a
small, collarlike ring on the upper stalk which
may disappear. **Flesh** very thin; yellow.
Spores ellipsoid, with an apical pore, smooth,
dextrinoid, 8–13 × 5.5–8μ. Deposit white.
Habitat singly or in groups or dense tufts on
rich organic matter and on compost, decaying
hay, and leaf piles; also in gardens and lawns
in the South. Common in the South. Found
widely distributed in many parts of North
America; indoors (greenhouses) in the North,
outdoors in warmer areas. **Season** May–
September, anytime indoors. **Not edible** —
reported as poisonous to some people.

Cystoderma cinnabarinum (Secr.) Fayod **Cap**
3–8cm (1¼–3in) across, convex; deep
cinnabar red to orange-brown; covered with
granular, pyramidal spines or scales, easily

brushed off. **Gills** adnate, crowded; white.
Stem 30–80 × 6–15mm (1¼–3 × ¼–½in),
concolorous with cap; covered below the veil
with a granular coating of scales; smooth above
the faint ring zone. **Flesh** white to reddish.
Odor not distinctive. **Taste** mild. **Spores**
ellipsoid, smooth, nonamyloid, 3.5–5 × 2–3μ.
Deposit white. **Habitat** in conifer woods.
Rather uncommon. Found widely distributed
in North America. **Season** August–October.
Not edible.

Cystoderma granulosum (Fr.) Fayod **Cap** 1–5cm
(½–2in) across, convex; brick red to deep red-
brown; surface granular-warty. **Gills** crowded,
adnate; white. **Stem** 20–60 × 3–6mm
(¾–2¼ × ⅛–¼in); of same color as cap;
sheathed up to ring with mealy granular
coating, smooth above ring; ring slight, soon
vanishing. **Flesh** white. **Odor** not distinctive.
Taste not distinctive. **Spores** ellipsoid, smooth,
nonamyloid, 3.5–5 × 2–3μ. Deposit white.
Habitat on soil or moss in mixed woods.
Widely distributed throughout North America.
Season August–October. **Not edible.**

Cystoderma fallax Smith & Singer **Cap** 3–5cm
(1¼–2in) across, convex-umbonate; dull rusty
brown to tawny ochre; covered with granulose
scales or finely powdery. **Gills** crowded,
adnate; white. **Stem** 30–60 × 3–10mm
(1¼–2¼ × ⅛–½in); dark red-brown and
granular below the ring, pallid above; ring
large, flaring. **Flesh** white. **Odor** pleasant.
Taste mild. **Spores** ellipsoid, smooth, amyloid,
3.5–5.5 × 2.8–3.6μ. Deposit white. **Habitat** on
moss or humus in conifer woods. Found in the
Pacific Northwest across to the Great Lakes.
Season August–October. **Not edible.**
Comment Cap surface turns black with KOH.

Leucocoprinus birnbaumii life-size

Cystoderma cinnabarinum life-size

Cystoderma granulosum almost life-size

Cystoderma fallax life-size

Phaeolepiota aurea ½ life-size

free, close, broad; pure white, then olive-gray in age. **Stem** 100–150 × 10–25mm (4–6 × ½–1in), solid, base enlarged into an oval bulb; creamy, with olive-gray stains in age; longitudinally lined with appressed hairs above and below the ring. **Veil** partial veil leaves a thin, membranous ring; pale pinkish gray and minutely hairy on both sides. **Flesh** thick; white. **Odor** slightly mealy when wet. **Taste** mild or slightly mushroomy. **Spores** ellipsoid to subglobose, smooth, nonamyloid, 4–5 × 3.5–4μ. Deposit white. **Habitat** scattered or in groups under elm or ash, in swamps. Found in Michigan. **Season** September–October. **Edibility not known. Comment** This mushroom was photographed in England.

Armillariella tabescens (Scop ex Fr.) Singer **Cap** 3–10cm (1¼–4in) across, convex then expanded to flat or irregular; margin uplifted in age; ochre-brown with darker, cottony scales; dry. **Gills** partially descending stem, nearly distant, narrow to broad; whitish soon becoming pinkish brown. **Stem** 70–200 × 5–15mm (2¾–8 × ¼–½in), stuffed to hollow, tapering toward the base; whitish or yellowish to brownish; scurfy, fibrous; no ring. **Flesh** thick near center; whitish to stained brown in age. **Odor** strong. **Taste** astringent. **Spores** ellipsoid, smooth, nonamyloid, 6.5–8 × 4.5–5.5μ. Deposit pale cream. **Habitat** in dense clusters attached to living or dead stumps and roots or over buried wood, especially oak. Common. Found widely distributed in eastern North America, west to Kansas and Texas. **Season** July–November. **Edible** when cooked, but has been known to cause stomach upset. **Comment** The spores in my specimens are rather small.

Honey Fungus *Armillariella mellea* (Vahl ex Fr.) Karsten **Cap** 3–15cm (1¼–6in) across, very variable, convex then flattened and centrally depressed or wavy; ochre, tawny to dark brown, often with an olivaceous tinge; covered in dark hairy scales, especially at the center. **Gills** attached or slightly descending stalk, nearly distant, narrow; whitish, then yellowish becoming pinky-brown and often spotted darker with age. **Stem** 60–150 × 5–15mm (2¼–6 × ¼–½in), stuffed to hollow; whitish becoming reddish brown. **Veil** partial veil leaving a thick whitish to yellow cottony ring on upper stem. **Flesh** white. **Odor** strong. **Taste** astringent. **Spores** ellipsoid, smooth, nonamyloid, 8–9 × 5–6μ. Deposit whitish. **Habitat** in small or large clusters at the bases of trees or near stumps. Common. Found widely distributed throughout North America. **Season** August–November. **Edible** but must be cooked. Some cases of severe stomach upset have been reported after eating this mushroom. Eat in small quantities the first time you try it. **Comment** This fungus spreads by long black cords called rhizomorphs, resembling bootlaces, which can be found beneath the bark of infected trees, on roots, or in the soil, where they can travel long distances to infect other trees. This is one of the most dangerous parasites on living trees, causing an intensive white rot and ultimately death. There is no cure, and the fungus is responsible for large losses of timber each year.

Phaeolepiota aurea (Matt. ex Fr.) Maire ex Konrad & Maublanc **Cap** 2–16cm (¾–6¼in) across, obtuse to convex, becoming flatter with a central umbo and the margin often hung with veil remnants; orange-tan to golden brown; dry, granular to powdery. **Gills** adnate to free, close, broad; pale yellow becoming tawny to orange-brown. **Stem** 40–150 × 10–40mm (1½–6 × ½–1½in), expanded toward the base; orange to buff or similar to cap; smooth above the ring, powdery or granular below. **Veil** partial veil sheathing stalk; same color as cap; granular underneath, smooth above; leaving persistent flaring to drooping ring. **Flesh** thick; pale or yellowish. **Spores** ellipsoid, smooth to minutely roughened, 10–14 × 5–6μ. Deposit yellowish brown to orange-buff. **Habitat** in groups or clusters on compost, rich soil, humus, or leaf litter under coniferous or deciduous trees. Quite rare but sometimes abundant. Found in Alaska and the Pacific Northwest. **Season** September–October. **Not edible** because it is mildly poisonous to some people.

Limacella illinita (Fr. ex Fr.) Maire **Cap** 2–7cm (¾–2¾in), round or ovoid becoming convex then flatter or with a broad umbo, the margin hanging with slimy veil remnants; white or creamy white; smooth, very sticky or slimy. **Gills** notched or free, close, broad; white. **Stem** 50–90 × 5–10mm (2–3½ × ¼–½in), tapering slightly toward the top; white; sticky or slimy. **Flesh** thin, soft; white. **Spores** globose to broadly ellipsoid, smooth, nonamyloid, 4.5–6.5 × 4–6μ. Deposit white. **Habitat** singly, scattered, or in groups in woods, swamps, fields, and sand dunes. Quite common. Found widely distributed in North America. **Season** August–October. **Edibility not known.**

Limacella guttata Konrad & Maublanc syn. *Limacella lenticularis* (Lasch) Earle **Cap** 6–15cm (2¼–6in) across, obtusely convex becoming broader with an umbo; margin incurved and cottony at first, then often splitting radially; background whitish, cuticle dull ochre to pale tan or creamy then somewhat cracked. **Gills**

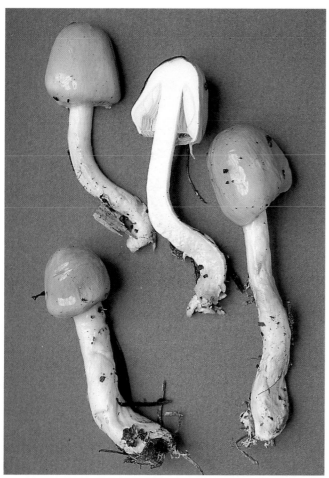

Limacella illinita nearly life-size

Limacella guttata ½ life-size

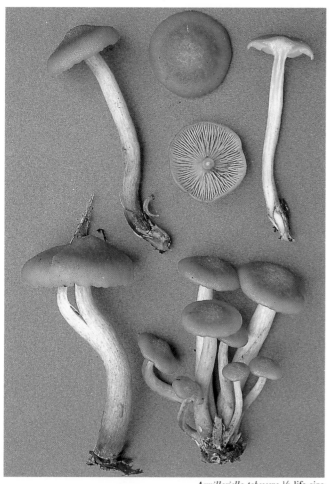

Armillariella tabescens ½ life-size

Honey Fungus *Armillariella mellea* ½ life-size

Armillaria zelleri ⅓ life-size

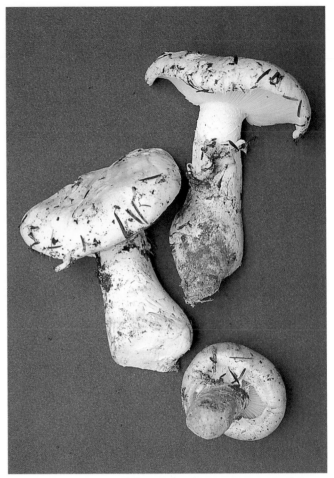

White Matsutake *Armillaria ponderosa* ½ life-size

Armillaria fusca ¾ life-size

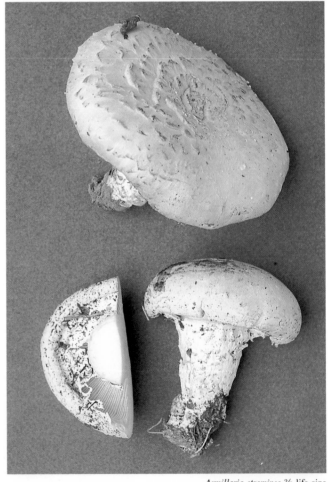

Armillaria straminea ⅔ life-size

Armillaria caligata ½ life-size

Armillaria zelleri Stuntz & Smith **Cap** 5–15cm (2–6in) across, obtuse becoming flat or broadly convex with an umbo and an incurved, cottony margin; orange, brown, olive, and yellow mixed; slimy and minutely hairy. **Gills** adnate, close to crowded, narrow to moderately broad; whitish staining rusty brown. **Stem** 40–130 × 10–30mm (1½–5 × ½–1¼in), tapering to a pointed base; white and cottony above ring, below ring the sheath breaking into orange scales and patches. **Veil** partial veil leaving membranous, ragged ring on upper stalk. **Odor** mealy, unpleasant. **Taste** mealy, unpleasant. **Spores** ellipsoid, smooth, nonamyloid, 4–5 × 3–4μ. Deposit white. **Habitat** scattered or in groups under pine and aspen. Sometimes abundant in Pacific Northwest. Found in northern North America and California and reported from Tennessee. **Season** July–August (December–February in California). **Edible** but not recommended.

White Matsutake *Armillaria ponderosa* (Pk.) Sacc. **Cap** 5–20cm (2–8in) across, convex becoming flatter with an inrolled, cottony margin becoming somewhat uplifted in age; white with flattened reddish-brownish scales and spots, particularly over the center; tacky becoming dry with streaks of brown fibers near the margin. **Gills** adnexed, crowded, narrow to broad; whitish staining pinkish brown. **Stem** 50–150 × 20–40mm (2–6 × ¾–1½in), hard, firm; white becoming pinkish brown from scales and patches of veil remnants; white and cottony above the ring. **Veil** partial veil leaves thick, soft, membranous ring on the upper stalk. **Flesh** firm; white. **Odor** distinctly fragrant. **Spores** broadly ellipsoid to globose, smooth, nonamyloid, 5–7 × 4.5–5.5μ. Deposit white. **Habitat** scattered to numerous under pine and in sandy soil, especially near coastal areas. Common. Found in northern North America and the Rockies. **Season** August–November (December–February in California). **Edible** — excellent. **Comment** Known among orientals as the White Matsutake. This is one of the most sought after edible mushrooms.

Armillaria fusca Mitchel & Smith **Cap** 4–7cm (1½–2¾in) across, convex becoming flat with an umbo and a margin hung with veil remnants; dingy gray-brown or dark fawn; smooth at first but streaked with fine hairs, then becoming appressed, scaly. **Gills** adnexed, close; white to grayish. **Stem** 60–70 × 10–15mm (2¼–2¾ × ½in), equal; white above the ring, below covered in fine, grayish-buff, woolly veil remnants. **Veil** white, partial veil leaving a ring on the upper stem and remnants on the margin. **Taste** mild. **Spores** ellipsoid, smooth, amyloid, 6–8 × 4–5μ. Deposit white. **Habitat** singly or scattered under spruce. Found in Colorado. **Season** June–August. **Edibility not known.**

Armillaria straminea (Krombh.) Kummer var. *americana* Mitchel & Smith **Cap** 4–18cm (1½–7in) across, conical to convex, becoming umbonate then flat, with incurved, cottony margin that straightens in age; straw yellow fading to whitish with conspicuous, flattened, bright yellow or darker scales arranged in concentric circles; dry. **Gills** sinuate, close, broad; whitish then lemon yellow. **Stem** 50–125 × 15–25mm (2–5 × ½–1in), sometimes with a thick bulb; smooth and white above the ring, whitish with shaggy yellowish scales below. **Veil** partial veil leaving thick, yellowish, cottony ring on upper stalk. **Taste** mild. **Spores** ellipsoid, smooth, weakly amyloid, 6–8 × 4–5μ. Deposit white. **Habitat** on the ground under aspen and in mixed woods. Often abundant. Found in the Pacific Northwest and the Rockies. **Season** July–October. **Edibility not known.** **Comment** There is also an albino form of this species.

Armillaria caligata (Viv.) Gilbert **Cap** 5–12cm (2–4¾in) across, broadly convex with margin sometimes uplifted in age and hung with veil remnants; creamish flesh showing beneath cinnamon-brown patches or scales; dry. **Gills** adnate, close, narrow to moderately broad; white, staining brownish with age. **Stem** 50–100 × 20–30mm (2–4 × ¾–1¼in); white above the ring, below cinnamon-brown zones and patches of veil remnants. **Veil** partial veil leaving membranous ring on the upper stem and brownish patches below. **Odor** fragrant, pleasant or foul, disgusting (see Comment). **Taste** slight, mild, or bitter (see Comment). **Spores** broadly ellipsoid, smooth, nonamyloid, 6–7.5 × 4.5–5.5μ. Deposit white. **Habitat** on the ground, sometimes in sandy soil, under hardwoods, particularly oak, in the East; under conifers in the West. Found widely distributed in North America. **Season** July–November. **Edible** — excellent. **Comment** Various forms of this mushroom exist. In Colorado and the West, found under spruce, it is usually fragrant and mild tasting; in the East, under hardwoods, I have found the foul, disgusting-smelling variety with the bitter taste.

Xerula rubrobrunnescens ½ life-size

Xerula furfuracea ⅓ life-size

Xerula megalospora ½ life-size

Tricholoma fulvum ⅔ life-size

Xerula rubrobrunnescens Redhead, Ginns, & Shoemaker **Cap** 2.5–8cm (1–3in) across, convex becoming flat with a distinct, broad umbo; rust to cinnamon with midbrown or honey patches; moist, smooth in the center, wrinkled elsewhere. **Gills** broadly adnate, moderately distant, 2 layers; white with rusty-colored edges, bruising darker. **Stem** 80–115 × 2–20mm (3–4¾ × 3/32–½in), enlarging slightly toward the base, sometimes twisted, long rooting base; white, staining rust when bruised or cut. **Odor** not distinctive. **Taste** not distinctive. **Spores** lemon-shaped to almond-shaped, with a prominent apiculus, finely roughened, nonamyloid, 13.5–16 × 8–9μ. Deposit white. **Habitat** singly or in groups on buried roots in mixed forests. Quite rare. Found in northeastern North America, south to Virginia. **Season** July–September. **Edibility not known.**

Xerula furfuracea (Pk.) Redhead, Ginns, & Shoemaker syn. *Collybia radicata* var. *furfuracea* Pk. **Cap** 2–12cm (¾–4¾in) across, broadly bell-shaped to flat with a distinct, broad umbo in a central depression; smoky brown to dark buff or honey-colored, occasionally with a few darker streaks; dry and slightly velvety becoming moist and greasy when wet, with a translucent edge; wrinkled or puckered on the disc, smooth elsewhere or lined at the margin. **Gills** broadly adnate to uncinate, subdistant, 2–3 layers; whitish, sometimes with hazel-colored edges. **Stem** 70–125 × 2–13mm (2¾–5 × ⅛–½in), slightly enlarged toward the base, rooting; pale grayish brown to pale gray or whitish; furfuraceous, often lined, powdery at the top, often hairy where the stem enters the soil; long white rooting section bruises rust. **Flesh** white, not staining. **Odor** not distinctive. **Taste** not distinctive. **Spores** broadly ovoid to ellipsoid, with a prominent apiculus, smooth, nonamyloid, 14–16 × 9.5–11μ. Deposit white. **Habitat** singly or in groups on buried hardwood roots. Common. Found in northeastern North America, south to West Virginia and west to Michigan. **Season** June–October. **Edible.**

Xerula megalospora (Clements in Clements & Pound) Redhead, Ginns, & Shoemaker. **Cap** 2–8cm (¾–3in) across, broadly conical becoming flatter and slightly depressed around a central umbo, with a lined margin that is sometimes scalloped; pale smoky brown to sepia or buff or dark cream, sometimes slightly darker in the center and at the edge; sticky when wet, becoming dry and almost polished, radially lined to wrinkled. **Gills** broadly adnate to uncinate, distant to subdistant, 2–4 layers; white. **Stem** 60–130 × 2–10mm (2¼–5 × 3/32–½in), slightly enlarged toward the base, rooting, sometimes a little twisted; white, dry, smooth, often silky and finely lined; pseudoroot stains rusty. **Flesh** white, not staining. **Odor** mildly or strongly of geraniums or carrots. **Taste** not distinctive. **Spores** lemon-shaped to almond-shaped, finely roughened, nonamyloid, 18–23 × 10–14μ. Deposit white. **Habitat** singly or in groups on buried roots in deciduous woods. Found in eastern North America, south to Louisiana and west to Michigan and Nebraska. **Season** June–October. **Edibility not known.**

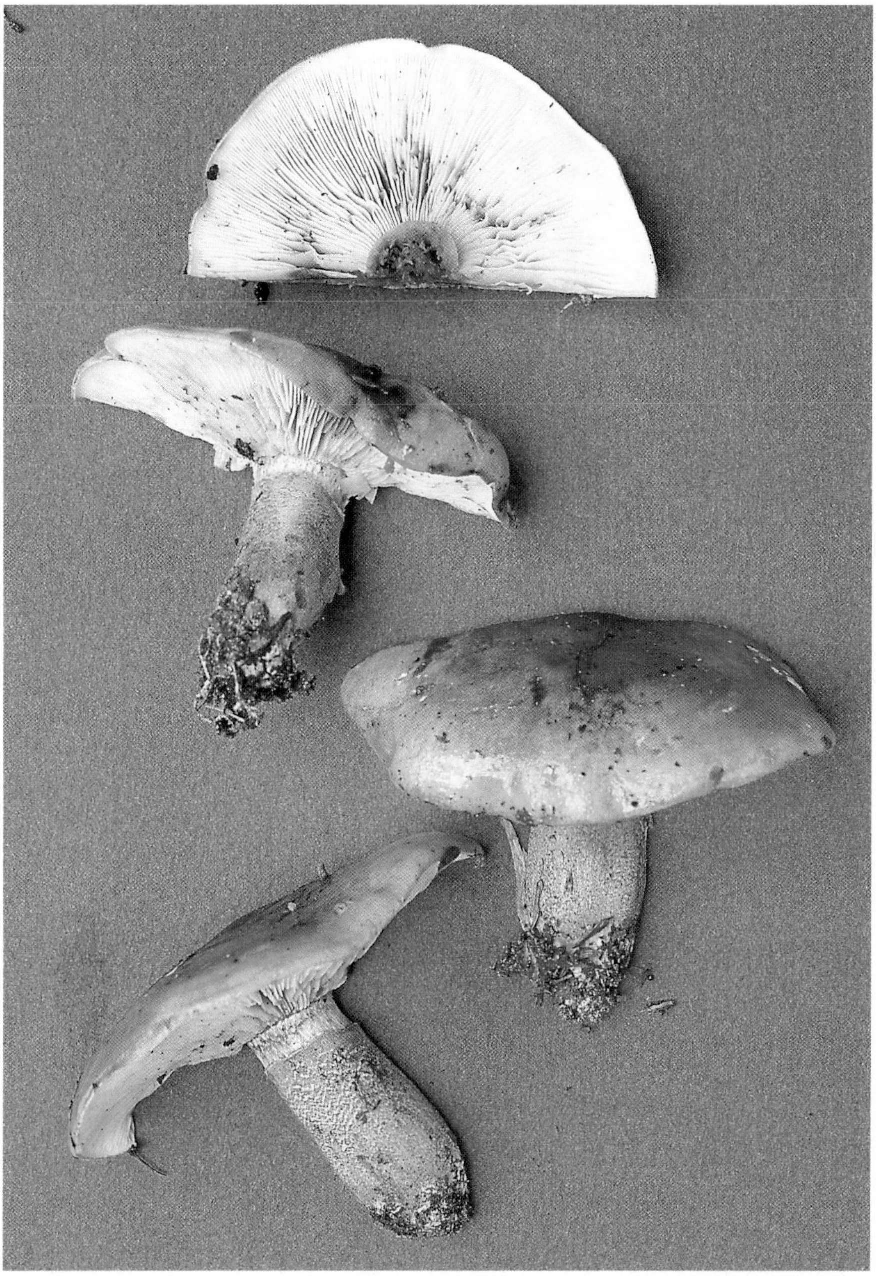

Tricholoma aurantium life-size

Tricholoma aurantium (Fr.) Ricken **Cap** 4–10cm (1½–4in) across, convex becoming flatter with an obtuse umbo and an inrolled margin at first; color varies from yellow-orange to tawny to rusty orange or orange-brown or orange-red sometimes splashed with green; sticky when moist, smooth then breaking into small scales; sometimes orange droplets on the margin when moist. **Gills** adnate or notched, close, narrow; white staining rusty brown. **Stem** 30–80 × 8–20mm (1¼–3 × 5/16–¾in), solid, firm; white background densely covered with rusty-orange scales below the ring zone; scaly or scurfy; small, cobweblike partial veil leaves a line on the stem. **Flesh** thick; white. **Odor** strongly of rancid meal. **Taste** disagreeable, of meal or rancid oil. **Spores** ellipsoid to subglobose, smooth, 4–6 × 3–5μ. Deposit white. **Habitat** singly, scattered, or in groups on the ground in mixed woods, under conifers, aspen, and madrone. Sometimes common and abundant. Found widely distributed in North America. **Season** July–October (November–February in California). **Not edible.**

Tricholoma fulvum (DC ex Fr.) Sacc. **Cap** 3–10cm (1¼–4in) across, expanded, convex with a slight umbo; the incurved margin often becoming wavy in age; honey brown to reddish brown; sticky becoming finely dry, streaky, scurfy, occasionally cracking. **Gills** sinuate or adnate, close, broad; yellowish, becoming spotted brownish in age. **Stem** 30–70 × 8–15mm (1¼–2¾ × 5/16–½in), sometimes slender or short and stout with the base rounded or abruptly tapering; yellow discoloring reddish brown in large patches then all over, but the top usually remaining yellow; silky-hairy and velvety at the top. **Flesh** pale yellow, discoloring brown around worm holes. **Odor** mealy. **Taste** mealy. **Spores** ellipsoid, smooth, nonamyloid, 5.7–7.2 × 4.3–5.2μ. Deposit white. **Habitat** in groups or dense clusters under hardwoods or in mixed hardwood and coniferous forests. Frequent. Found in Michigan and the Pacific Northwest. **Season** September–November. **Not edible.**

Tricholoma flavovirens ½ life-size

Tricholoma columbetta ⅓ life-size

Tricholoma odorum ½ life-size

Tricholoma intermedium ½ life-size

Tricholoma vaccinum ¾ life-size

Tricholoma flavovirens (Pers. Ex Fr.) Lundell **Cap**
5–12cm (2–4¾in) across, convex becoming
flat, margin inrolled becoming uplifted and
sometimes split; bright yellow, often with olive-
brown or brown disc; slimy-tacky when moist,
smooth or with adpressed scales in the center.
Gills notched or adnexed, close, broad; bright
sulphur or lemon yellow. **Stem** 30–80 ×
8–20mm (1¼–3 × ³⁄₁₆–¾in), solid, stout,
slightly bulbous toward base; pale yellow with
darker stains at base; dry, smooth. **Odor**
slight. **Taste** slightly mushroomy. **Spores**
ellipsoid, smooth 6–8×3–5μ. Deposit white.
Habitat scattered or in dense groups under
pine in sandy, grassy, or shrubby areas.
Sometimes common. Found widely distributed
in North America. **Season** August–November
(October–January in California).
Edible — good.

Tricholoma columbetta (Fr.) Kummer **Cap**
5–10cm (2–4in) across, convex then
expanding, with a wavy sometimes cracked
margin; whitish, sometimes with greenish,
pinkish, or violet-blue spots when old;
smooth, silky. **Gills** adnate to sinuate,
crowded; white. **Stem** 40–100×10–20mm
(1½–4 × ½–¾in), rooting; whitish, some blue-
green at base; smooth. **Flesh** firm; white.
Odor mild. **Taste** none. **Spores** oval,
5.5–7×3.5–5μ. Deposit white. **Habitat** singly
or scattered or in small groups in deciduous
and coniferous woods. Rare. Found in
Colorado. **Season** August–September. **Edible.**
Comment This seems to be a first record of
this species in North America.

Tricholoma odorum Pk. **Cap** 2–9cm (¾–3½in)
across, obtusely convex expanding to flatter
and occasionally becoming concave, often with
an umbo and incurved margin becoming wavy
at times; greenish yellow when young,
becoming yellow then fading to yellowish buff
or dark buff with the disc light brown; dry,
dull, smooth, naked. **Gills** sinuate or adnate,
close, broad; yellow to pale yellowish buff.
Stem 30–110×5–15mm (1¼–4½ × ¼–½in),
often twisted, with a bulbous or rounded base,
sometimes tapering abruptly; light greenish
yellow to very pale yellow in age, sometimes
bruising brownish-spotted at the base. **Flesh**
thick; pale greenish yellow then pale buff.
Odor of coal tar. **Taste** mealy. **Spores**
ellipsoid to fusiform, smooth, nonamyloid,
9.5–11.4×4.8–6.7μ. Deposit white. **Habitat** in
groups or dense clusters under hardwoods or in
mixed deciduous and coniferous woods. Quite
common. Found widely distributed in
northeastern North America, west to the Great
Lakes. **Season** August–October. **Not edible.**

Tricholoma vaccinum (Pers. ex Fr.) Kummer **Cap**
2–7cm (¾–2¾in) across, obtusely conic
expanding to convex then flatter and slightly
umbonate, with an incurved margin hanging
with hairy flaps of veil remnants, disappearing
later as margin expands; flesh-brown or rusty
tan and darker toward the center; dry,
minutely hairy, then breaking into woolly
scales. **Gills** adnate becoming notched, close,
broad; white at first, later pallid flesh-color.
Stem 30–80×6–22mm (1¼–3 × ¼–¾in),
hollow, the base usually rounded or abruptly
tapered; similar color to cap or paler; dry,
silky-hairy, sometimes with small brown scales.
Flesh often hollow in the stem; pallid to rosy.
Odor mealy. **Taste** bitter. **Spores** ovoid to
ellipsoid, smooth, nonamyloid, 6.2–7.6×
4.3–5.2μ. Deposit white. **Habitat** scattered or
in dense tufts under conifers, especially pine
and spruce. Common and often abundant.
Found widely distributed in northern North
America and California. **Season** July–
December. **Not edible** —possibly poisonous.

Tricholoma intermedium Pk. **Cap** 3.5–10cm
(1¼–4in) across, convex becoming flatter,
often slightly concave or with a rounded umbo;
light yellow with a more olive-tan center, often
fading in age; sticky becoming dry, smooth or
minutely scaly over the center. **Gills** sinuate,
close, broad; white. **Stem** 35–90×7–19mm
(1¼–3½ × ¼–¾in), solid becoming hollow,
base somewhat bulbous and rounded; white or
pale yellowish; silky-hairy, often with a light
bloom at the top. **Flesh** thick; white. **Odor**
mealy. **Taste** mealy. **Spores** ellipsoid, smooth,
nonamyloid, 5.7–6.7×3.8–4.8μ. Deposit white.
Habitat in groups under conifers or occasionally in mixed woods. Found in northern
North America. **Season** September–October.
Edibility not known.

Tricholoma saponaceum ½ life-size

Tricholoma saponaceum var. *ardosiacum* ¾ life-size

Tricholoma saponaceum (Fr.) Kummer **Cap** 4–10cm (1½–4in) across, convex at first, then expanded with a broad umbo; incurved margin becoming uplifted, wavy, or cracked in age; color very variable, from dark olive-gray to gray-brown, often with rusty or olivaceous tints and darker at the center; dry to moist, smooth becoming cracked with fine scales. **Gills** adnate to adnexed or notched, almost distant, broad; whitish sometimes with greenish tints or finely spotted reddish. **Stem** 50–100 × 10–30mm (2–4 × ½–1¼in), solid, variable but often thick in the middle and tapering to a rooting base; white with reddish or olivaceous tints; smooth or finely scaly. **Flesh** thick; white becoming more or less pink. **Odor** soapy. **Taste** mushroomy. **Spores** ellipsoid, smooth, nonamyloid, 5.2–7.6 × 3.3–4.8μ. Deposit white. **Habitat** singly, scattered, or in groups or tufts under conifers or hardwoods. Sometimes common. Found widely distributed in North America. **Season** July–November. **Not edible. Comment** *Tricholoma saponaceum* var. *ardosiacum* Bres. differs from the type in having a much darker, bluish-gray, almost black cap and sometimes gray or brown scales on the stem.

Tricholoma portentosum (Fr.) Quél. **Cap** 4–12cm (1½–4¾in) across, conical to bell-shaped expanding with a broad umbo; margin incurved at first, sometimes becoming uplifted in age; light gray to gray-black with violaceous or olivaceous tints; slimy-sticky, smooth with finely hairy, radiating, innate streaks. **Gills** adnexed or notched, close, broad; whitish becoming lemon yellow or graying. **Stem** 40–100 × 10–20mm (1½–4 × ½–¾in), stout,

solid; white, often becoming flushed lemon yellow; dry, streaked with hairs. **Flesh** quite thick; white or gray-tinged. **Odor** mealy. **Taste** mild or mealy. **Spores** ellipsoid, smooth, 5–6 × 3.5–5μ. Deposit white. **Habitat** scattered or in groups in sandy soil under conifers and oaks. Frequent. Found widely distributed in northern North America, the Rocky Mountains, and California. **Season** October–November (February in California). **Edible** — good.

Tricholoma pardinum Quél. **Cap** 5–15cm (2–6in) across, convex becoming flatter; whitish with pale-gray to dark-gray scales; dry with hairy scales. **Gills** notched or adnexed, close, moderately broad; whitish very occasionally flushed pinkish. **Stem** 50–150 × 10–20mm (2–6 × ½–¾in), firm, solid, sometimes enlarged at the base; white, sometimes tinged gray; dry, smooth. **Flesh** thick, firm; white. **Odor** mealy. **Taste** mealy. **Spores** ellipsoid, smooth, 8–10 × 5.5–6.5μ. Deposit white. **Habitat** singly, scattered, or in groups on the ground in mixed and deciduous woods, especially under fir. Sporadically common and abundant, otherwise infrequent. Found widely distributed in northern North America. **Season** August–October. **Poisonous.**

Tricholoma myomyces (Fr.) Lange **Cap** 1–7cm (½–2¾in) across, obtusely conic expanding to convex, then flat with a low umbo; margin incurved at first, then often wavy; dark drab gray to brownish gray or blackish gray, generally paler on the margin; dry, densely matted, and hairy on the disc and hairy to scaly elsewhere. **Gills** arcuate to sinuate, close,

broad; light gray, fading near the stem in age, very rarely discoloring with dull yellow spots. **Stem** 15–70 × 5–10mm (½–2¾ × ¼–½in), solid or hollow, generally rounded or abruptly tapered; white to pale gray; silky with white or gray hairs. **Veil** a cortina of white or gray hairs that leaves a faint, quickly disappearing zone on the stem. **Flesh** pale gray. **Spores** ellipsoid, smooth, nonamyloid, 6.7–7.6 × 4.3–4.8μ (4-spored form), 8.6–11.4 × 3.8–5.7μ (2-spored form). Deposit white. **Habitat** in groups or dense clusters under conifers in woods or on lawns. Frequent and sometimes abundant. Found widely distributed in northern North America. **Season** August–October. **Edibility not known.**

Tricholoma virgatum (Fr. ex Fr.) Kummer **Cap** 3–8cm (1¼–3in), conical to broadly conical or convex becoming flatter with a pointed umbo; brownish black or grayish initially with violaceous tints, paler toward the margin; dry, streaked with very fine black radiating hairs or scales. **Gills** adnexed or notched, close, broad; white or grayish-tinged flesh color. **Stem** 60–120 × 10–20mm (2¼–4¾ × ½–¾in), solid; white or tinged gray; smooth or minutely hairy. **Flesh** thin; white to grayish. **Odor** musty. **Taste** bitter and peppery. **Spores** ellipsoid, smooth, 6–7.5 × 5–6μ. Deposit white. **Habitat** singly, scattered, or in groups in mixed woods and under conifers. Frequent. Found widely distributed in North America. **Season** August–November. **Not edible** — possibly poisonous.

Tricholoma portentosum ⅔ life-size

Tricholoma pardinum ½ life-size

Tricholoma myomyces ¾ life-size

Tricholoma virgatum ⅓ life-size

Plums and Custard *Tricholomopsis rutilans* ⅔ life-size

Tricholomopsis platyphylla ⅓ life-size

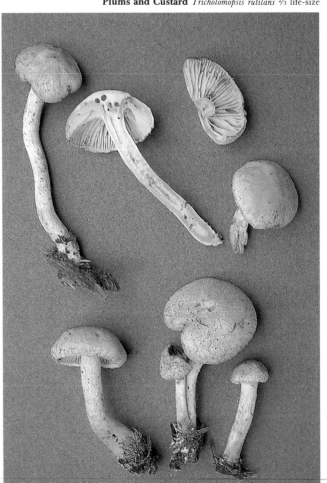

Tricholomopsis sulfureoides almost life-size

Tricholomopsis decora ¾ life-size

Calocybe carnea ½ life-size

Tricholomopsis formosa nearly life-size

Tricholomopsis flavissima ⅓ life-size

Calocybe carnea (Bull. ex. Fr.) Kühner syn. *Tricholoma carneum* (Bull. ex Fr.) Kühner & Romagnesi **Cap** 1.5–3.5cm (½–1¼in) across, flattened-convex; distinctive flesh-pink color flushed brown at the center; dry, smooth to minutely hairy. **Gills** attached or slightly descending the stem, crowded, narrow; white. **Stem** 20–40 × 2–7mm (¾–1½ × ³⁄₃₂–¼in), hollow; pinkish, fibrous, minutely felty. **Flesh** whitish. **Spores** ellipsoid, smooth, 4.5–5.5 × 2.5–3μ. Deposit white. **Habitat** on the ground, often in small clumps, in lawns and open places or woods. Uncommon. Found widely distributed in northern North America. **Season** August–October. **Edible.**

Plums and Custard *Tricholomopsis rutilans* (Schaeff. ex Fr.) Singer **Cap** 4–10cm (1½–4in) across, convex to bell-shaped, expanding to almost flat, often with a low, broad umbo; margin inrolled at first; yellow; dry, densely covered in reddish-purple downy tufts or scales, denser on the disc. **Gills** adnate or notched, crowded, narrow; rich egg yellow. **Stem** 50–100 × 5–20mm (2–4 × ¼–¾in), curved, stuffed becoming hollow; purplish scales. **Flesh** pale yellow or cream. **Odor** pungent, like rotten wood. **Taste** watery. **Spores** broadly ellipsoid, smooth, 5–7 × 3.5–5μ. Deposit white. Cheilocystidia present, thin-walled, voluminous. **Habitat** singly, several, or in groups on coniferous wood. Frequent. Found widely distributed in North America. **Season** June–October (November–February in California). **Edible** but of poor quality and not recommended.

Tricholomopsis platyphylla (Pers. ex Fr.) Singer **Cap** 3–12cm (1¼–4¾in) across, hemispherical soon expanded to almost flat with a sunken center and sometimes a low umbo, margin often splitting in age; dark gray-brown with an ochraceous tint, sometimes lighter; somewhat streaky surface from radial fibers sometimes forming minute scurfy scales toward the margin. **Gills** adnate, almost distant, broad; whitish to grayish. **Stem** 60–120 × 10–30mm (2¼–4¾ × ½–1¼in), tough, solid becoming hollow; white, often flushed with cap color, with thick white rooting mycelial strands; hairy-striate. **Flesh** white. **Odor** not

distinctive. **Taste** bitter. **Spores** subglobose, smooth, 7–9 × 5–7μ. Deposit white. Cheilocystidia present, swollen pear-shaped. **Habitat** singly or scattered on hardwood stumps, logs, debris, and pieces of buried wood. Common. Found widely distributed in eastern North America west to Ohio, and occasionally reported on the West Coast. **Season** May–July and occasionally until early October. **Edible** with caution as there have been a few reports of gastric upset.

Tricholomopsis sulfureoides (Pk.) Singer **Cap** 3–7cm (1¼–2¾in) across, convex becoming flatter and sometimes with a broad umbo; yellow with some paler patches at times, drying pale reddish brown; finely hairy becoming scaly, with scales darkening in age. **Gills** slightly decurrent or notched, close, broad; similar color to cap with a fringed edge. **Stem** 30–70 × 3–7mm (1¼–2¾ × ⅛–¼in), off-center, curved, stuffed becoming hollow; yellowish drying pale reddish brown; minutely hairy below. **Flesh** pale yellow, unchanging. **Spores** ovoid, smooth, 5.5–6.5 × 4.5–5μ. Deposit white. **Habitat** singly or in groups on coniferous logs, particularly eastern hemlock. Found in northeastern North America, south to North Carolina and west to the Great Lakes; also found in the Pacific Northwest. **Season** July–October. **Edibility not known.**

Tricholomopsis decora (Fr.) Singer **Cap** 3–6cm (1¼–2¼in) across, convex at first, becoming centrally depressed; deep golden yellow; covered in tiny brownish-black fibrillose scales, especially at the center. **Gills** adnate, crowded, narrow; yellow. **Stem** 30–60 × 3–10mm (1¼–2¼ × ⅛–½in), occasionally off-center; yellow; minutely scaly. **Flesh** deep yellow. **Spores** ellipsoid, smooth, 6–7.5 × 4.5–5μ. Deposit white. Pleurocystidia rare; cheilocystidia abundant. **Habitat** singly, scattered, or in groups on conifer logs. Common but not abundant. Found widely distributed throughout North America. **Season** July–October. **Edibility not known.**

Tricholomopsis formosa (Murr.) Singer **Cap** 5–8cm (2–3in) across, convex to flat with an incurved margin; background color dark buff;

covered with dark, brick-red, recurved scales. **Gills** adnate, crowded, edges fringed; whitish to yellowish. **Stem** 60–80 × 10–20mm (2¼–3 × ½–¾in), sometimes curved in age; dark brick red; hairy-scaly like the cap, longitudinally furrowed in age. **Flesh** firm. **Odor** unpleasant. **Taste** disagreeable. **Spores** pip-shaped, thin-walled, 6–7.2 × 4–5.7μ. Deposit white. Pleurocystidia absent; cheilocystidia present. **Habitat** on pine sawdust. Frequent. Found in Florida and the Gulf Coast. **Season** July–November. **Edibility not known.**

Tricholomopsis flavissima (Smith) Singer **Cap** 3–5cm (1¼–2in) across, broadly convex becoming flatter and centrally depressed; even chrome yellow all over but darker on the center disc; dry and finely hairy, sometimes becoming more scaly on the disc and sometimes margin fringed with clusters of tiny hairs. **Gills** adnate or notched, subdistant; strong yellow with brighter edges appearing gelatinous under the lens. **Stem** 20–50 × 3–6mm (¾–2 × ⅛–¼in), hollow, slightly curved; pale when young from thin, yellowish partial veil, the same color as cap; appressed, minute hairs. **Flesh** dull yellow. **Odor** faintly fragrant. **Taste** slightly peppery. **Spores** globose, 7–9 × 7–9μ. Deposit white. Pleurocystidia absent; cheilocystidia present. **Habitat** scattered or in clusters or tufts on coniferous wood. Frequent. Found in the Pacific Northwest. **Season** September–October. **Edibility not known. Comment** The two mushrooms at the top left of the photograph and the two at the bottom right are *Tricholomopsis flavissima;* the other two with the darker colors are *Tricholomopsis decora* (q.v.; nobody can be right all the time!).

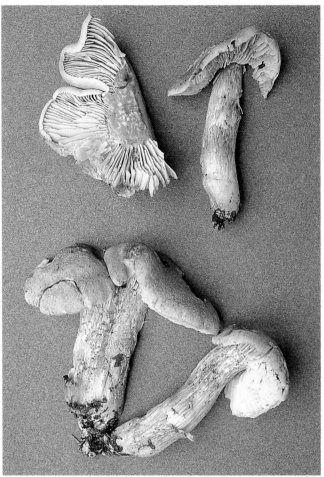

Porpoloma umbrosum ¾ life-size

Lyophyllum montanum ½ life-size

Porpoloma umbrosum (Smith & Wall.) Singer **Cap** 5–10cm (2–4in) across, convex-flattened; dull gray-brown; dry, smooth, often cracking with age. **Gills** adnate, crowded, broad; gray bruising reddish. **Stem** 25–50 × 15–20mm (1–2 × ½–¾in), tapered at base; gray-brown. **Flesh** pale gray bruising reddish. **Taste** mealy, cucumber-like. **Odor** strong, mealy, cucumber-like. **Spores** ellipsoid, amyloid, 7–9 × 3–4μ. Deposit white. **Habitat** under conifers. Found widely distributed in North America. **Season** July–September. **Edibility not known.**

Lyophyllum montanum Smith **Cap** 3–8cm (1¼–3in) across, convex to umbonate; grayish clay with a white hoary coating; glabrous to slightly greasy surface. **Gills** crowded, emarginate; grayish. **Stem** 30–80 × 10–15mm (1¼–3 × ½in), equal, gray-brown with hoary coating. **Flesh** firm; brownish to pallid. **Odor** pleasant. **Taste** mild. **Spores** long ellipsoid, 6.5–8 × 3.5–4μ. Deposit white. **Habitat** caespitose under spruce. Found close to melting snow in the Rocky Mountains. **Season** April–August. **Edible.**

Fried Chicken Mushroom *Lyophyllum decastes* (Fr. ex Fr.) Singer **Cap** 3–12cm (1¼–4¾in) across, convex then expanded becoming wavy; gray-brown to yellowish brown with silky or silvery streaks; smooth, soapy. **Gills** adnate to slightly decurrent, moderately broad; white to grayish. **Stem** 30–100 × 10–25mm (1¼–4 × ½–1in), stout, often eccentric; white at the top becoming brownish toward the base; fibrous,

tough. **Flesh** firm; white. **Odor** mild. **Taste** not distinctive. **Spores** subglobose, smooth, 5–7 × 5–6μ. Deposit white. **Habitat** in clusters on the ground in waste places, grassy areas, wood edges, and paths. Common and sometimes abundant. Found widely distributed in North America. **Season** June–October (overwinters in California). **Edible** — good. **Comment** This is a much sought after edible species but should be tried with caution as there have been some reports of gastric upset.

Lyophyllum connatum (Schum. ex Fr.) Singer **Cap** 3–7cm (1¼–2¾in) across, broadly convex to flat, margin often wavy; pure white with some small buff-colored scurfy patches. **Gills** slightly decurrent; white, gradually turning violet in FeSO₄. **Stem** 30–60 × 8–15mm (1¼–2¼ × ⁵⁄₁₆–½in), tapering toward the base; white. **Flesh** white, gradually turning violet in FeSO₄. **Spores** long ellipsoid, 6–7 × 3.5–4μ. Deposit white. **Habitat** in dense tufts in mixed woodlands. Found in Colorado and probably other areas. **Season** August. **Edibility not known. Comment** This species is a European species that I am reporting for the first time from Colorado.

Lyophyllum luteogriseascens Clemenceau & Smith **Cap** 3–6cm (1¼–2¼in) across, obtuse to convex expanding to almost flat, with a margin inrolled at first; creamy white staining yellowish and finally grayish all over; moist, smooth, cracking when dry. **Gills** broadly adnate, subdistant; whitish staining orangish

yellow then gray. **Stem** 20–40 × 10mm (¾–1½ × ½in), club-shaped; creamy white, but staining like the gills and sooty brown when dried; smooth, but with a faint bloom at the top. **Flesh** pallid to yellowish bruising orangish. **Odor** distinctly resembling fresh green corn. **Taste** mild. **Spores** broadly fusoid, smooth, nonamyloid, 6–10.2 × 3.6–6μ. Deposit whitish. **Habitat** singly in hardwood mixed forests. Found in Michigan and New Jersey. **Season** July. **Edibility not known. Comment** Note the distinctive way this mushroom bruises blackish.

Lyophyllum palustre (Pk.) Singer syn. *Collybia palustris* (Pk.) Smith **Cap** 1–2cm (½–¾in) across, bell-shaped flattening when expanded, usually with an umbo; grayish brown becoming paler when dry; striate almost to the paler center. **Gills** adnate or with a small decurrent tooth; pale buff. **Stem** 20–30 × 2–3mm (¾–1¼ × ³⁄₃₂–⅛in), long, thin; pallid; fragile. **Flesh** pale buff. **Odor** mealy. **Spores** ovoid, 5.5–8.5 × 4–5μ. Deposit white. **Habitat** often in groups on sphagnum bogs. Found in northeastern North America. **Season** July–September. **Edibility not known.**

Fried Chicken Mushroom *Lyophyllum decastes* ½ life-size

Lyophyllum connatum ½ life-size

Lyophyllum luteogriseascens almost life-size

Lyophyllum palustre life-size

Melanoleuca alboflavida ⅔ life-size

Melanoleuca alboflavida (Pk.) Murr. **Cap** 3–10cm (1¼–4in) across, broadly convex becoming flatter then often depressed with a low, broad umbo; margin incurved at first; yellowish brown to cream or whitish, the umbo remaining darker; smooth, dry to moist. **Gills** sinuate, crowded, narrow; whitish. **Stem** 30–100×4–10mm (1¼–4×³⁄₁₆–½in), slender with a small bulb at the base; whitish; cartilaginous, longitudinally lined with minute hairs. **Flesh** solid; white. **Spores** ellipsoid to ovoid, ornamented with small warts, 7–9×4–5.5μ. Deposit white. **Habitat** singly or in groups on the ground in deciduous or mixed woods and in open fields. Common. Found in northeastern North America, west to the Great Lakes. **Season** July–September. **Edible.**

Melanoleuca cognata (Fr.) Konrad and Maublanc **Cap** 5–13cm (2–5in) across, convex expanding to flat or slightly depressed in age with a broad umbo and sometimes a wavy margin; ochre-brown to gray-brown, fading in age; smooth, shiny, dry or slightly sticky. **Gills** sinuate, crowded, becoming broad; whitish to ochraceous cream to tan or pinkish cinnamon. **Stem** 60–110×10–16mm (2¼–4½×½–¾in), swollen at the base; color similar to cap or paler, base sometimes stained brownish; longitudinally lined. **Flesh** whitish to cream. **Odor** flowery. **Taste** sweetish. **Spores** ellipsoid, minutely warted, amyloid, 7–10×4.5–6μ. Deposit ochraceous cream. Cystidia abundant on gills. **Habitat** singly, scattered, or in small groups on wet soil in dense forests or wood edges under conifers. Not common. Found widely distributed in Colorado and the Southwest. **Season** May–September. **Edible. Comment** My specimens are rather old, faded, and pale.

Melanoleuca melaleuca (Pers. ex Fr.) Murr. **Cap** 2–8cm (¾–3in) across, convex then flattened, often slightly depressed with a central umbo; dark brown when moist, drying to buff, center often darker; smooth, dry, moist, hygrophanous. **Gills** sinuate, crowded, narrow; whitish to cream, tan, or salmon. **Stem** 35–130× 3–18mm (1¼–5×⅛–¾in), quite slender with a slightly bulbous base; whitish with dark, gray-brown hairs; longitudinally lined with hairs, often minutely scurfy at the top. **Flesh** thin; whitish in cap, flushed ochre-brown from stem base upward. **Odor** mild. **Taste** mild or slightly unpleasant. **Spores** ellipsoid, minutely ornamented, amyloid, 7–8.5×5–5.5μ. Deposit cream. Cystidia harpoonlike, frequent. **Habitat** scattered or in groups on the ground in lawns, pastures, paths, and open trails and also in woods. Common. Found widely distributed in North America. **Season** June–October (October–February in California). **Edible. Comment** Some mycologists claim this to be an aggregate species in which are placed all those brown collections to which a name cannot be put.

Melanoleuca cognata ¼ life-size

Melanoleuca melaleuca ½ life-size

CLITOCYBE *The gills are decurrent (running down the stem). The caps often have a central depression. Some have interesting and distinctive smells; in fact, the majority have a rather sweet smell which I can only describe as a clitocybe smell.*

Clitocybe robusta Pk. **Cap** 3–16cm (1¼–6¼in) across, flattened-convex with an inrolled margin, becoming flatter with a shallowly depressed disc and a wavy margin; whitish becoming pale tan; smooth, dry with a downy bloom, patchy and slightly sticky when wet. **Gills** adnate to decurrent, close, narrow becoming moderately broad; white to dark cream. **Stem** 45–100×10–35mm (1½–4× ½–1¼in), solid becoming hollow, base slightly bulbous; whitish; surface felty, dotted with fine hairs or fine scales at the top and base covered in fine white down. **Flesh** thick at the disc, firm when dry; white. **Odor** strong, unpleasant, like *Clitocybe nebularis* (Batsch ex Fr.) Kummer. **Taste** sweetish to rancid. **Spores** ellipsoid, smooth, nonamyloid, 5.5–7.5×3–4.5μ. Deposit creamy yellow. **Habitat** in groups or clusters on leaf debris under conifers or in deciduous woods. Found widely distributed in many parts of North America. **Season** August–October. **Not edible. Comment** I found this species growing in vast and numerous overlapping rings.

Clitocybe phyllophila (Fr.) Kummer **Cap** 3–10cm (1¼–4in) across, convex at first, later flatter or irregularly funnel-shaped with a wavy margin; whitish or watery flesh-colored with brownish patches when moist; opaque, covered in a dull white bloom. **Gills** decurrent, crowded, narrow to moderately broad; whitish to flesh-colored. **Stem** 25–65×5–15mm (1½–2½×¼–½in), stuffed becoming hollow, swollen at the base; whitish or light tan; finely hairy and base covered in white down. **Flesh** thicker at disc, firm when dry; whitish to tan. **Odor** strong, sweet. **Taste** mild, becoming unpleasant in age. **Spores** ovoid to ellipsoid, smooth, nonamyloid, 4–4.5×3–3.5μ. Deposit pale ochraceous clay. **Habitat** in groups or dense clusters among leaf litter and pine needles in mixed woods. Found in northeastern North America and the Pacific Northwest. **Season** September–October. **Not edible.**

Clitocybe dilatata Pers. ex Karsten **Cap** 3–15cm (1¼–6in) across, convex becoming flat with a swollen umbo; an incurved margin becoming irregular, upturned, and wavy; often misshaped from overlapping; gray then whitish or chalky; dry, smooth, downy. **Gills** adnate to decurrent, close, narrow then broad; whitish to buff. **Stem** 50–125×5–20mm (2–4¾×¼–¾in), solid becoming hollow, often curved, enlarged toward the base, which is sometimes united with several others; white, bruising darker at base; finely felty and furrowed, sometimes minutely scaly toward the base. **Flesh** firm, thicker on the disc; watery gray to whitish. **Odor** none. **Taste** unpleasant. **Spores** ellipsoid, smooth, nonamyloid, 4.5–6.5× 3–3.5μ. Deposit white. **Habitat** in groups or dense clusters, often overlapping, along roads or in sandy soil or gravel. Found in the Pacific Northwest and California. **Season** May–November. **Poisonous.**

Clitocybe robusta ⅔ life-size

Clitocybe phyllophila ⅓ life-size

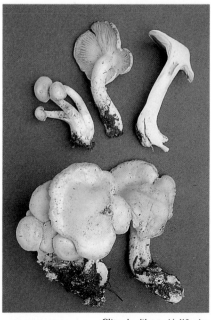

Clitocybe dilatata ¼ life-size

Clubfoot *Clitocybe clavipes* (Pers. ex. Fr.)
Kummer **Cap** 2–8cm (¾–3in) across, flattened
convex with a slight umbo at first, later
depressed; buff to gray-brown with an
olivaceous tint, paling toward the margin;
moist, smooth, with matted hairs and scurfy on
the disc. **Gills** deeply decurrent, nearly distant,
narrow to broad; pale creamy yellow. **Stem**
30–70×5–15mm (1¼–2¾×¼–½in), stuffed,
greatly swollen at the base and tapering
upward; whitish to ash; spongy, covered in
silky fibers. **Flesh** thick, spongy then rather
brittle when dry; white, but yellowish toward
the base. **Odor** strong, sweet. **Taste** mild.
Spores subglobose to ellipsoid, smooth,
nonamyloid, 5–8×3–5µ. Deposit white.
Habitat singly or in groups on the ground
under conifers or sometimes deciduous trees.
Common. Found widely distributed in North
America. **Season** July–November (December–
March in California). **Not edible. Comment**
If eaten in conjunction with alcohol a headache
or transient rash may occur.

Clitocybe fragrans (Fr.) Kummer **Cap** 1.5–4cm
(½–1½in) across, flattened convex sometimes
slightly depressed, with an inrolled margin
becoming somewhat wavy in age; hygrophanous, pale yellowish brown when wet, whitish
cream when dry, with a darker center; smooth,
finely lined at the margin. **Gills** adnate to
slightly decurrent, close, narrow to moderately
broad; whitish buff. **Stem** 30–60×3–6mm
(1¼–2¼×⅛–¼in), stuffed then hollow, often
curved and slightly enlarged toward the base;
whitish to palebuff; silky with fine hairs on
stem, felty with a few thin rhizoids at the base.
Flesh thin, soft, pliant; whitish to buff. **Odor**
of aniseed. **Taste** of aniseed. **Spores** ellipsoid,
smooth, nonamyloid, 6.5–8×3.5–4µ. Deposit
white. **Habitat** growing either scattered, in
groups, or in clusters under deciduous trees.
Found in northeastern and northwestern North
America and California. **Season**
July–September (December in California).
Edible but avoid due to possible confusion
with similar dangerously poisonous species.

Clitocybe odora (Fr.) Kummer **Cap** 3–9cm
(1¼–3½in) across, convex at first with a low,
broad umbo, later expanding and becoming
irregular and wavy at the margin; dingy green
to bluish green; grayish, bluish, or nearly
white; finely matted with silky hairs or
sometimes with a hoary bloom. **Gills** slightly
decurrent, close or crowded, broad; whitish
tinged with cap color. **Stem** 30–70×5–15mm
(1¼–2¾×¼–½in), solid becoming hollow,
sometimes curved and enlarged toward base;
whitish tinged with cap color; base spongy and
covered in fine whitish down. **Flesh** thin, firm;
whitish to pale tan. **Odor** strongly of aniseed.
Taste strongly aniseed. **Spores** ellipsoid,
smooth, nonamyloid, 6–7.5×3–4µ. Deposit
whitish pink. **Habitat** singly, scattered, or in
groups on leaf litter under hardwoods,
especially oak. Found widely distributed in
North America. **Season** July–September
(November–February in California). **Edible.**
Comment Said to be edible, but I would avoid
eating it.

Clubfoot **Clitocybe clavipes** ¾ life-size

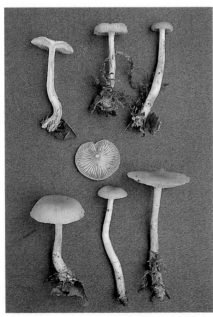

Clitocybe fragrans ½ life-size

Clitocybe odora ⅓ life-size

Clitocybe candida Bres. syn. *Leucopaxillus candidus*
(Bres.) Singer **Cap** 8–20cm (3–8in) across,
broadly convex to flat with the margin
incurved, becoming shallowly and broadly
funnel-shaped with the margin often wavy,
arched, and sometimes grooved at the edge;
whitish or pale cream, becoming slightly sordid
on the disc in older specimens; dull, dry, with
very finely matted hairs. **Gills** decurrent to
short decurrent, crowded, narrow; yellowish;
separable from the cap. **Stem** 40–90 ×
15–35mm (1½–3½ × ½–1¼in), sometimes enlarged and curved at the base; white to pale
buff; finely matted hairs at the top, becoming
more scurfy and hairy below. **Flesh** thin, but
thicker on the disc; whitish. **Odor** mildly
mealy. **Taste** mild. **Spores** ellipsoid, smooth,
weakly amyloid, 6–8.5 × 3–4.5μ. Deposit white.
No pleurocystidia. **Habitat** in groups or tufts,
sometimes forming rings, in mixed hardwood or
coniferous forests. Sometimes abundant. Found
in northern North America and New Mexico.
Season August–November. **Not edible.**

Clitocybe subalpina Bigelow & Smith **Cap** 5–12
cm (2–4¾in) across, convex becoming flatter
with a somewhat depressed disc; margin
incurved at first, but becoming lobed and
striate; dark pinky-brown or dingy nut brown;
surface has a slight bloom then smooth, moist,
and hygrophanous, but minute dry scales in
dry conditions. **Gills** adnate to decurrent, close
or crowded, narrow to quite broad; pale pinky-
brown or dull pinky-gray. **Stem** 30–150 ×
10–20mm (1¼–6 × ½–¾in), solid or hollow,
sometimes compressed; same color as gills;
fibrous with a bloom, then minutely hairy in
furrowed lines; base rooting, sometimes with
white rhizoids. **Flesh** thin, brittle; dull pinky-
brown fading whitish. **Odor** mushroomy or
none. **Taste** pleasant becoming a touch
astringent. **Spores** ellipsoid, smooth, nonamyloid, 4.5–5.1 × 3–3.9μ. Deposit pale pinky-
buff. **Habitat** in dense clusters near conifers on
sandy soil along roadsides. Found in Washington. **Season** August–October. **Not edible.**

Clitocybe avellaneialba Murr. **Cap** 5–20cm
(2–8in) across, flat becoming sunken to funnel-
shaped in age, with an inrolled margin that
becomes wavy; olive-brown to grayish or
blackish brown; smooth, finely felty or scaly.
Gills decurrent, close, narrow; whitish to
cream. **Stem** 50–180 × 10–30mm (2–7 ×
½–1¼in), stuffed, enlarged toward the base,
sometimes curved; brownish, finely felty and
furrowed. **Flesh** thicker on the disc, firm;
whitish. **Odor** not distinctive. **Taste** not
distinctive. **Spores** broadly spindle-shaped,
smooth, nonamyloid, 8–11 × 4–5.5μ. Deposit
white. **Habitat** in groups or clusters on or near
rotting logs and debris of alder and conifers.
Quite common. Found in the Pacific Northwest
and California. **Season** September–December.
Not edible.

Clitocybe candida ½ life-size

Clitocybe subalpina ½ life-size

Clitocybe avellaneialba ¼ life-size

Clitocybe subbulbipes ½ life-size

Clitocybe intermedia ⅓ life-size

Clitocybula familia ⅓ life-size

Clitocybe trullaeformis ⅓ life-size

Clitocybe truncicola ⅔ life-size

Clitocybe albirhiza ⅓ life-size

Grayling *Cantharellula umbonata* ½ life-size

Clitocybe subbulbipes Murr. **Cap** 1–5cm (½–2in) across, convex becoming flatter with a depressed disc, margin inrolled then becoming wavy; yellowy, honey brown fading to pale grayish, yellowy fawn, somewhat darker at the disc, and margin fading last; hygrophanous, smooth. **Gills** decurrent, close or crowded, narrow to moderately broad; dirty white to pale fawn. **Stem** 20–70 × 2–10mm (¾–2¾ × ³/₃₂–½in), hollow, sometimes compressed; pale honey or buff, whitish at the top; smooth, minutely felty; base with some rhizoids and whitish mycelium. **Flesh** thin, firm; pallid. **Odor** mildly sweet or none. **Taste** mushroomy, slightly unpleasant, or none. **Spores** ellipsoid, smooth, nonamyloid, 4.5–6 × 2.8–4μ. Deposit cream. **Habitat** singly, scattered, or in groups on hardwood logs. Found in northeastern North America. **Season** July–August. **Not edible.**

Clitocybe intermedia Kauffman **Cap** 1.5–4cm (½–1½in) across, broadly convex and shallowly depressed with an inrolled margin, becoming funnel-shaped with a wavy margin; yellowy brown to fawn to buff; hygrophanous, moist, smooth. **Gills** decurrent, close, narrow to moderately broad; pale buff or pale pinky-gray. **Stem** 10–65 × 2–6mm (½–2½ × ³/₃₂–¼in), stuffed then hollow, sometimes curved; same color as cap, smooth to finely hairy, sometimes with white mycelium around the base. **Flesh** thin, brittle; buff or fawn. **Odor** mealy. **Taste** mealy. **Spores** ellipsoid, smooth, amyloid, 6.5–8.5 × 4–5μ. Deposit white. **Habitat** scattered or in clusters on soil along roads and paths and in open woods. Found in northeastern North America and the Pacific Northwest. **Season** September–October. **Not edible.**

Clitocybula familia (Pk.) Singer **Cap** 1–4cm (½–1½in) across, bell-shaped then convex becoming flatter, with an incurved margin that spreads and finally becomes torn in age; grayish buff to brownish buff or dirty cream; smooth, moist. **Gills** adnate to nearly free, crowded, narrow; ash gray to whitish. **Stem** 40–80 × 1.5–3mm (1½–3 × ¹/₁₆–⅛in), fragile, gray or whitish, with flat white hairs on the base; smooth with a fine bloom. **Flesh** thin, fragile. **Taste** slightly disagreeable. **Spores** globose, smooth, amyloid, 3.5–4.5 × 3.5–4.5μ. Deposit white. **Habitat** in large clusters on

Leucopaxillus albissimus ½ life-size

Leucopaxillus tricolor ⅓ life-size

conifer logs. Often abundant. Found widely distributed in North America. **Season** August–October. **Edible.**

Clitocybe trullaeformis (Fr.) Quél. **Cap** 1–5cm (½–2in) across, very broadly convex quickly becoming flatter and then broadly funnel-shaped, sometimes with a slightly wavy margin; gray to pale gray with a dark gray, sometimes almost blackish disc and a pale margin; dry, velvety or with minute felty scales on the disc, sometimes becoming lined in age. **Gills** adnate to decurrent, close, narrow; whitish to creamy yellow. **Stem** 10–45 × 1–5mm (½–1¾ × ¹⁄₁₆–¼in), solid, tapering slightly toward the base; pale dirty fawn or pale greenish buff; smooth or with minute hairy furrows; base with some rhizoids and white mycelium. **Flesh** thin; whitish. **Odor** mealy. **Taste** mealy. **Spores** ellipsoid, smooth, nonamyloid, 5–6.5 × 3–4µ. Deposit white. **Habitat** scattered or in groups on leaf litter or debris under coniferous or deciduous trees. Found widely distributed throughout North America. **Season** May–October. **Not edible.**

Clitocybe truncicola (Pk.) Sacc. **Cap** 1–5cm (½–2in) across, broadly convex with an inrolled margin, becoming flatter then broadly depressed in age with an undulating margin; whitish becoming buff-colored; thickly covered in dense, white downy hairs. **Gills** adnate to short decurrent, close to crowded, narrow; whitish to creamy buff. **Stem** 10–40 × 2–10mm (½–1½ × ³⁄₃₂–½in), stuffed, with base sometimes slightly enlarged and curved; white to pale cream or pale pinky-buff; slight bloom at the top, fine dense hairs below. **Flesh** thin, firm; white. **Odor** not distinctive. **Taste** not distinctive. **Spores** subglobose to broadly ellip-

soid, smooth, nonamyloid, 3.5–4.5 × 2.5–3.5µ. Deposit white. **Habitat** scattered or in groups on hardwood stumps and fallen trunks. Found widely distributed in central and eastern North America. **Season** August–September. **Not edible.**

Clitocybe albirhiza Bigelow & Smith **Cap** 3–10cm (1¼–4in) across, obtusely convex becoming flatter with an inrolled margin, sometimes with a slight umbo, then becoming funnel-shaped with an upturned, wavy, sometimes notched margin; pale buff to watery brown with faint concentric bands or patches of tawny pinkish buff; at first covered with a whitish downy bloom, more or less smooth in age. **Gills** adnate to decurrent, close, narrow to broad; whitish then buff or same color as cap. **Stem** 30–80 × 5–20mm (1¼–3 × ¼–¾in), stuffed then hollow, sometimes fluted, base densely covered in white rhizoids; same color as cap; tough, smooth or downy when moist, becoming finely hairy and furrowed when dry. **Flesh** firm, somewhat thicker at the disc; pale buff. **Odor** slightly unpleasant. **Taste** bitter. **Spores** ellipsoid, smooth, nonamyloid, 4.5–6 × 2.5–3.5µ. Deposit white. **Habitat** growing in groups or dense clusters under conifers on mountains, between 2000m and 3000m (6500–10,000ft). Found in mountainous areas of western North America. **Season** June–July. **Not edible.**

Grayling *Cantharellula umbonata* (Fr.) Singer **Cap** 2–5cm (¾–2in) across, convex becoming funnel-shaped with a small central umbo and sometimes a wavy margin; grayish brown to smoky or violaceous gray; dry and minutely hairy. **Gills** decurrent, crowded, narrow, thickish, regularly forked; whitish bruising red

or yellow. **Stem** 25–80 × 3–7mm (1–3 × ⅛–¼in), tough, stuffed; whitish to graying; silky. **Flesh** white, bruising red where cut or handled. **Odor** scented. **Taste** mild. **Spores** subfusoid, smooth, amyloid, 8–11 × 3–4.5µ. Deposit white. **Habitat** scattered or in groups in hair-cap moss. Common. Found widely distributed in northeastern North America. **Season** August–November. **Edible.**

Leucopaxillus albissimus (Pk.) Singer **Cap** 3–10cm (1¼–4in) across, convex-flattened, margin incurved; white to cream color; glabrous, dry. **Gills** adnate to subdecurrent, crowded, narrow; white. **Stem** 50–80 × 5–15mm (2–3 × ¼–½in); white with dense mat of white mycelial hairs at base. **Flesh** white. **Odor** pleasant. **Taste** bitter. **Spores** ellipsoid, with low amyloid warts, 5.5–8.5 × 4–6µ. Deposit white. **Habitat** often in circles in coniferous woods. Found widely distributed throughout most of North America. **Season** August–October. **Not edible.**

Leucopaxillus tricolor (Pk.) Kühner **Cap** 8–30cm 3–12in) across, convex-flattened, margin inrolled at first; pinkish buff to tan. **Gills** adnate to subdecurrent, crowded, broad; slightly yellowish. **Stem** 40–100 × 10–25mm (1½–4 × ½–1in), bulbous at base; white with mycelium binding the soil at base. **Flesh** thick; white. **Odor** strong, unpleasant. **Taste** mild to unpleasant. **Spores** ovoid, with isolated warts, 6–8 × 4–5.5µ. Deposit white. **Habitat** under mixed hardwoods. Found from the western Great Lakes through eastern North America. **Season** July–September. **Edibility not known.**

Catathelasma imperialis ½ life-size

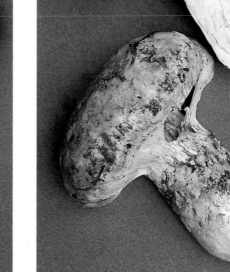

Leucopaxillus gentianus ⅓ life-size

Catathelasma ventricosa ½ life-size

Leucopaxillus gentianus (Quél.) Kotlaba syn. *Leucopaxillus amarus* (A. & S. ex Fr.) Kühner **Cap** 5–15cm (2–6in) across, convex flattened, with inrolled margin at first; deep vinaceous brown to nut brown; dry to slightly subtomentose. **Gills** adnate, crowded, narrow; white to cream. **Stem** 50–100×8–45mm (2–4×⁵⁄₁₆–1¾in), equal; white to slightly brownish. **Flesh** white. **Odor** strong, pungent. **Taste** bitter. **Spores** subglobose, with low warts, 4–5.5×4–5.5μ. Deposit white. **Habitat** under conifers. Common. Found in western North America. **Season** July–September. **Not edible.**

Catathelasma imperialis (Fr.) Singer **Cap** 12–40cm (4¾–16in) across, convex becoming flat with an incurved margin at first; blackish brown to dingy brown; slightly sticky becoming dry and breaking into small areas or patches over the middle. **Gills** decurrent, close, narrow becoming broad; yellowish to pale greenish gray. **Stem** 120–180×50–80mm (4¾–7× 2–3in), tapers to a pointed base; covered with a dingy yellow-brown or pinky-brown membranous sheath; dry. **Veil** partial double veil leaving double ring on upper stalk; top layer membranous and striate, bottom layer slimy. **Flesh** hard, thick; white. **Odor** mealy. **Taste** mealy. **Spores** cylindrical, smooth, amyloid, 11–14×4–5.5μ. Deposit white. **Habitat** singly or scattered in dense coniferous forests. Found in the Rocky Mountains and the Pacific Northwest. **Season** August–October. **Edible.**

Catathelasma ventricosa (Pk.) Singer **Cap** 7–20cm (2¾–8in) across, convex to broadly convex; dingy white to brownish or grayish; dry, smooth, and breaking up into patches with age. **Gills** decurrent, close to nearly distant, narrow to broad; whitish. **Stem** 50–150× 26–60mm (2–6×1–2¼in), stout but narrowing to a point and deep in soil; whitish to yellowish brown; dry. **Veil** partial double veil leaving flaring double ring; top layer hairy, bottom layer membranous and persistent. **Flesh** very thick, hard; white. **Taste** mildly unpleasant. **Spores** ellipsoid, smooth, amyloid, 9–12×4–5.5μ. Deposit white. **Habitat** scattered or in groups under conifers, especially spruce. Frequent. Found in northern North America south to Colorado, and in mountainous areas of northern California. **Season** August–October. **Edible** — good.

Laccaria montana Singer **Cap** 0.5–3.5cm (¼–1¼in) across, convex becoming flatter, then upturned with a tiny umbo; pale flesh-pink to pinkish brown; striate-furrowed at margin. **Gills** subdecurrent, thick, distant; pale pink. **Stem** 30–50×3–5mm (1¼–2×⅛–¼in); striate-fibrous; pinkish brown. **Flesh** white. **Odor** not distinctive. **Taste** not distinctive. **Spores** subglobose, 7.5–11.5(13)×6–11μ; spines 0.4–1μ long. Deposit white. Basidia 4-spored. **Habitat** stream banks and grassy areas, often outside woodlands. Found in high alpine and subalpine areas in western North America. **Season** August–October. **Edibility not known.**

Laccaria montana ¾ life-size

Laccaria ochropurpurea ⅓ life-size

Laccaria amethysteo-occidentalis ¾ life-size

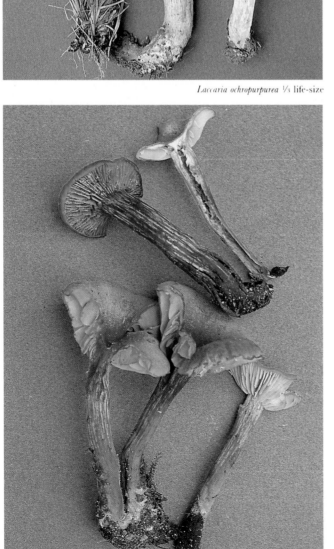

Laccaria proxima ½ life-size

Laccaria bicolor ¾ life-size

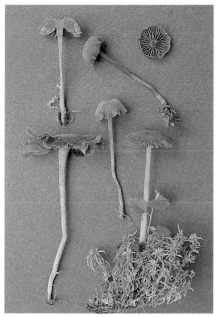

Laccaria amethystea ⅓ life-size

Laccaria laccata var. *pallidifolia* ⅓ life-size

Laccaria laccata ⅓ life-size

Laccaria nobilis ¼ life-size

Laccaria ochropurpurea (Berk.) Pk. **Cap** 5–15 (20)cm (2–6[8]in) across, convex then flattened and often upturned; purplish brown to purplish flesh or ochre, with surface dry and smooth or squamulose. **Gills** attached, broad, thick; purple, powdered with white spores when mature. **Stem** 50–150 × 10–20mm (2–6 × ½–¾in), solid, stout, tough and fibrous; colored as cap. **Odor** not distinctive. **Taste** not distinctive. **Spores** globose, spiny, 6–8 × 6–8μ. Deposit white to pale violet. **Habitat** in open grassy woods under oak and other deciduous trees. Common. Found east of the Great Plains. **Season** July–November. **Edible** — good. **Comment** Easily confused at first sight with some cortinarius species. Check for the white spores to confirm identity.

Laccaria amethysteo-occidentalis Mueller **Cap** 1–8cm (½–3in) across, convex becoming umbilicate; deep purple-violet, dull slate-violet to brownish purple or even buff when dry, hygrophanous. **Gills** adnexed, thick, broad, widely spaced; deep violet. **Stem** 20–100 × 3–10mm (¾–4 × ⅛–½in); purple violet with a violet woolly base; dry, striate. **Flesh** violet. **Odor** not distinctive. **Taste** not distinctive. **Spores** subglobose, 7.4–10.6 × 6.4–9.2μ; spines up to 1.5μ long. Deposit white. **Habitat** under conifers (Douglas Fir). Found in western North America. **Season** August–October. **Edible.** **Comment** This species is distinguished from the closely related and eastern *Laccaria amethystea* (q.v.) by its deeper, more purple colors and more ellipsoid spores.

Laccaria proxima (Boud.) Pat. **Cap** 2–8cm (¾–3in) across, convex-flattened; reddish brown to ochre when dry; usually rather coarsely scaly-scurfy at center. **Gills** adnexed, broad, thick; pallid to pale pink. **Stem** 30–130 × 3–5mm (1¼–5 × ⅛–¼in), fibrous; concolorous with cap, with white down at base. **Flesh** whitish ochre. **Odor** pleasant. **Taste** pleasant. **Spores** ovate, spiny, 7–9.5 × 6–7.5μ. Deposit white. **Habitat** under conifers and mixed woods. Uncommon. Found in eastern and northern North America. **Season** August–October. **Edible.**

Laccaria bicolor (Maire) Orton **Cap** 2–5cm (¾–2in) across, convex-flattened; ochraceous tan drying to pinkish ochre; smooth to slightly scaly. **Gills** adnate; pale lilac to pallid with age. **Stem** 50–140 × 4–10mm (2–5½ × 3/16–½in), equal; concolorous with cap, but with lilac tomentum at base; fibrillose. **Flesh** thin; whitish ochre. **Odor** pleasant. **Taste** pleasant. **Spores** ovoid, spiny, 7–9.5 × 6–7.5μ. Deposit white. **Habitat** in mixed woodlands. Rather uncommon. Found in eastern and western North America. **Season** August–October. **Edible** but not worthwhile.

Laccaria amethystea (Bull. ex Mérat) Murr. **Cap** 1–6cm (½–2¼in) across, convex with inrolled margin, then soon flattened; deep amethyst purple when wet, then pale lilac gray when dry; smooth to slightly rough-scaly on disc. **Gills** adnate, thick, distant, waxy; deep violet dusted with white spores. **Stem** 40–100 × 5–10mm (1½–4 × ¼–½in), equal; concolorous with cap; fibrous, tough; base with lilac down. **Flesh** thin; lilac. **Odor** pleasant. **Taste** pleasant. **Spores** globose, spiny, 9–11 × 9–11μ. Deposit white. **Habitat** in moist shady woodlands. Uncommon. Apparently restricted almost entirely to eastern North America. **Season** August–September. **Edible.**

Laccaria laccata (Scop. ex Fr.) Cke. **Cap** 1–6cm (½–2¼in) across, convex then flattened, often becoming finely wavy at the margin and centrally depressed; tawny to brick red drying paler to ochre-yellow; margin striate when moist, surface often finely scurfy. **Gills** adnate to short decurrent, distant, broad; pinkish, dusted white with spores when mature. **Stem** 20–80 × 3–10mm (¾–3 × ⅛–½in), often compressed or twisted; same color as cap; tough and fibrous. **Flesh** thin; reddish brown. **Odor** not distinctive. **Taste** slight. **Spores** globose to broadly ellipsoid, spiny, 7–10 × 6–8μ. Deposit white. **Habitat** scattered or in groups in sandy or pool soil in wasteland and under conifers. Quite common. Found widely distributed in North America. **Season** June–November (overwinters in California). **Edible. Comment** *Laccaria laccata* var. *pallidifolia* (Pk.) Pk. Differs from the type form in its very pallid, whitish gills and generally smaller stature.

Laccaria nobilis Smith and Mueller **Cap** 2–7.5cm (¾–2¾in) across, convex becoming flatter or occasionally uplifted, depressed, or deeply depressed, with a margin that sometimes becomes wavy; bright reddish orange to brownish orange, occasionally with a darker center; minutely hairy to scaly. **Gills** sinuate to adnate, close to distant, moderately broad to broad; pinkish flesh-color to orange-pink. **Stem** 25–110 × 4–10mm (1–4½ × 3/16–½in), slightly bulbous; same color as cap, with white mycelium at base; dry, longitudinally hairy, almost forming reticulate ridges, and with recurved scales near the top in mature specimens. **Flesh** thin; same color as cap. **Spores** subglobose to broadly ellipsoid, 7.4–9.7 × 6.4–8.7μ. Deposit white. **Habitat** singly or scattered at high elevations (but below tree line). Found in the Rocky Mountains and westward, and in New Mexico. **Season** July–November. **Edibility not known.**

Collybia acervata ⅓ life-size

Collybia confluens ⅓ life-size

Oak-loving Collybia *Collybia dryophila* ½ life-size

Laccaria altaica ½ life-size

Laccaria altaica Singer syn. *Laccaria striatula* of
many authors **Cap** 1–4cm (½–1½in) across,
convex then soon flattened; flesh-pink, pinkish
brown, to reddish; dry. **Gills** adnate, thick,
widely spaced; pinkish. **Stem** 20–50×2–4mm
(¾–2×³⁄₃₂–³⁄₁₆in); pinkish brown; fibrillose.
Odor not distinctive. **Taste** not distinctive.
Spores almost round, 9–12.5(13)×9–12.5
(13)μ. Deposit white. Basidia both 2- and 4-
spored. **Habitat** riverbanks and wet, marshy
areas. Rather common. Found in northern
North America. **Season** July–October. **Edible.**

COLLYBIA *The stems are tough and fibrous and
will usually bend without snapping. The gills are
crowded but never decurrent.*

Collybia acervata (Fr.) Kummer **Cap** 1–5cm
(½–2in) across, convex with an incurved
margin at first, becoming flatter; reddish
brown becoming paler, more pinkish buff to
whitish tan on drying; moist to dry, smooth.
Gills adnate to free, close, narrow; whitish to
dingy pale pink. **Stem** 30–100×2–5mm
(1¼–4×³⁄₃₂–¼in), hollow; reddish brown with

white hairs around the lower section and base;
dry, smooth, brittle. **Flesh** thin, flexible;
pinkish white. **Taste** bitter when cooked.
Spores ellipsoid, smooth, 5–6.5×2–2.5μ.
Deposit white. **Habitat** in dense clusters on
decaying or buried logs or stumps of coniferous
wood. Common among conifers. Found in
northern North America, Colorado, and Texas.
Season July–October (November in Texas).
Not edible.

Collybia confluens (Pers. ex Fr.) Kummer **Cap**
2–6cm (¾–2¼in) across, convex with incurved
margin, then soon flattened, thin, often wavy
margin; dull reddish brown fading to pale buff;
glabrous. **Gills** almost free, very crowded,
narrow; pallid. **Stem** 50–100×2–6mm
(2–4×³⁄₃₂–¼in), often flattened; reddish
beneath a dense whitish pubescence; tough;
base densely hairy. **Odor** odd, distinctive
(buggy?). **Taste** mild. **Spores** ellipsoid,
smooth, 6–8×2–4μ. Deposit white. **Habitat** in
dense clumps on fallen leaves or needles.
Found widely in most of North America.
Season July–November. **Edible** but worthless.

Oak-loving Collybia *Collybia dryophila* (Bull.
ex Fr.) Kummer **Cap** 2.5–5cm (1–2in) across,
convex and then flat, margin inrolled; ochre-
brown to reddish or pale tan; smooth, dry.
Gills adnate, crowded, narrow; white to pallid,
not staining. **Stem** 10–70×1.5–5mm
(½–2¾×¹⁄₁₆–¼in), smooth, brittle; reddish
brown to ochre, paler at base, which is bristly-
hairy. **Flesh** very thin; white. **Odor** not
distinctive. **Taste** not distinctive. **Spores**
ellipsoid, smooth, 5–6×2–3μ. Deposit white.
Habitat usually gregarious, under both oak
and pine. Rather common. Found in most of
North America. **Season** May–November.
Edible but not worthwhile (see Comment).
Comment This species has recently been split
into several distinct species, including the
usually spring fruiting *Collybia subsulphurea* (Pk.)
Bull., which differs in the general sulphur-
yellow color of its fruit body and in possessing
rhizomorphs, which are pinkish buff at the
stem base; the gills may be white or yellow. In
view of the confusion in this species it is
recommended that it not be eaten.

Collybia lentinoides (Pk.) Sacc. **Cap** 2–5cm
(¾–2in) across, convex with an incurved
margin; hygrophanous, dull buff-clay to pinkish

buff when dry, deeper brown when wet; dry.
Gills adnexed, crowded, minutely serrated
even when young; pale buff. **Stem**
20–60×3–6mm (¾–2¼×⅛–¼,) equal to
slightly swollen at base; whitish. **Flesh** white.
Odor mild. **Taste** mild. **Spores** ellipsoid,
smooth, dextrinoid, 5.4–7.5×3.2–4.4μ. Deposit
cream to pinkish cream. **Habitat** under eastern
white pine. Found in New Jersey. **Season**
August–September. **Not edible.**

Butter Cap *Collybia butyracea* (Fr.) Quél. **Cap**
3–7cm (1¼–2¾in) across, convex becoming
flattened with a distinct umbo; dark reddish
brown fading to light brown, pale reddish tan,
or pale orange-yellow; smooth, greasy (buttery)
to the touch. **Gills** adnate to free, close to
crowded, becoming moderately broad; whitish
to pale grayish pink. **Stem** 30–80×4–8mm
(1¼–3×³⁄₁₆–⁵⁄₁₆in), slightly swollen toward the
base, sometimes twisted, becoming hollow;
same color as cap or lighter; whitish at the
base with cottony hairs; smooth to lined. **Flesh**
watery, soft, thin, but thicker on disc; whitish
buff. **Odor** not distinctive. **Taste** not
distinctive. **Spores** ellipsoid, smooth,
6–7×3–3.5μ. Deposit yellowish. **Habitat**
scattered or in groups on soil and decaying
needles under conifers. Common in pine
plantations. Found in northern North America.
Season September–October. **Edible** but not
worthwhile.

Spotted Collybia *Collybia maculata* (A. & S. ex
Fr.) Kummer syn. *Rhodocollybia maculata*
(A. & S. ex Fr.) Singer **Cap** 4–10cm (1½–4in)
across, flattened-convex; white, soon becoming
spotted with tan-brown to rust spots on aging
or bruising, finally may become brown overall;
smooth, dry. **Gills** attached to almost free,
very crowded; white becoming spotted rust-
brown. **Stem** 50–100×8–12mm (2–4×
⁵⁄₁₆–½in), equal to slightly swollen at base,
hollow; similarly colored to cap; often with
rooting base. **Flesh** tough; white. **Odor**
pleasant. **Taste** bitter. **Spores** globose, smooth,
4–6×3–5μ. Deposit cream to pale pink.
Habitat often gregarious in mixed woodlands.
Found from Quebec to North Carolina. **Season**
July–November. **Not edible** due to toughness
and rather unpleasant flavor. **Comment**
Collybia maculata var. *scorzonerea* (Batsch: Fr.)
Gillet differs in having distinctly yellow gills.
Common.

Collybia lentinoides ¾ life-size

Butter Cap *Collybia butyracea* ¾ life-size

Spotted Collybia *Collybia maculata* ½ life-size

Collybia maculata var. *scorzonerea* ½ life-size

Collybia strictipes ⅓ life-size

Collybia velutipes life-size

Collybia velutipes (Curt. ex Fr.) Kummer syn.
Flammulina velutipes (Fr.) Karsten **Cap** 2–7cm
(¾–2¾in) across, convex becoming flatter or
with an umbo, the margin incurved at first;
orangish brown to reddish yellow or tawny,
darkening toward the center; smooth and
slimy. **Gills** adnate, close to nearly distant,
broad; pale yellow. **Stem** 25–75×3–5mm
(1–2¾ × ⅛–¼in), narrowing downward
slightly; yellowish at the top, dark brown and
densely velvety below; tough and cartilaginous.
Flesh thin; same color as cap. **Odor** pleasant.
Taste pleasant. **Spores** ellipsoid, smooth,
7–9×3–6μ. Deposit white. **Habitat** clustered
on deciduous logs, especially elm, poplar,
willow, and aspen. Found widely distributed in
North America. **Season** October–May
(July–August in mountains). **Edible.**
Comment The cultivated variety of this
species is sold commercially as Enotake, but it
looks very different from the wild mushroom.

Collybia strictipes Pk. **Cap** 2–6cm (¾–2¼in)
across, convex-flattened to slightly umbonate;
yellow to orange-yellow or ochre; smooth. **Gills**
narrow, crowded; pale yellowish to pinkish.
Stem 25–75×3–10mm (1–3 × ⅛–½in),
hollow; white to pale yellow; smooth, pruinose.
Odor pleasant. **Taste** pleasant. **Spores**
ellipsoid, smooth, 6–8×3–5μ. Deposit white.
Habitat in groups on leaf litter in mixed
hardwoods. Found in eastern North America to
the Great Lakes. **Season** July–October.
Edible.

Collybia alkalivirens Singer **Cap** 1–4cm
(½–1½in) across, broadly convex becoming
flatter or broadly umbonate; margin incurved
at first, then plane or recurved or wavy in age;
dark reddish brown to blackish brown, fading
to dingy pinky-brown or buff; smooth, moist to
dry, margin radially lined to furrowed. **Gills**
adnate, close to distant, narrow to broad; same
color as cap. **Stem** 25–75×1–3mm (1–2¾ ×
1/16–⅛in), hollow, often flattened above; dark
reddish brown or blackish brown; smooth,
fibrous-brittle. **Flesh** thin, pliant; same color as
cap. Turns greenish to dark olive with KOH.
Odor none. **Taste** very bitter. **Spores**
narrowly ellipsoid, smooth, nonamyloid,
greenish with KOH, 5.4–8.6×2.2–5.4μ.
Deposit white. **Habitat** singly or in groups on
soil, moss, or very rotten conifer wood. Often

Collybia alkalivirens ½ life-size

Collybia tuberosa ½ life-size

Cyptotrama chrysopeplum nearly life-size

Meadow Wax Cap *Hygrophorus pratensis* ¼ life-size

common. Found widely distributed in northern North America. **Season** May–July in the East, October–December in the Northwest. **Not edible.**

Collybia tuberosa (Bull. ex Fr.) Kummer **Cap** 5–10mm (¼–½in) across, convex becoming flat or rarely with an umbo; whitish or very weakly tinted buff; smooth, dry. **Gills** adnate, close, narrow; whitish. **Stem** 10–30 × 1mm (½–1¼ × ¹⁄₁₆in), often flexuous; whitish; dry, attached to reddish-brown, tuberlike body. **Flesh** very thin; whitish. **Odor** not distinctive. **Taste** not distinctive. **Spores** ellipsoid, smooth, 4–5.5 × 2–3μ. Deposit white. **Habitat** numerous on remains of decaying mushrooms. Found widely distributed in North America. **Season** August–November (overwinters in California). **Not edible.**

Cyptotrama chrysopeplum (Berk. & Curt.) Singer **Cap** 0.5–2cm (¼–¾in) across, convex to flat, with a lined to furrowed margin; lemon yellow to golden; dry, powdery to pitted. **Gills** adnate to decurrent, distant, broad; whitish to yellowish. **Stem** 10–50 × 1–3mm (½–2 × ¹⁄₁₆–⅛in), solid, tough; lemon yellow or paler; scaly with cottony blisters. **Spores** broadly ovoid to ellipsoid, smooth, 8–12 × 6–7.5μ. Deposit white. **Habitat** scattered on twigs, logs, and sticks of hardwoods. Common. Found in eastern North America, west to the Great Lakes and Texas. **Season** May–October (May–November in Texas). **Edibility not known.**

Crinipellis zonata (Pk.) Pat. **Cap** 1–2.5cm (½–1in) across, convex to nearly flat with a small cuplike depression on the disc; tawny

Crinipellis zonata ¼ larger than life-size

and cream; zones dry, densely covered in coarse, tawny hairs set in lines; dextrinoid. **Gills** free or nearly so, close, narrow; white. **Stem** 25–50 × 1–2mm (1–2 × ¹⁄₁₆–³⁄₃₂in), hollow; densely tawny-hairy. **Spores** ellipsoid, smooth, 4–6 × 3–5μ. Deposit white. **Habitat** on roots and debris of hardwoods. Found in eastern North America, west to Indiana and Texas. **Season** August–September. **Edibility not known.**

HYGROPHORUS *Cap often brightly colored, the caps and/or the stems are often glutinous, or they may be greasy or occasionally dryish. The gills are attached to the stem and often run down it (decurrent), and they are normally waxy to the touch. Some species blacken or discolor.*

Meadow Wax Cap *Hygrophorus pratensis* Fr. syn. *Camarophyllus pratensis* (Fr.) Kummer **Cap** 3–8cm (1¼–3in) across, convex then flattened with a broad umbo, becoming distorted and often cracking with age; salmon orange to tawny buff; dry and smooth with minute fibers. **Gills** deeply decurrent, distant, moderately broad, waxy; pale buff. **Stem** 20–70 × 5–15mm (¾–2¾ × ¼–½in), sometimes tapering toward base; whitish, paler than cap. **Flesh** thick, brittle; pale buff. **Odor** mild, mushroomy. **Taste** pleasant. **Spores** broadly ovoid to almost round, nonamyloid, 5.5–7.5 × 4–5.5μ. Deposit white. **Habitat** singly, scattered, or gregarious in open and grassy areas and in woods. Common. Found widely distributed in North America. **Season** May–December (November–March in California). **Edible** — excellent.

Hygrophorus camarophyllus life-size

Hygrophorus marginatus var. *marginatus* ⅔ life-size

Hygrophorus camarophyllus (Fr.) Dumée, Grandjean & Maire **Cap** 2–7cm (¾–2¾in) across, bluntly convex, slightly knobbed or flat with an umbo; browny gray with fine dark lines; sticky when wet, then dry and smooth, with a downy margin. **Gills** adnate, close to subdistant, moderately broad, very waxy; white or slightly grayish. **Stem** 25–130 × 10–20mm (1–5 × ½–¾in); pale, smoky gray-brown; silky with fine hairs near the top, smooth and hairless toward base. **Flesh** thick, fragile; white. **Odor** slight, faintly of coal tar. **Taste** mild. **Spores** ellipsoid, 7–9 × 4–5μ. Deposit white. **Habitat** scattered to gregarious. Found in northern North America. **Season** June–November. **Edible.**

Hygrophorus marginatus var. *marginatus* **Cap** 1–5cm (½–2in) across, conical to bell-shaped; bright yellow-orange to orange; smooth, moist. **Gills** adnate, widely spaced, thick; brilliant almost fluorescent orange, particularly on their edges. **Stem** 40–100 × 3–10mm (1½–4 × ⅛–½in); pale orange-yellow; smooth. **Flesh** brittle; orange. **Odor** pleasant. **Taste** pleasant. **Spores** 7–10 × 4–6μ. Deposit white. **Habitat** on soil and leaf litter. Common. Found in northwestern and eastern North America. **Season** June–October. **Edible** but not worthwhile.

Hygrophorus marginatus var. *olivaceus* Smith & Hesler **Cap** 2.5–4cm (1–1½in) across, cone-shaped becoming bell-shaped, then flatter with a sharp umbo; olive orange-brown to deep olive but paler orange near the margin; smooth

and moist with a translucent, striate margin. **Gills** ascending adnate, close to subdistant, broad; dark yellowy orange. **Stem** 30–50 × 2–10mm (1¼–2 × 3⁄32–½in), very fragile becoming hollow; yellowy green becoming paler; smooth. **Flesh** thin, fragile; dull greeny gray. **Odor** mild. **Taste** mild. **Spores** subellipsoid, nonamyloid, 6.5–8 × 4–6μ. Deposit white. **Habitat** singly on decayed conifer wood. Frequent. Found in eastern and western North America. **Season** July–September. **Edibility not known.**

Hygrophorus marginatus var. *concolor* Smith **Cap** 1–3cm (½–1¼in) across, obtusely cone-shaped becoming flatter and wavy in large, mature specimens; apricot to chrome yellow; smooth and moist with a thin, striate, spreading margin. **Gills** broadly adnate, subdistant, broad; chrome yellow. **Stem** 30–60 × 4–8mm (1¼–2¼ × 3⁄16–5⁄16in), sometimes narrower toward the base; bright yellow; smooth, moist. **Flesh** thin, brittle; same color as cap. **Odor** not distinctive. **Taste** not distinctive. **Spores** ellipsoid, nonamyloid, 7–9 × 4–5μ. Deposit white. **Habitat** scattered under hemlock. Common. Found in northeastern North America, west to Michigan. **Season** August–September. **Edibility not known.**

Hygrophorus parvulus Pk. **Cap** 1–3cm (½–1¼in) across, broadly obtuse to flat, sometimes with a depressed disc; amber or apricot yellow or orange, sometimes fading to paler yellow; smooth, moist and sometimes a little sticky; translucent striate. **Gills** decurrent, subdistant,

broad; cream to wax yellow. **Stem** 30–60 × 2–3mm (1¼–2¼ × 3⁄32–⅛in), hollow, fragile, sometimes narrower at the base; lemon to amber yellow, often tinged red or orangy pink on the lower half; smooth. **Flesh** thin, brittle, waxy; same color as cap. **Odor** mild. **Taste** mild. **Spores** ellipsoid, nonamyloid, 5–7.7 × 3.5–5μ. Deposit white. **Habitat** gregarious on soil and humus in deciduous and mixed woods. Common. Found in east and west North America. **Season** June–October. **Edibility not known.**

Hygrophorus lawrencei Smith and Hesler **Cap** 1–4cm (½–1½in) across, broadly convex to plane, with margin sometimes upturned in matured specimens; white; smooth and dry. **Gills** decurrent, distant, broad; white drying dirty, dark yellow. **Stem** 30–60 × 5–8mm (1¼–2¼ × ¼–5⁄16in), wider at the top, tapering slightly toward base; white; smooth and dry. **Flesh** fragile; white. **Odor** very strong, cedarlike. **Taste** rather medicinal, like cedar. **Spores** broadly ellipsoid, nonamyloid, 6.5–8 × 5–6μ. Deposit white. **Habitat** under conifers. Rare. Found in Oregon and New Jersey. **Season** October–January. **Edibility not known. Comment** This species is recognized by its most distinctive smell; possibly it is much more common than I have indicated, but it has not yet been recorded from other areas. The European species *Hygrophorus russocoriaceus* Berk. & Miller also has the strong cedarwood smell, but the spores are longer and narrower.

Hygrophorus marginatus var. *olivaceus* ½ life-size

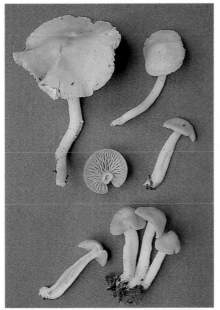

Hygrophorus marginatus var. *concolor* ⅓ life-size

Hygrophorus parvulus ⅔ life-size

Hygrophorus lawrencei ½ life-size

Hygrophorus virgineus ½ life-size

Hygrophorus niveus ½ life-size

Hygrophorus virgineus (Fr.) Fr. syn. *Camarophyllus virgineus* (Fr.) Kummer **Cap** 2–5cm (¾–2in) across, convex becoming flattened and depressed, with an incurved margin that becomes almost plane; white then ivory tinged ochraceous with age, especially at the center; moist then dry with a slight bloom, margin striate when moist. **Gills** decurrent, widely spaced, broad; white then tinged yellow. **Stem** 25–70 × 3–7mm (1–2¾ × ⅛–¼in), solid, sometimes twisted and tapering toward the base; white. **Flesh** soft, thick at center of cap; whitish. **Odor** not distinctive. **Taste** mild. **Spores** ellipsoid, nonamyloid, 8–10 × 5–7μ. Deposit white. **Habitat** gregarious on soil or moss in woods. Frequent. Found in northern North America. **Season** July–November. **Edible** — good. **Comment** *Hygrophorous niveus* (below) is similar but has a greasy cap.

Hygrophorus niveus Fr. **Cap** 1–3cm (½–1¼in), convex becoming flattened and depressed; white becoming ivory; smooth, greasy and striate to the disc when moist. **Gills** decurrent, distant, narrow; white to yellowish. **Stem** 20–70 × 2–6mm (¾–2¾ × ³⁄₃₂–¼in), becoming hollow, tapering slightly toward base; white; smooth, dry. **Flesh** thin, but thicker at center of cap; whitish. **Odor** not distinctive. **Taste** not distinctive. **Spores** ellipsoid, nonamyloid, 7–8.6 × 5.2–6.3μ. Deposit white. **Habitat** scattered to gregarious on humus and soil in deciduous and coniferous woods. Frequent. Found in northeast and northwest North America. **Season** August–January. **Edible** — good. **Comment** *Hygrophorus virgineus* (above) is very similar to this but has a dry, nongreasy cap.

Hygrophorus conicus ⅓ life-size

Hygrophorus conicus ¼ life-size

Hygrophorus flavescens ½ life-size

Hygrophorus vitellinus ⅔ life-size

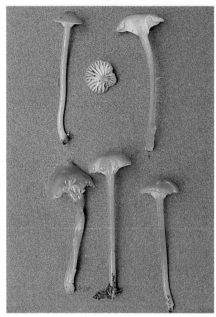

Hygrophorus cantharellus ⅔ life-size

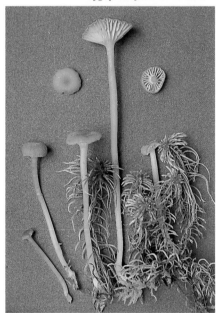

Hygrophorus turundus var. *sphagnophilus* ½ life-size

Hygrophorus squamulosus ½ life-size

Hygrophorus reai ½ life-size

Hygrophorus helobius ½ life-size

Hygrophorus conicus (Fr.) Fr. **Cap** 2–9cm (¾–3½in) across, conical or sometimes convex with a conical umbo and lobed margin; scarlet-orange to red, paler toward margin, often with olive tints, gray to black in age, bruising black when damaged; tacky when moist, smooth when dry, sometimes streaked with small fibers. **Gills** nearly free, close, broad; white tinted yellow to olive-yellow, bruising black. **Stem** 40–150×3–12mm (1½–6 × ⅛–½in), hollow, slightly twisted and lined; colored like cap but paler to whitish at base, bruising black. **Flesh** thin, fragile; same color as cap. **Odor** not distinctive. **Taste** not distinctive. **Spores** ellipsoid, nonamyloid, 10.8–11.7×5.6–6.4μ. Deposit white. **Habitat** single to gregarious in coniferous woods. Common. Found widely distributed in North America. **Season** July–October (east), October–April (west). **Edibility suspect** — avoid; possibly poisonous. **Comment** Distinctively blackening with age; old specimens may be completely black.

Hygrophorus flavescens (Kauffman) Smith & Hesler syn. **Cap** 2.5–7cm (1–2¾in) across, broadly convex becoming flatter in age, sometimes with a depressed disc; margin incurved at first, sometimes upturned and wavy in mature specimens; bright orange, paler toward margin; sticky, smooth, then dry and shiny, margin striate when moist. **Gills** adnexed, close to nearly distant, broad; yellow, waxy. **Stem** 40–70×6–14mm (1½–2¾ × ¼–½in), hollow, fragile, often compressed or fluted, easily splits; yellowy orange with whitish base; smooth and moist. **Flesh** thin, waxy; yellowish. **Odor** pleasant. **Taste** not distinctive. **Spores** ellipsoid, nonamyloid, 6.75–8×4.75–5.5μ. Deposit white. **Habitat** scattered to gregarious on soil and humus in deciduous, coniferous, and mixed woods, particularly in damp, mossy areas. Widely distributed in eastern North America, Pacific Northwest, and Texas. **Season** June–January. **Edible. Comment** *Hygrophorus chlorophanus* (p. 68) can easily be confused with this, but it has a really sticky coating, which this species does not, and is more yellow-colored.

Hygrophorus vitellinus Fr. **Cap** 1–2cm (½–¾in) across, convex soon expanding and often slightly depressed with an incurved margin; bright yellow; smooth with a lined margin. **Gills** decurrent, distant, broad; bright yellow. **Stem** 10–35×1–3mm (½–1¼ × ¹⁄₁₆–⅛in); bright yellow; smooth. **Flesh** thin; yellow. **Odor** delicate, slightly perfumed. **Taste** none. **Spores** ellipsoid-ovoid, 6–8×4.5–5μ. Deposit white. **Habitat** in mixed and coniferous woods. Found in Maine. **Season** September. **Edibility not known. Comment** Very close to *Hygrophorus nitidus* (p. 68). Needs further study.

Hygrophorus cantharellus (Schw.) Fr. **Cap** 0.5–4cm (¼–1½in) across, convex to flattened becoming depressed in the center, with a spreading to wavy margin; scarlet becoming vermilion, and ochre or yellowish in age; dry, silky then scurfy-scaly. **Gills** deeply decurrent, distant, broad; pale yellowish becoming deep egg yellow. **Stem** 30–90 × 1–4mm (1¼–3½ × ¹⁄₁₆–³⁄₁₆in), stuffed to hollow, fragile; orange at the top, same color as cap below; smooth, dry. **Flesh** thin; orange. **Odor** mild. **Taste** mild. **Spores** ellipsoid to subovoid, smooth, nonamyloid, 8.6–10.3 × 5.7–7μ. Deposit white. **Habitat** in groups or clusters in rich soil, bogs, or on decaying and moss-covered logs. Frequent. Found in eastern North America. **Season** July–October. **Edibility not known.**

Hygrophorus turundus var. *sphagnophilus* (Pk.) Smith & Hesler **Cap** 1–3.5cm (½–1¼in) across, broadly convex or flattened becoming broadly or deeply depressed with a decurved then wavy margin that sometimes spreads in age; red to scarlet, fading to yellow or brownish orange; dry, but minutely and densely woolly on the disc, becoming scaly. **Gills** adnate and sometimes becoming deeply decurrent, distant, broad or medium broad; red or orange or whitish to faintly yellow. **Stem** 40–120 × 1–3mm (1½–4¾ × ¹⁄₁₆–⅛in), stuffed solid or becoming tubular, wavy; bright red or yellowish, whitish toward the bottom where buried in moss; smooth, silky, fragile. **Flesh** thin or thickish, soft; same color as cap. **Odor** mild. **Taste** mild. **Spores** subellipsoid to subreniform (kidney-shaped), smooth, 10–14×5–9μ Deposit white. **Habitat** scattered or in groups in sphagnum bogs. Found throughout northern and eastern North America. **Season** June–October. **Edibility not known.**

Hygrophorus squamulosus Ellis & Ev. **Cap** 1.5–5cm (½–2in) across, obtuse to convex, with the disc often slightly depressed and the margin incurved; bright red to bright orange-yellow, margin yellowish pink and finally becoming bright yellow all over; dry or slightly moist, smooth when young, becoming distinctly scaly, especially near the margin. **Gills** adnate or with a slight decurrent tooth; close to subdistant, broad; reddish to pale yellow. **Stem** 30–50×3–6mm (1¼–2 × ⅛–¼in), hollow, often compressed; apricot yellow with white pith; smooth except for white bloom at the top. **Flesh** thick, firm; same color as cap fading to yellow. **Odor** not distinctive. Taste not distinctive. **Spores** subellipsoid, smooth, 6–8×4–5μ. Deposit white. No pleurocystidia; occasional cheilocystidia. **Habitat** scattered or in groups around rotten stumps and logs. Found in northeastern North America as far

Scarlet Waxy Cap *Hygrophorus coccineus* ½ life-size

south as Tennessee and in the Pacific Northwest. **Season** August–September. **Edibility not known.**

Hygrophorus reai Maire **Cap** 1–3cm (½–1¼in) across, convex with a minutely scalloped margin; bright flame red on the disc becoming brilliant scarlet toward the edge, fading to deep chrome yellow overall in age; smooth, sticky, faintly translucently lined toward the edge. **Gills** bluntly adnate soon seceding, subdistant, broad, 2 layers; whitish becoming pale yellow. **Stem** 30–50 × 2–4mm (1¼–2 × ³⁄₃₂–³⁄₁₆in), fragile, hollow; same color as cap; smooth, sticky, shining, often translucent. **Flesh** thin, brittle; same color as surface. **Odor** none. **Taste** very bitter. **Spores** ellipsoid, smooth, nonamyloid, 8.1–8.4×4.7–6.3μ. Deposit white. **Habitat** in groups under conifers. Found in New York, Michigan, and Washington. **Season** July–October. **Edibility not known.**

Hygrophorus helobius (Arnolds) Bon **Cap** 1–3cm (½–1¼in) across, flattish with the center often depressed and the margin sometimes wavy in age; scarlet paling to more orange; finely scaly. **Gills** deeply adnate to shortly decurrent; yellowish to light orange. **Stem** 15–30×2–4mm (½–1¼ × ³⁄₃₂–³⁄₁₆in), hollow; orange to slightly reddish; dry. **Spores** ellipsoid with a central waist, 8–12.5×4–6μ. Deposit white. **Habitat** in mossy woods. Rare. Found in New Jersey. **Season** August–September. **Edibility not known.**

Scarlet Waxy Cap *Hygrophorus coccineus* (Fr.) Fr. **Cap** 2–5cm (¾–2in), convex then expanded, bluntly umbonate; bright, intense scarlet, blood red, or orange-red; smooth, moist. **Gills** adnate, broad, thick, waxy, interveined; bright red to orange. **Stem** 30–70×3–8mm (1¼–2¾ × ⅛–⁵⁄₁₆in), hollow; orange-red, paler at base. **Flesh** yellow-orange. **Odor** not distinctive. **Taste** not distinctive. **Spores** ellipsoid, 7–10(11) × 4–5μ. Deposit white. **Habitat** gregarious on soil or humus in conifer or deciduous woods. Common. Found widely distributed over most of North America. **Season** July–October. **Edible. Comment** The similar *Hygrophorus miniatus* (Fr.) Fr. differs in the size and shape of its squamulose cap.

Hygrophorus chlorophanus ⅓ life-size

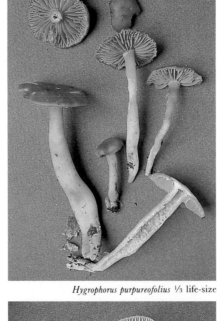

Hygrophorus purpureofolius ⅓ life-size

Hygrophorus nitidus ⅔ life-size

Hygrophorus perplexus ½ life-size

Hygrophorus laetus ½ life-size

Hygrophorus caespitosus ⅓ life-size

Hygrophorus chlorophanus (Fr.) Fr. **Cap** 2–4cm (¾–1½in) across, convex then expanded to flat; deep lemon yellow with disc sometimes more orange-yellow; sticky, with a striate margin. **Gills** adnexed, close, narrow; whitish to yellowish. **Stem** 30–70×4–8mm (1¼–2¾×³⁄₁₆–⁵⁄₁₆in), hollow; same color as cap; smooth and sticky. **Flesh** thin; yellowish. **Odor** mild. **Taste** mild. **Spores** ellipsoid, nonamyloid, 7–8.75×4–5μ. Deposit white. **Habitat** on soil in open woods. Uncommon. Found in northern United States and southeast to Tennessee. **Season** April–October. **Edibility not known. Comment** *Hygrophorus flavescens* (p. 67) is extremely similar, but the cap is more orange than yellow, and the stem, which is a bit slippery, lacks the really sticky coating that is characteristic of *Hygrophorus chlorophanus*.

Hygrophorus purpureofolius Bigelow **Cap** 1–5cm (½–2in) across, broadly bell-shaped becoming more convex then flat, with an incurved margin that sometimes becomes rather wavy in age; dark reddish orange when young, becoming a little more orange and sometimes paler in mature specimens; surface smooth, moist, and watery-looking becoming opaque. **Gills** broadly adnate to decurrent, close to subdistant, broad, waxy; mauvy-lavender to purple, then yellowish in age. **Stem** 25–70×4–9mm (1–2¾×³⁄₁₆–⅜in), hollow, brittle, compressed with a groove, curved, slightly enlarged toward the base; same color as cap, fading to whitish or yellowish. **Odor** not distinctive. **Taste** not distinctive. **Spores** ellipsoid, nonamyloid, 7–11×4–5.5μ. Deposit white. **Habitat** numerous or growing in dense tufts on humus in mixed woods, particularly maple. Infrequent. Found only in Massachusetts, as far as I know. **Season** August. **Edibility not known. Comment** The gills in my photograph only show a hint of lavender color.

Hygrophorus nitidus Berk. & Curt. **Cap** 1–4cm (½–1½in) across, broadly convex or flattened with a depressed disc and incurved margin; evenly apricot yellow or primrose yellow fading to pale creamy or whitish yellow; smooth and sticky with a striate margin. **Gills** arcuate then long decurrent, subdistant to distant, narrow becoming moderately broad, waxy; pale yellow, sometimes with deeper yellow edges. **Stem** 30–80×2–5mm (1¼–3×³⁄₃₂–¼in), hollow, fragile, sometimes slightly larger at the top; same color as cap; dry, smooth, and hairless. **Flesh** very thin, soft, fragile; yellowish. **Odor** not distinctive. **Taste** not distinctive. **Spores** ellipsoid, nonamyloid, 7–9.2×5.2–5.9μ. Deposit white. **Habitat** scattered to gregarious on humus in wet, mossy areas under hardwoods; also in bogs. Uncommon. Found in eastern North America and Washington. **Season** July–November. **Edibility not known. Comment** The faded white color with the

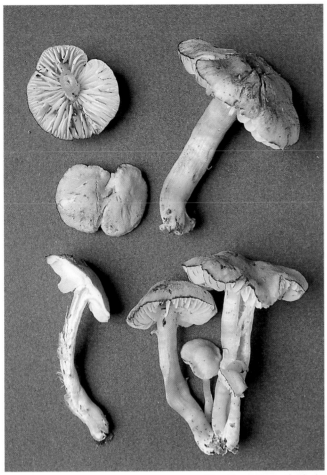

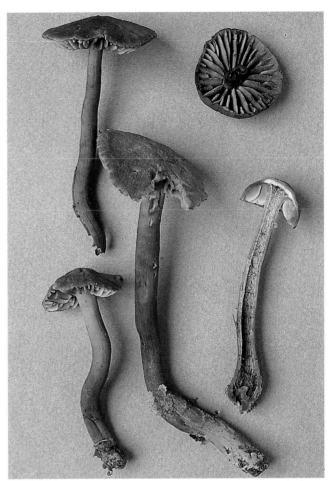

Hygrophorus nitiosus ¾ life-size   *Hygrophorus ovinus* ⅔ life-size

yellow edge to the cap is a distinctive
characteristic.

Hygrophorus perplexus Smith & Hesler **Cap**
1–3cm (½–1¼in) across, broadly cone-shaped
with the margin curved in to the gills,
expanding to broadly bell-shaped or flatter
with an obtuse umbo; brownish orange slowly
developing a greenish or orangish tinge on the
margin, then changing from orange-tan and
fading to buff-pink; very slimy and sticky with
a translucently lined margin. **Gills** ascending
and adnate with a tooth or expanded
depressed-adnate, close to subdistant, broad;
amber yellow becoming apricot yellow. **Stem**
30–50 × 2–5mm (1¼–2 × ³⁄₃₂–¼in), sometimes
slightly thicker below; pale watery gray with
yellowish-buff base, becoming yellow all over;
slimy and sticky all over. **Flesh** very thin,
fragile; same color as cap. **Odor** slight. **Taste**
none. **Spores** ellipsoid to ovoid, smooth,
6–8 × 4–5μ. Deposit white. **Habitat** in groups
and tufts under aspen and beech on thin sandy
soil. Found in Michigan and New Jersey.
Season June–August. **Edibility not known.**
Comment My photograph seems to have made
the mushrooms look too strongly orange.

Hygrophorus laetus (Fr.) Fr. **Cap** 1–3cm
(½–1¼in) across, convex to flat or depressed
at the disc with the margin sometimes
upturned; color variable from pale violet to
flesh, tawny-brown, orange to olive-orange,
often paler at the margin; smooth, slimy, and
sticky, with a striate margin. **Gills** decurrent,
subdistant but connected by veins, narrow to
moderately broad; variously colored like cap,

whitish to flesh-pink. **Stem** 30–100 × 2–6mm
(1¼–4 × ³⁄₃₂–¼in), hollow; pale yellow tinged
with cap color; smooth and very slimy. **Flesh**
thin, tough; same color as cap or lighter. **Odor**
faint or slightly fishy. **Taste** not distinctive.
Spores broadly ellipsoid, nonamyloid,
5–7.5 × 4–5μ. Deposit white. **Habitat** scattered
to gregarious in damp soil among moss or
bracken or in woods. Found widely distributed
in North America. **Season** May–December.
Edible but not worthwhile.

Hygrophorus caespitosus (Murr.) Murr. syn.
Camarophyllus caespitosus Murr. **Cap** 1–6cm
(½–2¼in) across, convex with the disc
flattened, then much flatter with a deeply
depressed disc and spreading or upturned
margin; creamy yellow to pale buff, sometimes
bright yellow and small, dark yellow or
brownish scales all over the cap; surface moist
then dry. **Gills** broadly adnate-decurrent,
subdistant to distant, broad; white to
yellowish. **Stem** 20–50 × 3–10mm (¾–2 ×
⅛–½in), spongy becoming hollow, often
tapering slightly toward the base; same color
as cap or darker; smooth and moist or dry.
Flesh thick; yellow. **Odor** not distinctive.
Taste mild or slightly of raw potato. **Spores**
ellipsoid, nonamyloid, 6.5–10 × 4–7μ. Deposit
white. **Habitat** in small groups or dense tufts
on mossy soil in coniferous or deciduous woods
and on clay banks. Uncommon. Found in
eastern North America. **Season** June–August.
Edibility not known.

Hygrophorus nitiosus Blytt **Cap** 2–5cm (¾–2in)
across, broadly convex with an umbo; snuff
brown; moist becoming dry with radiating

appressed hairs. **Gills** deeply emarginate,
subdistant, broad; whitish, becoming pinkish
or blackish where bruised. **Stem**
40–80 × 3–8mm (1½–3 × ⅛–⁵⁄₁₆in), hollow,
sometimes swollen below; pale brownish,
becoming pinkish brown where bruised and
drying blackish; dry, smooth, often compressed
and lined. **Flesh** quite thin; dingy grayish or
pinkish when cut. **Odor** nitrous. **Taste** rather
sour. **Spores** ellipsoid, smooth,
7.5–10 × 4.5–6μ. Deposit white. **Habitat** on
soil in deciduous woods. Found in Tennessee
and northern Massachusetts. **Season**
July–August. **Edibility not known.**

Hygrophorus ovinus (Fr.) Fr. **Cap** 2–6cm
(¾–2¼in) across, convex to cone-shaped,
expanding irregularly and often broadly
umbonate; dark sepia with yellowish-brown
margin; moist then dry and silky, becoming
cracked in age, sometimes forming scales on
the disc. **Gills** adnate, moderately close to
subdistant, broad; pale sepia at first, darkening
with age, bruises reddish. **Stem** 30–70 ×
6–15mm (1¼–2¾ × ¼–½in), hollow, often
compressed and curved; dark gray-brown
bruising pinky-red. **Flesh** thin, thicker at disc,
margin brittle; bruises pinky-red. **Odor**
nitrogenous when crushed. **Taste** nitrogenous.
Spores ellipsoid to subovoid, nonamyloid,
7–9 × 5–6μ. Deposit white. **Habitat** scattered
or gregarious on soil in mixed, deciduous, or
coniferous woods. Uncommon. Found in
eastern North America and California. **Season**
July–September. **Not edible. Comment** The
dark color and red bruising are distinctive
features of this species.

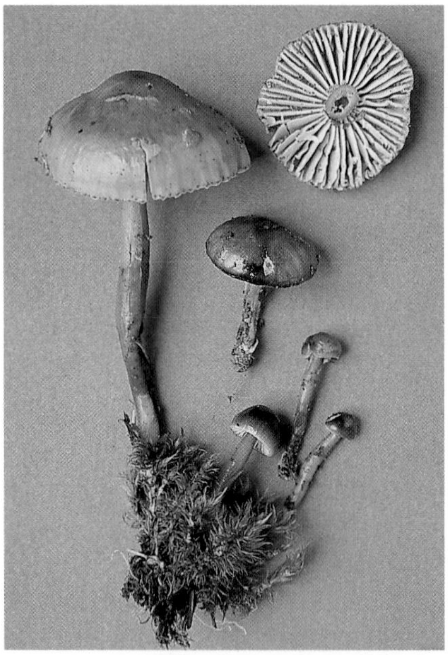
Hygrophorus unguinosus ⅓ life-size

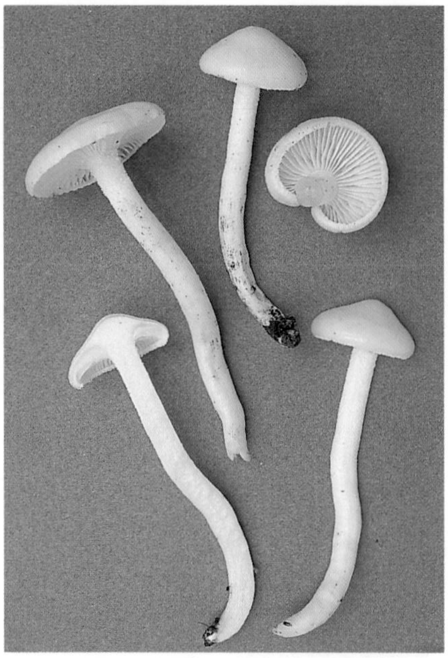
Hygrophorus eburneus ½ life-size

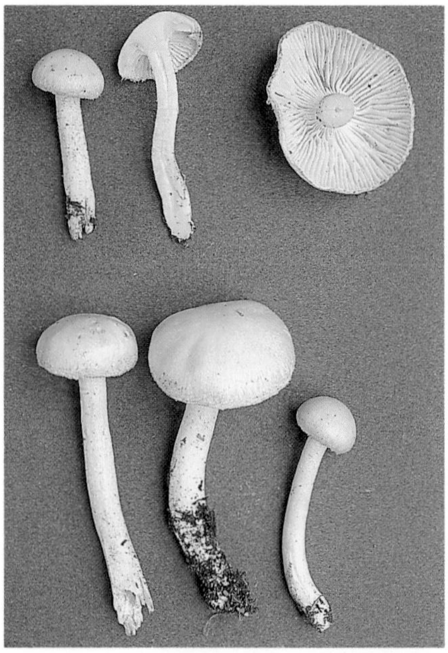
Hygrophorus chrysodon ½ life-size

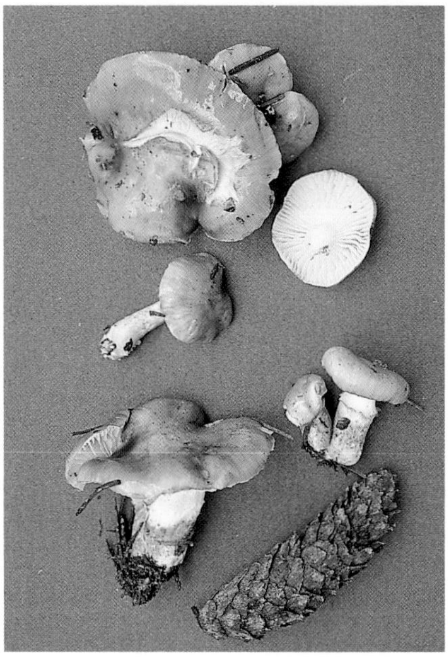
Hygrophorus olivaceoalbus ⅓ life-size

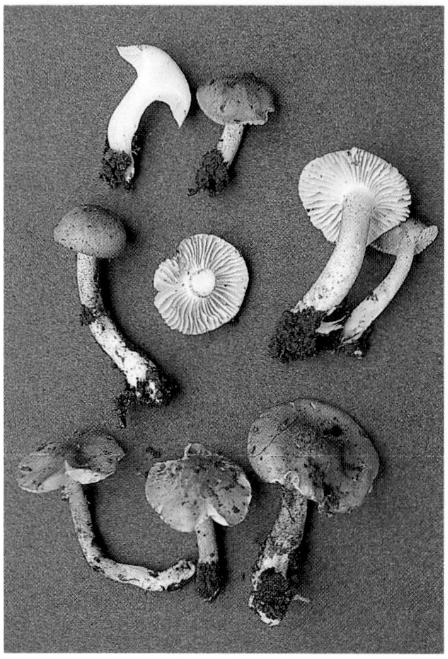
Hygrophorus pustulatus ½ life-size

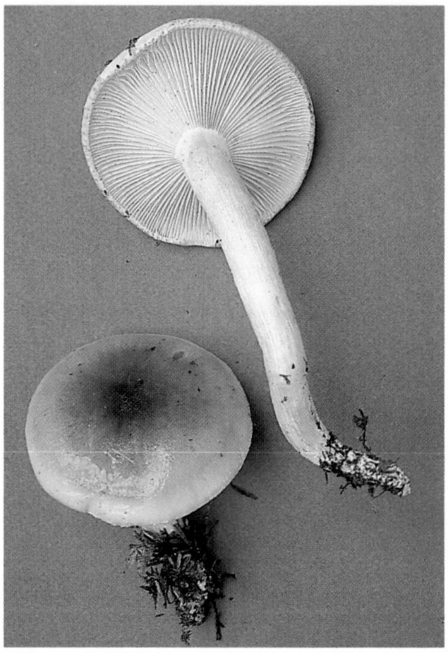
Tawny Almond Waxy Cap
Hygrophorus bakerensis ¼ life-size

Hygrophorus unguinosus (Fr.) Fr. **Cap** 2–5cm (¾–2in) across, convex or bell-shaped, then flatter with an umbo; blackish to gray-brown, then smoke gray when older; smooth, shiny, and sticky, translucent-striate on the margin. **Gills** adnate with a decurrent tooth, widely spaced but connected by veins, broad; whitish. **Stem** 30–80 × 3–6mm (1¼-3½ × ⅛–¼in), fragile, slightly wavy, sometimes tapering toward the base; same color as cap or paler; smooth and slimy. **Flesh** thin, fragile; pale gray or white. **Odor** not distinctive. **Taste** mild. **Spores** broadly ellipsoid, 7–8.5 × 4–5μ. Deposit white. **Habitat** scattered to gregarious on humus or soil in mixed and coniferous woods and in swamps. Frequent. Found widely distributed throughout North America. **Season** June–December. **Edibility not known.**

Hygrophorus eburneus (Fr.) Fr. syn. *Limacium eburneum* (Fr.) Kummer **Cap** 3–7cm (1¼–2¾in) across, convex becoming flattened with an umbo or with a depressed disc and raised margin; pure white; slimy, sticky, smooth. **Gills** decurrent, subdistant to distant, moderately broad, waxy; whitish. **Stem** 45–160 × 2–15mm (1¾–6¼ × ³⁄₃₂–½in), tapering toward base, slightly twisted; white; covered in mealy granules near the top when young; slimy. **Flesh** thick at center of cap; white. **Odor** faint but pleasant. **Taste** mild but pleasant. **Spores** broadly ellipsoid, no reaction with KOH on cap, 6–8 × 4–5μ. Deposit white. **Habitat** scattered to numerous in leaf litter under conifers and beech, also in thickets and grassy areas. Fairly common. Found in northern North America south to North Carolina and in California. **Season** August–January. **Edible** but not worthwhile. **Comment** *Hygrophorus niveus* (p. 65) and *Hygrophorus virgineus* (p. 65) lack the sticky stem.

Hygrophorus chrysodon (Fr.) Fr. syn. *Limacium chrysodon* (Fr.) Kummer **Cap** 3–8cm (1¼–3in) across, convex expanding to flatter and knobby with an inrolled margin; whitish; sticky when wet, shiny when dry, with many tiny, cottony yellow scales, mostly at the margin. **Gills** decurrent, distant, moderately broad, waxy; white with yellow edges. **Stem** 40–70 × 10–15mm (1½–2¾ × ½in); white, covered in minute yellow scales, usually at the top; sticky. **Flesh** thick; white, sometimes with a reddish tinge. **Odor** fungusy. **Taste** mild or slightly bitter. **Spores** ellipsoid, 7–9 × 4–5.5μ. Deposit white. **Habitat** singly to scattered on soil in coniferous and mixed deciduous woods. Found widely distributed in North America. **Season** July–January. **Edible. Comment** Easily recognized by its yellow scales.

Hygrophorus olivaceoalbus (Fr.) Fr. **Cap** 3–10cm (1¼–4in) across, convex or knobbed becoming flatter; brown to black at the disc, paler ash gray toward margin; slimy and sticky, streaked with distinct, fine dark gray fibers beneath the

Hygrophorus speciosus ⅔ life-size

Hygrophorus erubescens ½ life-size

pellicle. **Gills** adnate to decurrent, subdistant, moderately broad; white to ashy. **Stem** 80–140 × 10–30mm (3–5½ × ½–1¼in); white and smooth at the top above the ring, below the ring a double sheath which is very sticky on the outside with black fibers underneath; in age this breaks up into patches. **Veil** all over the cap and stalk, making it very slimy; partial veil is fibrous and black, leaving sheath on stem. **Flesh** soft; white. **Odor** not distinctive. **Taste** not distinctive. **Spores** ellipsoid, nonamyloid, 9–12 × 5–6μ. Deposit white. **Habitat** scattered or growing in dense tufts under conifers and redwood. Sometimes abundant. Found widely distributed in northern North America. **Season** July–December. **Edible** with caution.

Hygrophorus pustulatus (Pers. ex Fr.) Fr. **Cap** 2–5cm (¾–2in) across, convex with a depressed center and an inrolled margin that becomes arched in maturity; gray-brown with a darker disc; sticky, glutinous when wet, with dark granular scales, especially in the center. **Gills** decurrent, bluntly adnate, close to subdistant, narrow; white. **Stem** 40–80 × 3–8mm (1½–3 × ⅛–⁵⁄₁₆in), solid or stuffed at the top, slightly enlarged below; whitish flushed with cap color; lower section sticky from remnants of gelatinous universal veil, elsewhere dry with some dark gray pits. **Flesh** soft, quite thin; white. **Odor** not distinctive. **Taste** not distinctive. **Spores** ellipsoid to ovoid, smooth, 7.5–9.5 × 4.5–5.5μ. Deposit white. **Habitat** in groups under fir and redwood. Found in central and western North

America and in California. **Season** August–December. **Edible.**

Tawny Almond Waxy Cap *Hygrophorus bakerensis* Smith & Hesler **Cap** 4–15cm (1½–6in) across, convex-obtuse with incurved margin, becoming plane; margin whitish, center tawny olive, umber-brown to cinnamon buff, fading outward; glutinous when wet with appressed fibers below gluten. **Gills** adnate-decurrent, close; creamy white. **Stem** 70–140 × 8–25mm (2¾–5½ × ⁵⁄₁₆–1in), equal or tapering below; white to pale pinkish buff; pruinose at the top, without any gluten but often exuding small droplets of liquid at the top. **Flesh** thick; white, unchanging. **Odor** strong, fragrant, of bitter almonds according to some reports. **Taste** mild. **Spores** ellipsoid, smooth, 7–8(10) × 4–5(6)μ. Deposit white. **Habitat** under conifers. Common. Found in the Pacific Northwest. **Season** September–December. **Edible.**

Hygrophorus speciosus Pk. **Cap** 2–5cm (¾–2in) across, convex then expanding and sometimes knobbed; bright orange-red to orange or golden yellow, sometimes fading to more orangy yellow over the margin; smooth and very sticky. **Gills** decurrent, distant to subdistant, narrow; waxy white to yellowish. **Stem** 40–100 × 4–10mm (1½–4 × ³⁄₁₆–½in), solid, enlarging toward base; white stained yellow or orange by slimy veil or veil remnants. **Veil** thin and very sticky. **Flesh** soft; white to yellowish. **Odor** mild. **Taste** mild. **Spores** ellipsoid, nonamyloid, 8–10 × 4.5–6μ. Deposit

white. **Habitat** singly, scattered, or clustered on humus and needle duff in moist areas under conifers, particularly larch. Common and sometimes abundant. Found widely distributed in North America. **Season** August–October. **Edible.**

Hygrophorus erubescens (Fr.) Fr. **Cap** 3–10cm (1¼–4in) across, convex becoming flatter and broadly umbonate with an incurved margin; whitish or flesh-colored, with a pink tinge over the whole cap and darker red on the disc; purple-pink scales or fibers all over the cap, yellow-spotted in places; margin minutely woolly and often beaded with drops of moisture. **Gills** adnate to decurrent, subdistant, moderately broad; pink spotted red. **Stem** 40–70 × 6–12mm (1½–2¾ × ¼–½in), at first beaded with drops of moisture; top white, lower part pale reddish-brown and sometimes yellowish when bruised; scaly and covered in minute, close fibers. **Flesh** thin, but thicker on disc; white; staining yellowish where bruised. **Odor** not distinctive. **Taste** bitter. **Spores** ellipsoid, nonamyloid, 6.4–7.7 × 4.1–5.2μ. Deposit white. **Habitat** gregarious under conifers, particularly pine and spruce. Common at high altitudes. Found in western North America, especially in the mountains. **Season** September–November. **Not edible.**

Hygrophorus subalpinus ⅓ life-size

Hygrophorus monticola ½ life-size

Hygrophoru russula ⅔ life-size

False Chanterelle *Hygrophoropsis aurantiaca* ⅔ life-size

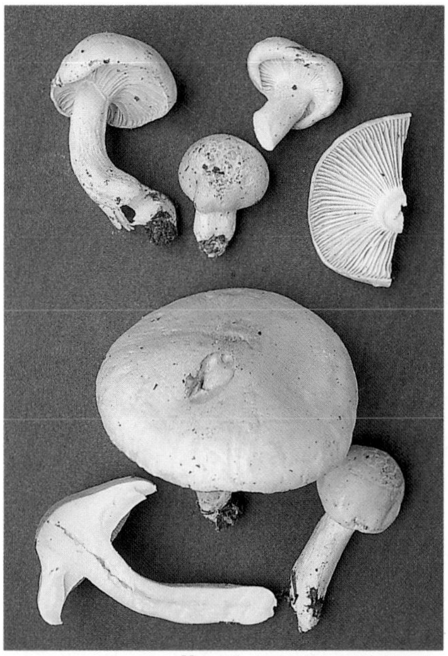

Hygrophorus pudorinus ¼ life-size

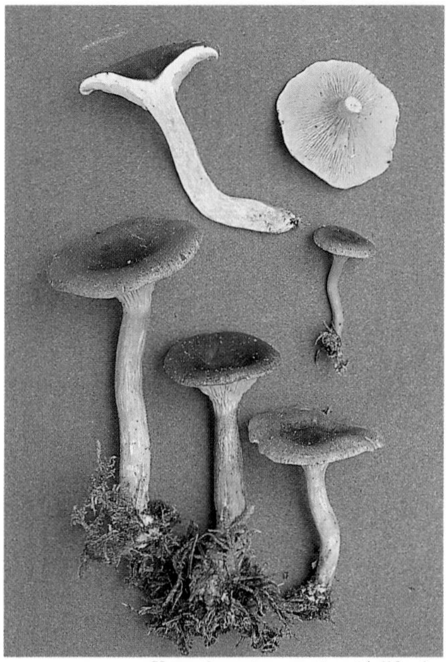

Hygrophoropsis aurantiaca ⅓ life-size

Hygrophorus subalpinus Smith **Cap** 4–12cm (1½–4¾in) across, broadly convex becoming flat with a slight umbo and sometimes a wavy margin with patches of broken veil attached; opaque and snow-white with a distinct sheen; smooth and sticky with a thin pellicle that is difficult to separate. **Gills** decurrent, close, narrow; same color as stem. **Stem** 30–50 × 10–30mm (1¼–2 × ½-1¼in) solid, thick, bulbous at the base; white. **Veil** hairy, white, leaving a thin ring in the center of the stem. **Flesh** thick; white. **Odor** mild. **Taste** mild. **Spores** ellipsoid, nonamyloid, 8–10 × 4.5–6μ. Deposit white. **Habitat** singly to gregarious under conifers and melting snowbanks. Frequent. Found in the Pacific Northwest, Rocky Mountains, and Michigan. **Season** May–October. **Edible. Comment** Giant forms may sometimes be found.

Hygrophorus monticola Smith & Hesler **Cap** 2–5cm (¾–2in) across, broadly convex becoming flat or depressed, with an inturned margin that becomes irregular in mature specimens; yellowy fawn with a darker, more orange-buff disc and paler toward the margin; sticky but soon dry. **Gills** decurrent, distant, broad, many forked halfway to margin; whitish, becoming flushed with cap color, only paler. **Stem** 30–60 × 3–25mm (1¼–2¼ × ⅛–1in), slightly thicker at top than base; white, though a few older specimens flushed with cap color. **Flesh** thick, firm, whitish. **Odor** cherry pits. **Taste** mild. **Spores** ellipsoid, nonamyloid, 10–14 × 5.2–7.5μ. Deposit white. **Habitat** gregarious on soil and moss in coniferous woods. Occasional. Found in the northeastern United States, Idaho, and Canada. **Season** August–November. **Edibility not known.**

Hygrophoru russula (Fr.) Quél. **Cap** 5–12cm (2–4¾in) across, broadly convex with an inrolled margin, expanding to nearly flat; pink to purple-red (on disc); sticky then dry and often streaked with small, purple-pink fibers, giving the cap a scaly appearance, sometimes bruising yellowy. **Gills** adnate, close to crowded, moderately broad, waxy; white then pale pink and spotted purple-red. **Stem** 20–70 × 15–35mm (¾–2¾ × ½–1¼in) dry, smooth, solid, sometimes tapering toward the base; white streaking pinkish. **Flesh** thick, firm; white tinged pink. **Odor** mild. **Taste** mild. **Spores** ellipsoid, nonamyloid, 6–8 × 3–5μ. Deposit white. **Habitat** scattered, gregarious, and sometimes in fairy rings under oak and sometimes conifers. Common and somewhat abundant in the East. Found widely distributed in North America. **Season** August–December (November–February in California). **Edible** — good.

False Chanterelle *Hygrophoropsis aurantiaca* (Wulf. ex Fr.) Maire **Cap** 2–8cm (¾–3in) across, convex to shallowly funnel-shaped, often remaining incurved at the margin; color variable, typically some shade of orange-yellow to brownish yellow or dark brown, often darker at the center and more yellowish orange at the edge; dry, downy to felty. **Gills** decurrent, close, narrow, and dichotomously forked; color varies from deep orange to yellowish. **Stem** 20–100 × 5–20mm (¾–4 × ¼–¾in), often enlarged toward the base and curved; same color as cap or darker; dry, somewhat hairy. **Flesh** thin, tough; yellowish to orangish. **Odor** mild, mushroomy. **Taste** mushroomy. **Spores** ellipsoid, smooth, dextrinoid, 5.5–8 × 2.5–4.5μ. Deposit white. **Habitat** singly, scattered, or in groups on the ground or on rotting coniferous wood. Common. Found widely distributed in North America. **Season** August–November (overwinters in California). **Not edible** as it is known to cause alarming symptoms (hallucination) in some cases. **Comment** Some authors feel that this species should be split: the almost white-capped form and the very dark brown-capped form would then probably be separate varieties. (Both illustrated.)

Hygrophorus pudorinus (Fr.) Fr. **Cap** 5–12cm (2–4¾in) across, convex to nearly flat with a broad umbo and a downy, inrolled margin that later expands; pale tan to pinky-buff; smooth and sticky. **Gills** subdecurrent, distant,

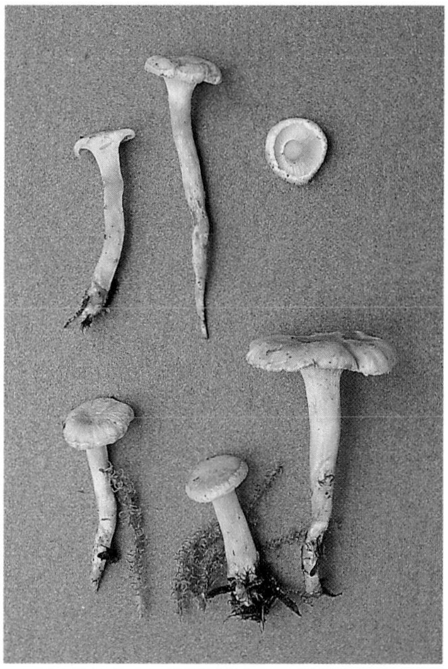

Hygrophoropsis morganii ¾ life-size

narrow, waxy; white tinged buff or pale salmon. **Stem** 40–100 × 8–20mm (1½–4 × 5/16–¾in), solid; white tinged pink with distinct white tufts of hair at the top, flattened hairs toward the bottom which become reddish with age and turn orange-yellow with KOH; dry. **Flesh** thick, firm; white tinged with pink. **Odor** none or faintly fragrant. **Taste** unpleasant, like turpentine. **Spores** ellipsoid, nonamyloid, 6.5–9.5 × 4–5.5μ. Deposit white. **Habitat** scattered to gregarious under conifers and sometimes in bogs. Common. Found widely distributed in northern North America, south to northern California. **Season** August–December (January in California). **Edible.**

Hygrophoropsis morganii (Pk.) Bigelow **Cap** 1–4cm (½–1½in) across, irregularly domed-funnel-shaped, with a margin remaining inrolled for a long time then expanding to become somewhat wavy; salmon to flesh-pink or ochre-pink; slightly sticky, then dry and minutely hairy. **Gills** decurrent, strongly forking; cream to very pale yellow-pink or flesh-pink. **Stem** 10–35 × 3–6cm (½–1¼ × ⅛–¼in), generally tapering to a long, rooting base in age; flesh-pink, slightly paler toward the top; dry, minutely scurfy. **Odor** fruity. **Spores** 3.5–5.5 × 2.5–3μ. Deposit whitish. Clamps present. **Habitat** in mixed or coniferous woods. Rare. Found in the Pacific Northwest. **Season** September–October. **Edibility not known.**

Fairy Ring Mushroom *Marasmius oreades* ½ life-size

Marasmius cohaerens ⅔ life-size

MARASMIUS *Caps usually small, often tiny, they are tough and leathery. If dried they will revive on wetting. The caps are not normally conical as are mycenas, which are often similar in size.*

Fairy Ring Mushroom *Marasmius oreades* (Bolton ex Fr.) Fr. **Cap** 2–5cm (¾–2in) across, convex-umbonate, with inrolled and slightly scalloped margin; pale tan to buff; glabrous to felty. **Gills** adnexed, broad, fairly distant; pale tan. **Stem** 30–80×3–5mm (1¼–3 × 1/16–¼in), tough, fibrous, often twisted; cream to pale buff; minutely felty. **Odor** pleasant to almondlike. **Taste** pleasant. **Spores** ellipsoid, smooth, 7–10×4–6μ. Deposit white. **Habitat** often in fairy rings in lawns, grassy meadows. Found throughout North America. **Season** May–October. **Edible** and good, although there are records of adverse effects with certain individuals.

Marasmius cohaerens (Pers. ex. Fr.) Cke. & Quél. **Cap** 1–3.5cm (½–1¼in) across, convex-campanulate; dark yellowish brown to cinnamon, darker at disc; smooth, dry, or subhygrophanous at margin; tough, pliable, dry cap texture revives when remoistened. **Gills** adnexed, distant, broad; yellowish white to brownish at margin. **Stem** 20–75 × 0.75–3mm (¾–2¾ × 1/32–⅛in), long, slender; white pruinose at apex, yellow-brown below and reddish brown in lower half, with a basal pad of pale yellow or white mycelium; dry, very smooth and polished, pliant to cartilaginous or horny with age. **Flesh** pallid brown. **Odor** somewhat pungent, earthy. **Taste** none to slightly alkaline and with a bitter aftertaste. **Spores** subfusiform, smooth, (6)7–9.8(11) × 3–5.5μ. Deposit white. **Habitat**

in dense clusters or gregarious on decaying leaves, twigs, etc., in deciduous woods. Found widely distributed in North America. **Season** July–October. **Edibility not known.** **Comment** A drop of alkali (KOH) applied to the reddish areas of the stem turns green. However, this appears to be the first record of this reaction in this species.

Marasmiellus nigripes (Schw.) Singer **Cap** 1–2 cm (½–¾in) across, convex; white translucent; overlapping with distinct radial ridges; rubbery in feel. **Gills** decurrent, distant, forked; pure white, bruising a touch reddish. **Stem**; 10–15 × 2mm (½ × 3/32in), shortish and rather stout, covered in dense white hairs; black or white at first then black from the base up. **Odor** slight. **Spores** triangular or star-shaped, 8–9u. Deposit white. **Habitat** on dead wood and twigs of many tree species. Found in northeastern North America, west to Michigan. **Season** August–October. **Edibility not known. Comment** The stems of my collection were only just beginning to blacken.

Marasmius scorodonius (Fr.) Fr. **Cap** 1–3cm (½–1¼in) across; broadly convex expanding to nearly flat, with an inturned margin that becomes wavy; reddish or yellowish brown becoming faded; dry, smooth, radially wrinkled. **Gills** adnate or nearly free, crowded, narrow, often forked; yellowish pink to pallid. **Stem** 15–60 × 1–3mm (½–2¼ × 1/16–⅛in), round to compressed; yellowish white toward the top, dark brown below; dry, smooth, shining, brittle. **Odor** of onions or garlic. **Taste** of onions or garlic. **Spores** ellipsoid, smooth, 7–10×3–5μ. Deposit white. **Habitat** scattered or in groups on debris, bark, twigs,

grass, needle duff. Frequent. Found in northern North America and California. **Season** July–October (overwinters in California). **Edible.**

Marasmius olidus Gilliam **Cap** 0.3–2cm (⅛–¾in) across, cushion-shaped or convex becoming flatter and often umbilicate, then flat to concave with a wavy margin; light brownish to yellowish brown tinged with pink; dry, dull, finely velvety, minutely wrinkled on the disc. **Gills** adnate, adnexed, or subdecurrent, distant, moderately numerous; light yellowish brown. **Stem** 12–30 × 1–2mm (½–1¼ × 1/16–3/32in), tapering slightly to the base, straight or curved, hollow; white at the top, pale yellowish or light yellow-brown in the middle, grayish brown tinged pink or blackish brown below; dry, dull, with a bloom above and hairy below, with white mycelium on the basal disc. **Flesh** thin, firm; yellowish white. **Odor** pungent, of garlic. **Taste** like onion or garlic. **Spores** narrowly club-shaped and often curved, 10.2–16.5×2.8–3.8μ. Deposit white. **Habitat** in deciduous woods. Found in Michigan and New Jersey. **Season** August–September. **Edibility not known.**

Marasmius cystidiosus (Smith & Hesler) Gilliam **Cap** 2–5cm (¾–2in) across, obtusely bell-shaped; pale tan to buff becoming pale, creamish, with a darker center; smooth, striate when moist. **Gills** adnate-seceding, crowded; pale buff. **Stem** 60–120×3–6mm (2¼–4¾ × ⅛–¼in), long, thin, sometimes twisted; whitish above, very pale gray-brown toward the white mycelium base; cartilaginous. **Odor** strong. **Taste** slight. **Spores** ellipsoid, 7–9×2.5–3.5μ. Deposit white. Pleurocystidia subfusoid, apex often ripplelike, 36–67 ×

Marasmiellus nigripes ⅔ life-size

Marasmius scorodonius life-size

Marasmius olidus ⅔ life-size

3–8μ. **Habitat** in clusters on decayed wood of deciduous trees. Not common. Found in northeastern North America. **Season** May–July. **Edibility not known.**

Marasmius strictipes (Pk.) Singer **Cap** 2–6.5cm (¾–2½in) across, broadly convex becoming flatter with a broad umbo, margin slightly upturned with age; yellowish to orange-yellow; smooth. **Gills** adnate, close to crowded, narrow; yellowish to pinky-fawn. **Stem** 25–75 × 3–10mm (1–2¾ × ⅛–½in), hollow; minutely hairy. **Flesh** dextrinoid. **Odor** sometimes of radish. **Taste** sometimes of radish. **Spores** ellipsoid, smooth, 6–8 × 3–5μ. Deposit white. **Habitat** scattered or in groups on fallen leaves under mixed woods or hardwoods. Found in northeastern North America, south to North Carolina and west to the Great Lakes. **Season** July–September. **Not edible.**

Marasmius sullivanti Montagne **Cap** 0.5–2.5cm (¼–1in) across, convex becoming flatter with a faintly lined margin; bright rusty red or orange-red; dull, unpolished. **Gills** free but reaching the stem, close, narrow but slightly swollen in the middle; white, with edges faintly pink at first. **Stem** 5–25 × 1–2mm (¼–1 × ¹⁄₁₆–³⁄₃₂in); whitish at the top, reddish brown becoming darker, almost black, toward the base; hornlike and rigid with a faint bloom, rough stiff hairs on the base. **Flesh** pallid. **Odor** none. **Taste** faintly bitter. **Spores** ellipsoid, smooth, 7–9 × 3–3.5μ. Deposit white. **Habitat** singly, scattered, or in groups on debris and humus in hardwood forests. Not common. Found in southeastern North America, north to New Jersey. **Season** July–August. **Edibility not known.**

Marasmius delectans Morgan **Cap** 1–4cm (½–1½in) across, convex to depressed at center; pale ochre then soon white; surface slightly sulcate in age. **Gills** adnate, distant, broad; white. **Stem** 20–80 × 1–3mm (¾–3 × ¹⁄₁₆–⅛in); dark brown, upper portion white; glabrous, with slightly swollen, hairy base. **Odor** not distinctive. **Taste** not distinctive. **Spores** ellipsoid, 6–8 × 3–5μ. Deposit white. **Habitat** on fallen leaves of hardwood trees. Found in eastern and central North America. **Season** July–August. **Edibility not known.**

Marasmius cystidiosus ¼ life-size

Marasmius strictipes ½ life-size

Marasmius sullivanti ⅔ life-size

Marasmius delectans ¾ life-size

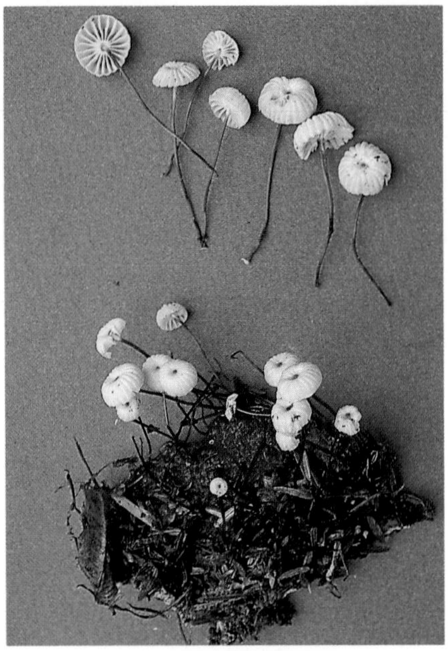
Marasmius rotula ½ life-size

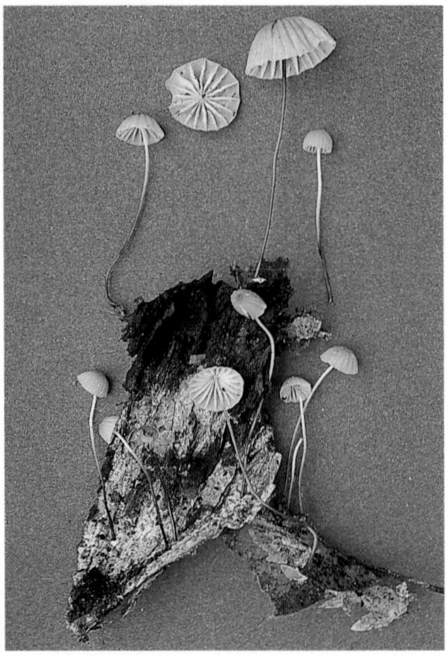
Orange Pinwheel *Marasmius siccus* ½ life-size

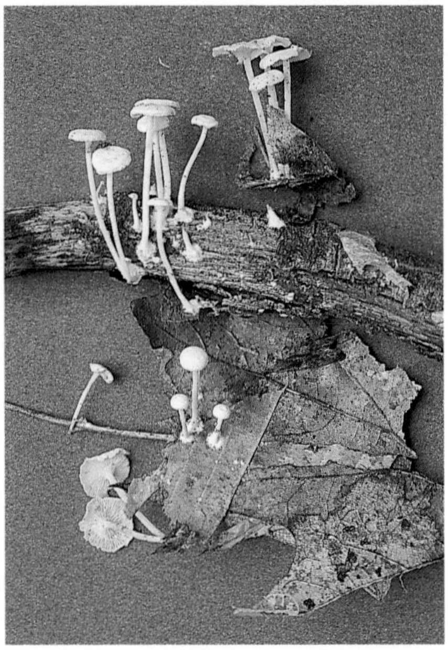
Resinomycena rhododendri ⅔ life-size

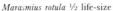

Marasmius rotula (Scop. ex Fr.) Fr. **Cap** 1.5–17mm (1/16–½in) across, hemispherical, then convex with a central depression (umbilicate), with deep radial grooves; pale yellowish brown, yellow to pallid; dry and minutely granular. **Gills** adnate to sinuate, often arching, distant, rather broad; yellowish white. **Stem** 15–85 × 1mm (½–3¼ × / 1/16in), equal; smooth, pale yellow-white above, light brown to reddish brown below; often blackish brown overall except apex when old; dry, shining to dull, cartilaginous or wiry. **Odor** none. **Taste** mild or with bitter aftertaste. **Spores** ellipsoid, 6–9.5(10) × 3–4.5μ. Deposit white to pale yellow. **Habitat** gregarious or in clusters on wood of deciduous (rarely conifer) trees or on leaf litter. Found widely distributed in much of North America, but commoner in the East. **Season** June–October. **Not edible.**

Orange Pinwheel *Marasmius siccus* (Schw.) Fr. **Cap** 0.5–3cm (¼–1¼in) across, bell-shaped with flattened or depressed center, with deep, wide, radial pleats; rust-orange to rust-brown or pale tawny; minutely velvety. **Gills** attached or free, distant, broad; pallid to buff; edges even. **Stem** 20–70 × 1–2mm (¾–2¾ × 1/16–3/32in), equal; deep brown from base upward, yellowish above; smooth, dry, polished. **Flesh** very thin; pallid, dextrinoid. **Odor** slight. **Taste** slight. **Spores** spindle- to club-shaped, smooth, (13)16–21 × 3–4.5(5)μ. Deposit white. **Habitat** scattered to gregarious on leaves, wood, twigs, etc., of deciduous trees. Very common. Found east of the Great Plains in North America. **Season** July–October. **Edibility not known.**

Resinomycena rhododendri (Pk.) Redhead & Singer **Cap** 0.4–1.5cm (3/16–½in) across, convex becoming flatter with a central depression and occasionally umbilicate; margin incurved at first, becoming uneven in age; white to pale yellowish white; dry to tacky, micaceous when dry, opaque to vaguely lined. **Gills** adnate to subdecurrent or arcuate-decurrent, close, moderately narrow; whitish; edges notched, sometimes beaded with resin. **Stem** 12–50 × 1–2mm (½–2 × 1/16–3/32in), hollow, equal or slightly enlarged above; whitish; sticky to tacky or dry with scattered heads of resin; tough and pliant drying to a horny texture; silky, white, radiating basal disc. **Flesh** tough-pliant; whitish. **Odor** not distinctive. **Taste** not distinctive. **Spores** ellipsoid or broadly cylindrical, with a prominent apiculus, smooth, amyloid, 5.4–8.5 × 2.4–4.1μ. Deposit whitish. **Habitat** scattered or in clusters on leaf litter, twigs, and woody debris in hardwood forests. Common. Found in eastern North America, south to Georgia and west to Illinois. **Season** June–September. **Not edible.**

Baeospora myriadophylla (Pk.) Singer **Cap** 1–4cm (½–1½in) across, convex becoming flat or sometimes sunken in the center, with an incurved margin becoming wavy and uplifted in age; lavender or brownish mauve, becoming more dingy yellowish brown or pale buff in age; smooth, moist. **Gills** adnate or nearly free, close to crowded, narrow; similar color to cap. **Stem** 10–50 × 1–3mm (½–2 × 1/16–⅛in), stuffed to hollow, pliant, tough; similar color to cap; dry, often grooved becoming smooth, with long, coarse hairs at the base. **Flesh** tough; grayish. **Spores** broadly ellipsoid, smooth, amyloid, 3.5–4.5 × 2–3μ. Deposit white. **Habitat** in small groups or clustered on decaying wood, especially poplar and hemlock. Found in northern North America. **Season** June–October. **Not edible.**

Omphalina wynniae (Berk. & Br.) Ito **Cap** 1–3cm (½–1¼in) across, broadly convex becoming flatter with an uplifted, lined margin; olive-brown to greenish yellow, paler overall in age; smooth, moist, hygrophanous. **Gills** decurrent, distant, broad; greenish yellow, brighter and paler in age. **Stem** 10–30 × 1–4mm (½–1¼ × 1/16–3/16in), equal; yellow or whitish; more or less smooth. **Spores** ellipsoid, 7–9 × 4–5μ. Deposit white. **Habitat** singly or in small clusters on wet, rotting conifers. Found in the Pacific Northwest. **Season** September–October. **Edibility not known.**

Omphalina ericetorum (Fr.) Lange **Cap** 1–2.5cm (½–1in) across, convex then soon flattened to funnel-shaped; from pale buff to deep brown or vinaceous; glabrous, strongly striate. **Gills** decurrent, widely spaced; pale yellowish. **Stem** 10–30 × 1–3mm (½–1¼ × 1/16–⅛in; pallid to yellow-brown; finely pubescent. **Spores** ovoid, 7–9 × 4–6μ. Deposit white to pale yellow. Basidia can be 1-, 2-, 3-, or 4-spored. **Habitat** attached to the lichen *Botrydina vulgaris*, which is the vegetative stage, the mushroom being the reproductive stage, the joint organism forming a basidio-lichen. Found widely distributed throughout northern North America across to the Pacific Northwest and up into the Arctic Circle. **Season** all year. **Edibility not known.**

Omphalina ectypoides (Pk.) Bigelow syn. *Clitocybe ectypoides* (Pk.) Sacc. **Cap** 2–6cm (¾–2¼in) across, sunken on the disc, funnel-shaped with a spreading, wavy margin; ochraceous; minutely scaly with brownish black hairs, moist. **Gills** decurrent, almost distant, narrow, some forked; yellowish. **Stem** 20–65 × 2–9mm (¾–2½ × 3/32–⅜in), sometimes eccentric, expanding toward the base; honey yellow staining olivaceous to brownish on handling; slightly wrinkled. **Flesh** thin. **Odor** mild. **Taste** mild. **Spores** ellipsoid, smooth, amyloid, 6.5–8 × 3.5–5μ. Deposit white. **Habitat** on rotting conifer logs, particularly hemlock. Common but not abundant. Found in northeastern North America, west to the Great Lakes and south to North Carolina; also in the Pacific Northwest. **Season** July–September. **Edibility not known.**

Gerronema chrysophylla (Fr.) Singer syn. *Omphalina chrysophylla* (Fr.) Murr. **Cap** 0.5–4cm (¼–1½in) across, flat with a shallowly depressed disc and incurved margin, becoming more funnel-shaped in age; ochre to yellowish

Baeospora myriadophylla ½ life-size

Omphalina wynniae ¾ life-size

Omphalina ericetorum ½ life-size

brown; moist to dry, minutely hairy to scaly.
Gills decurrent, distant, narrow to broad;
orange-yellow. **Stem** 20–50×1–3mm
(¾–2×¹⁄₁₆–⅛in), stuffed to hollow, somewhat
curved or flattened; orange-yellow, sometimes
tinged with brown, cottony white at the base;
smooth, moist. **Flesh** thin, flexible; orange.
Odor not distinctive. **Taste** not distinctive.
Spores ellipsoid to cylindrical, smooth,
nonamyloid, 8.5–15.5×4.5–6μ. Deposit
yellowish. **Habitat** scattered or in clusters on
rotting conifer logs and moss. Found widely
distributed throughout North America. **Season**
May–September. **Edibility not known.**

Omphalina pyxidata (Bull. ex Fr.) Quél. **Cap**
0.5–2cm (¼–¾in) across, convex, deeply
umbilicate; reddish brown, pinkish brown to
yellowish; smooth and deeply radially fluted.
Gills decurrent, widely spaced; brownish.
Stem 10–30×1–2mm (½–1¼×¹⁄₁₆–³⁄₃₂in),
paler than cap; smooth. **Spores** almond-
shaped, 7–10×4.5–6μ. Deposit white. **Habitat**
in grass in sandy soils, subalpine to alpine.
Found in western North America. **Season**
July–September. **Edibility not known.**

Gerronema strombodes (Berk. & Montagne) Singer
Cap 2.5–6cm (1–2¼in) across, convex
becoming centrally depressed; gray to gray-
brown, sometimes yellowish toward the
margin; smooth, darkly and innately striate.
Gills decurrent, distant, broad; yellowish white
to whitish. **Stem** 30–60×3–7mm
(1¼–2¼×⅛–¼in); whitish, yellowish, or
grayish; longitudinally hairy. **Taste** mild to
bitterish. **Spores** ellipsoid, smooth,
nonamyloid, 7.5–9×3–5.5μ. Deposit whitish.
Clamps present. **Habitat** on deciduous and
coniferous tree trunks. Rare. Found in West
Virginia and probably other areas. **Season**
August–September. **Edibility not known.**

Omphalina ectypoides ½ life-size

Gerronema chrysophylla ⅓ life-size

Omphalina pyxidata ¾ life-size

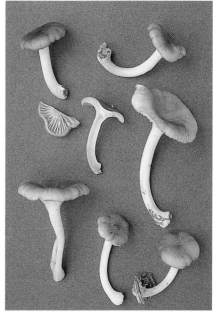

Gerronema strombodes ½ life-size

Strobilurus trullisatus ½ life-size

Asterophora lycoperdoides ¾ life-size

Rickenella fibula ¾ life-size

Xeromphalina campanella ½ life-size

Xeromphalina tenuipes ¾ life-size

Xeromphalina fulvipes ⅔ life-size

Strobilurus trullisatus (Murr.) Lennox **Cap** 0.5–2cm (¼–¾in), convex-depressed; whitish to pale pinkish brown; densely and minutely hairy. **Gills** adnexed, crowded; white to pale pink. **Stem** 25–50×1–15mm (1–2× ¹⁄₁₆–³⁄₃₂in), brittle; white above, yellowish rust below; minutely hairy-fuzzy. **Odor** not distinctive. **Taste** not distinctive. **Spores** 3–4.5×1.5–3μ. Deposit white. **Habitat** on old or partially buried cones of Douglas fir. Found in western North America. **Season** August–November. **Not edible.**

Asterophora lycoperdoides (Bull. ex Mérat) S. F. Gray syn. *Nyctalis asterophora* Fr. **Cap** 0.5–2cm (¼–¾in) across, subglobose; white becoming covered in a clay-buff coating of chlamydospores; dry, cottony. **Gills** adnate, distant, narrow, thick and sometimes forked, often malformed; whitish. **Stem** 20–30×3–10mm (¾–1¾×⅛–½in), stuffed to hollow; white, stained brownish in age; minutely downy. **Flesh** firm; whitish. **Odor** mealy. **Taste** mealy. **Spores** oval, smooth, nonamyloid, 5–6× 3.5–4μ. Deposit white. Chlamydospores subglobose, covered in long blunt projections, giving a star-shaped appearance, clay-buff in color, 13–16μ in diameter. **Habitat** growing on other mushrooms, especially lactarius and russula. Very common. Found widely distributed throughout North America. **Season** July–November. **Not edible.**

Rickenella fibula (Bull. ex Fr.) Raith. **Cap** 0.3–1.5cm (⅛–⅝in), convex with deep depression at center, striate; orange-yellow to buff. **Gills** deeply decurrent, distant; white. **Stem** 10–50×1–2mm (⅜–2×¹⁄₁₆–³⁄₃₂in); yellow-orange; smooth to finely hairy. **Odor** not distinctive. **Taste** not distinctive. **Spores** ellipsoid, 4–5×2–2.5μ. Deposit white. **Habitat** in moss. Found throughout most of North America. **Season** June–November. **Edibility not known.**

Xeromphalina campanella (Bat. ex Fr.) Kühner & Maire **Cap** 0.3–3cm (⅛–1¼in) across, convex usually with a sunken center, with prominent radial lines; bright tawny ochre to cinnamon; smooth. **Gills** decurrent, widely spaced; pale yellow to orange. **Stem** 10–50×0.5–3mm (½–2×¹⁄₃₂–⅛in), often with swollen base; red-brown at base, yellow at apex; smooth, base covered with dense tawny hairs. **Odor** not distinctive. **Taste** not distinctive. **Spores** elliptic, amyloid, 5–7×3–4μ. Deposit pale buff. **Habitat** densely clustered, often in many hundreds, over the surface of conifer stumps and logs. Common. Found widely distributed in North America. **Season** June–November. **Not edible.**

Xeromphalina tenuipes (Schw.) Smith **Cap** 2–7cm (¾–2¾in) across, broadly conical at first, quickly becoming flat to depressed with a very wavy margin; orange-brown with an olive-brown tinge when fresh; dry, velvety, becoming somewhat wrinkled with a striate margin. **Gills** adnate, distant; white becoming pale yellow. **Stem** 50–80×3–8mm (2–3×⅛–⁵⁄₁₆in), hollow, sometimes with a rootlike extension into the soil; similar color to the cap; velvety to minutely hairy. **Flesh** pliant, watery brown. **Spores** smooth, amyloid, 7–9×4.5–5μ. Deposit white. **Habitat** singly or in clusters on hardwood logs and stumps. Quite common. Found in eastern North America, west to the Great Plains. **Season** April–July. **Edibility not known.**

Mycena inclinata ⅔ life-size

Mycena haematopus ¾ life-size

Xeromphalina fulvipes (Murr.) Smith **Cap** 1–2.5cm (½–1in), convex to flattened; bright yellow-brown to paler at margin; glabrous. **Gills** adnate, crowded, yellowish. **Stem** 20–80 × 1–2.5mm (¾–3 × ¹/₁₆–⅛in); reddish brown to black at base; tomentose, hairy at base. **Odor** pleasant. **Taste** bitter. **Spores** long ovoid, smooth, 4.5–6 × 1.5–2μ. Deposit white. **Habitat** scattered on conifer debris. Uncommon. Found in the Pacific Northwest. **Season** all year except for dry periods. **Edibility not known.**

MYCENA *Very small mushrooms with long, thin, brittle stems; the caps are conical or bell-shaped. Some exude juice when the stem is broken. Some have strong characteristic smells. Check the gills to see if there is a dark edge. They grow on the ground in humus or on wood.*

Mycena inclinata (Fr.) Quél. **Cap** 1–3cm (½–1¼in) across, conical expanding to bell-shaped, with a prominent umbo and scalloped margin slightly overhanging gills; bay-colored, darker toward center, whitish toward margin; surface moist and smooth. **Gills** adnate, close to subdistant, broad; white or grayish becoming flesh-pink. **Stem** 50–100 × 2–4mm (2–4 × ³/₃₂–³/₁₆in); whitish at apex, deepening to dark red-brown toward the base, which is covered in fine white down. **Flesh** thin, fragile; whitish. **Odor** mealy or rancid. **Taste** mildly mealy. **Spores** ovoid, amyloid, 8–9 × 6–7μ. Deposit white. **Habitat** in dense tufts on decaying hardwood stumps, especially oak. Very common in eastern North America, rare along the western coast. **Season**

May–November. **Edibility not known.** **Comment** One of the most common of all mycenas, it is usually found in profuse clumps.

Mycena haematopus (Pers. ex Fr.) Kummer **Cap** 1–4cm (½–1½in) across, egg-shaped then conical to bell-shaped with distinct umbo and upturned margin; surface pale, reddish-pink bloom when dry, becoming dark gray or reddish brown when moist and paler gray toward the striate margin. **Gills** adnate, close to nearly distant, narrow to moderately broad; white becoming pale pink, sometimes darker at the flocculose edges. **Stem** 4–10 × 2–3mm (1½–4 × ³/₃₂–⅛in); gray-pink exuding a deep, blood-red latex when broken; covered with fine hairs, polished with age; often fused together to form tufts. **Flesh** thin, fragile; exuding dark red juice when cut. **Odor** not distinctive. **Taste** mildly bitter. **Spores** ellipsoid, amyloid, 7–10 × 5–6μ. Deposit white. **Habitat** singly or clustered on decaying wood. Common. Found widely distributed throughout North America. **Season** April–November and later if conditions are favorable. **Edible. Comment** One of the most easily recognized mycenas because of the dark red juice it exudes when damaged.

Mycena pseudoinclinata Smith **Cap** 1.5–2.5cm (½–1in) across, obtusely cone-shaped to bell-shaped, with the margin sometimes becoming plane and often splitting; slightly reddish gray-brown, darker at the center, becoming lighter toward the whitish margin; smooth, moist, striate. **Gills** ascending adnate, close to subdistant; white or pale gray, sometimes with a red tinge. **Stem** 25–60 × 1.5–3mm (1–2¼ × ¹/₁₆–⅛in); pale above,

Mycena pseudoinclinata ½ life-size

tawny brown below. **Flesh** fragile, thicker on disc; grayish to white. **Odor** slightly mealy. **Taste** slightly mealy. **Spores** broadly ellipsoid, strongly amyloid, 8–11 × 5–6μ. Deposit white. **Habitat** growing in dense tufts on decaying wood. Found in eastern North America. **Season** April–November. **Edibility not known. Comment** This mushroom differs from *Mycena inclinata* (above) in that the cystidia have smaller, fewer, and less distinct projections, and it has a larger spore quotient.

Mycena overholtzii ½ life-size

Mycena pura life-size

Mycena overholtzii Smith & Sol. **Cap** 2.5–5cm (1–2in) across, convex to nearly flat with an umbo; pale, dark or bluish gray; smooth, greasy, and striate. **Gills** adnate, subdistant, broad; whitish, staining gray when bruised. **Stem** 40–100 × 1.5–5mm (1½–4 × ¹⁄₁₆–¼in); pale gray and smooth, becoming reddish brown toward the thicker base, which is covered in dense white hairs. **Flesh** pale grayish. **Odor** fungusy. **Taste** distinct, strange. **Spores** ellipsoid, amyloid, 7–10 × 5–6µ. Deposit white. **Habitat** clustered on conifer logs, sometimes under snow, and in deciduous woods. Fairly common. Found in mountainous areas of western North America. **Season** April–July. **Edibility not known.**

Mycena pura (Pers. ex Fr.) Kummer **Cap** 2–5cm (¾–2in) across, convex with a broad umbo; margin sometimes split in large, mature specimens; color varying from shades of purplish lilac or pink to yellowish or white; radially lined at margin when moist, paler when dry. **Gills** adnate, close to almost distant, broad; whitish, gray, or pink. **Stem** 40–100 × 2–8mm (1½–4 × ³⁄₃₂–⁵⁄₁₆in), rigid, widening slightly at base, which is covered in fine, white fibers; whitish, flushed pink. **Flesh** purplish becoming white. **Odor** radishy. **Taste** mildly radishy. **Spores** subcylindric, amyloid, 5–9 × 3–4µ. Deposit white. **Habitat** scattered to gregarious on humus in hardwood and coniferous forests. Common. Found throughout North America. **Season** April–November. **Edible.** Traditionally eaten, but use with caution. **Comment** The most distinctive feature is the strong radishy smell.

Mycena stylobates (Fr.) Kummer **Cap**

0.4–1.5cm (³⁄₁₆–½in) across, convex to bell-shaped, with margin curving in slightly or sometimes flaring and recurved; pale white with a whitish margin; surface smooth or spiny, moist, translucent-striate. **Gills** attached by a line or very narrowly adnate, close becoming distant, narrow becoming ventricose or even very broad in age; pale gray becoming whitish. **Stem** 10–60 × 0.5–1mm (½–2¼ × ¹⁄₃₂–¹⁄₁₆in); covered with fine white fibrils and attached to a flat, circular, striate disc at the base; bluish gray when fresh, becoming pallid. **Flesh** thin; pale. **Odor** none. **Taste** none. **Spores** ellipsoid, amyloid, 6–8 × 3.5–4.5µ or 8–10 × 3.5–4.5µ. Deposit white. **Habitat** scattered or gregarious on fallen leaves, needles, bark, or debris. Found widely distributed in North America. **Season** May–November. **Edibility not known.**

Mycena rosella (Fr.) Kummer **Cap** 0.5–2cm (¼–¾in) across, obtusely cone-shaped, convex, sometimes becoming almost flat in age; pink or pinkish gray, margin fading and becoming more yellowish with age; surface moist, greasy, striate. **Gills** usually horizontal, adnate, close to subdistant, moderately broad; same color as cap with edges darker red. **Stem** 25–70 × 10–15mm (1–2¾ × ½in); pinky-gray; flexible, with base covered in rough white hairs; greasy but not sticky. **Flesh** thin; dirty pink to whitish. **Odor** not distinctive. **Taste** not distinctive. **Spores** ellipsoid, amyloid, 7–9 × 4–5µ. Deposit white. **Habitat** scattered or gregarious on needle beds in coniferous forests. Common. Found throughout North America. **Season** August–November. **Edibility not known.**

Mycena rutilantiformis Murr. **Cap** 2–7cm (¾–2¾in) across, convex expanding to broadly convex, sometimes with an uplifted, wavy or split margin; dark, dingy pinkish brown to dull pinky-buff, often becoming paler with yellowish tinges; smooth, moist and slightly greasy, lined. **Gills** sinuate to adnexed, close to subdistant, broad; dull pinky-fawn with dingy reddish-purple gill edges. **Stem** 30–80 × 5–10mm (1¼–3 × ¼–½in), fragile, hollow, slightly enlarged toward the base; pale grayish buff all over, yellowish to whitish in age. **Odor** radishy. **Taste** similar or bitter. **Spores** subovoid to ellipsoid, amyloid, 8–10 × 4–5µ. Deposit white. **Habitat** scattered or in groups on humus or rotten hardwood debris, often under oak or hickory. Found in northeastern North America, west to the Great Lakes, and in the Pacific Northwest. **Season** May–June, September–October. **Not edible.**

Mycena clavicularis (Fr.) Gillet **Cap** 1–2cm (½–¾in) across, convex, sometimes with a low, rounded umbo when mature and a straight margin spreading with age; gray-brown or bluish gray; surface bloomlike when young, wrinkled and naked in age, faintly striate when moist, moist to greasy. **Gills** adnate, subdistant to distant, narrow to moderately broad; reddish gray-brown with whitish edges. **Stem** 20–50 × 1–1.5mm (¾–2 × ¹⁄₁₆–³⁄₃₂in); same color as cap, though lighter at apex; very sticky. **Flesh** thin, gristly; gray or whitish. **Odor** not distinctive. **Taste** not distinctive. **Spores** ellipsoid, amyloid, 7–12 × 4–6µ. Deposit white. **Habitat** gregarious in coniferous forest. Found widely distributed in western and eastern states of

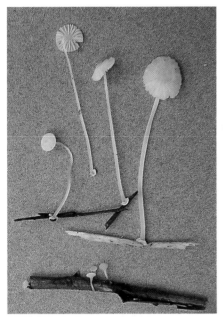

Mycena stylobates ¼ larger than life-size

Mycena rosella ¾ life-size

Mycena rutilantiformis ⅓ life-size

North America. **Season** April–October.
Edibility not known.

Walnut Mycena *Mycena luteopallens* (Pk.) Sacc.
Cap 1–1.5cm (½in) across, ovate to bell-
shaped, then soon flattened; bright yellow to
orange, fading to white; slightly downy then
smooth, translucent when wet. **Gills** adnexed,
almost distant; yellow to apricot with paler
margins; minutely hairy. **Stem** 40–100 ×
1–1.5mm (1½–4 × ¹⁄₁₆–¹⁄₃₂in); pale yellow with
prominent hairs at base. **Spores** ellipsoid,
amyloid, 7–9 × 4–5.5μ. Deposit white. **Habitat**
only found growing on the remains of hickory
and walnut shells. Very common. Found
throughout eastern North America, west to
Michigan. **Season** September–November.
Edibility not known. Comment A very pretty
species, easily recognized by its specific
habitat.

Mycena fusco-occula Smith **Cap** 0.5–1.5cm
(¼–½in) across, obtusely cone-shaped,
becoming bell-shaped in age; dark gray, pinky-
brown on the disk, more fawn-buff elsewhere
with a paler margin; smooth, moist, grooved,
striate. **Gills** adnate, close to subdistant,
narrow; whitish. **Stem** 40–80 × 1–1.5mm
1½–3 × ¹⁄₁₆–³⁄₃₂in), fragile; fawn-brown, paler
toward apex; smooth above the rough, hairy
base; moist and slightly sticky. **Flesh** thin,
pliant; pale. **Odor** mild. **Taste** mild. **Spores**
ellipsoid, amyloid, 10–12 × 5–6μ. Deposit
white. **Habitat** gregarious on needle beds
under hemlock and redwood. Found in
northwestern North America. **Season**
September–November. **Edibility not known.**

Mycena praedecurrens Murr. **Cap** 1–2cm
(½–¾in) across, broadly conic becoming
flatter with a shallow umbo and a slightly
scalloped, lined margin; grayish brown, slightly
darker on the disc and paler at the edge;
smooth, very slightly sticky when wet. **Gills**
long decurrent, distant; white with an ashy
tint. **Stem** 20–40 × 1–2mm (¾–1½ ×
¹⁄₁₆–³⁄₃₂in), stuffed, slightly enlarged at the top;
white at the top, pale grayish brown below;
almost smooth, slightly sticky when wet.
Spores ellipsoid, smooth, amyloid,
4–5 × 3–3.5μ. Deposit white. **Habitat** on moss
in damp, deciduous woods. Found in New
York and Connecticut. **Season**
August–September. **Not edible.**

Mycena clavicularis ⅔ life-size

Walnut Mycena *Mycena luteopallens* ⅔ life-size

Mycena fusco-occula ¾ life-size

Mycena praedecurrens ¾ life-size

Mycena leaiana ½ life-size   *Mycena strobilinoides* ⅔ life-size   *Mycena aurantiidisca* ¾ life-size

Mycena leaiana (Berk.) Sacc. **Cap** 1–5cm (½–2in) across, bell-shaped becoming convex, with center sometimes depressed; bright reddish orange becoming more yellow in age; slimy, shiny, smooth. **Gills** adnate, close to crowded, broad; dirty yellow-pink, staining orange-yellow when cut, with bright red-orange edges. **Stem** 30–70 × 1–3mm (1¼–2¾ × 1/16–1/8in), tough, fibrous; orange to yellow, paler near apex, exuding a little watery, orange juice; slimy and somewhat sticky with base covered in dense, coarse hairs. **Flesh** thickish, pliant; white beneath the orange cuticle. **Odor** faintly mealy. **Taste** slight. **Spores** ellipsoid, amyloid, 7–10 × 5–6μ. Deposit white. **Habitat** in dense clusters on deciduous wood. Common. Throughout central and eastern states of North America. **Season** June–September. **Edibility not known.**

Mycena strobilinoides Pk. **Cap** 1–2cm (½–¾in) across, very conical when young, becoming bell-shaped, often with a scalloped margin that flares in age; deep reddish orange fading to yellow; smooth, moist surface, with striate margin becoming grooved in maturity. **Gills** ascending adnate, subdistant, narrow; yellow to pinky-orange with scarlet edges. **Stem** 30–40 × 1–2mm (1¼–1½ × 1/16–3/32in); orange-yellow all over; covered with minute hairs toward apex, becoming smooth except at base, which is covered with long, coarse, orange fibers. **Flesh** thin, pliant; yellow. **Odor** not distinctive. **Taste** not distinctive. **Spores** ellipsoid, amyloid, 7–9 × 4–5μ. Deposit white. **Habitat** densely clustered on conifer needles, especially pine. Common in mountainous areas of northwest North America, rare in eastern North America. **Season** August–November. **Edibility not known.**

Mycena aurantiidisca Murr. **Cap** 0.7–2cm (¼–¾in) across, broadly cone-shaped, margin appressed against the stem when young and flaring slightly when mature; bright orange becoming more yellowish with white edges; smooth, moist, striate. **Gills** adnate, close to subdistant, narrow to slightly swollen in the middle; white gradually tinged with yellow. **Stem** 20–30 × 1mm (¾–1¼ × 1/16in), fragile, hollow; white becoming faintly yellow at base, which is very slightly covered in fine fibers. **Flesh** thin, fragile; orange to yellow. **Odor**

mild. **Taste** mild. **Spores** ellipsoid, nonamyloid, 7–8 × 3.5–4μ. Deposit white. **Habitat** gregarious under pine and Douglas fir. Fairly common. Found in northeastern North America. **Season** April–November. **Edibility not known. Comment** One of the most striking and lovely mycenas.

Mycena acicula (Schaeff. ex Fr.) Kummer **Cap** 0.2–1cm (3/32–½in), hemispherical, sometimes with a small umbo, margin flaring as cap expands; bright red when young, fading to orange, paler, more yellowish toward margin; faintly striate. **Gills** adnate, close to almost distant, moderately broad; pale yellow with whitish edge. **Stem** 10–60 × 1mm (½–2¼ × 1/16in); bright yellow, becoming paler toward the white, hairy base. **Flesh** very thin, brittle; orange-yellow. **Odor** none. **Taste** mild. **Spores** smooth, spindle-shaped, nonamyloid, 9–12 × 2.5–4μ. Deposit white. **Habitat** singly or in groups on woody debris in wet places. Found in eastern North America and on the West Coast. **Season** May–October. **Edibility not known.**

Mycena epipterygia (Fr.) S. F. Gray **Cap** 1–2cm (½–¾in) across, convex expanding to bell-shaped with a prominent umbo; margin often delicately toothed and flaring with age; fawn-yellow in the center, becoming paler toward the edge and turning white flushed with pink or gray in age; slimy, easily removed skin. **Gills** ascending adnate, almost distant, narrow; white or pale yellow. **Stem** 40–70 × 1–2mm (1½–2¾ × 1/16–3/32in), slimy; yellow fading to white; rough hairs at base. **Flesh** very thin; yellowish. **Odor** very slight, fragrant. **Taste** mild. **Spores** ellipsoid, amyloid 8–11 × 4.5–6μ. Deposit white to pale buff. **Habitat** scattered to numerous among grass or moss in deciduous or coniferous woods. Common. Found widely distributed in North America. **Season** August–November (or later in California). **Edible** but not worthwhile. **Comment** I have found it in great quantities in the Pacific Northwest. It is more difficult to recognize in dry conditions because the yellow color may fade and the sliminess dry out.

Mycena fragillima Smith **Cap** 1.5–3.5cm (½–1¼in) across, broadly bell-shaped becoming flatter, with the margin appressed against the stem at first, then often flaring in maturity; dark watery gray becoming paler and

brownish; surface has a faint bloom becoming polished; moist and striate. **Gills** adnate, close in large caps, subdistant to distant in smaller ones, narrow; pale gray with whitish edges. **Stem** 50–150 × 1.5–2mm (2–6 × 1/16–3/32in), very variable, fragile; pale gray or fawn with downy hairs at first, becoming polished; base covered with stiff, white hairs and sometimes slightly enlarged. **Flesh** thin and very fragile; pale gray or fawn to white. **Odor** not distinctive. **Taste** not distinctive. **Spores** subovoid, amyloid, 7–10 × 4–5μ. Deposit white. **Habitat** gregarious around clumps of ferns or fern debris. Found widely distributed in western North America. **Season** November–January. **Edibility not known.**

Mycena citrinomarginata Gillet **Cap** 1–3cm (½–1¼in) across, cone-shaped with margin close to the stem when young, broadly convex with margin often flaring in maturity; color varying shades of yellow; surface with a bloom becoming polished, moist, striate. **Gills** ascending adnate, distant to subdistant, narrow; whitish with yellow edges. **Stem** 30–80 × 1–2.5mm (1¼–3 × 1/16–1/8in), fragile, generally smooth; greeny gray or yellowish. **Flesh** thick on disc of large caps, otherwise thin, fragile; yellowish gray to white. **Odor** not distinctive. **Taste** not distinctive. **Spores** ellipsoid to subcylindric, 8–11 × 4–5.5μ (12–14 × 5–6μ in 2-spored variety). Deposit white. **Habitat** singly, scattered, or gregarious on moss, leaves, needles, or debris in deciduous and coniferous woods. Found widely distributed in western and eastern states of North America. **Season** April–November. **Edibility not known.**

Mycena rorida (Fr.) Quél. **Cap** 0.2–1cm (3/32–½in) across, broadly convex or hemispherical, often with a depressed disc and a straight margin that spreads and often becomes notched in age; colors vary from cream to yellow-brown or gray-brown; surface dry and finely scaly, furrowed and striate. **Gills** arcuate-adnate, becoming strongly decurrent, subdistant, narrow to moderately broad; white. **Stem** 10–30 × 1mm (½–1¼ × 1/16in), elastic; same color as cap; covered in a thick mucus sheath that gathers toward the base in large amounts. **Flesh** thin, quite fragile; pale. **Odor** not known. **Taste** not

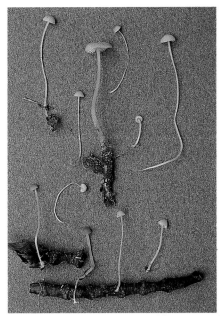

Mycena acicula life-size

Mycena epipterygia ½ life-size

Mycena fragillima ½ life-size

known. **Spores** narrowly ellipsoid, strongly amyloid, 8–10×4–5μ (9–12×4.5–6μ in 2-spored form). Deposit white. **Habitat** gregarious on needle beds under conifers and scattered on small sticks or branches of conifer wood. Not rare. Found in eastern and western states of North America. **Season** April–November. **Edibility not known.**

Mycena delicatella (Pk.) Smith **Cap** 0.3–1cm (⅛–½in) across, broadly cone-shaped to convex, with a slightly incurved margin becoming plane or flared in older specimens; watery milk-white becoming chalky white then slightly creamy; surface covered with minute, downy hairs, only becoming smooth in age, faintly striate. **Gills** free or narrowly adnate, close to subdistant, narrow; pure white. **Stem** 10–30×1mm (½–1¼×1/16in), pliant but gristly; white; densely downy all over, especially at the base, which is covered in stiff hairs that form a mat; stem appears smoother in maturity. **Flesh** thin, slightly fragile; white. **Odor** not distinctive. **Taste** not distinctive. **Spores** subcylindric, 7–9×2.5–3μ (9–12× 3–3.5μ in 2-spored form). Deposit white. **Habitat** scattered to gregarious on fallen twigs and conifer needles. Fairly common. Found in northern North America. **Season** July–November. **Edibility not known.**

Mycena oregonensis Smith **Cap** 0.2–1cm (3/32–½in) across, obtusely cone-shaped to bell-shaped with a papilla; margin flares and becomes wavy in age; bright orangy yellow to yellowish when older; faintly hoary then polished, moist, transparently grooved. **Gills** adnate with decurrent tooth, distant to subdistant though sometimes appearing close, narrow; orange with darker edges in young specimens. **Stem** 20–40×0.5–1mm (½–1½×1/16in), same color as cap; covered with short, downy, yellowish hairs and a few scattered yellow hairs on the base. **Flesh** thin, brittle; yellow. **Odor** not distinctive. **Taste** not distinctive. **Spores** subcylindric, nonamyloid, 7–8×2.5–3μ (9–10×3–4.5μ in 4-spored form). Deposit white. **Habitat** scattered to gregarious on oak leaves and pine needles of Douglas fir and spruce. Found in the Pacific Northwest. **Season** September–November. **Edibility not known. Comment** This is the smallest mushroom in the book. Not normally even 0.5cm (¼in) across.

Mycena citrinomarginata ⅔ life-size

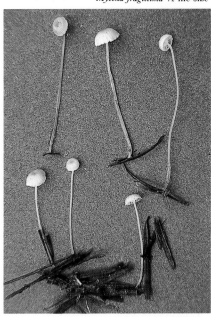

Mycena rorida life-size

Mycena delicatella ¾ life-size

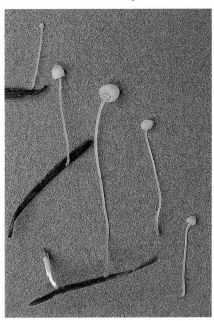

Mycena oregonensis twice life-size

Lactarius thyinos ¾ life-size

LACTARIUS *The milkcaps. They exude droplets of milky-looking fluid from the gills and flesh when damaged; note the milk color carefully and also whether it discolors after being exposed to air for some time. Taste a little drop on the tip of your tongue; it may be hot, bitter, or mild. The flesh of all parts of these mushrooms is granular in texture, so they break and crumble easily.*

Lactarius thyinos Smith **Cap** 3–9cm (1¼–3½in) across, convex with the disc soon becoming depressed, and then broadly funnel-shaped; carrot orange to salmon orange with concentric zones, weathering grayish in age; sticky then thinly slimy, smooth. **Gills** broadly adnate to decurrent, close becoming subdistant in age; bright orange becoming paler, with bruised areas staining dull red. **Stem** 40–80 × 8–20mm (1½–3 × 5⁄16–¾in), hollow, fragile; same color as gills or paler, staining dull red where cut; often with a whitish sheen above. **Flesh** thin; orange-buff when cut or dull red at base of stem. **Latex** cadmium orange slowly staining dull red. **Odor** faintly fragrant. **Taste** mild.

Spores subglobose to broadly ellipsoid, amyloid, 9–12 × 7.5–9μ; ornamented with a partial or broken reticulum and some isolated warts, prominences 0.5–0.7μ high. Deposit pale yellow. **Habitat** scattered or in groups in woods of thuja evergreens and in bogs and swamps. Common. Found widely distributed in northeastern North America. **Season** July–October. **Edible.**

Lactarius indigo (Schw.) Fr. **Cap** 5–15cm (2–6in) across, convex-depressed with an inrolled margin at first; indigo blue when fresh, fading to grayish, then having a silvery luster, with deep green areas where bruised; sticky, smooth, zoned. **Gills** adnate, close, broad; indigo blue or paler to yellowish from the maturing spores, staining green when bruised. **Stem** 20–80 × 1–25mm (¾–3 × 1⁄16–1in), hard becoming hollow, often tapered toward the base; indigo blue to silver blue, spotted at times; sticky but soon dry. **Flesh** whitish, promptly turning indigo blue when cut, staining greenish. **Latex** deep indigo blue, becoming dark green on exposure to the air. **Odor** mild. **Taste** mild or slightly bitter or

slightly acrid. **Spores** broadly ellipsoid to subglobose, amyloid, 7–9 × 5.5–7.5μ; ornamented with a complete or broken reticulum, prominences 0.4–0.5μ high. Deposit cream. **Habitat** scattered or in groups on soil in oak and pine woods. Common in the Southeast, rarer farther north. **Season** July–October. **Edible**.

Lactarius subpurpureus Pk. **Cap** 3–10cm (1¼–4in) across, convex then flat to depressed; with concentric bands of color wine-red to silvery ochre with age, staining green; smooth, slightly viscid when wet. **Gills** adnate; wine-red at first, then paler, bruising green. **Stem** 30–80 × 6–15mm (1¼–3 × ¼–½in); same color as the cap but with darker red spots; smooth, slightly sticky. **Flesh** white or pinkish, staining red then slowly green. **Latex** wine-red. **Odor** pleasant. **Taste** slightly peppery. **Spores** ellipsoid, amyloid, 8–11 × 6.5–8μ; ornamented with a reticulum. Deposit pale cream. **Habitat** frequent under pines. Found in eastern North America, south to North Carolina, west to Wisconsin. **Season** August–October. **Edible** for some people, but best avoided.

Lactarius rubrilacteus Smith & Hesler **Cap** 6–12cm (2¼–4¾in) across, broadly convex to flat with a depressed disc and inrolled margin, becoming broadly funnel-shaped with upturned and wavy margin; bright carrot orange or dull pinky-orange with green stains or patches in older specimens, arranged in concentric bands of color; smooth, sticky. **Gills** adnate to short decurrent, close to crowded, narrow to moderately broad; light, dull pinky-brown to dull purple-red, stained greenish in age. **Stem** 20–60 × 10–30mm (¾–2¼ × ½–1¼in), firm and rigid, hollow, slightly tapering toward the base; similar color to cap; sometimes pitted. **Flesh** dirty yellowish white, staining green with age. **Latex** dark scarlet to purple-red in young specimens, paler and more orangy red in older specimens, scanty. **Odor** very faintly aromatic. **Taste** mild or a little bitter. **Spores** broadly ellipsoid, amyloid, 7.5–9 × 6–7.5μ; ornamented with warts and bands forming a partial reticulum, prominences 0.2–0.4μ high. Deposit cream. **Habitat** scattered to gregarious under conifers, especially pine. Common. Found widely distributed in western North America. **Season** June–October. **Edibility not known.**

Lactarius chelidonium var. *chelidonioides* (Smith) Smith & Hesler **Cap** 3–8cm (1¼–3in) across, flattish with a distinctly depressed center, becoming shallowly funnel-shaped; dirty azure blue with orangy-brown patches or concentric rings in young specimens, becoming dull reddish brown staining green or olive in older specimens, often mottled with watery spots; smooth, sticky then dry. **Gills** decurrent, narrow, crowded; dingy yellow and darker, stained green or olive-brown and stained darker green in old caps. **Stem** 30–60 × 10–25mm (1¼–2¼ × ½–1in), hollow, sometimes enlarged toward the base; similar color to cap but paler; dry. **Flesh** azure blue toward top, more yellowish near gills. **Latex** dirty yellow to yellowish brown; scanty. **Odor** slightly sickly, like *Morchella esculenta* (q.v.). **Taste** slightly peppery after a few minutes. **Spores** ellipsoid with an oblique sterigmal appendage, amyloid, 7–9 × 5–6.5μ; ornamented with prominences 0.5–1μ high of irregular branches, forming at most a broken reticulum. Deposit pale buff. **Habitat** gregarious on soil in coniferous woods, particularly under pine. Frequent. Found in northeastern North America. **Season** August–October. **Edibility not known.**

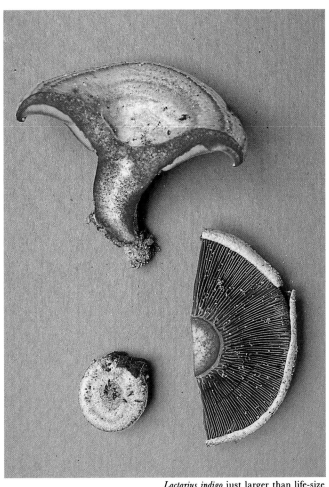

Lactarius indigo just larger than life-size

Lactarius subpurpureus ⅔ life-size

Lactarius rubrilacteus ½ life-size

Lactarius chelidonium var. *chelidonioides* ½ life-size

Lactarius deliciosus var. *deliciosus* ¾ life-size

Lactarius deliciosus var. *deliciosus* (Fr.) S. F. Gray
Cap 5–14cm (2–5½in) across, broadly convex
with a depressed disc and a distinctly inrolled
margin, becoming funnel-shaped with a wavy
margin in age; pale flesh or rosy buff tinged
greenish in places, with numerous purple-brick
or salmon-colored blotches arranged in narrow
concentric bands; sticky when moist. **Gills**
adnate-decurrent, crowded; pale salmon,
bruising pistachio green. **Stem** 30–70 ×
10–25mm (1¼–2¾ × ½–1in), stuffed then
hollow, pinched off at base; buff or red-orange
to salmon, sometimes with darker, spotlike
depressions, becoming green in places; brittle
with mycelium on base, pitted, with a distinct
bloom. **Flesh** rigid at first then fragile; cream,
yellow then carrot-colored. **Latex** orange
fading to orange-yellow then gray-green. **Odor**
slightly fruity. **Taste** mild or slightly bitter.
Spores ellipsoid, amyloid, 7–9 × 6–7μ;
ornamented with minute warts and ridges
forming a partial reticulum. Deposit bright

cream. **Habitat** scattered to gregarious under
conifers, especially pine. Often common. Found
widely distributed throughout North America.
Season August–October. **Edible — good.**
Comment In another variety of this species,
Lactarius deliciosus var. *deterrimus* Gröger, the
flesh stains wine-red.

Lactarius deliciosus Smith & Hesler var.
olivaceosordidus **Cap** 4–8cm (1½–3in), flatly
convex with an incurved margin, faintly striate,
becoming depressed at the disc; gray tinged
with yellow, becoming orange then more olive
to green overall, with faint concentric bands of
color; slightly sticky then dry. **Gills** decurrent,
close, narrow; orange becoming dirty yellow.
Stem 30–50 × 3–10mm (1¼–2 × ⅛–½in),
slightly larger toward the base; dull reddish
orange washed overall with green. **Flesh** thin;
orange near the gills, olive in the cuticle, and
yellowy in the middle. **Latex** muddy orange,
scanty. **Odor** normal. **Taste** mild or bitterish.
Spores subglobose to ellipsoid, amyloid,

8.5–10.5 × 6.5–8μ; ornamented with heavy
bands and crests forming a partial reticulum,
prominences about 0.5μ high. Deposit bright
cream. **Habitat** gregarious under Sitka spruce.
Found in the Pacific Northwest. **Season**
October–November. **Edibility not known.**

Lactarius subplinthogalus Coker **Cap** 3–5cm
(1¼–2in) across, flat with a broadly depressed
disc, becoming shallowly funnel-shaped, with a
pleated margin; dark fawn or drab gray;
smooth with a bloom, becoming finely wrinkled
in age. **Gills** decurrent, very distant, broad;
cinnamon buff bruising rosy apricot. **Stem**
30–80 × 7–15mm (1¼–3 × ¼–½in), solid
becoming hollow, tapering slightly toward the
base; same color as cap or slightly paler;
smooth, dry. **Flesh** soft, quite thick; whitish
bruising rosy apricot. **Latex** white, turning
dark rosy apricot in contact with flesh and
gills. **Odor** pleasant. **Taste** acrid. **Spores**
subglobose to broadly ellipsoid, amyloid,
7–8.7 × 7–8μ; ornamented with variable and
extremely prominent ridges, some branched,
not forming a reticulum, prominences 1.5–2.5μ
high. Deposit creamy ochre. **Habitat** singly or
scattered in woods. Found in the southeastern
United States. **Season** July–September. **Not
edible. Comment** Note the distinctive pleated
margin and the way the cut flesh turns pink
after a few minutes.

Lactarius gerardii Pk. **Cap** 4–12cm (1½–4¾in)
across, convex with a small umbo, becoming
flat or sunken with a wavy margin; dark brown
to yellowy brown, sometimes fading to golden
brown; dry, velvety, and minutely wrinkled.
Gills adnate to decurrent, very distant when
mature, broad, 4–5 layers; white to pale
cream. **Stem** 35–80 × 8–15mm (1¼–3 ×
⁵⁄₁₆–½in), stuffed becoming hollow, often folded
like a fan at the top; same color as cap; dry,
velvety. **Flesh** firm, thin; white, turning yellow
in 15 percent KOH. **Latex** white, unchanging,
does not stain flesh. **Odor** not distinctive.
Taste mild becoming slightly acrid. **Spores**
subglobose-broadly ellipsoid, amyloid,
8–9.5 × 6.5–8.5μ; ornamented with broad
bands forming a reticulum, prominences 0.5–
0.8μ high. Deposit white. **Habitat** scattered to
gregarious on soil in deciduous and coniferous
woods. Quite common. Found widely distributed in northeastern North America. **Season**
July–September or later. **Edible.**

Lactarius subgerardii Smith & Hesler **Cap**
1.5–3cm (½–1¼in) across; pale to dark smoky
brown, wrinkles darker; dry, velvety, and with
wrinkled surface. **Gills** adnate-decurrent,
subdistant to nearly close; white to cream.
Stem 15–30 × 3–4mm (½–1¼ × ⅛–³⁄₁₆in);
same color as cap; minutely velvety, dry. **Flesh**
thin; whitish. **Latex** white, unchanging,
quickly acrid. **Odor** not distinctive. **Taste**
acrid. **Spores** ellipsoid-ovoid, 7–9 × 6–7.5μ;
with warts 0.3–0.6μ high, with connectives
forming a partial reticulum, or none. Deposit
white. **Habitat** under hemlock. Uncommon.
Distribution uncertain; recorded from
Tennessee and New Jersey. **Season**
July–September. **Edibility not known.**

Lactarius deliciosus var. *olivaceosordidus* ¾ life-size

Lactarius subplinthogalus ½ life-size

Lactarius gerardii life-size

Lactarius subgerardii ¾ life-size

Lactarius lignyotus var. *canadensis* ¾ life-size

Deposit pale yellow. **Habitat** scattered to gregarious in conifer forests and mountain areas, particularly under balsam. Sometimes quite common. Found in the Pacific Northwest and California. **Season** July–November. **Edibility not known.**

Lactarius subvernalis var. *cokeri* Smith & Hesler **Cap** 3–6cm (1¼–2¼in) across, broadly convex, sometimes with a small central papilla, becoming flatter or shallowly depressed in age, with the margin often arched or lobed and wavy; whitish soon tinged with buff, sometimes pale smoky brown; dry and unpolished or with a bloom, becoming naked and minutely wrinkled around the disc. **Gills** broadly adnate to short decurrent, crowded, narrow; whitish to pinkish buff, spotted pinkish where injured. **Stem** 50–70×8–15mm (2–2¾×⅛–½in), stuffed, but becoming hollow in age; whitish, becoming dingy in age, staining pinkish where bruised; covered in a whitish bloom when young. **Flesh** thickish; white bruising pink. **Latex** white, unchanging, staining gills and flesh pinkish on cut or bruised surfaces. **Odor** mild. **Taste** bitter. **Spores** globose, 7–8× 7–8µ; ornamented with an almost complete reticulum of ridges and lines, prominences 0.6–1.5µ high. Deposit yellowish. **Habitat** scattered or in groups under hardwoods. Sometimes quite common. Found in eastern North America. **Season** July–October. **Edibility not known** but probably too bitter to eat.

Lactarius fumosus Pk. **Cap** 3–10cm (1¼–4in) across, broadly convex with a shallow depression at the disc, becoming flatter with the margin irregular, sometimes waved, lobed, or ribbed; pale yellowy brown or coffee-colored and tinged with smoky patches; dry, dull, at times minutely cracked in older specimens. **Gills** adnate to decurrent, crowded, narrow; pale becoming dirty yellow-buff, bruising reddish. **Stem** 40–110×6–15mm (1½–4½×¼–½in), stuffed with a pale pith that stains slower than gills; same color as cap or gills, with a whitish base; dry, dull. **Flesh** pale slowly changing to pink. **Latex** milk-white, unchanging, staining cut surfaces reddish. **Odor** slight. **Taste** variable; either peppery fading to mild, mild slowly becoming strongly acrid, or slowly and faintly burning. **Spores** globose to subglobose, amyloid, 7.4–8.7×7.4–8.7µ; ornamented with ridges and forked ridges forming a broken reticulum, prominences 0.6–2µ high. Deposit pinkish buff. **Habitat** in grassy soil in open woods. Found widely distributed in eastern North America. **Season** July–October. **Not edible. Comment** This mushroom is very similar to *Lactarius fuliginosus* (Fr.) Fr. of Europe.

Lactarius luteolus Pk. **Cap** 2.5–6cm (1–2¼in), across, convex; buff becoming brownish in age; dry, velvety, with a white bloom. **Gills** adnate to subdecurrent, close, narrow to moderately broad; white, becoming yellowish to brown when bruised. **Stem** 25–60×5–12mm (1–2¼×¼–½in), stuffed; whitish to buff staining brown; dry with a velvety bloom. **Flesh** whitish staining brown. **Latex** white or sometimes wheylike, plentiful, sticky. **Odor** mild becoming stronger, fetid. **Taste** mild. **Spores** ellipsoid, amyloid, 7–8.5×5.5–6µ; ornamented with isolated warts, prominences 0.3–0.8µ high. Deposit white to cream. **Habitat** on soil in deciduous and mixed woods. Found in eastern North America, west to Texas. **Season** June–November. **Edible.**

Lactarius lignyotus var. *canadensis* Smith & Hesler **Cap** 2–9cm (¾–3½in) across, convex-umbonate, then expanding to become umbilicate with a sharp central umbo, margin often crenate-sulcate; deep velvety blackish brown when young, becoming paler brown or yellowish brown with age; not hygrophanous. **Gills** adnate or very slightly decurrent, rather distant; white to pale tan, margins of the gills edged with brown. **Stem** 40–100×5–10mm (1½–4×¼–½in), rather long, even; slightly paler than cap, white at base. **Flesh** white, slowly stains pink when cut, as do cap and stem surface. **Latex** white staining pinkish and drying pinky-brown. **Taste** mild. **Spores** subglobose, 8–9.5×7–8µ excluding ornamentation, which forms a partial network of ridges, sometimes connected with fine lines and with large warts and spines 1–2.4µ high. Deposit pinkish buff. Cap surface a layer of short chains of inflated cells. **Habitat** under spruce in sphagnum bogs. Found in northeastern North America down to Vermont. **Season** August–October. **Edible** but not recommended. **Comment** The type variant lacks the dark margined gills and has spore ridges only up to 1µ high.

Lactarius fallax var. *concolor* Smith & Hesler **Cap** 2.5–8cm (1–3in) across, convex with a small umbo and inrolled margin, becoming broadly convex or flattened in age, with or without an umbo and with a wavy or ribbed margin; dark brownish black; dry, velvety, and wrinkled. **Gills** decurrent, crowded, narrow, many forked near stem; whitish, then yellowish. **Stem** 30–40×6–10mm (1¼–1½×¼–½in), solid becoming hollow, sometimes narrowing toward the base; same color as cap except paler at base; in some specimens the top of the stem is fluted where it attaches to the gills. **Flesh** quite thick, firm, brittle; white, staining brownish pink when bruised. **Latex** white, plentiful, unchanging. **Odor** mild. **Taste** mild or faintly peppery. **Spores** globose, amyloid, 9–11×9–11µ; ornamented with variable prominences 0.8–2µ high forming a broken or partial reticulum.

Lactarius fallax var. *concolor* ⅔ life-size

Lactarius subvernalis var. *cokeri* ¾ life-size

Lactarius fumosus ½ life-size

Lactarius luteolus life-size

Lactarius volemus ½ life-size

Lactarius corrugis ⅓ life-size

Lactarius volemus var. *flavus* ⅔ life-size

Lactarius hygrophoroides ¾ life-size

Lactarius piperatus ⅔ life-size

Lactarius piperatus var. *glaucescens* ⅓ life-size

Lactarius volemus (Fr.) Fr. **Cap** 5–10cm (2–4in) across, convex to expanded, often with a slight umbo; orange-brown; smooth to minutely velvety, dry. **Gills** adnate, crowded; pale cream bruising brown. **Stem** 50–100×8–20mm (2–4×⁵⁄₁₆–¾in); same color as cap; dry. **Flesh** brittle, firm; whitish, staining brown when cut. **Latex** white, abundant, sticky, discoloring brown. **Odor** fishy. **Taste** pleasant. **Spores** almost globose, with a well-developed amyloid reticulum, 7.5–10×7.5–9μ. Deposit white. **Habitat** in mixed deciduous woods. Common. Found widely distributed in eastern North America. **Season** June–September. **Edible** — good.

Lactarius corrugis Pk. **Cap** 4–13cm (1½–5in) across, convex becoming depressed, sometimes with a distinctly wrinkled margin; dark reddish brown to yellowish brown, sometimes paler at the margin; dry and velvety. **Gills** adnate to subdecurrent, close, quite broad, some forking; white to pale yellowy or buff, brown when bruised. **Stem** 40–110×15–25mm (1½–4½× ½–1in), solid; paler than cap, gray-brown sometimes tinged red-brown; dry, velvety. **Flesh** firm; white staining brown. **Latex** white, unchanging, abundant, staining tissues brown when cut. **Odor** slight. **Taste** mild. **Spores** globose, amyloid, 9–13×9–13μ; ornamented with an almost complete reticulum, prominences 0.4–0.7μ high. Deposit white. **Habitat** singly or in small groups on soil under hardwoods in deciduous, coniferous, or mixed woods. Found in eastern North America. **Season** July–October. **Edible**.

Lactarius volemus var. *flavus* Smith & Hesler **Cap** 5–9cm (2–3½in) across, flatly convex, becoming depressed from the uplifted margin; ivory yellow to buff yellow, becoming brownish where bruised; dry, velvety. **Gills** adnate, narrow to medium broad; whitish to cream. **Stem** 50–100×8–16mm (2–4×⁵⁄₁₆–½in), solid; cream to maize yellow; dry and velvety. **Flesh** firm; whitish to ivory. **Latex** white, unchanging, sticky, staining all parts brown. **Odor** strong (unpleasant). **Taste** mild. **Spores** globose to broadly ellipsoid, 7–8.5×6–7.5μ; broad and narrow bands together form a partial or complete reticulum, prominences 0.2–0.5μ high. Deposit whitish. **Habitat** on soil in deciduous or mixed woods. Rare. Found in southeastern North America. **Season** July–September. **Edible** — good.

Lactarius hygrophoroides Berk. & Curt. **Cap** 3–10cm (1¼–4in) across, convex with a depressed center, becoming flatter and sometimes funnel-shaped, with an incurved then spreading margin; orangy brown to reddish brown; dry and velvety, becoming wrinkled in age. **Gills** adnate-subdecurrent, distant, broad; white becoming cream or pale yellow fawn. **Stem** 30–50×5–15mm (1¼–2×¼–½in), stuffed or solid; same color as cap or paler; dry with a white bloom. **Flesh** firm; white. **Latex** white, plentiful, unchanging, not staining. **Odor** mild. **Taste** mild. **Spores** ellipsoid, amyloid, 7.5–9.5×6–7.5μ; ornamented with small warts connected by fine lines sometimes forming a broken reticulum, prominences 0.2–0.4μ high. Deposit white.

Habitat singly or scattered on soil in deciduous woods. Found widely distributed in eastern North America. **Season** June–September. **Edible** — excellent.

Lactarius piperatus (Fr.) S. F. Gray **Cap** 6–16cm (2¼–6¼in) across, convex with a concave to widely funnel-shaped center; white to cream stained with pinky-tan; dry, smooth to sometimes wrinkled. **Gills** adnate, crowded, narrow, often forked; white to pale cream or pinkish. **Stem** 20–80×10–25mm (¾–3× ½–1in), firm, solid; white; dry with a white bloom. **Flesh** thick; white, unchanging. **Latex** white, unchanging, or sometimes slowly drying yellow or staining gills yellowish. **Odor** slightly disagreeable. **Taste** very hot and acrid. **Spores** ellipsoid, amyloid, 5–7×5–5.4μ; ornamented with fine lines and tiny, isolated warts not really forming a reticulum, prominences 0.2μ high. Deposit white. **Habitat** scattered to numerous on soil in deciduous woods. Found widely distributed in eastern North America. **Season** July–September. **Not edible.** **Comment** *Lactarius piperatus* var. *glaucescens* (Crossland) Smith & Hesler is easily distinguished by observing the way the latex droplets dry greenish on the gills.

Lactarius allardii ⅓ life-size

most concepts of the species except for the greening milk.

Lactarius deceptivus Pk. **Cap** 5–20 cm (2–8in) across, convex becoming depressed then funnel-shaped, with an inrolled margin covered in thick, cottony tissue that sometimes partially covers the gills; whitish staining brown with yellowish patches; dry, smooth, but dull becoming scaly in age. **Gills** adnate-decurrent, close to subdistant, moderately broad; ivory to ochre, bruising tan. **Stem** 40–90 × 10–35mm (1½–3½ × ½–1¼in), hard, solid, sometimes tapering toward the base; white, staining brownish; dry, velvety, becoming scaly in age. **Flesh** firm, thicker at center of cap; white. **Latex** white, unchanging, staining flesh brownish. **Odor** mild becoming strong. **Taste** strongly acrid. **Spores** ovate to subglobose, amyloid, 9.5–12 × 7.5–10μ; ornamented with isolated warts and spines, prominences 0.5–1μ high. Deposit white to pale yellowy buff. **Habitat** gregarious on soil in mixed or deciduous woods near hemlock or oak, and in moss in coniferous woods. Common. Found widely distributed from central North America eastward to the Atlantic. **Season** June–September. **Not edible.** This species has been considered edible if thoroughly cooked to remove bitterness.

Lactarius peckii Burlingham **Cap** 5–15cm (2–6in) across, broadly convex with an inrolled margin, becoming shallowly depressed or occasionally funnel-shaped; dark bay to brick red, with darker zones of color when moist; dry and velvety when young, becoming shiny. **Gills** decurrent, close, narrow; pale buff, becoming reddish brown to dark brown when dried. **Stem** 20–60 × 10–25mm (¾–2¼ × ½–1in), stuffed becoming hollow, sometimes narrowing below; same color as cap or paler, sometimes spotted reddish brown when handled; covered with a white bloom at first. **Flesh** firm; pale pinkish brown. **Latex** white, unchanging, plentiful. **Odor** strong. **Taste** very acrid after 30 seconds. **Spores** globose, amyloid, 6–7.5 × 6–7.5μ; ornamented with a heavy partial or complete reticulum, prominences 0.3–0.8μ high. Deposit white. **Habitat** scattered or in groups along roads, trails, and streams and in grassy open oak woods. Found in the Eastern Seaboard states, the Southeast, and the Gulf Coast region. **Season** July–September. **Not edible.**

Lactarius atroviridis Pk. **Cap** 6–15cm (2¼–6in) across, convex becoming flatter with a central depression; inrolled margin sometimes ridged; dark olive green with darker olive spots arranged concentrically; scaly or covered with minute fibers, becoming cracked or patchy. **Gills** adnate to short decurrent, close, narrow becoming broad in age; white to pinky-buff becoming spotted green or brown, often staining greenish at the edges. **Stem** 20–80 × 10–25mm (¾–3 × ½–1in), becoming hollow; same color as cap or paler, spotted darker green with pitted, polished spots; dry, smooth. **Flesh** thick at disc, thinner at margin; white to pale pinky-buff. **Latex** white, unchanging or slowly becoming dull olive, staining the gills olive gray to greenish. **Odor** slight. **Taste** acrid. **Spores** ellipsoid, amyloid, 7–9 × 5.5–6.5μ; ornamented with a broken or partial reticulum and some isolated warts, prominences 0.2–0.5μ high. Deposit pale cream or yellowish. **Habitat** scattered to gregarious on soil in deciduous and conifer woods. Uncommon. Found in eastern North America. **Season** August–October. **Not edible.**

Lactarius allardii Coker **Cap** 6–15cm (2¼–6in) across, convex-depressed with an incurved margin, becoming broadly vase-shaped in age; light grayish pink, pinkish buff, or pale brick red, paler on the margin; dry, felty-velvety at first. **Gills** adnate to decurrent, close to subdistant, narrow; white to ivory to light buff, staining dull greenish to olive and becoming dingy brown where bruised. **Stem** 20–40 × 10–30mm (¾–1½ × ½–1¼in), hollow; white, then becoming same color as cap; hard. **Flesh** firm, thick on the disc; white, becoming pinky with a lavender tinge then olivaceous on exposure. **Latex** white, turning grayish olive then brown, plentiful, viscous. **Odor** mild, or strong in age. **Taste** acrid and disagreeable. **Spores** ellipsoid to subglobose, 8–10 × 5.5–8μ; many isolated warts, prominences 0.1–0.2μ high. Deposit white to yellowish. **Habitat** on soil in deciduous or mixed deciduous-coniferous woods. Found in central and eastern North America. **Season** July–October. **Not edible.**

Lactarius subvellereus Pk. **Cap** 7–15cm (2¾–6in) across, convex with a depressed center and strongly inrolled margin, then expanded, flattened; white, then slowly developing yellowish or ochre-brown stains; densely woolly, almost like suede or kid leather. **Gills** adnate to subdecurrent, narrow, very crowded, often forked; pale yellowish cream bruising brown. **Stem** 15–50 × 12–25mm (½–2 × ½–1in), tapered at base; concolorous with cap; tomentose or even downy. **Flesh** very firm; white to slightly yellowish. **Latex** white, but in the collection shown turning slowly green on a glass slide overnight (see Comment), copious. **Taste** hot. **Spores** ellipsoid, 7–8(9) × 5.5–6.5(7)μ; ornamented with fine, rather small warts less than 0.5μ in height connected by very faint lines. Deposit pale cream. **Habitat** in mixed woodlands. Of uncertain distribution but recorded in northeastern North America to Alabama. **Season** July–September. **Not edible** and possibly suspect. **Comment** This collection agreed well with

Lactarius subvellereus ¼ life-size

Lactarius deceptivus ⅓ life-size

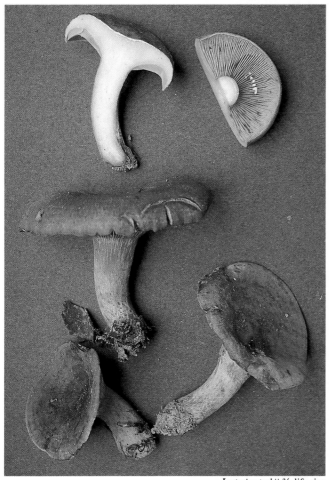

Lactarius peckii ¾ life-size

Lactarius atroviridis ⅓ life-size

Lactarius olympianus ¾ life-size

Reported from New Jersey and in the Pacific Northwest and Rocky Mountains. **Season** July–October. **Poisonous. Comment** The occurrence of this species in North America needs further study.

Lactarius yazooensis Smith & Hesler **Cap** 5–15cm (2–6in) across, convex, depressed with an inrolled and minutely downy margin, then becoming broadly funnel-shaped with a naked margin; creamy yellow and cinnamon buff with distinct alternating darker zones of orange-yellow, rusty orange, or dull orange-red, gradually becoming paler in age; sticky, smooth. **Gills** adnate becoming decurrent, crowded, becoming moderately broad; pale pinky-cinnamon becoming more vinaceous to fawn and slowly staining brownish. **Stem** 20–60×10–25mm (¾–2¼×½–1in), short, thick, hard; whitish, becoming discolored in age; dry. **Latex** milk-white, unchanging, plentiful. **Odor** usual. **Taste** exceedingly acrid after 30 seconds. **Spores** subglobose to broadly ellipsoid, 7–9×6–7.5μ; ornamented with short sparsely branched ridges and isolated warts not in a reticulum, prominences 1–1.5μ high. Deposit buff-yellowish. **Habitat** in groups and dense tufts on grassy areas under oak and other hardwoods. Common. Found in southeastern North America to Texas. **Season** July–October. **Not edible.**

Lactarius pubescens var. *pubescens* (Fr.) sensu lato **Cap** 2.5–9cm (1–3½in) across, broadly convex to flat with a depressed disc and sometimes developing a small navel, becoming saucer-shaped with a wavy, arched, incurved margin; pure white to cream and pale to bright salmon orange over the disc; slimy, dry, dull with a scaly layer that loosens and flakes in older specimens. **Gills** adnate to short decurrent, crowded, quite broad; whitish to pale yellow with salmon tints, bruising dingy clay color. **Stem** 20–55×3–12mm (¾–2⅛×⅛–½in), stuffed, usually tapering toward the base; whitish with pale yellow-brown or reddish tints upward from the base and often an orange-red girdle at the whitish apex, base often with dingy yellow-brown spots and white mycelium; silky with a bloom. **Flesh** elastic, firm, becoming soft; white. **Latex** white, scant in older specimens, unchanging. **Odor** faintly of geraniums. **Taste** bitter. **Spores** ellipsoid, amyloid, 6.5–8.5×5.5–6.5μ; ornamented with small warts often forming a reticulum. Deposit cream with salmon tint. **Habitat** gregarious in grass under birch. Found widely distributed west of Michigan, south to Arizona and Tennessee. **Season** August–October. **Poisonous.**

Lactarius pubescens Smith & Hesler var. *betulae* **Cap** 3–8cm (1¼–3in) across, convex with a depressed disc and inrolled margin, becoming flat then shallowly funnel-shaped with an arched margin; pale tan or cinnamon pink, slightly darker at the disc; disc slightly sticky becoming dry, densely covered in matted hair and coarsely bearded at the margin. **Gills** decurrent, quite close, narrow; white with a slight salmon tinge. **Stem** 30–80×10–18mm (1¼–3×½–¾in), solid then hollow; similar color as cap but slightly pinker; dry with a bloom. **Flesh** fragile, white to pinkish, particularly near the cuticle. **Latex** white quickly turning yellow, scant, staining yellow. **Odor** not distinctive. **Taste** slowly but slightly acrid. **Spores** ellipsoid, amyloid, 6.5–8×5.5–6.5μ; ornamented with ridges forming a well-developed, irregular reticulum, prominences 0.4–0.8μ high. Deposit white to cream. **Habitat** scattered to gregarious under birch. Found in northern North America. **Season** August–October. **Poisonous.**

Lactarius olympianus Smith & Hesler **Cap** 6–12cm (2¼–4¾in) across, broadly convex with a depressed disc and a wavy, incurved margin that becomes uplifted, making the cap funnel-shaped; tones of bright yellow and apricot orange with concentric bands of color; smooth, sticky. **Gills** adnate, close to subdistant, narrow; white to pale fawn or dingy yellow, bruising dirty orange or dark orange-brown. **Stem** 40–60×15–25mm (1½–2¼×½–1in), tapering slightly toward the base; whitish with a white bloom, bruising dingy ochre at the base. **Flesh** thin, fairly fragile; white. **Latex** white, unchanging but stains gills dark orange or dark orange-brown. **Spores** broadly ellipsoid, amyloid, 8–11×7.5–9μ; ornamented with short or long ridges, often branched, forming a broken or complete reticulum, prominences 0.5–1.5μ high. Deposit cinnamon buff. **Habitat** scattered to gregarious in conifer forests. Common. Found in the Rockies and the Pacific Northwest. **Season** June–October. **Not edible.**

Lactarius zonarius (St. Amans) Fr. sensu Neuhoff

Cap 5–10cm (2–4in) across, convex with a central funnel-shaped depression; margin inrolled and finely hairy at first, becoming broadly arched and hairless; pale yellow buff or straw becoming dingy orange or reddish ochre, although paler toward the margin, with several, darker concentric bands of color; hoary then naked, slightly slimy when wet. **Gills** decurrent, crowded, narrow, frequently forked near stem; milk-white to creamy yellow to bright ochre-yellow, bruising tawny or gray-brown. **Stem** 20–40×10–20mm (¾–1½×½–¾in), firm, solid becoming hollow, tapering toward the base; whitish to browny yellow with indistinct spots, bruising tawny; minutely pitted. **Flesh** hard, firm; white becoming faintly yellowy brown. **Latex** white, drying gray or brownish gray on gill edges. **Odor** slight to strong, fruity(?). **Taste** very hot, burningly acrid in throat. **Spores** subspherical to broadly ellipsoid, amyloid, 7–9×6.5–8μ; ornamented with warts and a partial reticulum, prominences about 1μ high. Deposit ochre or buff. **Habitat** scattered to gregarious under conifers and hardwoods. Uncommon.

Lactarius zonarius ½ life-size

Lactarius yazooensis ⅓ life-size

Lactarius pubescens var. *pubescens* ½ life-size

Lactarius pubescens var. *betulae* ¾ life-size

Lactarius payettensis ½ life-size

Lactarius scrobiculatus ¾ life-size

Lactarius croceus ⅔ life-size

Lactarius alnicola var. *alnicola* ⅓ life-size

Lactarius payettensis Smith **Cap** 8–16cm
(3–6¼in) across, broadly convex with a widely
depressed disc and an incurved margin; pale to
dark straw yellow; smooth in the disc, very
sticky and glutinous with coarse hairs or fibers,
particularly around the margin. **Gills**
decurrent, close to subdistant, narrow; cream
becoming dirty yellow, staining darker along
the margins. **Stem** 20–50 × 20–40mm
(¾–2 × ¾–1½in), hard, hollow, pinched off at
the base; white with large, smooth
brownish-yellow spots, bruising darker. **Flesh**
thick, firm, hard; pale becoming dirty brownish
yellow. **Latex** white, scant, staining gills and
flesh dull yellow to dark brownish yellow along
the edges. **Odor** faintly bitter and unpleasant.
Taste immediately and strongly acrid and
burning. **Spores** ellipsoid, mostly nonamyloid
but some amyloid warts, 8.2–9.6 × 7.4–7.8μ;
ornamented with a sparse network of lines
mostly not united into a complete reticulum,
prominences 0.3–0.8μ high. Deposit pale
cream. **Habitat** scattered under spruce, fir, and
alder. Found in Idaho and Colorado. **Season**
June–September. **Not edible. Comment** Very
similar to *Lactarius alnicola* var. *alnicola* (below).

Lactarius scrobiculatus (Fr.) Fr. **Cap** 7–20cm
(2¾–8in) across, broadly convex with a
depressed disc and long inrolled, hairy margin,
becoming flatter and broadly funnel-shaped
with a smooth margin; pale ochre-yellow to
yellow-orange, darker in the center with faint
concentric bands of color, bruising dingy
brown; very sticky, slimy when dry, scaly,
often in rings. **Gills** adnate to decurrent, quite
crowded, broad; whitish with a faint yellow or
pink tint, bruising pale pinky-brown. **Stem**
30–60 × 15–35mm (1¼–2¼ × ½–1¼in),
sometimes tapering to a rootlike base; tawny
with glazed, yellow-brown spots and some
white mycelium on the base; finely downy and
pitted. **Flesh** rigid; white. **Latex** white,
plentiful, quickly changing to sulphur yellow.
Odor fruity. **Taste** burningly acrid. **Spores**
broadly ellipsoid, amyloid, 7.1–8.6 × 5.9–6.8μ;
ornamented with warts, some paired, and fine
lines making a sparse reticulum, prominences
0.5–1μ high. Deposit bright ochre-yellow with
a slight flesh tint. **Habitat** scattered to
gregarious under conifers, particularly in
mountain areas. Rare. Found in Oregon.
Season September–October. **Not edible.**
Comment My collection had burningly acrid
milk, and I feel it should probably be
recognized as *Lactarius scrobiculatus* var.
scrobiculatus (Fr.) Fr., but the presence of this
variety is not confirmed in North America.

Lactarius croceus Burlingham **Cap** 5–10cm
(2–4in) across, broadly convex with a
depressed disc and inrolled margin, expanding
to shallowly depressed and broadly
funnel-shaped; bright yellow to orangy or
peachy yellow, sometimes with faint concentric
bands of color; sticky at first, then dry. **Gills**
adnate-decurrent, close to subdistant,
moderately broad; creamy to honey yellow or
buff, bruising orange-yellow. **Stem**
30–60 × 10–20mm (1¼–2¼ × ½–¾in), stuffed
becoming hollow; same color as cap or paler;
smooth or sometimes hairy at the base. **Flesh**
brittle; whitish staining orange-yellow. **Latex**
white changing to orange-yellow, stains gills
and flesh. **Taste** bitter to acrid. **Spores**
ellipsoid, nonamyloid, 7.5–10 × 5.5–7.5μ;
ornamented with widely spaced bands,
sometimes forming a partial reticulum,
prominences 0.3–0.6μ high. Deposit yellowish.
Habitat scattered on soil in deciduous woods.
Found in northeastern North America. **Season**
July–August. **Not edible.**

Lactarius colorascens life-size

Lactarius alnicola var. *alnicola* Smith **Cap**
8–18cm (3–7in) across, convex with a
depressed disc and inrolled margin, expanding
to become funnel-shaped; yellow-ochre all over,
although somewhat paler toward the edge,
with some concentric bands of color; sticky to
glutinous, with matted, minute hairs beneath
the gluten near the margin. **Gills** decurrent,
crowded, narrow, many layers, forked near the
stem; whitish becoming cream then yellowy
buff to yellowy brown, staining yellow when
bruised. 30–60 × 20–30mm (1¼–2¼ ×
¾–1¼in), hard, solid becoming stuffed; whitish
above, more pale yellowy brown below, and
tawny mycelium at the base; dry, distinctly
pitted. **Flesh** thick, hard; whitish. **Latex** white,
scanty, staining cut flesh yellow. **Odor** strong.
Taste extremely and immediately acrid.
Spores broadly ellipsoid, amyloid, 7.4–8.7 ×
6.7–7.4μ; ornamented with warts and ridges
up to 0.6μ high, with many connectives or
forming a partial reticulum. Deposit almost
white. **Habitat** gregarious in woods,
particularly under alder and conifers. Quite

common in the northern Rocky Mountains.
Found in western North America. **Season**
July–September. **Not edible. Comment** My
collection has the smell of *Lactarius alnicola* var.
pungens Smith & Hesler, but otherwise its
characteristics are identical to *Lactarius alnicola*
var. *alnicola*.

Lactarius colorascens Pk. **Cap** 2.5–5cm (1–2in)
across, nearly flat, becoming centrally
depressed; whitish when young, becoming
brownish red with age; moist, smooth. **Gills**
adnate or slightly decurrent, close, thin;
whitish, becoming yellowish with age. **Stem**
25–35 × 3–5mm (1–1¼ × ⅛–¼in), solid, even;
whitish, becoming brownish red with age.
Flesh pallid. **Latex** white, changing to sulphur
yellow on exposure to air. **Odor** normal. **Taste**
bitter, hottish. **Spores** broadly ellipsoid,
6–7.5 × 5–6μ; ornamented with warts and
short ridges, sometimes forming a broken
reticulum, prominences up to 0.5μ high.
Deposit whitish. **Habitat** in mixed woods.
Found in eastern North America. **Season**
August–September. **Edibility not known.**

LACTARIUS

Lactarius chrysorrheus ⅔ life-size

Lactarius vinaceorufescens ¾ life-size

Lactarius uvidus ⅔ life-size

Lactarius caespitosus ¾ life-size

Lactarius chrysorrheus Fr. **Cap** 3–8cm (1¼–3in) across, convex with a funnel-shaped depression and thin, incurved margin which straightens out in age; pale salmon to rosy to ochre-buff, with darker rings of watery blotches or narrow, concentric bands; smooth, moist to slightly sticky, becoming dry. **Gills** adnate to decurrent, crowded, with numerous layers, sometimes forked near the stem; pale orange-buff, tinged pink. **Stem** 30–70 × 10–20mm (1¼–2¾ × ½–¾in), stuffed becoming hollow; whitish to pale buff, often flushed pinkish below; sometimes uneven on the surface, with stiff hairs at base. **Flesh** thin; whitish, becoming sulphur yellow when cut. **Latex** white rapidly turning yellow, abundant. **Odor** slight. **Taste** slowly but distinctly acrid and somewhat hot. **Spores** broadly ellipsoid, amyloid, 6–8 × 5.5–6.5μ; ornamented with sparse ridges forming a broken reticulum but also many isolated warts and ridges 0.5–1μ high. Deposit pale yellow. **Habitat** in groups in hardwood and mixed forests and under conifers. Found widely distributed throughout North America. **Season** July–November. **Poisonous. Comment** This mushroom is easily recognized by the way the white milk turns yellow in a few seconds.

Lactarius vinaceorufescens Smith **Cap** 4–10cm (1½–4in) across, convex with an inrolled margin, then expanded and depressed at center; pale yellow-buff to cinnamon pink, with paler, watery zones or spots, becoming darker overall with age; smooth and sticky when wet. **Gills** adnate, crowded; vinaceous buff staining pinkish. **Stem** 40–60 × 10–20mm (1½–2¼ × ½–¾in); same color as cap; smooth, with stiff, short hairs at base. **Flesh** white staining yellow, all parts exuding latex. **Latex** white turning bright yellow. **Odor** none. **Taste** (of latex) slightly peppery but more acrid after color change. **Spores** subglobose, 6.5–8(9) × 6–7μ; ornamentation of nodulose ridges and isolated warts, partially reticulate, up to 0.5–0.8μ. Deposit dull white to yellowish. **Habitat** gregarious under mixed conifers and hardwoods. Found widely distributed in northeastern North America, west to Wisconsin. **Season** August–October. **Edibility not known** — best avoided.

Lactarius uvidus (Fr.) Fr. **Cap** 3–10cm (1¼–4in) across, flatly convex with a small umbo and incurved margin, becoming flatter with a broad depression and an arched or irregular margin; very pale fawn, lilac, buff, or violet-gray, becoming a little darker in age; smooth, sticky or slimy if very wet. **Gills** adnate to decurrent, close, moderately broad; creamy white, bruising dull lilac to tan. **Stem** 30–70 × 10–16mm (1¼–2¾ × ½in), becoming hollow; dirty white to pale, dull buff, staining rusty or yellow-brown at the base; smooth, slimy when young, shiny when dry. **Flesh** whitish staining dull lilac to pinky-brown. **Latex** milk-white becoming dingy cream, staining broken surfaces dull lilac. **Odor** typical. **Taste** mild becoming slightly bitter then faintly acrid. **Spores** oval, amyloid, 8–11 × 7–8.5μ; ornamented with widely spaced bands and ridges and isolated particles, no reticulum, prominences 0.5–1μ high. Deposit pale yellow. **Habitat** scattered to gregarious on low ground under mixed aspen, birch, and pine. Quite common. Found widely distributed in northeastern North America, west to Colorado. **Season** July–October. **Not edible** — probably poisonous.

Lactarius caespitosus Smith & Hesler **Cap** 4–10cm (1½–4in) across, convex with an inrolled margin, becoming shallowly depressed with an upturned margin; tawny brown to gray- or olive-brown; smooth, sticky. **Gills** adnate becoming decurrent, close, narrow becoming broad, 2 or 3 layers; whitish then pinkish brown. **Stem** 30–70 × 10–30mm (1¼–2¾ × ½–1¼in), hollow; off-white to pale brown or brownish gray; slimy and sticky when fresh, shiny when dry. **Flesh** thick, brittle; white, unchanging when first cut, but turning yellow after several hours. **Latex** white, scanty, unchanging but slowly staining flesh yellow and gills brownish. **Odor** pleasant. **Taste** mild then acrid. **Spores** broadly ellipsoid, amyloid, 9.5–11 × 7.5–9μ; ornamented with prominences 0.2–0.5μ high forming a partial or incomplete reticulum. Deposit pale buff. **Habitat** scattered to gregarious or growing in dense tufts on soil under conifers in wet places. Common in the spruce-fir zone of the Rocky Mountains. Found in western North America. **Season** June–October. **Edibility not known.**

Lactarius kauffmanii Smith & Hesler **Cap** 5–15cm (2–6in) across, flatly convex with a depressed disc and inrolled, minutely hairy margin which lifts upward in age; blackish brown becoming more reddish gray-black, and very occasionally with concentric bands of color; slimy and sticky, smooth, sometimes streaked under the slime. **Gills** adnate to short decurrent, close, narrow becoming broad, forking near the stem; pale pinky-buff to pale ochraceous salmon flushed cinnamon or with brown patches. **Stem** 50–100 × 10–30mm (2–4 × ½–1¼in), becoming hollow, sometimes thicker in the middle; pale pinky-brown or tan; slimy, sticky and shiny, smooth, pitted. **Flesh** violet-brown near the cuticle, pale pinky-brown near the gills. **Latex** white, unchanging, slowly staining the gills an olive- or gray-brown. **Odor** none. **Taste** acrid. **Spores** subglobose to ovate, amyloid, 9–9.5 × 7.5–8.5μ; ornamented with a distinct partial reticulum with irregular meshing, prominences 0.2–0.7μ high. Deposit whitish. **Habitat** on soil in coniferous woods. Quite common. Found widely distributed in northwestern North America. **Season** July–November. **Not edible.**

Lactarius kauffmanii ¾ life-size

Lactarius pseudomucidus ⅔ life-size

Lactarius mucidus var. *mucidioides* ½ life-size

Lactarius hibbardae ¾ life-size

Lactarius vietus ⅔ life-size

Lactarius pseudomucidus Smith & Hesler **Cap** 4–10cm (1½–4in) across, flatly convex becoming shallowly depressed to broadly funnel-shaped, with an inrolled margin that becomes wavy; dark charcoal brown; smooth and very slimy. **Gills** adnate to decurrent, close to subdistant, narrow to quite broad; white with a blue-gray tinge, spotting yellowish to brown in age. **Stem** 40–100 × 5–12mm (1½–4 × ¼–½in), hollow, very fragile, enlarging toward base; dingy brown or gray on surface and paler within; very slimy when fresh. **Flesh** thin, limber; grayish. **Latex** thin, milk-white, changing very slowly, spotting gills yellowish to brown. **Odor** slight. **Taste** slowly, increasingly acrid. **Spores** broadly ellipsoid, amyloid, 7–9 × 6–7μ; ornamented with heavy bands with branches forming a virtually distinct reticulum, prominences 0.5–1.2μ high. Deposit white. **Habitat** singly or gregarious under conifers. Common. Found widely distributed in the Pacific Northwest. **Season** August–October. **Not edible.**

Lactarius mucidus var. *mucidioides* Smith & Hesler **Cap** 3–9cm (1¼–3½in) across, convex with an incurved margin, becoming flatter with an unevenly depressed disc and sometimes a lobed margin; dull pinky-brown or dark fawn, becoming paler toward the margin and in age; smooth, sticky. **Gills** adnate to short decurrent, close, narrow, often forked near stem; pale pinky-buff, becoming more yellowy in age, staining gray-green with latex. **Stem** 50–90 × 10–17mm (2–3½ × ½–¾in), stuffed then hollow, enlarged at base to 25mm (1in) and sometimes swollen in the middle; same color as gills; smooth, sticky, rather waxy, sometimes compressed and fluted. **Flesh** thin, firm; pale pinky-buff. **Latex** white drying pale gray-green. **Odor** none. **Taste** acrid. **Spores** broadly ellipsoid, amyloid, 8–9 × 6.7–7.5μ; ornamented with a partial to well-developed reticulum with angular meshes and variable ridges, prominences 0.2–0.6μ high. Deposit cream. **Habitat** on soil under conifers. Found widely distributed in northeastern North America. **Season** July–October. **Not edible.**

Lactarius hibbardae Pk. **Cap** 2–8cm (¾–3in) across, flatly convex then somewhat depressed, sometimes with a papilla; margin even or lobed and often cracking or scaly; gray tinged dark pinkish brown; dry, minutely hairy or scurfy. **Gills** adnate to short decurrent, close to crowded, narrow; cream to pale ochre then dull pinkish cinnamon. **Stem** 20–50 × 4–10mm (¾–2 × 3⁄16–½in), quickly becoming hollow; same color as gills or cap (eventually), with a whitish base; dry, dull with a bloom. **Flesh** whitish, but tinged with color of cap beneath the cuticle. **Latex** white or wheylike in age, unchanging but globules drying creamish; stains white paper yellow or ochre. **Odor** mild, slightly fragrant, like anise or coconut. **Taste** distinctly, although sometimes slowly, acrid and hot. **Spores** broadly ellipsoid, 6.5–9 × 5–6.5μ; ornamented in the form of bands with branches forming a partial or distinct reticulum, prominences 0.2–0.4μ high. Deposit white to creamish. **Habitat** on soil or among sphagnum moss in coniferous and mixed woods. Found in northeastern North America. **Season** June–October. **Edibility not known** — probably too hot to eat.

Lactarius vietus (Fr.) Fr. **Cap** 3–8cm (1¼–3in) across, shallowly convex with a depressed center, becoming slightly vase-shaped with bluntly notched margin in age; darkish, reddish brown with faint concentric bands of color, becoming paler, more pinky-fawn and opaque white at the margin; smooth, moist,

Lactarius mucidus var. *mucidus* ¾ life-size

slightly sticky when wet. **Gills** broadly adnate to decurrent, close, moderately broad, several layers; very pale pinky-fawn, spotting grayish or brownish when cut. **Stem** 30–60 × 8–18mm (1¼–2¼ × 5⁄16–¾in), fragile; similar color to cap but paler; dry, naked. **Flesh** pale, becoming yellowy gray-brown near the cuticle and pale pinky-fawn near the gills. **Latex** milk-white, changing to olive-gray, staining gills olive-gray to gray-brown. **Odor** slight. **Taste** acrid. **Spores** ellipsoid, amyloid, 6.8–8.9 × 5.2–6.4μ; ornamented with very small warts and some connecting lines but only forming a partial reticulum, prominences 0.1–0.3μ high. Deposit white to cream. **Habitat** solitary to gregarious or growing in dense clumps on decaying conifer wood or hardwoods in wet areas near birch. Sometimes quite common. Found widely distributed in northern North America. **Season** August–October. **Not edible.**

Lactarius mucidus var. *mucidus* (Burlingham) Smith & Hesler **Cap** 3–9cm (1¼–3½in)

across, broadly convex to flat, with a wide depressed disc and sometimes an acute umbo; dingy chocolate or gray-brown to pinky-buff, paler toward the margin; smooth and persistently very sticky. **Gills** adnate or subdecurrent, close, medium broad; white to cream, staining blue-green to gray with latex. **Stem** 40–80 × 7–10mm (1½–3 × ¼–½in), solid eventually becoming hollow, tapering slightly toward the top; same color as cap but paler; smooth, sticky. **Flesh** thick, firm though thin in margin; white. **Latex** white, drying blue or greeny gray on gills. **Odor** mild. **Taste** acrid. **Spores** ellipsoid, amyloid, 7.5–9 × 6–7μ; ornamented with a partial reticulum and variable ridges and warts in a zebralike pattern, prominences 0.2–0.5μ high. Deposit white. **Habitat** scattered or in groups on the soil or in needle duff under conifers. Quite common. Widely distributed in north central and northeastern North America. **Season** June–November. **Not edible.**

Lactarius affinis ⅓ life-size

Lactarius subflammeus ¾ life-size

Lactarius griseus ¾ life-size

Lactarius glyciosmus ⅔ life-size

Lactarius rufus ⅔ life-size

Lactarius rufus var. *parvus* ¾ life-size

Lactarius affinis Pk. **Cap** 8–10cm (3–4in) across, convex with a depressed disc; pale ochre to yellow to pinkish cinnamon; smooth and sticky. **Gills** decurrent or nearly adnate, quite broad; whitish tinged with yellow. **Stem** 25–80×14–24mm (1–3×½–1in), stuffed then hollow; same color as cap; smooth. **Flesh** firm to soft; white. **Latex** milk-white, unchanging, not staining. **Taste** acrid. **Spores** ellipsoid, amyloid, 8–10×7.1–7.7μ; ornamented with a partial or broken reticulum, prominences 0.5–1μ high. Deposit white. **Habitat** scattered or in groups in pastures or mixed woods and under conifers. Widely distributed in northeastern North America. **Season** July–October. **Not edible.**

Lactarius subflammeus Smith & Hesler **Cap** 3–7cm (1¼–2¾in) across, convex with a depressed disc, becoming flat with an arched, translucently striate margin; scarlet to yellow-orange; smooth, naked and slimy, sticky. **Gills** broadly adnate to decurrent, close, moderately broad; whitish tinged with color of cap. **Stem** 40–90×8–15mm (1½–3½×5⁄16–½in), hollow, fragile, enlarging toward the base; similar color to cap but paler; hoary then naked, often with an uneven surface. **Flesh** thin, fragile; watery pinky-buff to dull orange-buff. **Latex** milk-white, unchanging, not staining. **Odor** slight. **Taste** slowly acrid. **Spores** broadly ellipsoid, amyloid, 7.8–9.5×6.4–7.8μ; ornamented with short ridges and warts sometimes forming a partial reticulum, prominences 0.5–1μ high. Deposit white. **Habitat** under pine in conifer forests or sand dunes. Common. Found widely distributed in the Pacific Northwest. **Season** September–November. **Not edible.**

Lactarius griseus Pk. **Cap** 1.5–5cm (½–2in) across, convex quickly becoming deeply depressed or funnel-shaped with a papilla; gray-brown, with the reddish or pinky-brown depression becoming paler with age and then more ochraceous; dry and covered in the middle with small scales. **Gills** adnate to decurrent, close to subdistant, medium broad, numerous; white becoming yellowish. **Stem** 20–65×3–6mm (¾–2½×⅛–¼in), fragile, hollow, tapering slightly toward the top; pale to grayish, with the top the same color as the gills; stiff, rough, white hairs at base. **Flesh** thin; white to yellowish. **Latex** white, scanty; dries yellowish in droplets and stains white paper yellow, gills not staining. **Odor** slight. **Taste** slowly, slightly acrid or mild in mature specimens. **Spores** subglobose to ellipsoid, amyloid, 7–8×6–7μ; ornamented with warts and ridges, no reticulum, prominences up to 1μ high. Deposit yellowish. **Habitat** scattered or in groups on moss, humus, decaying logs, sometimes in sphagnum bogs, and under conifers. Common. Widely distributed in eastern North America. **Season** June–October. **Not edible.**

Lactarius glyciosmus (Fr.) Fr. **Cap** 2–6cm (¾–2¼in) across, convex with an inrolled margin, becoming depressed at the disc then shallowly funnel-shaped with an upturned and sometimes wavy margin; lilac gray to dull buff, occasionally with faint concentric bands of color; dry and faintly hairy, sometimes slightly scaly in age. **Gills** decurrent, crowded, narrow, some forked; yellowy to pale flesh-buff then lilac gray. **Stem** 20–50×10–15mm (¾–2×½in), soft, fragile; same color as cap or paler; dry, faintly downy or almost smooth. **Flesh** thin; buff-colored. **Latex** white, unchanging, not staining. **Odor** fragrant, reminiscent of coconut. **Taste** slowly, slightly acrid and hot. **Spores** broadly ellipsoid, amyloid, 6.5–8.5×5–7μ; ornamented with small warts connected by thin ridges forming an incomplete network, prominences 0.5–0.8μ high. Deposit pale cream to buff. **Habitat** scattered to gregarious under birch or alder and conifers. Often common in wet seasons. Found widely distributed in northern North America. **Season** August–November. **Edible.**

Lactarius rufus (Fr.) Fr. **Cap** 3–12cm (1¼–4¾in) across, broadly convex with a shallow depression and sometimes a small umbo and an incurved margin, becoming flatter then broadly funnel-shaped; yellowish to brick-red cinnamon becoming paler in age; hoary becoming dull and dry. **Gills** adnate to short decurrent, crowded, narrow; whitish becoming pale pinky-yellow to pinky-tan. **Stem** 50–110×9–17mm (2–4½×⅜–¾in), stuffed, slightly enlarged toward the base; same color as cap except for whitish base; powdery and finely downy becoming dry. **Flesh** whitish to pale pinky-brown. **Latex** milk-white, plentiful, unchanging. **Odor** very slight, developing a smell like bouillon cubes as it dries. **Taste** immediately very acrid. **Spores** broadly ellipsoid, amyloid, 7.5–10×6–7.5μ; ornamented with ridges forming a partial reticulum and isolated warts, prominences about 0.5μ high. Deposit cream with a pale pinky tinge. **Habitat** scattered on soil under pine, in sand dunes, or in sphagnum bogs. Quite common. Found widely distributed throughout North America. **Season** July–November. **Poisonous. Comment** *Lactarius rufus* var. *parvus* Smith & Hesler is darker in color when wet, and the habitat seems to be confined to conifer logs.

Lactarius luculentus var. *laetus* ¾ life-size

Lactarius mutabilis ½ life-size

Lactarius hepaticus ⅔ life-size

Lactarius oculatus life-size

Lactarius luculentus var. *laetus* Smith & Hesler
Cap 3–6cm (1¼–2¼in) across, very broadly
convex with a papilla, becoming flat with a
broadly depressed center; bright orange all
over or slightly paler at the margin, becoming
a little duller in older specimens; smooth,
opaque when damp, slightly sticky when wet
but soon dry. **Gills** adnexed, narrow, crowded;
pale reddish clay, sometimes slowly staining
brownish. **Stem** 40–50×5–7mm
(1½–2×¼in), hollow, firm becoming fragile,
slightly bigger toward the base; same color as
gills or cap; smooth. **Flesh** thin, firm; whitish
to pale yellow. **Latex** milk-white, plentiful,
unchanging, slowly staining gills brown. **Odor**
typical. **Taste** mild, but slowly becoming bitter
then acrid. **Spores** broadly ellipsoid, amyloid,
7.8–9.7×6.7–7.7μ; ornamented with a broken
reticulum and some very small isolated warts
and ridges, prominences 0.3–0.6μ high.
Deposit cream. **Habitat** gregarious under
conifers and mountain alder. Found in the
Pacific Northwest. **Season** September–
October. **Not edible. Comment** Possibly this
mushroom is synonymous with *Lactarius
mitissimus* (Fr.) Fr.

Lactarius mutabilis Pk. **Cap** 3–8cm (1¼–3in)
across, at first almost round, then convex and
often slightly depressed, then expanded; dark
brown or walnut brown in concentric bands
and zones when wet; slightly sticky, but soon
dry and smooth or with a very faint bloom.
Gills adnate or subdecurrent, close, narrow to
medium broad; whitish tinged yellowish or
stained dingy red. **Stem** 30–80×6–10mm
(1¼–3×¼–½in), stuffed; same color as the
cap, with the top paler; smooth, dry. **Flesh**
thick on the disc, thinner at the edge; same
color as the cap. **Latex** white or watery white,
unchanging, sometimes staining gills pinkish.
Odor normal or faintly aromatic. **Taste** mild.
Spores ellipsoid, warty, 8–9.1×6.4–7.7μ.
Deposit whitish to yellowish. **Habitat** on soil in
coniferous and mixed woods. Found in eastern
North America. **Season** June–September.
Edibility not known.

Lactarius hepaticus Plowright **Cap** 3–9cm
(1¼–3½in) across, convex becoming flatter
then depressed in the center, sometimes with a
central papilla and a spreading or wavy
margin crimped with tiny lobes; rusty red or
dull chestnut, with papilla often darker; moist
but not sticky, smooth, opaque. **Gills**
decurrent, close, narrow; pale buff to pinkish or
with a mauvy bloom. **Stem** 40–90×4–10mm
(1½–3½×³⁄₁₆–½in), solid; about the same
color as the cap; dry, fragile, with a velvety
bloom when young and hairy at the base.
Flesh thin but brittle; pale, dingy pinkish buff.
Latex milk-white, unchanging, staining white
paper yellow. **Odor** not distinctive. **Taste**
slowly and strongly acrid. **Spores** ellipsoid,
7.5–9×6–7μ; ornamented with large warts
forming a partial reticulum, prominences
0.3–0.4μ high. Deposit cream. **Habitat** singly
or in groups in damp moss under pine.
Common. Found in northern North America.
Season July–October. **Not edible.**

Lactarius oculatus (Pk.) Burlingham **Cap**
1–3.5cm (½–1¼in) across, flat with a
depression on the disc and a central papilla,

eventually becoming shallowly funnel-shaped;
dark reddish brown or dark reddish cinnamon
around the papilla and similar but paler over
the rest of the cap; tacky and shining when
fresh, but soon drying. **Gills** decurrent, close,
broad; very pale pinky-cream staining light
pinkish cinnamon, becoming duller all over in
age. **Stem** 20–30×2–4mm (¾–1¼×
³⁄₃₂–³⁄₁₆in), hollow, sometimes slightly enlarged
toward base; dull pinkish cinnamon; naked,
moist, watery-fragile. **Flesh** watery-fragile;
similar color to surface. **Latex** wheylike,
unchanging, scanty; young caps stain white
paper yellow. **Odor** slight. **Taste** very slowly
and faintly peppery. **Spores** ellipsoid,
7.5–9.5×6–6.5μ; ornamented with a few
isolated warts, prominences 0.3–0.6μ high.
Deposit yellowish. **Habitat** scattered or in
groups in sphagnum bogs. Sometimes common.
Found in northeastern North America as far
south as Georgia. **Season** August–September.
Edibility not known.

Lactarius imperceptus Beardslee & Burlingham
Cap 3–8cm (1¼–3in) across, broadly convex
or bell-shaped, with a slight umbo and a
striate margin in mature specimens; pale to
dark fawn; sticky when wet, smooth. **Gills**
slightly decurrent, close, broad, narrowed and
sometimes forked at the stem; white, tinted
pale rose, sometimes staining brownish. **Stem**
30–80×7–15mm (1¼–3×¼–½in), solid;
pinkish buff with soft, dense, whitish hairs on
the base. **Flesh** bruises sulphur yellow. **Latex**
watery, white, changing slowly to sulphur
yellow. **Odor** none. **Taste** bitter becoming
peppery. **Spores** broadly ellipsoid, amyloid,
7–10×6–8μ; ornamented with heavy bands
forming a partial reticulum, prominences
0.5–1.5μ high. Deposit white to pale cream but
rather variable. **Habitat** on soil or occasionally
rotting logs under oak, pine, maple, or sweet
gum. Found in Florida and widely distributed
east of the Great Plains. **Season** July–October.
Not edible.

Lactarius imperceptus ¾ life-size

Burnt-sugar or **Curry Milkcap** *Lactarius aquifluus* life-size

Lactarius theiogalus ⅓ life-size

Lactarius obscuratus var. *obscuratus* ½ life-size

Curry-scented Milkcap *Lactarius camphoratus*
⅓ life-size

Burnt-sugar or **Curry Milkcap** *Lactarius aquifluus* Pk. syn. *Lactarius helvus* sensu European authors **Cap** 3–15cm (1¼–6in) across, convex and slightly umbonate with margin inrolled at first, then expanding to flattened or slightly funnel-shaped; dull grayish brown to fawn, yellowish, or cinnamon, often flecked with darker squamules; surface roughened, fibrillose, or with tiny scales. **Gills** adnexed to slightly decurrent, narrow and crowded, forking near stem; pale buffy ochre. **Stem** 35–70 × 12–15(20)mm (1¼–2¾ × ½[¾]in), hollow, often rather long and slender; concolorous with cap but paler and with a white base; surface often powdery or slightly downy. **Flesh** thin, fragile; whitish. **Latex** watery, almost clear. **Odor** very faint at first, then increasing as the fungus dries to become very strong of curry powder, bouillon cubes, or burnt sugar, very persistent. **Taste** mild or slightly bitter. **Spores** ellipsoid, (6)6.5–8(9) × 4.5–5.5(7.5)μ; with small warts mostly joined by slender ridges in a poorly developed network. Deposit whitish with slight salmon-buff tint. **Habitat** in moss in mixed woods. Found widely distributed in eastern and northern North America. **Season** July–October. **Slightly poisonous. Comment** The spores of this collection were rather smaller than usual, but otherwise the fungus agreed in all respects.

Lactarius theiogalus (Fr.) S. F. Gray **Cap** 2–7cm (¾–2¾in) across, shallowly convex becoming flat then funnel-shaped, often with an acute umbo and a spreading margin; dull orangy brown to dark apricot or dull rust, sometimes with faint concentric lines of color; moist, smooth. **Gills** decurrent, crowded, narrow; pale yellowy or pinky-cinnamon, becoming darker and pink-spotted in age. **Stem** 40–60 × 10–15mm (1½–2¼ × ½in), fragile, slightly enlarged toward the base; same color

as cap or paler, staining yellow from the milk; with a bloom at first, stiff, rough hairs at the base. **Flesh** firm but brittle; whitish to pale browny pink, staining yellow when cut. **Latex** white, scant, turning yellow slowly. **Odor** not distinctive. **Taste** mild or slowly, slightly hot and acrid. **Spores** subglobose to broadly ellipsoid, amyloid, 7.5–8.5 × 6.4–7.5μ; ornamented with blunt warts and irregular ridges sometimes connected by five lines, prominences 0.4–0.9μ high. Deposit white to creamy. **Habitat** in leaf mold, sphagnum, or on soil under hardwoods or in mixed forests, usually near birch. Found widely distributed throughout north central and northeastern North America and in Alaska. **Season** August–October. **Not edible.**

Lactarius obscuratus var. *obscuratus* (Lasch) Fr. **Cap** 0.6–1.6cm (¼–½in) across, convex to flatter with a sunken disc, sometimes nearly funnel-shaped, often with a small umbo and thin, translucent, striate margin; cinnamon brown with olive tints, especially in the darker center, becoming brighter, more reddish and losing the olive tone; smooth or very faintly wrinkled. **Gills** adnate to subdecurrent, quite close, fairly broad; yellowish dull cream. **Stem** 17–21 × 3–4mm (⅝–¾ × ⅛–³⁄₁₆in), soon hollow, slightly swollen at base; color similar to cap but more reddish brown, especially toward the base. **Flesh** thin; reddish to reddish ochre, and distinctly olive under umbo and cuticle when moist. **Latex** clear white, plentiful, unchanging. **Odor** weak. **Taste** mildly acrid. **Spores** ellipsoid, amyloid, 6–7.7 × 5–6.2μ; ornamented with a broken reticulum and warts joined by a well-developed network of ridges. Deposit whitish. **Habitat** in bogs under alder. Infrequent. Found in the Pacific Northwest. **Season** September–October. **Not edible.**

Curry-scented Milkcap *Lactarius camphoratus* (Bull. ex Fr.) Fr. **Cap** 3–5cm (1¼–1in) across, convex, then with a depression and often with a small umbo; margin inrolled at first, often furrowed; red-brown to bay or dark brick, sometimes with a purplish tinge; smooth and dull, not viscid. **Gills** decurrent, crowded, narrow; pale pinkish cinnamon. **Stem** 30–50 × 4–7mm (1¼–2 × ³⁄₁₆–¼in), equal to tapered downward; concolorous with the cap or darker. **Flesh** pallid brown. **Latex** thin, watery. **Odor** mild at first, but as fungus ages and dries becoming very strong of curry powder. **Taste** mild. **Spores** subglobose, 7.5–8.5 × 6.5–7.5μ; isolated warts with occasional connecting ridges. Deposit cream. **Habitat** in mixed woods. Common. Found widely distributed in northeastern North America, west to Minnesota, south to West Virginia. **Season** July–October. **Edible** — can be used dried as a seasoning.

Russula mutabilis ⅓ life-size

Russula densifolia ⅓ life-size

RUSSULA *Many species have brightly colored caps: red, purple, yellow, or green. The flesh of the whole mushroom is granular and brittle and will easily crumble. The gills, usually white at first, are neat and geometric in appearance. The spore color varies from pure white (A–B) to cream (C–E) to yellowish (F–G) to yellow-ochre (H); take a spore print by leaving a cap on paper or glass overnight. The taste is important; it may be hot, bitter, or mild. Check the amount that the cap cuticle peels. Note any special smells.*

Firm Russula *Russula compacta* Frost **Cap** 3–18cm (1¼–7in) across, convex with center flattened or depressed, margin smooth; whitish cream at first, soon turning yellow-orange to ochre with age or on bruising, finally completely rust brown; surface dry, dull, and can be scurfy-granular at center, often cracking; smooth and viscid when wet. **Gills** adnate, fairly crowded; white to pale cream-yellow, bruising rust-brown. **Stem** 20–100×12–30mm (¾–4×½–1¼in), becoming hollow, even; concolorous with cap but usually discolors less. **Flesh** very firm, brittle; white, flushing yellow with age or where eaten by insects. **Odor** strong, fishy, and unpleasant. **Taste** a little unpleasant. **Spores** broadly ovate, 7.5–9(9.9)×(5.5)6.3–7(8.6)μ; with blunt, conic warts up to 1.2μ high, usually isolated but with some faint connectives forming an incomplete network. Deposit white (A–B). **Habitat** in mixed woodlands. Common. Found in eastern North America, west to Michigan. **Season** August–September. **Edible** but rather poor. **Comment** This very firm, large russula is easily distinguished by the increasingly unpleasant odor and the color change. The superficially similar *Russula nigricans* (below) lacks the odor and finally turns black.

Russula mutabilis Murr. **Cap** 3–5cm (1¼–2in) across, convex to depressed, margin striate-tuberculate; dull orange-brown to pale tawny, dull red where injured; surface viscid, not peeling readily. **Gills** adnate, broad, forked at base; pale yellow. **Stem** 30–50×10–13mm (1¼–2×½in); dull ochre bruising deep blood red. **Flesh** firm; pale yellow. **Odor** fetid to aromatic. **Taste** unpleasant, acrid, or astringent. **Spores** subglobose, 8–10.5×7.5–9μ; warts 0.9–1.4μ high, large and blunt, isolated. Deposit cream (A–C). **Habitat** in grass, sandy soil under oak. Found in Florida and New Jersey. **Season** July–August. **Not edible. Comment** This appears to be the first published picture of this little-known species, which is distinctive by the whole mushroom drying blood red with age.

Short-stem Russula *Russula brevipes* Pk. **Cap** 6–20cm (2¼–8in) across, convex with depressed center, with inrolled, nonstriate margin; white stained dull ochre to brown; dry, minutely woolly. **Gills** adnate to slightly decurrent, crowded; white staining ochre-brown. **Stem** 25–70×25–40mm (1–2¾×1–1½in); white stained brown; dry, smooth. **Flesh** thick, brittle; white. **Odor** not distinctive. **Taste** slowly acrid, unpleasant. **Spores** broadly ellipsoid to subglobose,

8–11×6.5–10μ; ornamented with a broken to complete reticulum, with warts 0.7–1.7μ high. Deposit white to cream (A–C). Habitat in mixed woods. Found widely distributed throughout North America. **Season** July–October. **Not edible. Comment** *Russula brevipes* var. *acrior* Schaeff. differs in its green-tinted gills and apex of the stem; other details similar. *Russula cascadensis* Schaeff. has a more acrid flavor and smaller size and grows under conifers in the Northwest.

Russula adusta Fr. **Cap** 5–15cm (2–6in) across, flattened, then depressed at center, margin strongly inrolled; white to pale buff; thick, viscid when wet, cuticle adnate. **Gills** subdecurrent, very crowded, narrow; white to cream. **Stem** 40–100×20–40mm (1½–4×¾–1½in), hard; white bruising reddish brown. **Flesh** white, slowly pinkish when cut, then brown. **Odor** of sour wine. **Taste** mild. **Spores** broadly ellipsoid, 7–9×6–8μ; warts small, under 0.5μ high, connectives fine, abundant, well-developed reticulum. Deposit white (A). **Habitat** usually under conifers. Uncommon. Found throughout northeastern North America. **Season** July–October. **Edibility suspect** — best avoided; other members of this group are poisonous.

Russula densifolia Secr. **Cap** 5–15cm (2–6in) across, broadly convex, soon flattened and then funnel-shaped; white, then dull brown and finally black; viscid when wet. **Gills** slightly decurrent, very crowded (7–12 gills per cm/½in); white to pale cream. **Stem** 30–60×10–30mm (1¼–2¼×½–1¼in), tapered at base, hard; white becoming blackish. **Flesh** firm; white soon staining red, then gray to black. **Odor** not distinctive. **Taste** hot to very hot. **Spores** broadly ellipsoid, 7–9×6–7μ; warts only 0.5μ high or less, with partial reticulum. Deposit white (A). **Habitat** in mixed woods. Often common. Found in eastern North America up into Michigan. **Season** July–September. **Edibility suspect** — not recommended; other members of this group have caused poisonings.

Russula nigricans Fr. **Cap** 10–20cm (4–8in) across, broadly convex, soon funnel-shaped; dirty white soon staining brown and finally black as if burnt, bruising reddish. **Gills** adnate, thick, widely spaced; pallid to straw, eventually black. **Stem** 30–80×10–40mm (1¼–3×½–1½in); white then soon stained like cap. **Flesh** white, becoming blood red then gray to black. **Odor** fruity. **Taste** slowly hot. **Spores** broadly ellipsoid, 7–8×6–7μ; warts under 0.5μ high, with partial reticulum. Deposit white (A). **Habitat** under deciduous trees. Common. Found in northern North America. **Season** July–September. **Edibility suspect** — not advisable; other members are poisonous.

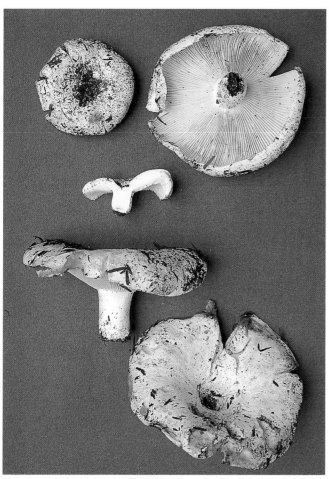

Short-stem Russula *Russula brevipes* ½ life-size

Russula adusta ½ life-size

Firm Russula *Russula compacta* ½ life-size

Russula nigricans ⅓ life-size

Almond-scented Russula *Russula laurocerasi* ½ life-size

Russula subfoetens ¾ life-size

Russula fragrantissima ⅓ life-size

Russula granulata ½ life-size

Russula polycystis life-size

Russula polyphylla ⅓ life-size

Almond-scented Russula *Russula laurocerasi*
Melz. **Cap** 4–10cm (1½–4in) across, convex
then expanded or even depressed at center,
margin conspicuously furrowed; bright
yellow-ochre, tawny, or yellow-brown; viscid
when wet. **Gills** adnate, well spaced, broad;
yellowish white to pale orange. **Stem**
25–100 × 15–25mm (1–4 × ½–1in), slightly
tapered at base, rigid but easily breaking;
dirty yellowish white to pale orange, stained
brownish; dry, dull. **Flesh** brittle, with
irregular cavities in stem; white. **Odor** more or
less marked of bitter almonds or marzipan.
Taste unpleasant, rather hot. **Spores**
subglobose, (6.8)7–8(10.7) × 6–7(9)μ; very
distinctive with well-developed, sometimes
branching "wings" up to 2μ high connecting a
large proportion of the warts. Deposit pale egg
yellow (C–D). **Habitat** in mixed deciduous
woods. Fairly common. Found in eastern
Canada, south to North Carolina, west to
Michigan. **Season** July–September. **Not
edible. Comment** The similar species *Russula
fragrantissima* (below) is usually much more
fetid-smelling, also darker and larger. This and
other similar species lack the enormous
"wings" of the spores of *Russula laurocerasi*.

Russula subfoetens Smith **Cap** 5–10cm (2–4in)
across, rounded then with a depressed center;
dull honey yellow to brownish; margin coarsely
tuberculate-striate, viscid when wet. **Gills**
adnate, cream-yellow, often brown-spotted.
Stem 50–100 × 10–25mm (2–4 × ½–1in),
narrowing near base, firm; pale honey yellow.
Flesh pale straw, yellowing when cut, and
turning bright golden in KOH. **Odor** slightly
unpleasant, fetid. **Taste** hot in cap cuticle but
mild in flesh. **Spores** oval-ellipsoid, 7–9 ×

5–6μ; warts 0.3–0.7μ high, few connectives.
Deposit cream (C–D). **Habitat** in mixed
woods. Found in eastern North America, west
to Michigan, south to North Carolina. **Season**
July–September. **Not edible.**

Russula fragrantissima Romagnesi **Cap** 7–20cm
(2¾–8in) across, subglobose, slowly expanding
with incurved margin; thick, fleshy; pale yellow
to yellow-brown or tawny; very viscid when
wet, shiny when dry, tuberculate-striate at
margin. **Gills** adnate, close, narrow at front;
pale yellow, often beaded with moisture on
margin when young; stained brown where
injured. **Stem** 70–150 × 15–60mm (2¾–6 ×
½–2¼in), equal, firm, soon hollow;
colored as cap, staining darker brown; dry,
dull. **Flesh** hard; white. **Odor** fetid to
almondlike, rarely very fragrant in North
American collections (see Comment below).
Taste oily-acrid. **Spores** broadly elliptic,
6–9 × 5.5–7.7μ; warts up to 1μ high, with
partial to complete reticulum. Deposit pale
orange-yellow (C–D). **Habitat** in mixed
deciduous woods. Common. Found in
northeastern North America, west to Michigan,
south to North Carolina. **Season** July–
October. **Not edible. Comment** This species
as defined in America differs markedly from
European collections.

Russula granulata (Pk.) Pk. **Cap** 4–10cm
(1½–4in) across, convex with incurved margin,
disc flattened with small crustlike patches and
granules; pale orange-yellow to tawny; margin
tuberculate-sulcate; viscid when wet. **Gills**
adnate, close; pale yellow, often spotted brown.
Stem 30–75 × 10–20mm (1¼–2¾ × ½–¾in),
equal; often stained darker brown below;
smooth. **Flesh** yellowish staining brown. **Odor**

unpleasant. **Taste** oily-acrid. **Spores** ellipsoid,
5.7–8 × 4.4–6.3μ; warts up to 1μ high, isolated
with very few connectives. Deposit pale orange-
yellow (D–E). **Habitat** in mixed woodlands.
Frequent. Found in northeastern North
America to Michigan and Tennessee. **Season**
July–September. **Not edible.**

Russula polycystis Singer **Cap** 4–10cm (1½–4in)
across, soon flattened; pale creamy yellow, buff
to ivory or straw; viscid when wet, surface
often cracking into small areolate patches;
cuticle thick and rubbery or elastic, peels
almost to center. **Gills** close; pale creamy-
yellow. **Stem** 25–50 × 10–20mm (1–2 ×
½–¾in); white to pale yellow; pruinose-scurfy
surface. **Flesh** firm; white. **Odor** none. **Taste**
slowly acrid. **Spores** subglobose,
6.5–7.5 × 5–6.5μ; warts 0.5–1μ high, almost
no connectives. Deposit cream (B–C). **Habitat**
in mixed woods. Uncommon. Recorded from
Tennessee, Virginia, and New York, but
probably widespread. **Season** July–August.
Not edible.

Russula polyphylla Pk. syn. *Russula magnifica* Pk.
Cap 8–20cm (3–8in) across, soon a flattened
funnel shape; white to ivory, with darker
brown crusty squamules at center; viscid when
wet. **Gills** very close, thin; white with pinkish
flush. **Stem** 50–120 × 15–35mm (2–4¾ ×
½–1¼in); white staining pale brownish pink.
Flesh firm; white. **Odor** very pungent alkaline,
even when dried. **Taste** strong, unpleasant,
alkaline. **Spores** ovoid-ellipsoid, 7–9 × 5.5–7μ;
warts 0.1–0.4μ high, isolated, spores often
appear almost smooth. Deposit white (A–B).
Habitat in beech and oak woods. Very
uncommon. Found in New York and New
Jersey. **Season** July–August. **Not edible.**

Russula integra ½ life-size

Russula brunneola ½ life-size

Russula vinosobrunnea ½ life-size

Russula occidentalis ⅓ life-size

Russula graveolens ¼ life-size

Russula curtipes ⅓ life-size

Russula versicolor ⅓ life-size

Russula integra Fr. **Cap** 6–15cm (2¼–6in) across, soon flattened-depressed; firm, fleshy; reddish brown to dull brown; smooth. **Gills** moderately spaced, broad; deep yellow. **Stem** 40–60 × 15–25mm (1½–2¼ × ½–1in), equal, firm; white or slightly yellowish at base. **Flesh** white. **Odor** pleasant. **Taste** mild. **Spores** broadly ovoid, 8–8.5 × 7μ; warts up to 1μ high, with almost no connectives. Deposit deep ochre (G). **Habitat** under beech and oak. Not common. Found in western and eastern North America. **Season** August–October. **Edible** — good.

Russula brunneola Burlingham **Cap** 5–12cm (2–4¾in) across, subglobose then soon almost flat; deep sepia brown, gray-brown to olive brown, sometimes with paler, yellowish areas; smooth, dry, often pruinose when young, margin finely striate-tuberculate. **Gills** moderately crowded, broad, prominently forking near stem; pale cream to slightly brownish. **Stem** 40–80 × 10–30mm (1½–3 × ½–1¼in), equal, firm; yellowish white or flushed pale purple to reddish brown at base; dry, dull. **Flesh** firm; white. **Odor** pleasant. **Taste** mild. **Spores** ellipsoid, 6–9.5 × 4–7.5μ; warts up to 0.7μ high, mostly isolated, occasionally a few fine connectives. Deposit white to palest cream (A–B). **Habitat** conifer and mixed woods. Common. Found in eastern North America and west to Washington. **Season** July–September. **Edible** — good.

Russula vinosobrunnea (Bres.) Romagnesi **Cap** 5–13cm (2–5in) across, convex; fleshy, firm; deep purple-brown to blackish purple or even olivaceous; cap cuticle hardly peeling. **Gills** rather close, strongly interveined; yellow-ochre, edge often purple. **Stem** 35–80 × 10–30mm (1¼–3 × ½–1¼in), equal; white, flushed rose or reddish in part. **Flesh** thick, firm; white. **Odor** pleasant. **Taste** mild. **Spores** ovoid, 7–10 × 6–8μ; warts up to 1.5μ high, connectives irregular, rarely forming a reticulum. Deposit clear ochre (F–G). **Habitat** under mixed woods. Found in western North America. **Season** September–November. **Edible.**

Russula occidentalis Singer **Cap** 6–13cm (2¼–5in) across; usually purplish with yellow-green center but very variable in color,

sometimes with bluish-green shades at center and bluish purple at margin; viscid when wet; cuticle peels halfway. **Gills** subcrowded; pale yellow, blackening at margin with age. **Stem** 50–80 × 15–25mm (2–3 × ½–1in), firm; white, often grayish with age, bruising reddish then brown. **Flesh** firm; white, turning slowly reddish when cut, then gray to black. **Odor** none. **Taste** mild. **Spores** subglobose to ellipsoid, 8–10 × 7.5–9μ; ornamented with large protuberances, some connected by lines. Deposit cream (B). **Habitat** under fir and hemlock. Common. Found in the Washington Cascades area. **Season** September–November. **Edible.**

Russula graveolens Romell **Cap** 4–8cm (1½–3in) across, convex to flattened; reddish brown to snuff brown or ochre, sometimes with hints of violet or carmine at margin; cuticle not peeling easily. **Gills** fairly crowded; pale ivory. **Stem** 40–80 × 10–25mm (1½–3 × ½–1in); white, staining yellow-brown when handled. **Flesh** firm; white. **Odor** fishy or crabmeat-like. **Taste** mild. **Spores** ovoid, 7.5–8.8 × 6.2–7.5μ; warts up to 1.1μ high, isolated. Deposit pale ochre (D–E). **Habitat** under deciduous trees. Recorded from western North America, but distribution uncertain owing to confusion in the *Russula xerampelina* group (q.v.). **Season** August–October. **Edible. Comment** When FeSO₄ is applied, the typical dark green reaction of this group occurs.

Russula curtipes Møller & J. Schaeff. **Cap** 3–10cm (1¼–4in) across, soon flattened; often concentrically pigmented, red-brown to vinaceous red or purplish, center often paler cream; surface dull. **Gills** fairly close, broad; medium ochre. **Stem** 30–50 × 10–15mm (1¼–2 × ½in), usually shorter than cap diameter; white. **Flesh** firm; white. **Odor** mild. **Taste** mild. **Spores** ellipsoid, 7–9.5 × 6–7.5μ; with low, blunt warts and a partial network. Deposit deep ochre (F–H). **Habitat** in beech woods. Found in New Jersey; possibly elsewhere but not recorded. **Season** July–August. **Edible.**

Russula versicolor Schaeff. **Cap** 1.5–8cm (½–3in) across, soon flattened; fragile; usually pale, a mixture of colors including buff, greenish, yellow to pink or vinaceous purple; margin usually sulcate, viscid when wet; peeling almost fully. **Gills** pale yellow. **Stem**

Russula odorata ⅓ life-size

15–50 × 10–25mm (½–2 × ½–1in), fragile; dull white staining yellowish ochre. **Flesh** soft; white. **Odor** pleasant. **Taste** mild to rather hot. **Spores** subglobose, 6–10 × 4–7μ; warts up to 0.4μ high, isolated or with a few connectives, sometimes even reticulate. Deposit medium yellow (E–F). **Habitat** in wet, boggy areas. Probably widely distributed in North America although little-known. **Season** August–November. **Not edible.**

Russula odorata Romagnesi **Cap** 3–5cm (1¼–2in) across, convex becoming flatter, soon depressed; dull purplish red, brown, rosy, or grayish red, fading to buff; fragile; margin sulcate; cuticle peels almost completely. **Gills** thin, strongly interveined; deep ochre-yellow. **Stem** 25–80 × 10mm (1–3 × ½in), fragile; white bruising slightly yellow. **Flesh** white. **Odor** rather aromatic. **Taste** mild to slightly hot. **Spores** ovoid, 6.7–8.5 × 5.7–7μ; warts up to 1μ high, with partial reticulum. Deposit deep ochre (G–H). **Habitat** in mixed deciduous woods. Found in western North America. **Season** September–November. **Edibility not known.**

Russula brunneoviolacea ¾ life-size

Russula olivaceoviolascens ½ life-size

Russula turci ½ life-size

Russula zelleri ¾ life-size

Russula elegans ½ life-size

Russula fragilis ½ life-size

Russula gracilis ½ life-size

Russula brunneoviolacea Craw. **Cap** 6–9cm (2–3in) across, convex to plane; livid violet to vinaceous brown, olive-brown to occasionally even entirely greenish; dry, smooth; pigment of cuticle usually distinctly speckled, patchy, when viewed through lens (10×), peels two-thirds. **Gills** subdistant; cream. **Stem** 35–60×10–15mm (1¼–2×½in), equal; white staining slightly yellowish brown at base. **Flesh** brittle; white. **Odor** pleasant. **Taste** mild. **Spores** ovoid, 6–8×5.5–6.8μ; warts up to 0.8μ high, with many fine connectives forming partial to complete reticulum. Deposit pale orange-yellow (C–E). **Habitat** in mixed woods. Frequent. Found widely distributed in eastern North America, west to Michigan. **Season** July–September. **Edible.**

Russula olivaceoviolascens Gillet **Cap** 2–4cm (¾–1½in) across, convex when very young, soon with a central depression; a mixture of greenish violet; smooth, viscid and shining when wet; peels easily. **Gills** broad; cream. **Stem** 50–70×10–15mm (2–2¾×½in), clavate, very fragile; white. **Flesh** soft; white. **Odor** fruity. **Taste** hot. **Spores** ellipsoid, 7–8×5–6μ; warts up to 0.5μ high, reticulum almost complete. Deposit cream (B–C). **Habitat** in marshy places. Rare. Found in northwestern North America. **Season** September. **Edibility not known.**

Russula turci Bres. **Cap** 3–10cm (1¼–4in) across; dark purple to vinaceous brown; viscid when wet, smooth, dull when dry; cuticle peeling one-quarter. **Gills** close; pale yellow. **Stem** 30–70×10–25mm (1¼–2¾×½–1in), equal; white or rarely flushed rose. **Flesh** soft; white. **Odor** of ink or iodine at stem base when scratched. **Taste** mild. **Spores** ovoid, 7–9×6–8μ; warts up to 0.5μ high, reticulum nearly complete. Deposit deep yellow-orange (G). **Habitat** under conifers. Found in western North America. **Season** September–October. **Edible.**

Russula zelleri Burlingham **Cap** 3–10cm (1¼–4in) across, convex then plane; deep vinaceous brown to purple vinaceous, browner at center; smooth, viscid when wet; cuticle separable up to half the radius. **Gills** rather distant; pale buff. **Stem** 50–60×10–20mm (2–2¼×½–¾in), equal; white. **Flesh** rather fragile; white. **Odor** pleasant. **Taste** mild. **Spores** ovoid, 7.5–10×6–8.5μ; warts up to 0.8μ high, isolated. Deposit pale ochre (D–F). **Habitat** under spruce. Rather rare. Found in Oregon. **Season** September–December. **Edible.**

Russula elegans Bres. **Cap** 2–6cm (¾–2¼in) across, convex; deep carmine red to purplish red or red-brown, often with discolored cream patches; peels halfway. **Gills** fragile; white to cream, yellowing with age. **Stem** 25–50×5–12mm (1–2×¼–½in); white staining yellow. **Flesh** soft; white, yellowing. **Odor** fruity. **Taste** mild. **Spores** ellipsoid, 7–8.5×6–7.5μ; warts 0.3–0.6μ high, with connectives forming a partial reticulum. Deposit clear yellow (F–G). **Habitat** under oak in grass. Found in New Jersey. **Season** July–September. **Edible.**

Russula fragilis (Pers. ex Fr.) Fr. **Cap** 2–5cm (¾–2in) across, convex then flattened-depressed; very variable in color, usually purplish to violet with darker, blackish to olive center, paler at margin, often with mixture of colors; fragile, cuticle three-quarters peeling. **Gills** adnexed, usually with minutely serrate margin; white to very pale cream. **Stem** 25–60×5–15mm (1–2¼×¼–½in), slightly clavate below, fragile; white. **Flesh** white. **Odor** pleasant, slightly fruity. **Taste** soon extremely hot. **Spores** subglobose, 7.5–9×6–8μ; with warts up to 0.5μ high joined by fine lines forming an almost complete network. Deposit whitish (A–B). **Cap cystidia** cylindrical to club-shaped, with 0–2 septa, react strongly with SV. **Habitat** under mixed deciduous trees. Probably common. Found throughout northeastern North America. **Season** July–September. **Not edible.**

Russula gracilis Burlingham **Cap** 3–6cm (1¼–2¼in) across, soon flattened-depressed; pale lilac rose to salmon, mixed with pale greenish gray or buff; viscid when wet; cuticle separable almost to center. **Gills** rather close; cream-buff. **Stem** 30–50×8–20mm (1¼–2×⁵⁄₁₆–¾in), equal; white or rarely tinted pink. **Flesh** soft; white. **Odor** pleasant. **Taste** acrid. **Spores** ellipsoid, 7.5–9×6–7μ; warts 0.3–0.5μ high, isolated or with a few connectives. Deposit medium ochre (D–E). **Habitat** in wet, boggy areas in mixed woods. Fairly common. Found in eastern North America. **Season** July–September. **Not edible.**

Russula sardonia var. *viridis* just over life-size

Russula sardonia ½ life-size

Russula sardonia Fr. **Cap** 4–10cm (1½–4in) across, convex; hard and fleshy; violet to purplish red or brownish red, greenish in var. *viridis*; not peeling. **Gills** very slightly decurrent, narrow, arching, forked; pale primrose yellow. **Stem** 30–80 × 10–20mm (1¼–3 × ½–¾in), equal, firm; usually flushed entirely pale lilac to grayish rose; very finely pruinose. **Flesh** white. **Odor** slightly fruity. **Taste** acrid. **Spores** ovoid, 7–9 × 6–8μ; warts up to 0.5μ high, connected to form a partial reticulum. Deposit cream-yellow (E–F). **Habitat** under pine. Frequent. Found in western North America. **Season** August–November. **Edibility suspect** — not advisable.

Purple-bloom Russula *Russula mariae* Pk. syn. *Russula alachuana* Murr. **Cap** 2–7(10)cm (¾–2¾[4]in) across, convex then flattened, margin striate to sulcate; deep purple-violet, wine-red, ruby-red to coral-pink,

often mixed with greenish tints, rarely entirely greenish yellow; dry and velvety to pruinose. **Gills** adnate, crowded, often forking; white then soon pale yellow-buff (C–D). **Stem** 20–60 × 5–20mm (¾–2¼ × ¼–¾in), equal to tapered at base and apex, firm; white to pinkish or entirely deep pinkish purple; surface dull, pruinose to granular. **Flesh** white. **Odor** oily. **Taste** mild to slightly acrid or oily. **Spores** subglobose, amyloid, 6.5–8.5(9) × 5.5–7.5(8)μ; ornamentation up to 1.2μ high, consisting of long irregular crests and ridges, flattened spines, and a few conical warts, forming a partial to complete reticulum. Deposit pale yellow (C–D). **Cap cystidia** in clumps, lance-shaped, without septa. **Habitat** common especially near oak. Found widely in eastern North America, west to Michigan. **Season** June–October. **Edible** — good.

Blackish-red Russula *Russula krombholtzii* Shaffer syn. *Russula atropurpurea* (Krombh.)

Britz. non Pk. **Cap** 4–10cm (1½–4in) across, convex then flattened with slight depression; dark blackish purple at center, paler, more blood red at margin, often mottled with paler, discolored areas; smooth, slightly viscid when wet. **Gills** adnexed, crowded; palish cream. **Stem** 30–60 × 10–20mm (1¼–2¼ × ½–¾in), fairly firm, later softer and easily broken; white, often becoming grayish with age. **Flesh** white. **Odor** rather fruity, of apples. **Taste** from almost mild to rather hot. **Spores** ovoid, 7–9 × 6–7μ; with warts joined by fine ridges to form a well-developed but not quite complete network. Deposit whitish (A–B). **Cap cystidia** abundant, cylindrical to somewhat club-shaped, without septa. **Habitat** usually under oak or other deciduous trees. Common. Found widely throughout northeastern North America, west to Michigan. **Season** June–October. **Edible.**

Russula mordax Burlingham **Cap** 5–13cm (2–5in) across, convex to slightly centrally depressed; deep madder-brown to blood-red brown; viscid when wet; cuticle separable halfway. **Gills** crowded, equal, many forking at base; yellowish. **Stem** 40–80 × 15–30mm (1½–3 × ½–1¼in), equal; white washed with red or reddish brown below, paler at apex. **Flesh** firm; white. **Odor** nil. **Taste** very acrid. **Spores** ovoid, 7–9.5 × 6–7.5μ; warts up to 0.8μ high, connected by fine partial reticulum. Deposit ochraceous (D–F). **Habitat** under fir. Found in Washington State. **Season** September–November. **Not edible.** **Comment** Singer suggests that this species may be the same as *Russula badia* Quél., but since Burlingham specifically disagreed with this interpretation, I choose her name instead.

Russula vinosa Lindblad syn. *Russula obscura* Romell **Cap** 5–15cm (2–6in) across; pale to dark blood red or livid purple; peeling only at margin. **Gills** somewhat distant; pale buff. **Stem** 40–150 × 15–30mm (1½–6 × ½–1¼in); white, flushing grayish black when bruised or with age. **Flesh** white, blackening. **Odor** not distinctive. **Taste** mild. **Spores** ovoid, 8–10 × 6.5–9μ; warts up to 0.5μ high, no connectives. Deposit ochre (E–F). **Habitat** in conifer woods and boggy areas. Found in northeastern North America. **Season** July–August. **Edible.**

Purple-bloom Russula *Russula mariae* ⅔ life-size

Blackish-red Russula *Russula krombholtzii* ⅔ life-size

Russula mordax ½ life-size

Russula vinosa ⅓ life-size

Russula xerampelina var. *elaeodes* ½ life-size

Russula sericeonitens ½ life-size

Russula xerampelina ⅓ life-size

Russula xerampelina var. *barlae* ½ life-size

Variable Russula *Russula variata* ⅓ life-size

Russula cyanoxantha ⅓ life-size

Russula nauseosa ⅓ life-size

Russula xerampelina Fr. **Cap** 5–15cm (2–6in) across, convex then flattened; thick, fleshy; dark red, vinaceous, brownish or with greenish tints (see variations below); soon dull, dry; peeling only at margin. **Gills** quite distant, broad; pale buff. **Stem** 40–130 × 15–30mm (1½–5 × ½–1¼in), equal; typically rose to white, staining brownish with age or bruising. **Flesh** firm; white, browning. **Odor** fishy, or crablike when crushed. **Taste** mild. **Spores** ellipsoid, 8–10 × 7–8.5μ; warts up to 1μ high, few connectives. Deposit deep yellow (E–F). **Habitat** in mixed woods. Found throughout North America. **Season** July–October. **Edible. Comment** Several varieties, probably good species in their own right, have been described: var. *barlae* Quél. is orange-yellow, var. *elaeodes* (Bres.)? is small with greenish cap and has smaller spores.

Russula sericeonitens Kauffman **Cap** 4–6cm (1½–2¼in) across, convex then depressed; deep violet-purple, almost black at center; smooth, silky, shining; peels easily. **Gills** subdistant, broad; white. **Stem** 30–70 × 10–15mm (1¼–2¾ × ½in), equal; white. **Flesh** soft; white. **Odor** none. **Taste** mild. **Spores** ovoid, 7–8 × 6–7μ; warts up to 1.2μ high, isolated. Deposit white (A). **Habitat** in mixed woods. Occasional. Found in eastern North America, west to Michigan. **Season** July–September. **Edible.**

Variable Russula *Russula variata* Ban. apud Pk. **Cap** 5–15cm (2–6in) across, convex then flattened, center depressed; very variable in color, shades of greenish yellow mixed with pink or purple; dry to slightly greasy. **Gills** adnate to slightly decurrent, crowded,

conspicuously forked; white. **Stem** 30–100 × 10–30mm (1¼–4 × ½–1¼in); white to dull cream; dry, dull, often wrinkled lengthwise. **Flesh** brittle; white. **Odor** unpleasant. **Taste** disagreeable, often acrid. **Spores** ovoid, 7–11.4 × 5.5–9.5μ; warts 0.3–1μ high, no reticulum present. Deposit white (A). **Cap cystidia** scattered, small, short, and narrow, ends slightly swollen. **Habitat** in oak woods. Often abundant. Found widely distributed in North America. **Season** July–October. **Edible. Comment** Chemical reaction to FeSO₄ on flesh is almost nil to slightly greenish.

Russula cyanoxantha (Schw.) Fr. **Cap** 5–15cm (2–6in) across, at first globose, then expanded, depressed at disc; extremely variable in color, usually shades of lilac to purple, often intermixed with green or entirely green in var. *peltereaui* Maire; dry to greasy when moist; cuticle peeling halfway. **Gills** narrow, crowded, not usually forking (see *Russula variata* for comparison), flexible and greasy to touch; white to pale cream. **Stem** 50–100 × 15–30mm (2–4 × ½–1¼in), firm to spongy; white, sometimes flushed purplish. **Flesh** white. **Odor** not distinctive. **Taste** mild. **Spores** ellipsoid, 7–9 × 6–7μ; warts up to 0.5μ high, isolated. Deposit white (A). **Habitat** in mixed woods, rather uncommon. Found widely throughout North America. **Season** August–September. **Edible** — good. **Comment** Chemical reaction to FeSO₄ on flesh is almost nil to slightly greenish.

Russula nauseosa (Pers. ex Secr.) Fr. **Cap** 3–8cm (1¼–3in) across, convex soon plane; fragile; margin tuberculate-sulcate; very varied in color, dark olive, vinaceous brown, reddish, grayish rose, buff, the center usually darker;

cuticle peeling almost completely. **Gills** thin, brittle; ochre. **Stem** 30–80 × 5–10mm (1¼–3 × ¼–½in), fragile; white, browning with age. **Flesh** white. **Odor** pleasant. **Taste** mild to slightly hot. **Spores** ovoid, 7–11 × 6–9μ; warts up to 1.25μ high, isolated, no connectives. Deposit deep ochre (G–H). **Habitat** under conifers. Probably widely distributed over North America. **Season** August–November. **Edibility not known.**

Russula olivacea ½ life-size

white. **Stem** 35–90 × 15–30mm (1¼–3½ × ½–1¼in), soon becoming hollow; creamy white; dry. **Flesh** firm at first then soon soft, brittle; white. **Odor** not distinctive. **Taste** mild. **Spores** ellipsoid, 5.5–10 × 4.5–7.8μ; warts 0.2–0.3μ high, isolated or with fine lines. Deposit pale yellow-ochre (D). **Habitat** in mixed woods, often under oak. Common. Found in eastern North America, west to Michigan. **Season** July–September. **Edible** — good. **Comment** Compare with the very similar *Russula virescens* (below), which differs in its very sparse facial gill cystidia and white spore print.

Russula virescens Fr. **Cap** 5–12cm (2–4¾in) across, convex then flattened; thick-fleshed; dull green to olive; surface breaking up into small, flattened patches, or granules like a mosaic. **Gills** crowded, brittle; cream. **Stem** 40–90 × 20–40mm (1½–3½ × ¾–1½in), equal; white, browning slightly. **Flesh** very firm; white. **Odor** pleasant. **Taste** mild, nutty. **Spores** ellipsoid to subglobose, 7–9 × 6–7μ; warts up to 0.5μ high, with partial reticulum. Deposit white (A–B). **Habitat** in mixed woods, in particular beech. Found throughout northern and eastern North America. **Season** July–September. **Edible** — delicious. **Comment** The very similar *Russula crustosa* (above) has abundant facial gill cystidia and cream spore print.

Russula modesta Pk. **Cap** 3–10cm (1¼–4in) across, convex then flat to slightly convex, slightly striate at margin when mature; pale grayish green, olive, or olive-buff; dry, dull, usually densely pruinose; cuticle peeling one-quarter to one-half. **Gills** adnate, crowded to slightly spaced; pale buff. **Stem** 25–60 × 10–25mm (1–2¼ × ½–1in), equal; white; dry. **Flesh** solid then spongy; white. **Odor** not distinctive. **Taste** mild. **Spores** ovoid, 6–7(8.5) × 4.5–6(6.8)μ; warts up to 0.8μ high, isolated or with a very partial reticulum. Deposit cream (B–C). **Habitat** often gregarious under mixed woods. Found in New England and southward to West Virginia, Virginia, and Tennessee. **Season** July–September. **Edible.**

Russula redolens Burlingham **Cap** 2.5–8cm (1–3in) across, flattened to depressed; deep bluish green, gray-green often with a whitish bloom at margin; viscid when wet; cuticle peels halfway or more. **Gills** crowded, forking near stem; pale yellow. **Stem** 30–80 × 10–15mm (1¼–3 × ½in); white; dry, dull. **Flesh** white, sometimes slightly graying. **Odor** strong when old of parsley or celery. **Taste** mild to unpleasant (not hot). **Spores** ellipsoid, 6–8 × 4.5–6μ; warts up to 1μ high, isolated. Deposit pale cream (B). **Habitat** under broad-leaved and conifer trees. Rather uncommon. Apparently widespread in eastern North America. **Season** August–September. **Edibility not known.**

Russula heterophylla Fr. **Cap** 5–10cm (2–4in) across, convex then flat with a small central depression; shades of green to greenish yellow to ochre; smooth to radially veined; peels about halfway. **Gills** crowded, flexible, forked near stem; white. **Stem** 50–100 × 10–20mm (2–4 × ½–¾in); white. **Flesh** white. **Odor** pleasant. **Taste** mild. **Spores** subglobose, 5–7 × 4–6μ; warts only 0.2–0.6μ high, mostly isolated. Deposit white (A). **Habitat** in deciduous woods. Found in eastern North America. **Season** July–September. **Edible. Comment** Often confused with *Russula aeruginea* Lindblad ex Fr., which has cream-yellow spores (D–E). It differs from *Russula variata* (p. 119) in having a strong salmon-colored reaction to iron salts (FeSO₄).

Russula crustosa ½ life-size

Russula olivacea Fr. **Cap** 6–16cm (2¼–6¼in) across, convex then soon flattened with depressed center; mixture of purplish red to livid purple or olive to brown; dry, slightly roughened, often concentrically cracking. **Gills** distant, broad, interveined; pale yellow, margins usually colored like cap at edge. **Stem** 50–100 × 15–40mm (2–4 × 1½in), equal; white flushed rose entirely or mostly near apex. **Flesh** firm; white. **Odor** pleasant. **Taste** mild, nutty. **Spores** ovoid, 8–11 × 7–9μ; warts up to 1.5μ high, isolated, few connectives. Deposit deep yellow-orange (G–H). **Habitat** under beech. Uncommon. Found in both western and eastern North America. **Season** July–August. **Edible** — very good. **Comment** FeSO₄ crystals rubbed on the stem give a green reaction. When phenol is applied to the stem it turns bright blackberry juice color in 10–15 minutes.

Russula crustosa Pk. **Cap** 5–15cm (2–6in) across, convex then rather flat; pale blue-green to pale yellowish green or even ochre-yellow; dry, dull, surface cracking all over like mosaic. **Gills** attached, crowded to almost distant;

Russula virescens ¾ life-size

Russula modesta ½ life-size

Russula redolens ½ life-size

Russula heterophylla ½ life-size

Russula viridofusca ½ life-size

Russula simulans ½ life-size

Russula raoultii ¾ life-size

Russula crassotunicata ¾ life-size

Russula earlei ½ life-size

Russula farinipes ½ life-size

Russula perlactea ⅔ life-size

Russula seperina ¾ life-size

Russula viridofusca Grund **Cap** 5–12cm
(2–4¾in) across, broadly convex then
depressed at disc, margin tuberculate-striate;
olive-green to ochre-brown, usually darker at
center; viscid when wet, dull when dry; peels
up to halfway. **Gills** close; pale yellow. **Stem**
45–85 × 18–25mm (1¾–3¼ × ¾–1in); white,
bruising bright rust-brown when injured. **Flesh**
firm; white bruising brown. **Odor** not
distinctive. **Taste** mild. **Spores** ovoid, 10–13 ×
8–10μ; warts up to 1μ high, isolated. Deposit
cream (B). **Habitat** under hemlock. Found in
Washington State. **Season** September–
October. **Edibility not known.**

Russula simulans Burlingham **Cap** 4–10cm
(1½–4in) across, soon flattened-convex; leaf-
green to pale yellow-green, often with faint
flush of lilac or violet; smooth; cuticle peels
halfway. **Gills** close, forking near stem; white.
Stem 50–80 × 10–20mm (2–3 × ½–¾in);
white. **Flesh** white. **Odor** not distinctive.
Taste soon slightly acrid. **Spores** ovoid,
8.5–9.5 × 6–7μ; warts 0.5–1μ high, with
partial connectives. Deposit pure white (A).
Habitat in deciduous woods. Uncommon.
Found in northeastern North America. **Season**
July–August. **Edible. Comment** This species
is very similar to *Russula variata* (p. 119) but
differs in its gills only forking at the base and
in having numerous partial gills; also its cuticle
peels more easily.

Russula raoultii (Quél.) Singer **Cap** 2–6cm
(¾–2¼in) across, fragile; lemon yellow to
yellow-buff; dry, smooth; peels halfway. **Gills**
close, broad; white. **Stem** 20–40 × 5–10mm
(¾–1½ × ¼–½in), clavate; white staining
grayish brown. **Flesh** soft; white. **Odor**
pleasant. **Taste** hot. **Spores** ovoid, 7–8.5 ×
5.5–7μ; warts up to 0.75μ high, almost
complete reticulum. Deposit white (A).
Habitat in deciduous woods. Uncommon.
Found in eastern North America. **Season**
August–September. **Not edible.**

Russula crassotunicata Singer **Cap** 3–8cm
(1¼–3in) across, deeply convex then expanded
with depressed disc; white to yellowish white
or pale buff, staining yellowish brown; viscid
when moist, usually dry and felty; cuticle thick
and rubbery, almost totally separable. **Gills**
adnate, quite distant, narrow; pale yellow,
staining yellow-brown when injured. **Stem**
35–50 × 9–20mm (1¼–2 × ⅜–¾in), equal,

solid then spongy; white, staining yellow-
brown; dry, dull, almost velvety. **Flesh** firm;
white, staining when cut. **Odor** puffball- or
coconut-like. **Taste** acrid to burning. **Spores**
broadly ellipsoid, 8.5–11.5 × 7–9μ; warts up to
1.2μ high, isolated. Deposit white (A). **Habitat**
under conifers. Found in northern North
America from Michigan to Washington.
Season August–September. **Edibility not
known** — not recommended.

Russula earlei Pk. **Cap** 4–10cm (1½–4in) across,
soon flattened-depressed, often irregular in
outline, margin slightly striate when old; pale
straw yellow to ochre-tan; smooth, viscid when
wet, waxy; cuticle peels to one-quarter. **Gills**
adnate, widely spaced, thick, almost waxy; pale
yellow. **Stem** 25–40 × 5–15mm (1–1½ ×
¼–½in), equal, spongy within; yellowish;
smooth. **Flesh** pale yellow. **Odor** not
distinctive. **Taste** mild. **Spores** small,
subglobose, 6–7 × 4–5μ; warts isolated, minute,
usually less than 0.25μ high. Deposit white
(A). **Habitat** usually under mixed deciduous
trees. Apparently quite common in eastern
states but rarely reported. **Season** August–
September. **Edibility not known. Comment**
This species is remarkable in appearance,
looking like a cross between a hygrophorus and
a lactarius, very little like a russula. It is the
only member of the primitive group the
Archaeinae found in the United States; other
members are from Africa and Madagascar.

Russula farinipes Romell **Cap** 3–6cm (1¼–2¼in)
across, flattened, thinnish but tough and
almost elastic, margin slightly sulcate-
tuberculate; pale yellow to straw; cuticle hardly
peeling. **Gills** rather distant, narrow; pale
yellow. **Stem** 30–60 × 10–15mm (1¼–2¼ ×
½in), hard and rigid; white to straw yellow;
distinctly pruinose-farinose above. **Flesh**
brittle; yellowish. **Odor** fruity. **Taste** very hot.
Spores ovoid, 6.5–8 × 5–7μ; warts up to 0.75μ
high, isolated. Deposit white (A). **Habitat**
usually under deciduous trees. Rather rare.
Found in eastern to southeastern North
America. **Season** July–August. **Edibility
suspect** — not recommended.

Russula perlactea Murrill **Cap** 3–8cm (1¼–3in)
across, soon flattened-depressed; pure white
with a creamy yellow center; dry, smooth.
Gills close; broad; white. **Stem**
25–80 × 10–15mm (1–3 × ½in), fragile; white.

Flesh white. **Odor** pleasant. **Taste** instantly,
painfully hot. **Spores** ovoid, 9–10 × 7–8.5μ;
warts 0.7–1.3μ high, mostly isolated with a few
fine connectives. Deposit white (A). **Habitat** in
mixed woods. Found in New York and New
Jersey. **Season** September–October. **Not
edible. Comment** A beautiful species easily
recognized by the combination of white cap
and hot taste.

Russula seperina Dupain **Cap** 5–10cm (2–4in)
across; purplish red to ochraceous at the
center; smooth, margin striate; peels halfway.
Gills close; pale ochre, blackening.
Stem 50–60 × 10–20mm (2–2¼ × ½–¾in);
white, bruising pinkish red then black. **Flesh**
white, reddening, then gray or black. **Odor**
slightly fishy. **Taste** mild. **Spores** ellipsoid,
8–9.5 × 6.7–8μ; warts 0.5–1μ high, with well-
developed connectives. Deposit deep ochre
(G–H). **Habitat** in mixed woods. Found in
northern New York State. **Season** August.
Comment This is a European species, but the
specimens illustrated agreed in all essential
details with European descriptions.

Russula bicolor ⅔ life-size

Russula ochroleucoides ½ life-size

Russula ochroleucoides Kauffman **Cap** 6–12cm (2¼–4¾in) across, convex, soon flattened; straw yellow to ochraceous, dull ochre at center; dry, pulverulent. **Gills** close, narrow; white. **Stem** 40–60 × 15–20mm (1½–2¼ × ½–¾in), equal, rigid; subpruinose, white. **Flesh** hard; white. **Odor** faint, pleasant. **Taste** slowly bitter, unpleasant. **Spores** ovoid, 8.5–10 × 7–8μ; warts 0.4–0.8μ high, nearly complete reticulum. Deposit white (A). **Habitat** under mixed deciduous trees. Common. Found widely in eastern North America. **Season** June–October. **Edible.** **Comment** Similar to *Russula flavida* (below), but the latter has brighter colors (stem is also yellow) and darker spores.

Russula bicolor Burlingham **Cap** 4–8cm (1½–3in) across, convex then flattened; surface copper red mixed with yellow-orange or pale ochre; smooth, viscid when wet; cuticle separable for one-quarter of radius. **Gills** subcrowded, broad; white. **Stem** 30–70 × 10–20mm (1¼–2¾ × ½–¾in), spongy; white. **Flesh** white. **Odor** not distinctive. **Taste** acrid.

Spores ovoid, 8–10 × (6)7–8μ; warts less than 0.5μ high, almost no connecting lines. Deposit white (A). **Habitat** particularly under birch. Found in both eastern and western North America. **Season** August–September. **Edible.**

Yellow Swamp Russula *Russula claroflava* Grove syn. *Russula flava* Romell **Cap** 4–10cm (1½–4in) across, convex then a little flattened; bright yellow to pale lemon yellow; smooth, viscid when wet, otherwise dry, firm. **Gills** adnexed to almost free, often forked near stem; palish yellow, graying with age. **Stem** 40–100 × 10–20mm (1½–4 × ½–¾in), soft but not fragile; white, turning gray with age or bruising (very slowly). **Flesh** white turning gray. **Odor** pleasant. **Taste** mild to slightly acrid. **Spores** ovoid, (7.5)9–10 × (6)7.5–8μ; warts up to 1μ high, joined by numerous fine lines to form a fairly well-developed network. Deposit pale ochre (F). **Habitat** in wet, swampy places in mixed woods. Found widely in North America. **Season** July–September. **Edible — good.**

Russula flavida Frost & Pk. **Cap** 3–8cm (1¼–3in) across, convex then plane; firm, fleshy; chrome yellow to orange-yellow; dry, velvety pruinose. **Gills** rather crowded; white then cream. **Stem** 35–80 × 10–15mm (1¼–3 × ½in), equal, firm; colored like cap or a little paler. **Flesh** white. **Odor** pleasant. **Taste** pleasant. **Spores** subglobose, 5–7 × 5.5–8.5μ; warts up to 0.6μ high, with partial to complete reticulum. Deposit yellow (D–E). **Habitat** in deciduous woodlands. Occasional. Found in southeastern North America from New York south to Florida and west to Texas. **Season** July–September. **Edible.**

Russula amoenolens Romagnesi **Cap** 3–10cm (1¼–4in) across, globose then expanded-depressed, margin striate-tuberculate; deep grayish yellow to brown, often spotted with darker reddish brown; viscid when wet; cuticle thin, peels one-quarter to one-half the radius. **Gills** adnate, rather crowded; pale yellow-ochre. **Stem** 35–70 × 10–27mm (1¼–2¾ × ½–1in), firm, soon hollow, brittle; yellowish white, soon stained strongly yellowish brown at base. **Odor** rancid, cheesy. **Taste** oily, unpleasant, slowly very hot. **Spores** ellipsoid, 6–8.5 × 4.5–7μ; warts up to 1μ high, with few connectives, rarely a partial reticulum. Deposit pale orange-yellow (C–D). **Habitat** in mixed woods. Found in northeastern North America. **Season** July–October. **Not edible.**

Russula simillima Pk. **Cap** 3–8cm (1¼–3in) across, convex-flattened, margin tuberculate; pale ochre to yellow-brown or buff; smooth; cuticle peels up to two-thirds. **Gills** close; pale buff-yellow. **Stem** 25–60 × 5–20mm (1–2¼ × ¼–¾in), equal; pale buff-yellow. **Flesh** yellowish white. **Odor** not distinctive. **Taste** rather acrid. **Spores** broadly ellipsoid, 6.5–9 × 5.5–7μ; warts 0.4–1.0μ high, with partial to complete reticulum. Deposit white (A–B). **Habitat** in beech woods. Frequent. Found in eastern North America. **Edible.** **Comment** Singer places this as a subspecies of *Russula fellea* (Fr.) Fr., a European species differing only in its odor of household geraniums and slight microscopic differences.

Yellow Swamp Russula *Russula claroflava* ⅔ life-size

Russula flavida ½ life-size

Russula amoenolens ½ life-size

Russula simillima ⅔ life-size

Russula decolorans ⅔ life-size

white. **Odor** slightly oily or fetid. **Taste** mild or slightly acrid. **Spores** ovoid, 7.5–9 × 5.5–7.5μ; warts 0.4–0.6μ high, with an incomplete reticulum. Deposit cream (D–E). **Habitat** in mixed hardwoods. Quite common. Found in northeastern North America, west to Michigan, south to North Carolina. **Season** July–September. **Not edible.**

Russula puellaris Fr. **Cap** 2.5–5cm (1–2in) across, convex then flat with a central depression, fragile, margin sulcate; livid purple, vinaceous, rosy or brick red, often pale and washed out, often darker at center; cuticle peels halfway or more. **Gills** moderately spaced; pale ochre. **Stem** 25–60 × 5–15mm (1–2¼ × ¼–½in), equal to clavate, soft; white, usually staining entirely yellow-ochre, as does cap. **Flesh** fragile; white. **Odor** pleasant. **Taste** mild. **Spores** ovoid, 6.5–9 × 5.5–7μ; warts up to 1.25μ high, with very few or no connectives. Deposit pale yellow (D–E). **Habitat** in mixed deciduous woods. Frequent. Found in eastern North America. **Season** July–September. **Edible.**

Russula ballouii Pk. **Cap** 2–5cm (¾–2in) across, broadly convex then flattened or slightly depressed in the center; yellow-ochre to tawny; surface dry and cracking into many minute mosaic-like patches. **Gills** adnate, close; pale cream. **Stem** 20–50 × 8–10mm (¾–2 × ⁵⁄₁₆–½in), equal; colored like cap; base with cracked surface. **Flesh** fragile; creamy white. **Odor** not distinctive. **Taste** mild. **Spores** ovoid, 8–10 × 7–8μ; warts up to 0.5μ high, partial reticulum. Deposit white (A). **Habitat** in mixed deciduous woods. Rather rare. Found widely distributed in eastern North America. **Season** August–September. **Edibility not known.**

Bare-toothed Russula *Russula vesca* Fr. **Cap** 5–10cm (2–4in) across, globose then convex and flattened; fleshy, firm; variable in color but usually a pinkish buff or brown, often with pale wine or olive tints; the cap skin peels halfway, thus exposing the slightly toothed white margin below. **Gills** adnexed, rather crowded, narrow, forked near stem; whitish to very pale cream. **Stem** 30–100 × 15–25mm (1¼–4 × ½–1in), firm, often tapered at base; white. **Flesh** white. **Odor** not distinctive. **Taste** mild, nutty. **Spores** ovoid, 6–8 × 5–6μ; small warts up to 0.5μ high, sometimes with short lines attached or joining pairs. Deposit white (A). **Cap cystidia** cylindrical to spindle-shaped, without septa, hardly reacting to SV. **Cap hyphae** with cylindrical or tapering terminal cells or sometimes a long, tapering, thick-walled hair; supporting rectangular cells. **Habitat** under deciduous trees. Occasional. Found in northeastern North America. **Season** July–September. **Edible — good. Comment** Gills and stem surface rapidly turn deep salmon pink when rubbed with an iron salt (FeSO$_4$).

Russula montana Schaef. **Cap** 3.5–7cm (1¼–2¾in) across, convex-flattened, margin rarely striate; deep red to grayish red or reddish brown, sometimes with discolored areas; smooth; cuticle peels up to two-thirds cap radius. **Gills** fairly close, fragile; white. **Stem** 25–50 × 10–35mm (1–2 × ½–1¼in), equal to clavate; white. **Flesh** soft; white. **Odor** none or slightly fruity. **Taste** strongly acrid. **Spores** subglucose, 7–10 × 6–8μ; warts up to 0.4μ high, with more or less complete reticulum. Deposit white to slightly cream (A–B). **Habitat** under conifers. Found in Colorado. **Season** July–August. **Not edible. Comment** Similar to *Russula silvicola* (p. 130) in the eastern states, but differing in its duller gray-red to brown-red cap.

Russula decolorans (Fr.) Fr. **Cap** 4–11cm (1½–4½in) across, subglobose then broadly convex, becoming flat with a slight depression, margin slightly sulcate when old; reddish orange to yellow-orange or brick, stains darker brown or blackish when injured; smooth, hardly peeling. **Gills** strongly interveined; pale ochre, blackening. **Stem** 45–100 × 10–25mm (1¾–4 × ½–1in), often clavate; white, graying. **Flesh** white, strongly graying. **Odor** not distinctive. **Taste** mild. **Spores** ovoid, 9–14 × 7–12μ; warts up to 1.5μ high, very few connectives. Deposit deep yellow (E–F). **Habitat** under conifers. Found in northern North America. **Season** August–October. **Edible.**

Russula pectinatoides Pk. **Cap** 3–8cm (1¼–3in) across, convex to centrally depressed, margin strongly striate-tuberculate; yellowish brown to dull straw color or cinnamon; viscid when wet; cuticle peels halfway or more. **Gills** thin; white to pale cream. **Stem** 25–50 × 5–10mm (1–2 × ¼–½in), hollow, equal; white to pale yellowish or brown where bruised. **Flesh**

Russula pectinatoides ¼ life-size

Russula puellaris ¾ life-size

Russula ballouii ⅔ life-size

Bare-toothed Russula *Russula vesca* ⅓ life-size

Russula montana ¾ life-size

Russula pulchra life-size

Russula pulchra Burlingham **Cap** 4–10cm (1½–4in) across, soon flattened to slightly depressed at center; surface scarlet red to cerise, becoming pale peach or cream at center; dry, pulverulent, often minutely cracking; cuticle hardly peeling. **Gills** equal, forked at stem; white. **Stem** 30–100×15–25mm (1¼–4×½–1in), equal to pointed at base, firm to spongy. **Flesh** white. **Odor** nil. **Taste** mild. **Spores** ellipsoid, 8.7–10×7–7.5μ; warts up to 0.8μ high, with a few connectives, no real reticulum. Deposit pale cream (B–D). **Habitat** under mixed deciduous trees, especially beech and oak. Quite common. Found in eastern North America. **Season** July–September. **Edible. Comment** Distinguished by means of its dry, pruinose, bright red cap cracking into areolae, combined with its mild taste.

Russula persicina Krombh. **Cap** 4–13cm (1½–5in) across, soon flattened and depressed; fleshy; bright red to scarlet or blood, even garnet red, often decolored at center; glabrous; cuticle hardly peels. **Gills** thick, rather crowded; pale cream. **Stem** 25–80×10–25mm (1–3×½–1in), fleshy; white, sometimes with a flush of pink, browning slightly with age. **Flesh** white. **Odor** distinctive. **Taste** mild at first, then somewhat acrid. **Spores** ovoid, 6.5–9.2×5.7–7.5μ; warts up to 1μ high, isolated, with almost no connectives. Deposit cream (C–D). **Habitat** under beech and conifers. Apparently widespread. Found in eastern North America but rarely recorded. **Season** August–October. **Not edible. Comment** This is undoubtedly one of a number of species that have gone under the catchall name of *Russula emetica* (p 130), from which it differs in spore color and ornamentation.

Russula aquosa Leclair **Cap** 3–12cm (1¼–4¾in) across, broadly convex then flattened-depressed, margin tuberculate-sulcate; pale cherry red to pinkish purple, often pallid, watery brown at center; viscid when wet; cuticle peels halfway to completely. **Gills** distant, broad; white. **Stem** 40–100×10–25mm (1½–4×½–1in), usually quite clavate, very fragile; dirty white. **Flesh** white. **Odor** faint, suggesting coconut. **Taste** slightly hot. **Spores** subglobose, 7–8.5×6–7μ; warts up to 0.75μ high, partial reticulum. Deposit white (A). **Habitat** in swamps and marshes. Found throughout northeastern North America. **Season** July–October. **Edibility suspect** — not recommended.

Russula humidicola Burlingham **Cap** 3–6cm (1¼–2¼in) across, soon depressed at center, margin soon striate-tuberculate; salmon, reddish yellow, to darker red at center; glabrous, viscid when wet; cuticle peeling almost entirely. **Gills** close, narrow; pale cream. **Stem** 30–50×5–10mm (1¼–2×¼–½in); white. **Flesh** spongy; white. **Odor** not distinctive. **Taste** mild. **Spores** ovoid, 7–8×5.5–6.5μ; warts 0.4–1.0μ high, with almost full reticulum. Deposit medium yellow (E–F). **Habitat** in mixed woods. Uncommon. Found in eastern North America. **Season** July–September. **Edible.**

Russula persicina ¾ life-size

Russula aquosa ⅓ life-size

Russula humidicola ½ life-size

Russula silvicola ¾ life-size

Russula emetica ¾ life-size

Russula silvicola Shaffer **Cap** 2–8cm (¾–3in) across, convex to flattened; bright red to pinkish red, reddish orange; smooth, dry; peeling easily. **Gills** close, broad; white. **Stem** 20–80×4–15mm (¾–3×³⁄₁₆–½in), clavate; white. **Flesh** soft; white. **Odor** fruity. **Taste** very hot. **Spores** broadly ovoid, 6–10.7× 5.3–9μ; warts up to 1.2μ high, partial to complete reticulum. Deposit white (A). **Habitat** in mixed woods, often on rotten logs. Very common. Found throughout northern and eastern North America. **Season** July–October. **Not edible.**

Russula emetica (Schaeff. ex Fr.) S. F. Gray **Cap** 5–10cm (2–4in) across, convex, margin sulcate; bright scarlet red, cherry, or blood red; peels easily. **Gills** broad; pure white to pale cream. **Stem** 50–100×10–25mm (2–4×½–1in), clavate, fragile; pure white. **Flesh** white. **Odor** pleasant, fruity. **Taste** very hot. **Spores** ovoid, 9–11×7.5–8.5μ; warts large, conical 1–1.25μ high, almost complete reticulum present. Deposit white (A). **Habitat** in swampy conifer woods. Found in northern North America. **Season** July–October. **Not edible.**

Russula cessans Pearson **Cap** 3–8cm (1¼–3in) across, soon flattened; deep crimson to purplish red, often darker at the center; smooth, dry; peels halfway. **Gills** fairly close; pale ochre. **Stem** 30–50×10–20mm (1¼–2×½–¾in), equal; white. **Flesh** white. **Odor** not distinctive. **Taste** mild. **Spores** broadly ovoid, 8–9×7–8μ; warts 0.6–1μ high, many connectives forming a partial reticulum. Deposit deep ochre (G). **Habitat** under eastern white pine. Found in New Jersey. **Season** September–November. **Edible. Comment** This collection agrees very well with the original description of Pearson and other British authors. Recent descriptions by European authors seem to include a much wider range of cap colors; they are possibly describing a mixture of species.

Russula subfragiliformis Murr. **Cap** 5–8cm (2–3in) across, convex; red to pinkish red; smooth, slightly viscid; peeling almost completely. **Gills** widely spaced; deep ochre. **Stem** 20–50×10–15mm (¾–2×½in), equal; white to partly pink. **Flesh** white. **Odor** none. **Taste** mild. **Spores** ellipsoid, 6–8×5–7.5μ; warts 0.5–1μ high, few connectives. Deposit medium ochre (E–F). **Habitat** under hardwoods. Uncommon. Found in eastern North America from New Jersey to Florida. **Season** July–August. **Edible.**

Russula rubescens Beardslee **Cap** 5–9cm (2–3½in), convex then flattened, thin-fleshed, striate at margin; scarlet red; dull, dry. **Gills** rather crowded, forking near base; pale cream. **Stem** 30–80×5–10mm (1¼–3×¼–½in), equal; white, turning red when scratched, then black. **Flesh** white then gray-black. **Odor** pleasant. **Taste** mild. **Spores** ovoid, 7–10×6–8μ; warts up to 1.5μ high, connectives few or absent. Deposit pale yellow (E–F). **Habitat** in wet, deciduous woodlands, especially oak. Locally common. Found in eastern North America. **Season** August–September. **Edibility not known. Comment** The very similar *Russula nigrescentipes* Pk. differs in its more crowded, white gills, and blackening stem without the initial red change.

Russula silvicola ¾ life-size

Russula cessans ⅔ life-size

Russula subfragiliformis ½ life-size

Russula rubescens ½ life-size

Russula pseudolepida ¾ life-size

Stem 45–100×8–24mm (1¾–4×⁵⁄₁₆–1in), clavate; white flushed strongly old rose to bright pink; dry, dull. **Flesh** white. **Odor** not distinctive. **Taste** mild to bitter. **Spores** ovoid, 6.5–9.5×5.6–8.4µ; warts up to 1.2µ high, isolated to partially reticulate. Deposit white (A–B). **Habitat** in mixed woods. Found throughout northern and eastern North America. **Season** July–October. **Not edible.**

Russula roseipes (Secr.) Bres. **Cap** 4–7cm (1½–2¾in) across, convex then flatter with a central depression; rosy pink to orange-rose, often with tiny whitish spots, fading with age; dry, dull, pruinose. **Gills** subdistant; white. **Stem** 30–60×5–10mm (1¼–2¼×¼–½in), clavate; white speckled with rose-pink. **Flesh** soft; white. **Odor** pleasant. **Taste** mild. **Spores** ovoid, 7.5–9.5×6–8µ; warts below 0.5µ high. Deposit deep yellow (E–F). **Habitat** under deciduous trees. Common. Found in eastern North America. **Season** July–September. **Edible.**

Russula paludosa Britz. **Cap** 6–15cm (2¼–6in) across, convex, margin tuberculate with age; bright yellow-orange to brick red or tawny; smooth, viscid when wet. **Gills** crowded, broad, forking near stem; pale yellow, graying when injured. **Stem** 45–120×14–45mm (1¾–4¾×½–1¾in), equal; white or flushed pinkish; dry. **Flesh** white. **Odor** not distinctive. **Taste** mild to quite acrid. **Spores** ovoid, 7.5–11×6.5–9µ; warts up to 1.5µ high, with partial reticulum. Deposit pale orange-yellow (E–F). **Habitat** often in sphagnum moss in conifer woods. Found in eastern North America and west to Michigan. **Season** August–September. **Edible.**

Rosy Russula *Russula sanguinea* (Bull. ex St. Amans) Fr. syn. *Russula rosacea* (Pers. ex Secr.) Fr. **Cap** 5–10cm (2–4in) across, convex then flattened or depressed; blood to cherry red or rose, often streaked or spotted with pale pink or cream to white areas or entirely pale cream-pink. **Gills** usually slightly decurrent, crowded, forking near stem; pale cream-ochre. **Stem** 40–100×10–30mm (1½–4×½–1¼in); white, tinted pink or red, often with yellow stains. **Flesh** spongy, brittle; white. **Odor** faintly fruity. **Taste** slightly to strongly acrid. **Spores** ovoid, 7–10×6–8µ; warts up to 1µ high, with very few connecting lines. Deposit pale to deep ochre (C–F). **Cap cystidia** cylindrical to narrow club-shaped, often with a bulbous end, with 0–4 septa. **Habitat** under pine. Found widely distributed in North America. **Season** July–October. **Not edible.**

Russula rhodopoda Zvára. **Cap** 3–10cm (1¼–4in) across, convex; fleshy; blood red to deep garnet red, or even brownish red; very viscid and shiny as if lacquered; cuticle easily separable. **Gills** rather crowded, narrow; pale cream to yellowish. **Stem** 30–80×10–25mm (1¼–3×½–1in), equal; white, flushed red or pink over most of lower half. **Flesh** white. **Odor** weak, fruity. **Taste** mild. **Spores** broadly ovoid, 7.5–9.5×6.7–7µ; warts up to 0.5µ high, with partial to complete reticulum. Deposit pale ochre-yellow (D–E). **Habitat** under conifers. Found in the Pacific Northwest. **Season** September–October. **Edibility not known.**

Russula peckii ½ life-size

Russula pseudolepida Singer **Cap** 3–8cm (1¼–3in) across, firm; bright scarlet to pinkish red; dry, subpruinose to velvety; hardly peeling. **Gills** rather crowded, occasionally forking; cream-white. **Stem** 30–60×10–15mm (1¼–2¼×½in), firm, rigid, equal; white flushed in part with vivid rose-red; dry. **Flesh** firm; white. **Odor** nil. **Taste** mild. **Spores** subglobose, 7–9.5×6.3–8µ; warts up to 1.3µ high, almost no connectives. Deposit pale yellow (D–E). **Habitat** in mixed woods. Found in eastern North America. **Season** August–September. **Edibility not known.** **Comment** This species differs from *Russula lepida* Fr. mostly in spore characters; the latter has spores with low (up to 0.5µ high) warts and rather extensive reticulum.

Russula peckii Singer **Cap** 3–10cm (1¼–4in) across, subglobose then convex then flattened-depressed; deep blood red to ruby red, often paler at disc; dry and dull, minutely velvety; peels about halfway. **Gills** close, broad, margin clearly serrated-crenulate; yellowish white.

Russula roseipes ¾ life-size

Russula paludosa ⅓ life-size

Rosy Russula Russula sanguinea ⅔ life-size

Russula rhodopoda ⅔ life-size

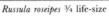

Sweetbread Mushroom or **The Miller**
Clitopilus prunulus ¼ life-size

Clitopilus cretatus ⅔ life-size

Pink-Spored Agarics pp. 134–139.
See Key B, p. 12

This is a small section in the book but one that is readily separated from the white-spored section. As mushrooms in this group mature, the gills will tend to discolor pink. The spore deposit will vary from pale to brownish pink.

Sweetbread Mushroom or **The Miller**
Clitopilus prunulus (Scop. ex Fr.) Kummer **Cap** 3–12cm (1¼–4¾in) across, convex then expanded and often very irregular, margin inrolled, wavy; white to pale cream; dry, felty, like kid leather. **Gills** decurrent; white then pink. **Stem** 15–25×4–12mm (½–1 × ³⁄₁₆–½in), often off-center; concolorous with cap. **Flesh** firm; white. **Odor** strongly mealy, of bread dough or cucumber. **Taste** similar. **Spores** pointed-elliptic, with longitudinal ridges and angular in end view, 9–11(12) × 4–5(7)μ. Deposit pink. **Habitat** in grass in open woodlands. Found throughout most of North America. **Season** June–September. **Edible** — good. **Comment** The spores are absolutely characteristic for this genus.

Clitopilus cretatus (Berk. & Br.) Sacc. **Cap** 0.5–3cm (¼–1¼in) across, flat with a depression, becoming irregularly funnel-shaped with a frequently lobed margin; pale pink to cream or buff with pale cinnamon tinges; smooth, silky-floury. **Gills** adnexed with a decurrent tooth, not crowded; medium; whitish. **Stem** almost none to 2cm (¾in), lateral or eccentric; similar color to cap; smooth. **Spores** ellipsoid, spindle-shaped, with 6–8 longitudinal grooves, 6.5–8.5 × 3.5–5μ. Deposit pink. **Habitat** growing on rotten wood. Rare. Found in the Pacific Northwest. **Season** September–October. **Edibility not known**.

Wood Blewit *Lepista nuda* (Bull. ex Fr.) Cke. syn. *Clitocybe nuda* (Bull. ex Fr.) Bigelow & Smith **Cap** 4–15cm (1½–6in) across, flattened-convex becoming depressed, wavy on the margin; bluish lilac then more brownish and drying paler; smooth, moist to dry. **Gills** adnate to adnexed or notched, close, moderately broad; bluish lilac fading in age to almost buff. **Stem** 30–80 × 10–25mm (1¼–3 × ½–1in), often somewhat bulbous at the base; bluish lilac, with downy purple mycelium at the base; dry, minutely hairy. **Flesh** thick, quite soft; bluish lilac. **Odor** strongly perfumed. **Taste** fragrant, sweet to slightly bitter. **Spores** ellipsoid, minutely spiny, 6–8 × 4–5μ. Deposit pinkish buff. **Habitat** scattered or in groups (often rings or arcs) on compost or organic debris in open areas, gardens, under bushes, and along paths. Common and abundant. Found widely distributed in North America. **Season** August–December (November–March in California). **Edible** — excellent.

Entoloma trachyospermum var. *purpureoviolaceum* Largent **Cap** 1.5–6cm (½–2¼in) across, convex to broadly convex, becoming flatter with an obscure or broad umbo, with a margin decurved then wavy; dark brownish gray with purplish or reddish tinges, fading to more orange-gray on the disc and light brownish orange on the margin; smooth, sometimes minutely velvety, greasy and slippery to the touch, hygrophanous in radial streaks. **Gills** finely adnexed with a decurrent tooth, mostly subdistant but sometimes crowded to close, moderately broad; gray to bluish gray, becoming brownish pink in age. **Stem** 10–90 × 2–8mm (½–3½ × ³⁄₃₂–⁵⁄₁₆in), equal or club-shaped; light bluish gray or grayish violet, base whitish tinged buff in age; greasy or slightly sticky, longitudinally lined, hygrophanous and rippled, base covered in long, soft hairs. **Flesh** pale buff, becoming grayish blue to dark blue on exposure to the air. **Odor** not distinctive or pleasant. **Taste** not distinctive or pleasant. **Spores** ovoid to warty, angular, 5.5–8 × 5.5–7μ. Deposit pinkish. **Habitat** scattered or in groups in needle duff or mossy humus in coniferous forests. Found in Washington. **Season** September–November. **Edibility not known**.

Entoloma sinuatum (Bull. ex Fr.) Kummer **Cap** 5–15cm (2–6in) across, convex with a broad hump, becoming flatter and sometimes wavy in age, with a downcurved margin; dirty cream to dull brownish or grayish; smooth, slightly slippery when wet, sometimes with a faint bloom. **Gills** adnate, close to almost distant, broad; pale grayish yellow, becoming pinkish in maturity. **Stem** 40–150 × 10–25mm (1½–6 × ½–1in); pale grayish; slightly hairy. **Flesh** thick near stalk, firm; white. **Odor** odd, like bad meal or fishy. **Taste** nasty. **Spores** subglobose, angular, with many oil drops, 7–10 × 7–9μ. Deposit salmon pink. **Habitat** scattered or in groups on the ground under conifers and hardwoods. Found widely distributed in North America. **Season** August–September. **Poisonous**. **Comment** Formerly known as *Entoloma lividum*, which has now been split off as a separate species because of the lack of yellow color showing in the gills.

Entoloma griseum (Pk.) Hesler **Cap** 4–5cm (1½–2in) across, convex with a small umbo, becoming flatter, sometimes with a depression, and somewhat upturned in age, umber or grayish, fawn brown; hygrophanous with a lined margin when wet and minutely scaly all over. **Gills** adnexed, crowded, medium broad; grayish white to nut brown. **Stem** 60–100 × 3–4mm (2¼–4 × ⅛–³⁄₁₆in), hollow, fragile, slightly tapering upward; silky white with dingy brown tinges. **Flesh** thin; dark. **Odor** mild. **Taste** mealy. **Spores** angular-spheroid, generally 5–6 sided, 9–12 × 7–9μ. Deposit cinnamon. No pleurocystidia or cheilocystidia. **Habitat** on soil and sphagnum in very damp woods. Found in northeastern North America, south to Tennessee. **Season** August–September. **Not edible**.

Wood Blewit *Lepista nuda* ½ life-size

Entoloma trachyospermum var. *purpureoviolaceum* ⅔ life-size

Entoloma sinuatum ⅔ life-size

Entoloma griseum ½ life-size

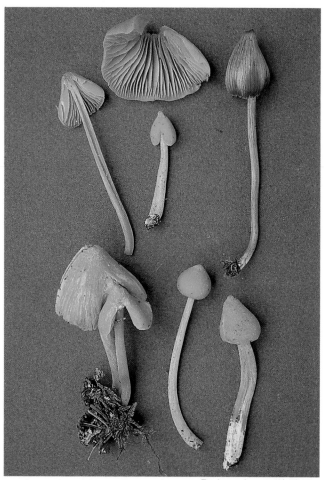

Entoloma salmoneum ⅔ life-size

Entoloma murraii ⅔ life-size

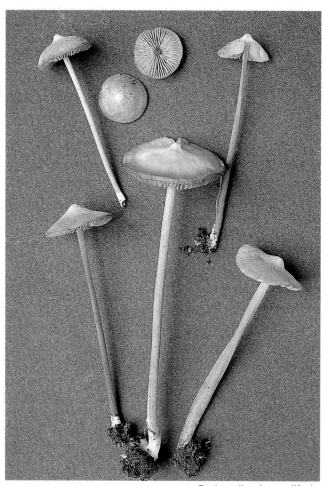

Entoloma alboumbonatum life-size

Aborted Entoloma *Entoloma abortivum* ½ life-size

Entoloma salmoneum (Pk.) Sacc. **Cap** 1–4cm (½–1½in) across, bell-shaped to cone-shaped with a distinctly pointed tip; salmon orange, fading with age; smooth, moist. **Gills** adnexed, subdistant, broad; same color as cap. **Stem** 40–100×2–6mm (1½–4×³⁄₃₂–¼in), hollow, sometimes compressed; salmon orange tinged greenish; with a faint bloom at the top and lined with fine hairs. **Flesh** thin, fragile; orange to salmon. **Odor** mild. **Taste** mild. **Spores** angular or almost square in section, 10–12×10–12μ. Deposit salmon pink. No pleurocystidia. **Habitat** scattered on moist or mossy soil, humus, or leaf litter under conifers or hardwoods. Found in northeastern North America, south to North Carolina and west to Ohio. **Season** June–October. **Poisonous** to some people.

Entoloma murraii (Berk. & Curt.) Sacc. **Cap** 1–3cm (½–1¼in) across, conical, bell-shaped with spikelike nipple at center; yellow to yellow-orange; silky, smooth. **Gills** adnate, distant, broad; pale yellow, then pinkish as spores mature. **Stem** 50–100×1.5–5mm (2–4×³⁄₃₂–¼in); dry, yellow. **Flesh** thin; pale yellow. **Odor** pleasant. **Taste** pleasant. **Spores** 4-sided, smooth, 9–12×8–10μ. Deposit salmon pink. **Habitat** in damp woods and swamps. Frequent. Found from Maine to Alabama and west to the Great Lakes. **Season** July–October. **Edibility uncertain** — best avoided.

Entoloma alboumbonatum Hesler **Cap** 1–3.5cm (½–1¼in) across, broadly cone- or bell-shaped with a distinct umbo and an incurved margin; dull pinky-buff to pale pinky-cinnamon with a grayish tinge, the umbo whitish; hygrophanous, shining with silky hairs in radial lines. **Gills** narrowly adnate-ascending, close, medium broad; pale whitish to pale pinky-cinnamon. **Stem** 50–80×1–3mm (2–3×¹⁄₁₆–¹⁄₈in), firm, tough, hollow, slightly enlarged at the base; grayish brown, paler toward the top, whitish mycelium at the base; dry, smooth but minutely hairy toward the top, splits easily lengthwise. **Taste** faintly mealy. **Spores** generally 4-sided, 8.5–10×7–8.5μ. Deposit pinkish. Pleurocystidia and cheilocystidia present. No caulocystidia. **Habitat** growing on the ground or in humus under deciduous trees and hemlock. Found in eastern North America. **Season** June–August. **Not edible**.

Aborted Entoloma *Entoloma abortivum* (Berk. & Curt.) Donk **Fruit body** in two forms, one of which is parasitized. The parasitized form is described last. **Normal form** — **Cap** 4–10cm (1½–4in) across, convex with inrolled edge and central umbo; gray to gray-brown; smooth, dry, innately fibrillose. **Gills** sinuate, crowded; pale gray, becoming pink with mature spores. **Stem** 30–100×5–15mm (1¼–4×¼–½in), slightly clavate, often not central to cap; white to gray, base with white coating of mycelium. **Flesh** white. **Odor** mealy, cucumber-like. **Taste** similar. **Spores** ellipsoid, angular, 6-sided, 8–10×4.5–6μ. Deposit salmon pink. **Aborted form** — An irregular lumpy white ball often in large numbers along with the typical form; when cut open a spongy context lined with pink veins and patches is revealed. Size from 2–10cm (¾–4in). **Habitat** Common. Found widely in eastern North America west to Texas. **Season** August–October. **Edible** with caution.

Leptonia incana (Fr.) Quél. **Cap** 1–4cm (½–1½in) across, convex, depressed at center; green to greenish yellow; glabrous or minutely squamulose at center. **Gills** adnate, distant, broad; light green. **Stem** 20–40×1–3mm (¾–1½×¹⁄₁₆–¹⁄₈in); bright green bruising blue. **Flesh** pale green bruising blue. **Odor** of mice.

Taste mild. **Spores** angular-ellipsoid, 11–14×8–9μ. Deposit pink. **Habitat** in grassy areas. Found widely distributed east of the Great Plains. **Season** August–October. **Not edible**.

Alboleptonia sericella (Fr.) Largent & Benedict syn. *Entoloma sericellum* var. *sericellum* (Fr.) Kummer **Cap** 1–2.5cm (½–1in) across, convex becoming expanded, sometimes with a depressed disc, margin sometimes slightly lined in age; white becoming pinkish; dry, sometimes felty to scaly on the disc. **Gills** adnate often with a decurrent tooth, close to subdistant, broad to moderately broad; white then pale pinkish tawny. **Stem** 15–60×1–3mm (½–2¼×¹⁄₁₆–¹⁄₈in), solid or becoming hollow; whitish becoming dingy or vinaceous in age; smooth, silky, with a bloom toward the top. **Flesh** thin, fragile; white. **Odor** mild. **Taste** mild. **Spores** nodulose, 8.5–11×6–7.5μ. Deposit pinkish. **Habitat** growing on soil under conifers and hardwoods. Found in central and eastern North America and in Washington. **Season** July–September. **Not edible**.

Entoloma strictipes (Pk.) Hesler **Cap** 1–4cm (½–1½in) across, broadly cone-shaped becoming flatter, with an umbo; nut brown or darker; margin has short lines which then extend to the center, surface scaly. **Gills** adnate becoming sinuate, close, quite broad; whitish to fleshy pink. **Stem** 60–105×2–3mm (2¼–4¼×³⁄₃₂–¹⁄₈in), hollow, slightly tapering upward; whitish to pale dingy brown, with white mycelium on the base; smooth. **Flesh** thin, fragile; pale brown. **Odor** none. **Taste** slightly sweet. **Spores** generally 5–6 sided, 9–14.5×7–9μ. Deposit cinnamon pink. No pleurocystidia. **Habitat** on sphagnum moss. Frequent. Found in eastern North America from New York to West Virginia. **Season** July–August. **Not edible**.

Rhodocybe hirneola (Fr.) Orton **Cap** 1–3cm (½–1¼in) across, convex becoming flatter, often depressed; gray to dirty gray-brown, frosted whitish on the crown. **Gills** curvingly decurrent; gray to brownish. **Stem** 20–35×10–30mm (¾–1¼×½–1¼in), slender; gray to gray-brown; sometimes with spots or patches, base covered in white hairs. **Spores** finely warty, 6.5–9×5–6.5μ. Deposit gray-brown. **Habitat** in deciduous and mixed woods. Rare. Found in Washington. **Season** September–October. **Edibility not known**.

Leptonia incana ²⁄₃ life-size

Alboleptonia sericella ½ life-size

Entoloma strictipes ⅓ life-size

Rhodocybe hirneola ¾ life-size

Fawn-colored Pluteus *Pluteus cervinus* ½ life-size

Pluteus flavofuligineus life-size

Pluteus atromarginatus just over life-size

Pluteus petasatus ¾ life-size

Pluteus admirabilis ½ life-size

Fawn-colored Pluteus *Pluteus cervinus* (Fr.)
Kummer **Cap** 3–15cm (1¼–6in) across,
convex with a flattened center, then expanding;
gray-brown to dark brown with darker innate
fibers radiating from center; smooth, often
slightly squamulose at center. **Gills** free,
broad, crowded; pallid then soon pink. **Stem**
50–120 × 5–15mm (2–4¾ × ¼–½in); white but
with darker brownish fibrils below. **Flesh**
white. **Odor** not distinctive. **Taste** not
distinctive. **Spores** smooth, 5.5–7 × 4.5–5µ.
Deposit deep salmon pink. Pleurocystidia
abundant, prominent, spindle-shaped with
hornlike projections at tip. **Habitat** on dead
timber of hardwoods or conifers. Found
throughout North America. **Season**
July–October. **Edible** but poor quality.

Pluteus flavofuligineus Atkinson **Cap** 2–8cm
(¾–3in) across, bell-shaped then flattened;
deep blackish yellow to yellow-green, then
paler, more yellow-ochre as it expands; with
granular or velvety surface. **Gills** broad,
crowded; pallid then pink. **Stem**
40–100 × 4–8mm (1½–4 × ³⁄₁₆–⁵⁄₁₆in), equal;
pinkish, then yellow with age. **Flesh** yellowish.
Odor not distinctive. **Taste** not distinctive.
Spores ovoid, smooth, 6–7 × 4.5–5.5µ. Deposit
pink. Pleurocystidia abundant, with acute
point. **Habitat** on fallen hardwoods. Found
mostly in northeastern North America. **Season**
July–October. **Edible**.

Pluteus atromarginatus (Konrad) Kühner **Cap**
3–10cm (1¼–4in) across, convex-flattened;
blackish brown with darker streaks and fibrils;
smooth to slightly squamulose at center. **Gills**
broad, crowded; pallid then pink-salmon with
edges dark brown. **Stem** 50–100 × 5–10mm
(2–4 × ¼–½in); white with dark brown fibers
on surface. **Flesh** white. **Odor** pleasant. **Taste**
pleasant. **Spores** ellipsoid, smooth, 6.5–8 ×
4.5–5µ. Deposit brownish pink. Pleurocystidia
present and with horns at their apex. **Habitat**
on fallen conifer logs. Found throughout most
of North America. **Season** July–September.
Edible.

Pluteus petasatus (Fr.) Gillet **Cap** 4–10cm
(1½–4in) across, convex to obtuse; white to
ivory with dark brownish fibrils or scales at
center. **Gills** free, crowded, rather obtuse,
blunt at outer margin of cap; white for a long
time before turning pink. **Stem**
40–100 × 7–15mm (1½–4 × ¼–½in), slightly
broader at base; white to streaked below with
darker fibrils. **Flesh** white. **Odor** pleasant.
Taste pleasant. **Spores** ovoid, smooth,
6–7.5 × 4.5–5µ. Deposit dull pink. **Habitat** on
rotten logs, stumps, or sawdust piles. Found
throughout much of North America. **Season**
July–October. **Edible**.

Pluteus admirabilis Pk. **Cap** 1–3cm (½–1¼in)
across, convex then soon flat, umbonate; bright
yellow, then tawny with age; glabrous. **Gills**
free, close; pallid then pink. **Stem** 30–60 ×
1–2.5mm (1¼–2¼ × ¹⁄₁₆–¹⁄₈in), fragile; pale
yellow; smooth. **Odor** not distinctive. **Taste**
not distinctive. **Spores** broadly ellipsoid,
smooth, 6–7 × 4.5–5.8µ. Deposit dull pink.
Habitat often in groups on decayed hardwood
logs. Common. Found in the Great Lakes area
eastward and south. **Season** July–October.
Edible. **Comment** Cap cuticle is formed of
globose cells. The similar *Pluteus leoninus*
(Schaeff. ex Fr.) Kummer sensu Lange, with
filamentous cuticle, has a white stem.

Brown-Spored Agarics pp. 139–189.
See Key C, p. 12

CORTINARIUS *Mushrooms in this, the largest
genus in America and also the world, all have rust-
colored spores and a cortina (a fine cobweb between the
stem and cap edge). In addition, there is a universal
veil, which covers the whole mushroom when it is
developing; remnants of this may often be seen on the
stem and the cap edge. It is essential to have an
immature specimen and note the color of the young
gills; they will discolor rusty on maturity as the spores
form. The genus is usually divided into seven
subgenera: MYXACIUM The cap and stem are
sticky from glutinous universal veil. The cap cuticle is
often bitter to taste. PHLEGMACIUM Only the
cap is sticky. The stem frequently has a bulb, often
with a prominent edge (marginate).
SERICEOCYBE Caps are often silky or shaggy,
not sticky or changing color when wet.
CORTINARIUS Only two of this subgenus are
listed in Professor Moser's key; both are large and
deep violet all over. LEPROCYBE This subgenus
contains most of the olivaceous and yellowish (large)
cortinarii; many are probably deadly poisonous. They
give a luminous reaction under ultraviolet light.
TELAMONIA This subgenus contains most of the
small brown species. They are characterized by the
way the wet mushroom dries out a different color
(hygrophanous). DERMOCYBE This subgenus
generally has long, narrow stems and conical, fibrous
caps; they are often strongly colored olive-green or red.*

*The spores of cortinarius mushrooms vary greatly in
shape, and the descriptions in the text give an
indication of this by noting the quotient. The quotient
represents the spore's length divided by its width.
Thus, if the quotient is 1, this indicates that the length
and width are the same, meaning the spore is round; a
quotient of 2 means the spore is twice as long as it is
wide..*

Cortinarius pseudosalor Lange subgenus *Myxacium*
Cap 3–9cm (1¼–3½in) across, broadly conical
to convex; ochre-brown to deeper in the center,
rather variable; shiny and glutinous. **Gills**
adnexed; pallid milky coffee, or with a touch of
violet, then more rusty brown. **Stem** 40–100 ×
6–14mm (1½–4 × ¼–½in), equal or narrower
near the base; purply violet, whitish at the
base and apex; with a thick coating of gluten.
Flesh pallid. **Odor** slight. **Taste** mild. **Spores**
broadly almond-shaped, rough, warty,
12–15.5 × 6.7–9µ, quotient 1.75. Deposit rusty
brown. Gills with fat club-shaped cystidia.
Habitat in coniferous or broad-leaved woods.
Rather common. Found in northern North
America. **Season** July–September. **Not edible.**

Cortinarius pseudosalor ½ life-size

Cortinarius corrugatus ½ life-size

Cortinarius collinitus ½ life-size

Cortinarius corrugatus Pk. subgenus *Myxacium*
Cap 5–10cm (2–4in) across, obtuse to
campanulate, then expanding; tawny ochre to
rust-brown; surface viscid, coarsely and
prominently radially wrinkled or corrugated.
Gills adnexed; violaceous when very young but
soon cinnamon brown. **Stem** 60–120 ×
10–20mm (2¼–4¾ × ½–¾ in), with a small
rounded bulb at base; tawny below, paler
above; distinctly viscid over lower half; cortina
fine, white, soon disappearing. **Flesh** firm;
white to buff. **Odor** pleasant. **Taste** pleasant.
Spores warty, 12–15 × 7–9μ, quotient 1.7.
Deposit rust-brown. **Habitat** scattered in moist
deciduous woods, especially beech. Not very
common. Found east of the Great Plains.
Season August–October. **Not edible.**
Comment Although this species is often placed
in the subgenus *Phlegmacium*, the wrinkled cap
and viscid stem place it in *Myxacium*.

Cortinarius collinitus Fr. syn. *Cortinarius mucosus*
var. *coeruliipes* Smith subgenus *Myxacium* **Cap**
2–12cm (¾–4¾in) across, hemispherical then
broadly conical; orangy brown to rusty ochre;
glutinous. **Gills** adnexed; silvery grayish when
young, then rusty brown. **Stem** 50–140 ×
10–20mm (2–5½ × ½–¾in), equal or
narrowing slightly at the base; violet or purple
or whitish; covered in thick coating of gluten,
which may dry into horizontal bands. **Flesh**
pallid with a touch of violet. **Odor** slight.
Taste slight. **Spores** almond-shaped, rough,
12–18 × 6.5–8m, quotient 2.0. Deposit rusty
brown. No clamp connections; no cheilocystidia. **Habitat** in conifer woods. Occasional.
Possibly widespread in North America. **Season**
September–October. **Not edible. Comment**
This is a very variable mushroom and seems to

have often been muddled with other close
species.

Cortinarius mucifluus Fr. subgenus *Myxacium*
Cap 3–8cm (1¼–3in) across, convex then
expanding to plane; straw yellowish at first,
then tawny fibrous or olive-brown; glutinous.
Gills adnexed, equal; pallid buff then rusty.
Stem 60–110 × 7–12mm (2¼–4½ × ¼–½in),
equal; white; covered in glutinous veil which
sometimes dries up into bands. **Flesh** pallid to
creamy. **Odor** slight. **Taste** mild. **Spores**
lemon-shaped, warty, 12.5–17 × 6.5–8μ,
quotient 2.05. Deposit rusty brown. No clamps;
no cheilocystidia. **Habitat** in swampy, wet
areas in conifer woods. Occasional. Found in
northern North America. **Season** September–
October. **Not edible**.

Cortinarius arvinaceus Fr. subgenus *Myxacium*
Cap 4–8cm (1½–3in) across, hemispheric
for a long time, unexpanded with margin
inrolled, sometimes with an umbo; yellow-
ochre, more reddish ochre with age; glutinous.
Gills adnate; pallid, yellowish then rusty.
Stem 50–120 × 7–15mm (2–4¾ × ¼–½in),
usually thicker at the apex, attenuating
downward; fibrous and flaky beneath the
cortina of slime. **Flesh** whitish. **Odor** slight.
Taste mild. **Spores** long elliptical, roughened,
10–14 × 6–7.5μ, quotient 1.8. Deposit rusty
brown. Cheilocystidia bladder-shaped; no
clamps. **Habitat** in coniferous and mixed
woods. Uncommon. Found in the Rocky
Mountains. **Season** August–September. **Not
edible. Comment** This is a European species
not previously recorded as far as I know in
North America. Moser says the gills should
be gray-brown, but Cooke shows yellowish

gills in his illustration.

Cortinarius trivialis Lange subgenus *Myxacium*
Cap 4–11cm (1½–4½in) across, convex then
expanded with umbo; ochraceous tawny to bay
brown; very viscid. **Gills** adnate; pallid clay or
with a slight hint of violet at first, then ochre
to rusty. **Stem** 50–120 × 10–20mm
(2–4¾ × ½–¾in), equal; whitish above, orangy
brown below, with zones of whitish scales; very
glutinous. **Flesh** generally white, but reddish
brown in stem base or where damaged. **Odor**
slight. **Taste** mild. **Spores** almond-shaped,
roughened, 11–12.5 × 5–6.5μ, quotient 2.0.
Deposit rusty brown. **Habitat** in mixed and
deciduous woods, especially damp areas.
Uncommon. Found in North America from
Michigan westward. **Season** September–
October. **Not edible. Comment** Distinguished
from similar species by having some clamp
connections but no cheilocystidia.

Cortinarius oregonensis Smith subgenus *Myxacium*
Cap 2–4.5cm (¾–1¾in) across, broadly
convex becoming flat; pallid buff, buffy yellow
in the center, more violet near the edge;
glutinous. **Gills** adnate; pallid pinkish violet at
first, then pallid, eventually more rusty. **Stem**
40–60 × 6–10mm (1½–2¼ × ¼–½in), slightly
broader at the base; pallid violet at first, then
pallid whitish. **Flesh** watery violet to brownish.
Odor slight. **Taste** mild. **Spores** ellipsoid,
slightly roughened, 6.5–7.6 × 3.7–4.6μ,
quotient 1.70. Deposit rusty brown. **Habitat**
under conifers. Uncommon. Found in
Washington and Oregon. **Season** October–
November. **Not edible. Comment** My
material seems a good fit except that Smith's
stem is much longer, 6–10cm (2¼–4in).

Cortinarius mucifluus ⅔ life-size

Cortinarius arvinaceus ½ life-size

Cortinarius oregonensis almost life-size

Cortinarius trivialis ⅔ life-size

Spotted Cort *Cortinarius iodes* ½ life-size

Spotted Cort *Cortinarius iodes* ½ life-size

Cortinarius sphaerosporus ½ life-size

Cortinarius delibutus life-size

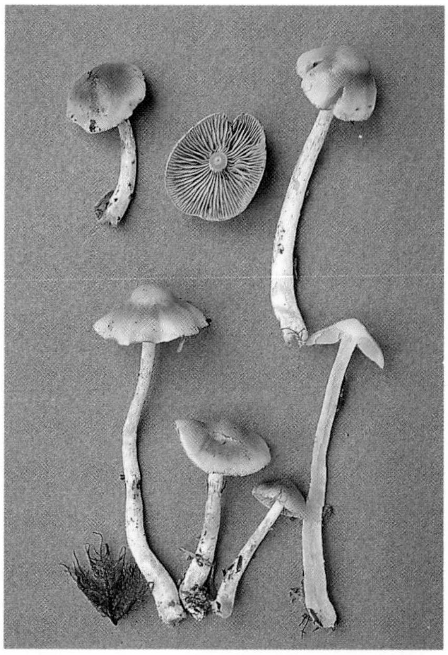

Cortinarius liquidus ⅓ life-size

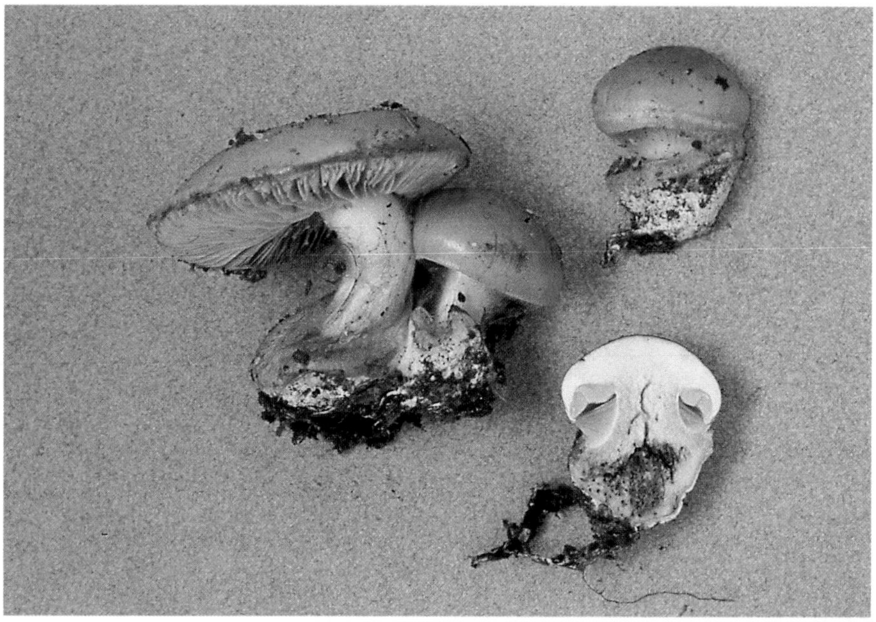

Cortinarius elegantoides ¾ life-size

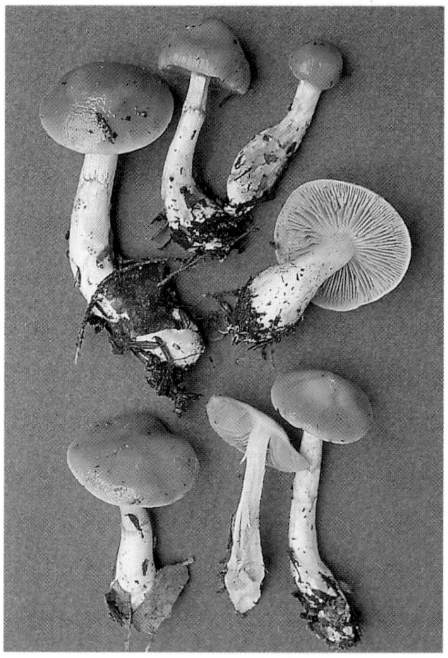

Cortinarius vibratilis ½ life-size

Spotted Cort *Cortinarius iodes* Berk. & Curt. subgenus *Myxacium* **Cap** 2–5cm (¾–2in) across, convex expanding to bell-shaped; deep purple-violet to slightly yellowish on disc, often spotted irregularly with yellowish spots; smooth, very viscid in wet weather. **Gills** adnate, somewhat broad; violet then gray-cinnamon. **Stem** 40–75 × 5–15mm (1½–2¾ × ¼–½in), often irregular and swollen at base; colored as cap; smooth, viscid below like cap, with faint ring zone at apex from cobwebby veil. **Flesh** soft; violet becoming pallid. **Odor** pleasant. **Taste** (of cuticle) mild. **Spores** ellipsoid, minutely roughened, 7–10 × 5–7μ, quotient 1.4. Deposit rust-brown. **Edibility** like all Cortinarius species — not recommended. **Habitat** common and abundant through eastern North America, rare in the Pacific Northwest. **Season** August–September. **Comment** The very similar *Cortinarius iodeoides* Kauffman differs in its bitter cap surface and longer, narrower spores.

Cortinarius sphaerosporus Pk. subgenus *Myxacium* **Cap** 3–7cm (1¼–2¾in) across, convex then plane; yellow; glutinous. **Gills** adnate; violet at first, eventually bright cinnamon. **Stem** 50–100 × 5–10mm (2–4 × ¼–½in), base swollen; white with patches of yellow gluten from the veil adhering, often slightly violaceous at the apex. **Flesh** pallid with a hint of violet, especially when very young. **Odor** slight. **Taste** mild. **Spores** subglobose, rough, 6–8 × 5.5–6.2μ, quotient 1.2. Deposit rusty brown. **Habitat** found in wet coniferous or wet mixed woods. Uncommon in eastern North America, rare in the Pacific Northwest. **Season** August–October. **Not edible. Comment** *Cortinarius delibutus* (below) is extremely similar, but it has slightly larger spores and its gills only mature to a clay color; possibly they are synonymous.

Cortinarius delibutus Fr. subgenus *Myxacium* **Cap** 1.5–4cm (½–1½in) across, hemispherical then convex; golden yellowish with a hint of violet near the edge; glutinous. **Gills** adnate; pallid violet, then clay-colored. **Stem** 40–80 × 5–15mm (1½–3 × ¼–½in), base swollen, clavate; whitish but with yellow patches from the veil; glutinous. **Flesh** pallid. **Odor** pleasant. **Taste** mild. **Spores** subglobose to globose, warty, 7.5–8.5 × 6.5–7.8μ, quotient 1.17. Deposit rusty brown. **Habitat** in coniferous and deciduous woods. Rare. So far found only in Colorado and Maine. **Season** September. **Not edible. Comment** This is a European species that may turn out to have a wider distribution. Extremely similar to *Cortinarius sphaerosporus* (above), which has smaller spores and gills more cinnamon-colored in age; but possibly further study will point to these two species being synonymous.

Cortinarius liquidus Fr. subgenus *Myxacium* **Cap** 1.5–3cm (½–1¼in) across, conical then almost plane with a large umbo; orange on the umbo, brownish around it, and then silvery creamy near the edge; glutinous with pale yellow veil, slightly striate at the extreme edge. **Gills** adnate to adnexed; pale creamy buff at first, then more cinnamon. **Stem** 50–90 × 2–5mm (2–3½ × 3/32–¼in), very long and narrow, attenuating upward; silky white; glutinous at first, soon hollow. **Flesh** pallid buff. **Odor** slight. **Taste** bitter, probably only the gluten. **Spores** ovoid, only lightly roughened, 6.5–7.6 × 4.9–5.6μ, quotient 1.35. Deposit rusty brown. Clamps not seen; cheilocystidia absent. **Habitat** in a swampy area of conifers, alders, etc. Locally abundant. Found in Washington. **Season** October. **Not edible. Comment** To find this mushroom is really exciting. It was originally described by Fries from Sweden in 1838 and then illustrated in his *Icones* in 1884, and as far as I can tell, until now it has not been seen in modern times.

Cortinarius elegantoides Kauffman subgenus *Phlegmacium* **Cap** 4–7cm (1½–2¾in) across, convex then plane; deep yellowish orange; glutinous. **Gills** adnate; pallid yellow at first, then rusty. **Spores** 40–80 × 10–25mm (1½–3 × ½–1in), with a large marginate bulb; yellowish except the bulb, which is clothed in an orange veil. **Flesh** yellow. **Odor** slight. **Taste** bitter. **Spores** lemon- to almond-shaped, very rough, 15–18 × 7.5–9μ, quotient 2.0. Deposit rusty brown. **Habitat** in deciduous or mixed woods. Uncommon. Found in eastern North America to Michigan. **Season** September–October. **Not edible. Comment** Note the very large spores.

Cortinarius vibratilis Fr. subgenus *Myxacium* **Cap** 2–5cm (¾–2in) across, hemispherical then convex to domed; orange, slightly hygrophanous, lighter when dry, darker when wet; glutinous. **Gills** adnate; pallid reddish ochre then more cinnamon. **Stem** 40–70 × 4–10mm (1½–2¾ × 3/16–½in), often swollen at the base, but sometimes tapering and appearing rootlike; white discoloring orangy in places, sometimes showing a ring zone; glutinous. **Flesh** whitish or a touch ochre. **Odor** slight. **Taste** very bitter. **Spores** ellipsoid, roughened, 6–8.4 × 4–5μ, quotient 1.5. Deposit rusty brown. **Habitat** in coniferous or deciduous woods. Frequent. Found in northern North America. **Season** September–November. **Not edible. Comment** The small size and very bitter taste help to identify this species.

Cortinarius calochrous ¾ life-size

Cortinarius elegantior ½ life-size

Cortinarius rapaceus life-size

Cortinarius glaucopus ⅔ life-size

Cortinarius volvatus ¾ life-size

Cortinarius glaucopus (olive type) ½ life-size

orange, eventually fading to dull yellowish brown; glutinous. **Gills** adnate; yellowish at first, then tawny ochre. **Stem** 50–120 × 10–20mm (2–4¾ × ½–¾in), with a marginate bulb; off-white with yellowish tints; fibrous. **Flesh** yellowish. **Odor** slight. **Taste** mild. **Spores** lemon-shaped, very rough and warty, 13.5–16 × 8–9.5μ, quotient 1.65. Deposit rusty brown. **Habitat** in spruce woods. Rare. Found in Colorado. **Season** August. **Not edible. Comment** No reaction with KOH on cap or flesh.

Cortinarius rapaceus Fr. subgenus *Phlegmacium* **Cap** 3–8cm (1¼–3in) across, hemispherical then convex; ivory to creamy ochre; sticky. **Gills** adnexed; pallid with just a hint of pinkish violet at first, then a little more brownish. **Stem** 40–70 × 9–12mm (1½–2¾ × ⅜–½in), usually with a marginate bulb; white with white veil and cortina. **Flesh** white. **Odor** nice, slight. **Taste** slight. **Spores** lemon-shaped, warty, 8.4–9 × 4.8–5.6μ, quotient 1.65. Deposit rusty brown. **Habitat** under conifers. Rare. Found in Washington. **Season** October. **Not edible. Comment** A European species; the closest American species is *Cortinarius albidus* Pk., which has larger spores, 9–11 × 5–6.5μ.

Cortinarius glaucopus (Schaeff. ex Fr.) Fr. sensu Moser subgenus *Phlegmacium* **Cap** 6–13cm (2¼–5in) across; orangy yellow, margin greenish or olivaceous, or more olive-greenish all over; glutinous, markedly innately fibrous. **Gills** adnate; strong bluish violet or more grayish violet. **Stem** 40–80 × 15–30mm (1½–3 × ½–1½in), with a bulb which is sometimes marginate, sometimes not; pallid yellowish, upper part violaceous. **Flesh** pallid yellowish white, usually tinged violaceous, at least in the upper stem. **Odor** slightly of meal. **Taste** slightly of meal. **Spores** ovoid to almond-shaped, warty, 7.5–9(10) × 4.5–6.3μ, quotient 1.6. Deposit rusty brown. **Habitat** rare in deciduous woods in eastern North America, more common (the olivaceous form) in conifer woods in the Pacific Northwest. **Season** August–October. **Not edible. Comment** As will be seen from my description there are differences from collection to collection, some having more or less olive colors, others having more or less violet colors; the spore sizes may also vary a little. Possibly when the American species have been further studied, more than one species will be named from this complex.

Cortinarius volvatus Smith subgenus *Phlegmacium* **Cap** 3–9cm (1¼–3½in) across, convex; drab grayish ochre with hints of violet, especially at the margin; fibrillose, the buttons especially with patches of the whitish universal veil, glutinous. **Gills** adnate; violet then dark grayish lavender or purplish, remaining so for a long time. **Stem** 40–70 × 6–15mm (1½–2¾ × ¼–½in), with a distinctly rounded to marginate bulb, the edge of which has distinct remains of the whitish veil, giving it an appearance of having a volva; above dull ochreous; densely fibrous. **Flesh** dull grayish yellow with streaks of purple. **Odor** slight. **Taste** a touch bitter. **Spores** ovoid, roughened, 7.5–9 × 4.7–5.6μ, quotient 1.6. Deposit rusty brown. **Habitat** under conifers, especially spruce. Rare. Found in the Pacific Northwest. **Season** October–November. **Not edible. Comment** The volva-like remnants on the bulb are a very distinct characteristic.

Cortinarius calochrous Fr. S. F. Gray subgenus *Phlegmacium* **Cap** 4–7cm (1½–2¾in) across, convex; chrome yellow fading to creamy yellow; glutinous. **Gills** adnexed; pale pinky-violet, remaining so for a long time. **Stem** 30–60 × 7–15mm (1¼–2¼ × ¼–½in), with an abrupt marginate bulb; whitish with violet shades at the apex and yellow veil on the bulb. **Flesh** whitish with a touch of violet near the stem apex. **Odor** slight. **Taste** slight. **Spores** almond-shaped, warty, 9–11 × 5.5–6.5μ, quotient 1.65. Deposit rusty brown. **Habitat** in coniferous or mixed woods. Rare. Found in Colorado and east of the Great Plains. **Season** August–September. **Not edible. Comment** KOH on the cap discolors reddish pink. In Europe American specimens showing the violet tints on the stem would be named *Cortinarius calochrous* var. *caroli* (Vel.).

Cortinarius elegantior Fr. subgenus *Phlegmacium* **Cap** 5–13cm (2–5in) across, hemispheric expanding to flat; orangy yellow or reddish

Cortinarius crassus ¾ life-size

Cortinarius multiformis ½ life-size

Cortinarius sodagnitus ¾ life-size

Cortinarius cedretorum ⅓ life-size

Cortinarius crassus Fr. sensu Smith subgenus
Phlegmacium **Cap** 10–20cm (4–8in) across,
slightly convex or flat; buff-colored then
cinnamon brown; soon dry, smooth. **Gills**
adnexed; pallid buff at first, then cinnamon
brown. **Stem** 50–80 × 15–40mm (2–3 ×
½–1½in), equal; whitish; fibrillose. **Flesh** off-
white with brownish areas. **Odor** slight. **Taste**
mild. **Spores** lemon-shaped, lightly roughened,
10–11.6 × 6–6.7μ, quotient 1.7. Deposit rusty
brown. **Habitat** under conifers and possibly
maple. Found in the Pacific Northwest, in
Colorado and other parts of the Rockies, and
in the Great Lakes region. **Season** August–
October. **Not edible. Comment** The flesh goes
yellow with KOH. I could not see cheilocystidia nor could Smith, but in Europe cheilocystidia (albeit looking like basidia) are found.

Cortinarius multiformis (Fr.) Fr. subgenus
Phlegmacium **Cap** 4–10cm (1½–4in) across,
convex then expanding to plane; ochraceous
buff to ferruginous orange; viscid, white hoary
layer when young. **Gills** adnate; pallid cream
at first, then buff, at length rusty. **Stem**
40–90 × 10–20mm (1½–3½ × ½–¾in), at first
with a marginate bulb, which usually becomes
just a swollen clavate bulb as the stem
expands; white then ochre. **Flesh** pallid with
touches of ochre. **Odor** slight, sweetish. **Taste**
mild. **Spores** ovoid, lightly roughened,
8–10 × 5–6μ, quotient 1.65. Deposit red-brown.
Habitat in conifer woods. Occasional. Found
in northern North America and down the
Rockies. **Season** August–October. **Not edible.**
Comment No reaction to KOH.

Cortinarius sodagnitus Henry subgenus
Phlegmacium **Cap** 4–7cm (1½–2¾in) across,
convex; strong violet at first, discoloring with
age to brownish ochre; glutinous. **Gills**
adnexed; violet then gradually pinkish rust.
Stem 50–90 × 10–16mm (2–3½ × ½–¾in),
with a smallish marginate bulb; violet
especially at the apex. **Flesh** white with a hint
of violet at the stem apex. **Odor** slight. **Taste**
mild. **Spores** ellipsoid to lemon-shaped, warty,
10.5–12 × 5.9–6.5μ, quotient 1.9. Deposit rusty
red. **Habitat** under spruce. Occasional. Found
in Colorado and probably other states. **Not**
edible. Comment With KOH the cap skin
turns bright red.

Cortinarius cedretorum Maire subgenus
Phlegmacium **Cap** 8–15cm (3–6in) across,
convex with an inrolled margin for a long time;
pallid yellow at first, later the center becoming
brick reddish, in age almost purple-brown;
very glutinous at first. **Gills** adnexed; pallid
yellowish then rusty. **Stem** 50–120 × 13–25mm
(2–4¾ × ½–1in) at the apex, with a large
rounded marginate bulb up to 50mm (2in)
across; whitish with pale lavender touches,
more yellow on the bulb. **Flesh** lavender under
the cuticle, otherwise pallid whitish. **Odor** not
distinctive. **Taste** pleasant. **Spores** almond-
shaped, warty, 10–13 × 6.5–8μ, quotient 1.6.
Deposit rusty brown. **Habitat** under spruce
and fir. Found in the Pacific Northwest.
Season September–October, later in the
South. **Edibility not known.**

Cortinarius fraudulosus ⅓ life-size

Cortinarius corrugis ⅓ life-size

Cortinarius variecolor ¾ life-size

Cortinarius percomis ¾ life-size

Cortinarius dionysae ½ life-size

Cortinarius variecolor Fr. subgenus *Phlegmacium*
Cap 5–13cm (2–5in) across, convex; a mixture
of violet and pale rusty red; glutinous
especially at first, later it may become dry.
Gills adnate; violet to pallid clay at first, then
more rusty. **Stem** 50–120 × 10–18mm
(2–4¾ × ½–¾in), base swollen to clavate;
violet. **Flesh** whitish with patches of violet, or
more brownish when old and wet. **Odor**
sweetish, pleasant. **Taste** mild. **Spores**
almond- to lemon-shaped, warty,
8.5–11.6 × 5–6.5μ, rather variable in size,
quotient 1.75. Deposit rusty brown. **Habitat**
under conifers. Occasional. Found in the
Pacific Northwest and possibly other areas.
Season October–November. **Not edible.**
Comment With KOH the flesh goes yellow.

Cortinarius fraudulosus Britz. subgenus
Phlegmacium **Cap** 5–10cm (2–4in) across,
convex; ochre to foxy red, often with white
spots; never glutinous although it may be a
little viscid, but usually dry from the
beginning, and markedly, innately fibrous.

Cortinarius communis ¾ life-size

Cortinarius porphyropus ½ life-size

Gills adnexed; pale cream at first, then a little darker. **Stem** 60–120 × 9–20mm (2¼–4¾ × ⅜–¾in), base usually swollen, clavate; whitish, covered in thick white veil to about three-quarters of the way up, which often forms a pseudo ring. **Odor** sweetish. **Taste** slight. **Spores** fat, lemon-shaped, very rough, warty, 12.5–15 × 8–9μ, quotient 1.6. Deposit rusty brown. **Habitat** in spruce woods. Rare. Found in Colorado. **Season** August–September. **Not edible. Comment** A European species; this is one of those contradictory dry phlegmacia.

Cortinarius corrugis Smith subgenus *Phlegmacium* **Cap** 6–12cm (2¼–4¾in) across, convex then flattened with slight dome; pinkish cinnamon; viscid when wet, becoming radially wrinkled. **Gills** adnate; pallid, pinky creamy buff, then darker. **Stem** 60–90 × 12–20mm (2¼–3½ × ½–¾in), mostly with a clavate bulbous stem base; pallid then brownish; fibrillose. **Flesh** pallid, whitish. **Odor** slight. **Taste** slight. **Spores** ellipsoid, roughened, 8.5–10 × 4.5–5.5μ, quotient 1.85. Deposit rusty brown. **Habitat** under conifers. Rare. Found in Washington. **Season** September–October. **Not edible. Comment** My specimens agree with Smith's description except that mine have clavate stems and my spores are a little fatter.

Cortinarius percomis Fr. subgenus *Phlegmacium* **Cap** 4–8cm (1½–3in) across, margin inrolled when young; strong bright yellow to greenish yellow; viscid when wet. **Gills** adnate; distinctively yellow when young, becoming more rusty yellow in age. **Stem** 40–80 × 10–18mm (1½–3 × ½–¾in), the base swollen or with a small submarginate bulb; bright yellow, veil yellow. **Flesh** yellow in all

parts. **Odor** strong, sweet, distinctive. **Taste** slight. **Spores** lemon-shaped, rough, 9–10.5 × 3.5–5.5μ, quotient 2.15. Deposit rusty brown. **Habitat** under conifers. Found in the Pacific Northwest and Colorado. **Season** September–October. **Not edible. Comment** The spores of my collection are rather small.

Cortinarius dionysae Henry subgenus *Phlegmacium* **Cap** 4–8cm (1½–3in) across; gray lilac to blue or brownish; innately fibrous, viscid when wet. **Gills** adnexed; gray-violet to purple when young, more rusty brown in age. **Stem** 35–80 × 7–18mm (1¼–3 × ¼–¾in), base swollen, clavate; pallid gray-violet, lighter than the cap; base shows yellowish hairs. **Flesh** whitish with violet touches especially near the apex of the stem. **Odor** of meal. **Taste** mild. **Spores** ovoid to lemon-shaped, rough, 9–11 × 5–6μ, quotient 1.8. Deposit rusty brown. **Habitat** in mixed woods. Found in Massachusetts. **Season** August–September. **Not edible. Comment** The gills of my collection are a little darker than described, but otherwise it fits Henry's species well. This is a first record for North America.

Cortinarius communis Pk. subgenus *Sericeocybe* **Cap** 2–6cm (¾–2¼in) across, convex; pallid creamy buff with a gray bloom at first, gradually becoming more yellowish then more reddish brown with age and as it is bruised; young specimens have white remnants of the cortina near the edge. **Gills** adnexed with a slight decurrent tooth; pale cream, more ochre-rusty in age. **Stem** 40–60 × 10–14mm (1½–2¼ × ½in), swollen at the base; whitish cream, bruising yellowish then reddish fawn; fibrous. **Flesh** pallid cream with touches of

yellowish. **Odor** pleasant. **Taste** slight. **Spores** ellipsoid, warty, 9–10.2 × 5.4–6.1μ, quotient 1.65. Deposit rusty brown. **Habitat** in conifers or mixed woods. Rare in the East — my collection was in Washington. **Season** September–October. **Not edible. Comment** The mushroom illustrated fits Peck's description of *Cortinarius communis* well, but since he referred to the cap as "subviscid" the species was placed in subgenus *Phlegmacium*. My specimens are not really subviscid, but nevertheless I feel I have Peck's fungus. I have therefore transferred it to subgenus *Sericeocybe*.

Cortinarius porphyropus Fr. subgenus *Phlegmacium* **Cap** 3–7cm (1¼–2¾in) across, convex then flat; pallid ochre to buffy brown; sticky at first. **Gills** adnexed; purple, bruising deeper purple. **Stem** 40–90 × 5–12mm (1½–3½ × ¼–½in), base swollen bulbous; violet at the apex, whitish elsewhere, bruising purple. **Flesh** pallid but bruising purplish. **Odor** slight. **Taste** mild. **Spores** ellipsoid, warty, 9.3–10.9 × 5.3–6.3μ, quotient 1.75. Deposit rusty brown. **Habitat** in frondose woods, especially beech and birch. Occasional. Found in New England and New York northward. **Season** September–October. **Not edible. Comment** The way all parts bruise dark purple is distinctive.

Cortinarius balteatus ½ life-size

Cortinarius herpeticus life-size

Cortinarius scaurus ¾ life-size

Cortinarius subscaurus ½ life-size

Cortinarius obliquus ½ life-size

Cortinarius violaceus ½ life-size

Cortinarius balteatus Fr. subgenus *Phlegmacium*
Cap 4–12cm (1½–4¾in) across, convex,
margin involute; hazel to foxy brown, violet
especially near the margin; sticky only at first,
then dry and shiny. **Gills** adnate; whitish
cream then buff. **Stem** 40–70 × 20–40mm
(1½–2¾ × ¾–1½in), base often swollen; white,
but with ochre veil remnants around the base.
Flesh white. **Odor** mild, musty. **Taste** mild.
Spores almond-shaped, warty, 10–12 × 5.5–7μ,
quotient 1.75. Deposit rusty brown. **Habitat** in
conifer woods, or with huckleberries or
blueberries. *Vaccinium* — rather rare. Found in
northeastern North America, Colorado,
and the Pacific Northwest. **Season**
September–October. **Not edible.**

Cortinarius herpeticus Fr. subgenus *Phlegmacium*
Cap 3.5–8cm (1¼–3in) across; liverish brown
or with hints of olive or yellowish colors,
margin yellowish or olive, the whole cap often
showing spots or blotches of darker colors;
glutinous then only viscid, having a slightly
hygrophanous appearance. **Gills** adnexed;
olive-yellow or with hints of slate-violet. **Stem**
40–80 × 10–25mm (1½–3 × ½–1in), with a
rather variable bulb sometimes marginate,
sometimes not; whitish, usually violet at the
apex; fibrillose. **Flesh** white with brown tints
near the cap and violet tints near the stem
apex. **Odor** strong. **Taste** slight. **Spores**
ellipsoid, warty, 8–9.5 × 5.5–6.5μ, quotient
1.45. Deposit rusty brown. **Habitat** with moss
and lichen in coniferous woods. Uncommon.
Found in the Pacific Northwest, Michigan, and
Nova Scotia. **Season** September–November.
Not edible. Comment Moser has said that
Cortinarius montanus Kauffman is a synonym for
Cortinarius herpeticus. The name *Cortinarius
scaurus* (below) must also be considered for this

species, but in general, in America, the
collections have been designated *Cortinarius
herpeticus.* However, variants between the two
certainly exist, and it would seem to me to be
possible that *scaurus* and *herpeticus* are part of
the same complex species.

Cortinarius scaurus Fr. subgenus *Phlegmacium* **Cap**
2–6cm (¾–2¼in) across, convex; browny
olive-green with brown spots near the margin;
greasy rather than glutinous. **Gills** adnate;
yellowish to olive-greenish. **Stem** 50–130 ×
8–12mm (2–5 × ⁵⁄₁₆–½in), the base bulbous
sometimes only slightly; whitish or flushed
greenish or violaceous; fibrous. **Flesh** pallid
with possible hints of violet. **Odor** not
distinctive. **Taste** slight. **Spores** almond-
shaped, rough, 9.5–11.5 × 6–6.5μ, quotient 1.7.
Deposit rusty brown. **Habitat** in conifer woods,
usually with sphagnum moss, often at high
altitudes. Found in the Pacific Northwest and
New York. **Season** August–October. **Not
edible.**

Cortinarius subscaurus (Moser) Moser subgenus
Phlegmacium **Cap** 3–8cm (1¼–3in) across,
convex then flattened; rich olive-brown with
darker flecks; the texture is greasy rather than
glutinous. **Gills** adnate; purple at first, then
rusty brown. **Stem** 50–90 × 8–14mm (2–3½ ×
⁵⁄₁₆–½in), base bulbous; whitish violet at first,
then more brown; fibrous. **Flesh** violet at first.
Odor not distinctive. **Taste** pleasant. **Spores**
ovoid, rough, 10.5–13.3 × 6–6.5μ, quotient 1.9.
Deposit rusty brown. **Habitat** similar to
Cortinarius scaurus, in conifer woods with
sphagnum. Found in New York. **Season**
August–September. **Not edible. Comment**
Cortinarius scaurus (above) is very close, but the
gills are not purple at first. This is the first

report of this species in North America.

Cortinarius obliquus Pk. subgenus *Sericeocybe* **Cap**
3–8cm (1¼–3in) across, broadly convex;
violaceous, purplish but with rusty tones
coming through; dry, not hygrophanous, silky
fibrillose. **Gills** adnate to adnexed; purple at
first, then cinnamon brown. **Stem** 30–80 ×
8–20mm (1¼–3 × ⁵⁄₁₆–¾in), rather short with
an abrupt marginate bulb; whitish, tinged
violet; bulb heavily sheathed in matted white
veil. **Flesh** whitish, mottled purple, especially
when young. **Odor** slight, pleasant. **Taste** just
a touch hot. **Spores** ovoid, warty, 8–10 ×
5–6.5μ, quotient 1.55. Deposit rusty brown.
Habitat in deciduous woods. Uncommon.
Found in northeastern North America, west to
Wisconsin and south to Missouri. **Season**
August–October. **Not edible. Comment**
Rather variable. Smith has reported larger,
longer-stemmed collections. My collection in
the photograph is not as violet as they are
usually described.

Cortinarius violaceus (Fr.) S. F. Gray subgenus
Cortinarius **Cap** 5–15cm (2–6in) across, convex
to slightly umbonate; dark violet; dry, scaly.
Gills adnate, widely spaced, broad; dark
violet. **Stem** 60–120 × 10–25mm (2¼–4¾ ×
½–1in); concolorous with cap; dry, fibrillose;
cortina faint, leaving a ring zone on stem.
Flesh firm; dark violet. **Odor** not distinctive.
Taste not distinctive. **Spores** ellipsoid, warty,
13–17 × 8–10μ, quotient 1.65. Deposit rust-
brown. Large flask-shaped cystidia. **Habitat**
usually in mixed woods. Not common. Found
throughout most of North America. **Season**
September–October. **Not edible. Comment**
A very distinctive large mushroom with deep
purplish cap and stem.

Cortinarius squamulosus ⅔ life-size

Cortinarius squamulosus Pk. subgenus *Telamonia*
Cap 3–10cm (1¼–4in) across, hemispherical
at first, then convex with an incurved margin,
finally flat; reddish ochre when young, then
darkening to chocolate brown with hints of
purple; the surface breaking up into tiny dark
brown scales. **Gills** adnate; purple at first, then
deep blackish brown. **Stem** 80–150 × 10–20mm
(3–6 × ½–¾in) at the apex, with an enormous
clavate bulb below, larger than the cap width
in young specimens; ochre-brown with dark
brown fibers and scales; stem has a ringlike
zone from the brownish ochre veil, violet
tomentum on base. **Flesh** mottled buff and
brown, with purple at the stem apex in young
specimens. **Odor** pleasant, spicy. **Taste** mild.
Spores broadly ellipsoid to subglobose, rough,
7–8 × 5.5–6.5µ, quotient 1.25. Deposit dark
rusty brown. **Habitat** in broad-leaved woods.
Occasional. Found in New England and west
to Wisconsin. **Season** August–September. **Not
edible. Comment** The enormous bulbous stem
is most distinctive.

Cortinarius anomalus (Fr. ex Fr.) Fr. subgenus
Sericeocybe **Cap** 3–5cm (1¼–2in) across,
hemispheric then broadly convex; often tinged
gray-violet at first, then grayish buff; the
surface covered with fine hairs that give a silky
sheen. **Gills** adnate to emarginate; when
young, violaceous or with violaceous tints, then
rusty brown. **Stem** 40–90 × 7–18mm
(1½–3½ × ¼–¾in), the base usually swollen,
sometimes almost bulbous; at first with violet
tints, especially at the apex, then pallid, with
distinct pale grayish or yellowish bands of veil
material. **Flesh** pallid with strong purplish
tints, especially inside the stem apex. **Odor**
faint. **Taste** slight. **Spores** subglobose,
roughened, 7–9 × 6–7µ, quotient 1.25. Deposit
rusty brown. **Habitat** in coniferous and mixed
woods, especially with birch. Frequent. Found
over most of North America. **Season**
September–October. **Not edible. Comment**
The copious veil remnants on the stem often
show a distinct yellow color; this form has been
given the name *Cortinarius lepidopus* Cke., but

Moser keeps it under the name *Cortinarius
anomalus.*

Cortinarius caninus (Fr.) Fr. subgenus *Sericeocybe*
Cap 3–10cm (1¼–4in) across, convex, domed;
at first buff with a hint of violet, then more
rusty brown; dry, smooth. **Gills** adnexed;
violet when young, then pallid rusty brown.
Stem 60–120 × 8–18mm (2¼–4¾ × ⁵⁄₁₆–¾in),
clavate, swollen below; violet near the apex at
first, then white with delicate bands or touches
of brown veil. **Flesh** pallid, with violet near the
stem apex. **Odor** slight. **Taste** mild. **Spores**
subglobose to broadly pip-shaped, roughened,
7–8.1 × 5.7–6.8µ, quotient 1.2. Deposit rusty
brown. **Habitat** in coniferous or broad-leaved
woods. Occasional. Found in northern North
America. **Season** August–October. **Not
edible. Comment** My specimens are rather
small compared with the specimens reported
by Smith.

Cortinarius traganus Fr. syn. *Cortinarius pyriodorus*
Kauffman subgenus *Sericeocybe* **Cap** 4–12cm
(1½–4¾in) across, hemispherical then convex
with center dome; bright lilac fading to rusty
ochre in age; not hygrophanous, finely silky,
shiny. **Gills** adnexed; cinnamon even when
young. **Stem** 60–120 × 10–25mm (2¼–4¾ ×
½–1in) at apex, the large base swollen to
40mm (1½in) across; violet like the cap and
thickly covered in violet veil fibrils, the base
sometimes whitish, cortina violet. **Flesh**
distinctly marbled buff with steaks of cinnamon
which make it look hygrophanous. **Odor**
strong, of overripe pears, fragrant. **Taste** a
touch bitter. **Spores** ellipsoid, roughened,
7.4–9 × 4.5–6µ, quotient 1.55. Deposit rusty
brown. **Habitat** in coniferous woods.
Occasional. Found in northern North America.
Season September–October. **Not edible.
Comment** In the East this species is usually
named *Cortinarius pyriodorus.*

Cortinarius pholideus Fr. subgenus *Sericeocybe* **Cap**
4–10cm (1½–4in) across, hemispherical at
first, eventually flattish with a broad umbo; the
dry surface is ochraceous buff, slightly darker
in the center; covered in small sepia fibrillose
scales. **Gills** adnate; with a touch of violet at
first, then clay-buff, eventually pale rusty
brown. **Stem** 50–120 × 8–15mm (2–4¾ ×
⁵⁄₁₆–½in); flushed violaceous at first, especially
near the apex, dull buff; covered below the ring
zone in small brown scales often breaking up
into snakelike patterns. **Flesh** purplish near the
apex when young, otherwise buff. **Odor** faint,
pleasant. **Taste** slight. **Spores** broadly
ellipsoid, rough, 6.5–8.5 × 5–6µ, quotient 1.35.
Deposit rusty brown. **Habitat** in mixed woods,
especially with birch. Occasional. Found in
New England and west to Wisconsin; rare in
the Pacific Northwest. **Season** September–
October. **Not edible. Comment** The tiny
scales on cap and stem are very unusual in
cortinarius mushrooms.

Cortinarius anomalus ⅔ life-size

Cortinarius caninus ⅔ life-size

Cortinarius traganus ½ life-size

Cortinarius pholideus ½ life-size

Cortinarius lilacinus ⅔ life-size

Cortinarius azureus ⅔ life-size

Silvery-violet Cort *Cortinarius alboviolaceus* ½ life-size

Cortinarius cyanites ⅓ life-size

Cortinarius pulchrifolius ½ life-size

Cortinarius pulchrifolius var. *odorifer* ½ life-size

Cortinarius lilacinus Pk. subgenus *Sericeocybe* **Cap** 5–10cm (2–4in) across, hemispheric, domed, thick-fleshed in the center; violet persisting but more violet-ochre in the center; covered in purplish fibers, dry, not hygrophanous. **Gills** adnexed; pallid violet. **Stem** 50–100 × 12–20mm (2–4 × ½–¾in), with a large clavate bulb in the button stage, but as the stem expands this is seen only as a slight clavate swelling; whitish with a hint of violet. **Flesh** mottled violet and cream. **Odor** slight. **Taste** mild. **Spores** ellipsoid, warty, 8.5–10 × 5.5–6µ, quotient 1.61. Deposit rusty brown. **Habitat** in deciduous or mixed woods. Rather common. Found in eastern North America, west to Wisconsin. **Season** August–October. **Not edible. Comment** Often confused with *Cortinarius alboviolaceus* (below), but *lilacinus* is much more violet, less shiny, with purplish, mottled flesh and slightly larger spores.

Cortinarius azureus Fr. subgenus *Sericeocybe* **Cap** 3–7cm (1¼–2¾in) across, hemispherical at first, then convex; violet with reddish ochre in the center; covered in fine fibers that give it a shiny appearance. **Gills** adnate; deep purple to violet. **Stem** 60–130 × 7–10mm (2¼–5 × ¼–½in), swollen at the base; violet to purple with matted white fibers overlaying it, with gray or ochre veil fragments. **Flesh** purplish. **Odor** slight. **Taste** slight. **Spores** ovoid to subglobose, warty, 7.5–10 × 6–7.5µ, quotient 1.3. Deposit rusty brown. **Habitat** in deciduous and mixed woods. Rare. Found in northeastern North America. **Season** August–September. **Not edible. Comment** This is a European species rather close to *Cortinarius anomalus*

(p. 152) but with much more violet color, especially in the cap.

Silvery-violet Cort *Cortinarius alboviolaceus* (Pers. ex Fr.) Fr. subgenus *Sericeocybe* **Cap** 3–10cm (1¼–4in) across, convex then expanding, often with umbo, bell-shaped, with incurved margin; pale silvery blue to grayish violet; surface smooth, silky fibrillose, dry, margin with fine cobwebby veil stretching to stem when young. **Gills** attached, crowded; pale violet, then cinnamon brown when old. **Stem** 40–80 × 5–15mm (1½–3 × ¼–½in), with a clavate, bulbous base; color similar to cap; sheathing white veil clothes lower stem below faint apical ring zone. **Flesh** firm; white, with pale violet particularly in stem apex. **Odor** not distinctive. **Taste** not distinctive. **Spores** ovoid to lemon-shaped, minutely roughened, 7.5–10 × 5–6µ, quotient 1.6. Deposit rusty brown. **Habitat** often in large numbers under beech and oak. Common. Found throughout North America. **Season** August–October. **Edible** but not recommended. **Comment** The very easily confused *Cortinarius subargentatus* Orton differs in its lack of sheathing white veil on stem.

Cortinarius cyanites Fr. subgenus *Sericeocybe* **Cap** 7–15cm (2¾–6in) across, convex; a mixture of violaceous purple and reddish brown colors; a touch slimy at first, then dry, covered in brownish fibrils or scales. **Gills** adnate; violet, keeping their color for a long time. **Stem** 90–150 × 20–35mm (3½–6 × ¾–1¼in), with a very large bulb before the stem extends and the cap expands, eventually becoming swollen as in my picture; pallid purplish; fibrillose. **Flesh** purple especially near the exterior,

turning red after 2 or 3 minutes when cut or bruised. **Odor** distinctive. **Taste** slight or a touch bitter. **Spores** lemon-shaped, warty, 8.8–11.5 × 5–6.5µ, quotient 1.7. Deposit rusty brown. **Habitat** in coniferous and deciduous woods. Uncommon. Found in the Pacific Northwest and Virginia and probably in other eastern states. **Season** August–October. **Not edible. Comment** Cut your specimen in half and rub the knife over the cut surface to bruise it; in a few minutes it will turn blood red in patches if it is this species.

Cortinarius pulchrifolius Pk. subgenus *Sericeocybe* **Cap** 5–12cm (2–4¾in) across, convex; dull whitish center, reddish; covered in fine, silky white fibrils from the copious veil and cortina, not hygrophanous. **Gills** adnexed; purple then dark purple-brown. **Stem** 50–100 × 8–16mm (2–4 × ⅜–⅝in), with a large, fat, oniony bulb up to 30mm (1¼in) wide; covered in white fibers and heavy patches of white universal veil. **Flesh** purple. **Odor** slight. **Taste** pleasant. **Spores** pip-shaped, warty, 9–11 × 5.7–7.3µ, quotient 1.55. Deposit rusty brown. **Habitat** in coniferous and mixed woods. Occasional. Found in northeastern North America, south to Virginia. **Season** August–September. **Not edible. Comment** *Cortinarius pulchrifolius* var. *odorifer* Hesler differs in its odor, which is strong, pleasant, and sweet, similar to ripe pears, and in its spores, which are larger, 10–12 × 6.3–7.5µ. It is found from Tennessee to New York.

Cortinarius cotoneus ⅔ life-size

Cortinarius bolaris ½ life-size

Cortinarius cotoneus Fr. subgenus *Leprocybe* **Cap** 4–9cm (1½–3½in) across, convex, flat when mature with an umbo; olive-green or darker olive-brown; not hygrophanous, dry, covered in minute fibrillose scales. **Gills** adnate; yellowish olive at first, then dark cinnamon. **Stem** 40–80 × 7–14mm (1½–3 × ¼–½in), slightly clavate; pale olivaceous yellow with remnants of olive veil, often looking rusty from spore deposit. **Flesh** olive-brownish, paler as it dries. **Odor** radishy, especially when crushed. **Taste** also a bit radishy. **Spores** ovoid, rough, 6.7–8.5 × 5.5–6μ, quotient 1.25. Deposit rusty brown. **Habitat** in conifer woods. Rare. Found in Washington. **Season** September–October. **Not edible. Comment** In Europe this species is found under broad-leaved trees.

Cortinarius bolaris (Fr.) Fr. subgenus *Leprocybe* **Cap** 3–8cm (1¼–3in) across, convex-expanded to slightly umbonate; with red fibrillose squamules on a pale buff ground color. **Gills** adnate, crowded, broad; pallid then cinnamon brown. **Stem** 50–100 × 5–12mm (2–4 × ¼–½in); pale buff, covered below the veil zone with reddish fibrils, bruising yellowish red. **Flesh** pallid. **Odor** not distinctive. **Taste** not distinctive. **Spores** ovoid, warty, 7–8 × 5–6μ, quotient 1.35. Deposit rust-brown. **Habitat** under conifers or beech and oak trees. Sometimes common. Found in eastern North America. **Season** August–October. **Not edible** — possibly poisonous. **Comment** Easily recognized by the red squamules on cap and stem.

Cortinarius gentilis (Fr.) Fr. subgenus *Leprocybe* **Cap** 2–5cm (¾–2in) across, conic or convex, umbonate in age; hygrophanous, bright red-

brown when wet, drying yellowish ochre; fibrous. **Gills** adnexed, deep; bright reddish cinnamon. **Stem** 30–80 × 3–6mm (1¼–3 × ⅛–¼in), equal; the same color as the cap but with remnants of yellow veil; fibrous. **Flesh** yellowish, red. **Odor** slight, but definite when crushed, of radish. **Taste** slight. **Spores** broadly ovoid, rough, 7.5–8.5 × 5.5–6.5μ, quotient 1.33. Deposit rusty brown. **Habitat** in conifer woods. Uncommon. Found in northern North America, more common in the Pacific Northwest. **Season** August–October. **Poisonous.**

Cortinarius raphanoides Fr. subgenus *Leprocybe* **Cap** 2–6cm (¾–2¼in) across, conic, umbonate; buffy brown to olivaceous brown; silky, fibrillose. **Gills** adnexed; buffy olive when young, then browny cinnamon. **Stem** 50–70 × 5–9mm (2–2¾ × ¼–⅜in), base swollen, coated in thick mycelium; olive-buff-brown; fibrous with olivaceous-brown veil remnants. **Flesh** concolorous. **Odor** of radish, especially when crushed. **Taste** slight. **Spores** broadly ellipsoid to pip-shaped, warty, 8–9 × 5.5–6.5μ, quotient 1.4. Deposit rusty brown. **Habitat** in mixed and conifer woods (in Europe this species is found in association with birch). Occasional. Found in Washington and probably other states. **Season** September–October. **Not edible. Comment** Reminiscent of the subgenus *Dermocybe*, and indeed it used to be in that subgenus previously.

Cortinarius limonius (Fr. ex Fr.) Fr. syn. *Cortinarius whitei* Pk. subgenus *Leprocybe* **Cap** 2–8cm (¾–3in) across, hemispheric to broadly conic to convex; hygrophanous, when damp foxy red, when dry apricot yellow; when very

young covered in yellowish fibers, in age only the edge fibrous. **Gills** adnexed; yellow to rusty yellow. **Stem** 60–120 × 10–20mm (2¼–4¾ × ½–¾in), usually tapering toward the base; heavily covered in yellowish fibrous veil, discoloring reddish toward the base. **Flesh** yellowish, more reddish near base of the stem. **Odor** strong, rather rusty. **Taste** mild. **Spores** broadly ovoid to subglobose, rough, 7–8.5 × 5.5–7μ, quotient 1.25. Deposit rusty brown. **Habitat** in coniferous woods. Rather common. Found in northeastern North America from Virginia northward. **Season** August–September. **Not edible** — probably poisonous. **Comment** The descriptions of *Cortinarius limonius* of Europe and that of *Cortinarius whitei* of North America both seem to fit the four collections I have made of this mushroom extremely well; I have therefore synonymized them.

Cortinarius callisteus (Fr.) Fr. subgenus *Leprocybe* **Cap** 3–8cm (1¼–3in) across, convex expanding to flattish or with a central flat area; yellowish, yellow-ochre; the surface is smooth but eventually begins to break up into minute scales. **Gills** adnexed; yellow then reddish ochre from the spores. **Stem** 50–80 × 6–12mm (2–3 × ¼–½in), the clavate bulb up to 30mm (1¼in) wide; yellowish; fibrous, with a distinct ring from the universal veil when young. **Flesh** pallid, yellowish to deeper orange-yellow, especially in the lower stem. **Odor** slight. **Taste** slight. **Spores** subglobose, very warty, 7–8.5 × 6–7μ, quotient 1.2. Deposit rusty brown. **Habitat** in conifer woods. Uncommon. Found in most of North America. **Season** August–September. **Not edible.**

Cortinarius gentilis ⅔ life-size

Cortinarius raphanoides ¾ life-size

Cortinarius limonius ½ life-size

Cortinarius callisteus ⅔ life-size

Cortinarius annulatus ⅔ life-size

Cortinarius melinus ½ life-size

Cortinarius torvus ⅓ life-size

Cortinarius laniger ⅔ life-size

Cortinarius annulatus Pk. **Cap** 3–9cm (1¼–3½in) across, broadly convex; bronze yellowish in color; dry, not hygrophanous, covered in minute tawny yellow scales. **Gills** adnate; pallid ochre-yellow at first, eventually bright, rusty-colored. **Stem** 40–80 × 7–18mm (1½–3 × ¼–¾in), base slightly swollen; covered in golden-yellow universal veil most of the way up from the base to the ringlike remnants, whitish above. **Flesh** pallid yellow, the base of the stem orangy yellow. **Odor** strong, plantlike or radishlike. **Taste** mild. **Spores** subglobose, distinctly rough and warty, 6–7 × 5–6μ, quotient 1.2. Deposit rusty brown. **Habitat** in mixed woods. Uncommon. Found on the eastern side of North America, west as far as Wisconsin and from Missouri north. **Season** July–September. **Not edible.** **Comment** I am uncertain about which subgenus this distinctive mushroom belongs to; possibly it will eventually be included in subgenus *Leprocybe* after further study.

Cortinarius melinus Britz. subgenus *Leprocybe* **Cap** 1.5–10cm (½–4in) across, convex then plane; yellowish brown to reddish brown; dry, not hygrophanous, finely tomentose-scaly. **Gills** adnate; deep yellowish then more rusty. **Stem** 35–65 × 5–10mm (1¼–2½ × ¼–½in), base slightly swollen; yellowish brown with patches of yellow universal veil. **Flesh** yellowish brown to pallid. **Odor** of radish, especially when crushed. **Taste** mild. **Spores** broadly ellipsoid to pip-shaped to subspherical, warty, 7.5–8.7 × 5.8–6.8μ, quotient 1.3. Deposit rusty brown. **Habitat** under spruce. Rare. Found in Colorado. **Season** August. **Not edible.** **Comment** This is a European species recorded for the first time in North America.

Cortinarius torvus (Bull. ex Fr.) Fr. subgenus *Telamonia* **Cap** 3–10cm (1¼–4in) across, convex then flattened; brown, sometimes with purplish tints, covered in darker innate radiating fibrils. **Gills** adnate, rather distant; deep purplish at first, then umber. **Stem** 40–60 × 10–15mm (1½–2¼ × ½in), swollen at base; clay-buff flushed violet above the membranous sheathing white veil, which gives the appearance of a stocking on the stem. **Flesh** buff flushed violet in upper stem. **Taste** slightly bitter or stinging. **Odor** heavy and sweet. **Spores** ellipsoid, minutely rough, 8–10 × 5–6μ, quotient 1.69. Deposit rusty brown. **Habitat** deciduous and mixed woods. Common. Found widely in eastern North America. **Season** September–October. **Edibility not known.**

Cortinarius laniger Fr. subgenus *Telamonia* **Cap** 4–10cm (1½–4in) across, domed,

campanulate; heavy and thick-fleshed; ochreous cinnamon-colored; broken up with fibrils, only a touch hygrophanous. **Gills** adnexed; bright cinnamon. **Stem** 80–100 × 10–20mm (3–4 × ½–¾in), with a large, swollen clavate bulb; pallid cinnamon with creamy fibers and white veil remnants often forming a distinct thick central belt. **Flesh** creamy marbled with cinnamon patches. **Odor** said not to have an odor, but mine smelled distinctly of pelargonium (household geranium). **Taste** mild. **Spores** ellipsoid to pip-shaped, rough, 8.5–10.5 × 5.5–6.5μ, quotient 1.6. Deposit rusty brown. **Habitat** in conifer woods. Occasional. Found in northern North America. **Season** September–October. **Not edible. Comment** The bright, cinnamon-colored gills are very distinct.

Cortinarius armillatus (Fr.) Fr. subgenus *Telamonia* **Cap** 5–15cm (2–6in) across, convex then bell-shaped, finally flattened; deep tawny red when moist, drying pale reddish orange to

Cortinarius armillatus ½ life-size

ochre; minutely fibrillose to smooth. **Gills** adnate, distant, broad; pale cinnamon then rust-brown. **Stem** 70–150 × 10–20mm (2¾–6 × ½–¾in), clavate to bulbous at base; whitish to pale brown with the remains of the veil forming 1–3 cinnabar-red bands around stem; cortina at stem apex white. **Flesh** thick, firm; pallid, darker in the stem. **Odor** slightly of radish. **Taste** bitter. **Spores** broadly ellipsoid to almond-shaped, roughened, 7–12 × (5)6–7(7.5)μ, quotient 1.45. Deposit rusty brown. **Habitat** under birch or pine. Common. Found widely from Canada to New Jersey, west to Idaho. **Season** August–October. **Not edible** — probably poisonous. **Comment** This is one of the few easily recognized species in its genus. In the Pacific Northwest *Cortinarius haematochelis* (Bull. ex Fr.) Fr. is found, similarly ringed but with spores 5.5–7 × 5–6.5μ.

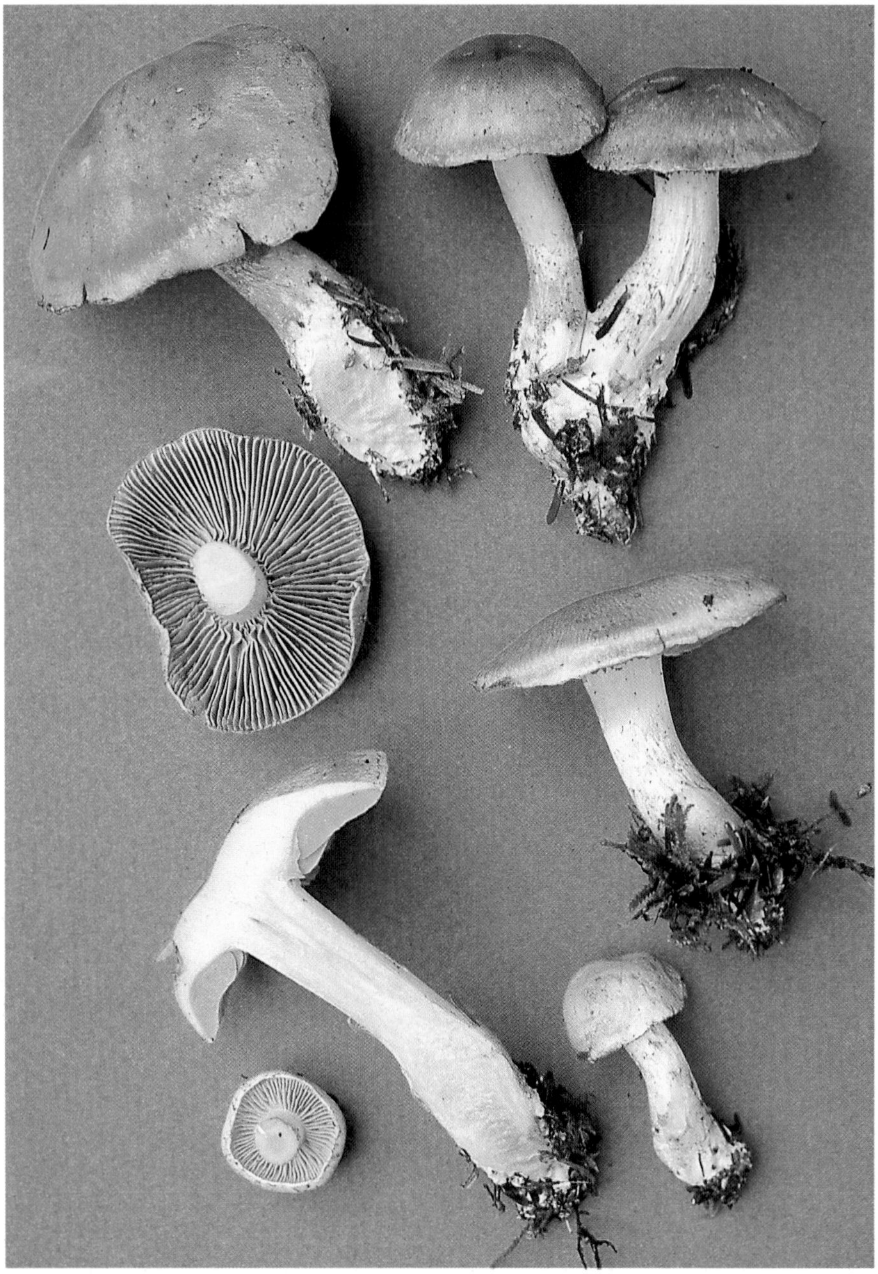

Cortinarius pseudobolaris ¾ life-size

Cortinarius pseudobolaris Maire sensu Smith subgenus *Leprocybe* **Cap** 4–9cm (1½–3½in) across, convex becoming flat; pallid whitish ochre at first, soon becoming dull orange-yellow; bruising bright chrome yellow then reddish; greasy to the touch when wet, on first sight you might consider it one of the dryish members of subgenus *Phlegmacium*; lightly fibrillose. **Gills** adnexed; pallid brown then rusty, bruising chrome yellow. **Stem** 40–80 × 8–14mm (1½–3 × 5/16–½in), base swollen, sometimes fused into small clumps (caespitose); pallid whitish, bruising yellow then turning more orange. **Flesh** white, discoloring chrome yellow, especially in the stem base. **Odor** slight, pleasant. **Taste** slight. **Spores** ellipsoid, rough, 6.1–7.3 × 4.2–4.6μ, quotient 1.5. Deposit rusty brown. **Habitat** in coniferous and mixed woods. Occasional. Found in the Pacific Northwest. **Season** September–November. **Not edible. Comment** Moser has synonymized this species with *Cortinarius rubicundulus* (Rea) Pearson, but my specimens do not have the yellow veil that Moser describes. However, my specimens do fit the Smith description of *Cortinarius pseudobolaris* and they were collected in the same area as Smith's. I have therefore referred this collection to the latter name.

Cortinarius privignoides Henry subgenus *Telamonia* **Cap** 3–8cm (1¼–3in) across, convex; ochre to reddish brown; shiny with fine fibers, the edge with white veil remnants; only slightly hygrophanous. **Gills** adnate, for a long time whitish denticulate; pallid brown at first, then rusty brown at the edge. **Stem** 40–70 × 10–20mm (1½–2¾ × ½–¾in), with a spongy egg-shaped bulb up to 40mm (1½in) across; covered in white fibers. **Flesh** pallid with the slightest hint of violet at first, later touches of ochre, especially near the base. **Odor** slight. **Taste** mild. **Spores** longish ovoid, warty, 8–10 × 3.75–5.7μ, quotient 1.9. Deposit rusty brown. **Habitat** in mixed woods. Occasional. Found in New York and very possibly elsewhere. **Season** August. **Not edible. Comment** This is a European species; my American collections fit Henry's species very well. The spores are exceptionally long and narrow.

Cortinarius privignus Fr. subgenus *Telamonia* **Cap** 4–7cm (1½–2¾in) across, conical to campanulate; hygrophanous, the center dark red-brown when wet, drying to yellow-brown; smooth and slightly shiny except the edge, which is coated with fine white veil fragments. **Gills** adnexed; pallid milky coffee to pale cinnamon at first, then much darker cocoa brown. **Stem** 40–90 × 7–15mm (1½–3½ × ¼–½in), base swollen, clavate; white covered in matted fibers. **Flesh** pallid with red-brown patches. **Odor** strongish, plantlike. **Taste** mild. **Spores** ellipsoid to pip-shaped, warty, 7.5–9.5 × 5.5–6.3μ, quotient 1.45. Deposit rusty brown. **Habitat** in hemlock and mixed woods. Uncommon. Found in northern North America, north of Tennessee. **Season** July–September. **Not edible. Comment** The dried material blackened slightly.

Cortinarius cacao-color Smith subgenus *Telamonia* **Cap** 5–10cm (2–4in) across, hemispherical then expanding with a broad umbo and a decurved margin; hygrophanous, dark chocolate brown when wet, drying to a lighter reddish brown, but with some streaks of the darker color still showing; finely silky, almost punctate with fine scales. **Gills** adnate; dark cocoa brown from the beginning. **Stem** 70–150 × 10–25mm (2¾–6 × ½–1in), broadest at the apex, narrowing downward; pallid, with darker areas showing through; covered in long fibers. **Flesh** rather mottled, lighter in the center and darker near the edge or when waterlogged. **Odor** strong, resembling rhubarb(?). **Taste** slight. **Spores** broadly ellipsoid, warty, 7.5–8.5 × 5.5–6.5μ, quotient 1.35. Deposit rusty brown. **Habitat** under conifers. Seemingly not rare. Found on the Olympic Peninsula. **Season** September–October. **Not edible. Comment** All parts of the mushroom, but especially the gills, turn dark grayish black on drying.

Cortinarius injucundus Fr. subgenus *Telamonia* **Cap** 4–8cm (1½–3in) across, broadly convex; hygrophanous, red-brown when wet, pallid when dry, sometimes with a hint of violet; the edge matted with patches of white veil. **Gills** adnate; with violet tints at first, then red-brown. **Stem** 50–80 × 7–16mm (2–3 × ¼–½in), the base swollen; whitish or pale violet at the apex, with a copius white cortina. **Flesh** pallid with violet tints, especially when young. **Odor** not distinctive. **Taste** slight. **Spores** pip-shaped, rough, 10–11.2 × 5.5–6.5μ, quotient 1.75. Deposit rusty brown. **Habitat** under conifers. Found in Colorado and northeastern North America. **Season** August–October. **Not edible. Comment** Cortinarius mushrooms in this group are extremely difficult to name with any certainty, and much work is still to be done before the position will become clearer. It is possible that *Cortinarius rimosus* Pk. is another name for this mushroom.

Cortinarius privignoides ½ life-size

Cortinarius privignus ⅓ life-size

Cortinarius cacao-color ⅓ life-size

Cortinarius injucundus ½ life-size

Cortinarius evernius ½ life-size

Cortinarius glandicolor ½ life-size

adnexed; dark blackish brown, more rusty dark brown as spores develop. **Stem** 30–80 × 4–10mm (1¼–3 × ³⁄₁₆–½in), equal or swollen a little at the base; dark blackish brown, blacker in age, with white veil patches usually forming a ring or rings. **Flesh** blackish brown. **Odor** slight. **Taste** mild. **Spores** ovoid, rough, 8.5–10 × 5.5–6.5μ, quotient 1.55. Deposit rusty brown. **Habitat** in coniferous woods. Uncommon. Found in northern North America. **Season** September–October. **Not edible. Comment** My specimens are not very good; the stems are rather short and do not show good bands of white veil.

Cortinarius saturatus Lange subgenus *Telamonia* **Cap** 4–9cm (1½–3½in) across, hemispherical, then flat with a broad umbo; hygrophanous, reddish brown when wet, ochre when dry; smooth with silky white veil remnants on the margin. **Gills** adnate; pallid buff then rusty brown. **Stem** 60–90 × 10–16mm (2¼–3½ × ½in), base sometimes swollen; silky white from veil, showing slightly brown from damp flesh. **Flesh** white to brownish mottled when wet. **Odor** slight. **Taste** mild. **Spores** subglobose to broadly ovate, rough, 6.5–7.3 × 5.5–6.3μ, quotient 1.25. Deposit rusty brown. **Habitat** under conifers. Rare. Found in Maine, possibly elsewhere. **Season** September. **Not edible. Comment** This is a European species not reported previously in North America. The subglobose spores are most distinctive.

Cortinarius duracinus Fr. subgenus *Telamonia* **Cap** 3–10cm (1¼–4in) across, convex then expanding to domed, usually with an inrolled margin; hygrophanous, cinnamon brown when wet, reddish ochre when dry; surface smooth. **Gills** adnate; creamy-colored at first, then reddish buff. **Stem** 40–120 × 6–15mm (1½–4¾ × ¼–½in), tapering downward, usually to a rooting point; white; smooth. **Flesh** white. **Odor** nutty to radishy when crushed. **Taste** slight. **Spores** ovoid, lightly roughened, 9–11 × 5–6.5μ, quotient 1.75. Deposit rusty brown. **Habitat** in broad-leaved or mixed woods. Uncommon. Found in northeastern North America, west to Michigan, and in Colorado. **Season** August–September. **Not edible. Comment** Note the rooting stem and hygrophanous cap.

Cortinarius hinnuleus Fr. subgenus *Telamonia* **Cap** 3–6cm (1¼–3in) across, conic-campanulate, sometimes with a pointed umbo; hygrophanous, dark reddish or yellowish brown when wet, pallid ochre when dry, often with rusty to blackish stains. **Gills** adnate, very broad; yellow-brown then rusty-colored. **Stem** 50–70 × 4–8mm (2–2¾ × ³⁄₁₆–⁵⁄₁₆in), equal; yellowish brown, usually with one or more white veil belts; longitudinally fibrous. **Flesh** very thin on the cap; yellowish. **Odor** either slight or strong and disgusting (earthy). **Taste** slight. **Spores** ovoid, warty, 7–9 × 5–7μ, quotient 1.35. Deposit rusty brown. **Habitat** in coniferous and mixed woods. Rare. Found in northeastern North America and Michigan. **Season** August–October. **Not edible. Comment** Some authors are adamant about the strong, disgusting smell; others, like myself, do not detect it.

Cortinarius obtusus Fr. subgenus *Telamonia* **Cap** 2–4.5cm (¾–1¾in) across, broadly conic, often with an umbo; hygrophanous, deep amber brown when wet, drying light buff; striate, especially when young, texture smooth and glabrous. **Gills** adnexed; cinnamon. **Stem** 40–60 × 3–7mm (1½–2¼ × ⅛–¼in), equal or narrower toward the base; silky with fine white fibrils over a cinnamon underlayer. **Flesh** reddish cinnamon when wet, drying more

Cortinarius evernius Fr. subgenus *Telamonia* **Cap** 3–9cm (1¼–3½in) across, conic then expanding to flat with an umbo; hygrophanous, dark purplish brown when wet, drying much lighter, ochre; the margin silky from remains of the veil. **Gills** adnexed, rather distant, very broad; dark purplish with white edges, then rusty red. **Stem** 80–150 × 7–15mm (3–6 × ¼–½in), equal; purplish-colored but overlaid with a layer of whitish universal veil. **Flesh** distinctly purple in the stem. **Odor** scented, possibly of radish. **Taste** slight. **Spores** ellipsoid, rough, 8–11 × 5–6μ, quotient 1.75. Deposit rusty red. **Habitat** in conifer woods, especially near boggy areas. Fairly common. Found in northern North America. **Season** August–October. **Not edible. Comment** The dark purplish cap and the very long white-and-purple stem are distinctive.

Cortinarius glandicolor Fr. subgenus *Telamonia* **Cap** 2–5cm (¾–2in) across, broadly conic, then expanded with an umbo or boss; hygrophanous, black-brown when wet, drying chestnut brown, blackening in age; silky, the margin white from veil remnants. **Gills**

(continued on page 164)

Cortinarius saturatus ½ life-size

Cortinarius duracinus ⅔ life-size

Cortinarius hinnuleus ⅔ life-size

Cortinarius obtusus ⅓ life-size

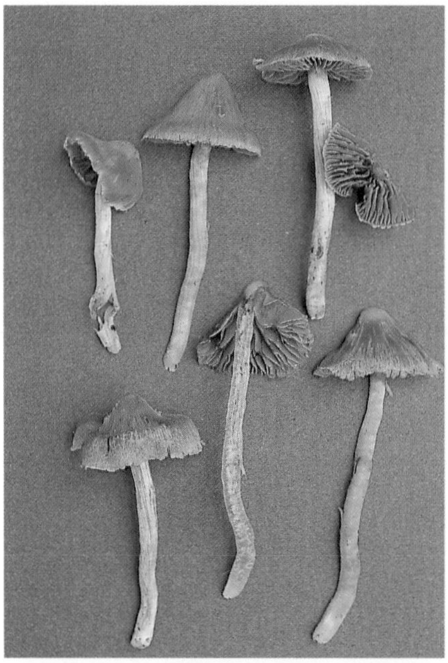
Cortinarius jubarinus ½ life-size

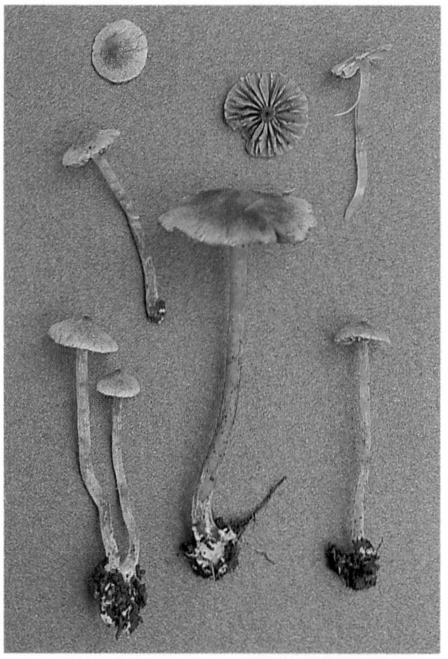
Cortinarius pulchellus life-size

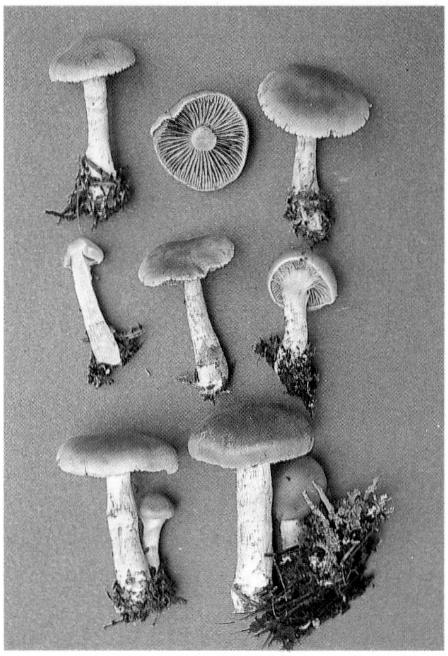
Cortinarius praestigiosus ⅓ life-size

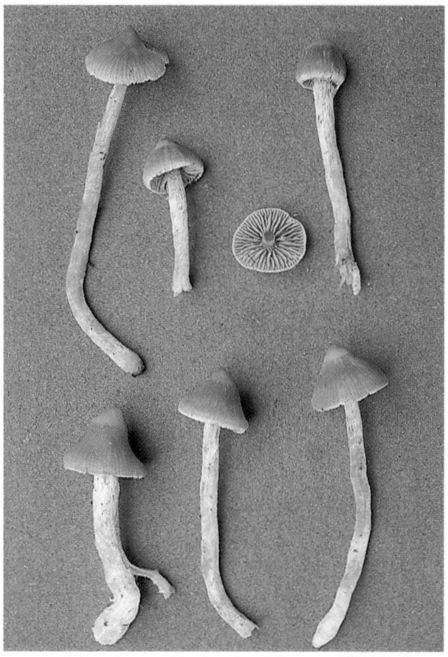
Cortinarius fulvescens ⅔ life-size

(*continued from page 162*)
ochraceous. **Odor** distinct, similar to radish. **Taste** slight. **Spores** ovoid, warty, very variable in length, 7.7–9.7 × 5–6μ, quotient 1.6. Deposit rusty brown. **Habitat** in conifer woods. Occasional. Found in northern North America. **Season** August–October. **Not edible. Comment** Variable in size; sometimes very small collections will be found.

Cortinarius jubarinus Fr. subgenus *Telamonia* **Cap** 2–4cm (¾–1½in) across, campanulate with a large pointed umbo; hygrophanous, cinnamon-colored when wet, ochre when dry; silky, shiny when dry, margin white from veil when very young. **Gills** adnate; cinnamon. **Stem** 30–80 × 3–5mm (1¼–3 × ⅛–¼in), equal; pale brown with whitish silky covering. **Flesh** pale yellowish cinnamon in stem, soon hollow. **Odor** earthy. **Taste** slight. **Spores** ellipsoid, finely roughened, 7–8.5 × 4.5–5μ, quotient 1.65. Deposit rusty brown. **Habitat** in coniferous or mixed woods. Rare. Found in northern North America. **Season** September–October. **Not edible.**

Cortinarius pulchellus Lange syn. *Cortinarius bibulus* Quél. subgenus *Telamonia* **Cap** 1–2cm (½–¾in) across, conic, then flat with an umbo; dull violaceous, fading to wine-brown; hygrophanous. **Gills** adnexed; violet at first, then rusty brown. **Stem** 20–50 × 2–4mm (¾–2 × 3/32–3/16in), equal or slightly swollen at the base; dull violaceous with a light coating of grayish fibrils. **Flesh** violaceous. **Odor** slight. **Taste** mild. **Spores** ellipsoid, roughened, 9–11 × 5.5–7μ, quotient 1.6. Deposit rusty brown. **Habitat** under alder. Occasional. Found in the Pacific Northwest. **Season** September–October. **Not edible.**

Cortinarius praestigiosus (Fr.) Moser subgenus *Telamonia* **Cap** 1.5–4.5cm (½–1¾in) across; reddish brown with whitish bloom, edge lighter; hygrophanous. **Gills** adnate to almost free; yellowish ochre then rusty. **Stem** 30–60 × 4–10mm (1¼–2¼ × ¼–½in), pallid with red-brown veil belts. **Flesh** marbled white and brown. **Odor** strong, radishy. **Taste** radishy. **Spores** ovoid, rough, 8–9 × 4–5μ, quotient 1.9. Deposit rusty brown. **Habitat** in coniferous woods. Found in Maine. **Season** September–October. **Not edible. Comment** Identified in Moser, this species needs further study in North America to confirm its identity.

Cortinarius fulvescens Fr. sensu Favre subgenus *Telamonia* **Cap** 1–3cm (½–1¼in) across, conical with a pointed umbo; hygrophanous, coppery red-brown when wet, drying ochre; smooth, striate near the margin. **Gills** adnexed; creamy buff then yellowy brown. **Stem** 30–90 × 2–5mm (1¼–3½ × 3/32–¼in), very long and thin, equal; off-white; longitudinally fibrous; copious cortina. **Flesh** pallid, brownish. **Odor** distinct, possibly a touch of iodine. **Taste** slight. **Spores** ellipsoid, rough, 8.5–11 × 4.5–6.5μ, quotient 1.75. Deposit rusty brown. **Habitat** in mossy coniferous (fir) woods at high elevations. Rare. Found in Washington. **Season** October. **Not edible. Comment** A European species of alpine regions.

Cortinarius badius Pk. subgenus *Telamonia* **Cap** 1–2.5cm (½–1in) across, conical to broadly conical, developing an umbo as it expands; hygrophanous, reddish brown, darker when very wet; surface appearing rather smooth to the naked eye, slightly silky at the margin. **Gills** adnexed; creamy gray-colored at first, then more reddish brown. **Stem** 20–60 × 2–3mm (¾–2¼ × 3/32–⅛in), not bulbous but attenuating from the base upward, hollow; covered in matted white fibers. **Flesh** brownish when wet, white when dry. **Odor** slight. **Taste** slight. **Spores** long ellipsoid, warty, 10.5–13.5 × 7–8.5μ, quotient 1.55. Deposit rusty red. **Habitat** in coniferous and mixed woods, especially under birch. Infrequent. Found over most of North America. **Season** September–October. **Not edible. Comment** At first sight this mushroom looks like a psathyrella, but a spore print will tell you that you have a cortinarius.

Cortinarius decipiens Fr. subgenus *Telamonia* **Cap** 2–4.5cm (¾–1¾in) across, sharply conical, eventually flat with an umbo; dull vinaceous buff, much darker blackish vinaceous in the center; covered at first with a hoary layer of fine fibrils giving it a shiny luster. **Gills** adnate; ochraceous tawny at first, then a very bright reddish cinnamon. **Stem** 30–70 × 2–5mm (1¼–2¾ × 3/32–¼in), equal; flesh brownish or pallid with a slight flush of purple at the apex; matted with white fibrils, sometimes showing slight zones. **Flesh** brown. **Odor** distinct when crushed, radishy(?). **Taste** slight. **Spores** broadly ellipsoid or pip-shaped, warty, 8–9 × 5–6.3μ, quotient 1.5. Deposit rusty brown. **Habitat** in conifer woods. Rare. Found in California and Michigan. **Season** September–November. **Not edible. Comment** My specimens are not conical as is typical in this species.

Cortinarius castaneus Fr. subgenus *Telamonia* **Cap** 2–5cm (¾–2in) across; very hygrophanus, dark chestnut brown when wet, edge white; silky. **Gills** adnexed; cinnamon brown. **Stem** 20–40 × 4–8mm (¾–1½ × 3/16–5/16 in), equal; slightly tapering at base; dirty whitish. **Flesh** whitish brown where really wet. **Odor** pleasant. **Taste** slight. **Spores** ovoid to pip-shaped, rough, 8–10 × 4.5–6μ, quotient 1.7. Deposit rusty brown. **Habitat** in coniferous or mixed woods. Found in northern California and New York. **Season** August–October. **Not edible. Comment** This mushroom was photographed in England.

Cortinarius scandens Fr. subgenus *Telamonia* **Cap**
(*continued on page 166*)

Cortinarius badius ¾ life-size

Cortinarius decipiens almost life-size

Cortinarius castaneus ¾ life-size

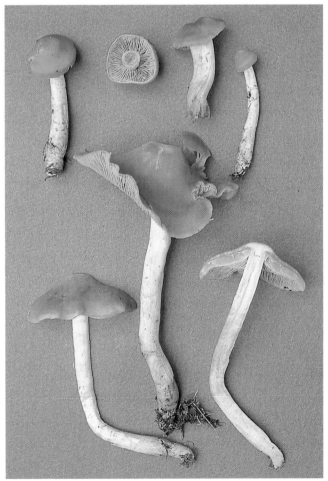

Cortinarius scandens ½ life-size

Cortinarius paleaceus 1/3 life-size

Cortinarius hemitrichus 1/2 life-size

Cortinarius rigidus 1/2 life-size

Cortinarius alnetorum 2/3 life-size

Cortinarius incisus 1/2 life-size

geraniums. **Taste** slight. **Spores** ellipsoid to pip-shaped, lightly roughened, 7.2–8.2 × 5–5.6(6)μ, quotient 1.45. Deposit rusty brown. **Habitat** under conifers in boggy places, often with sphagnum moss. Fairly common. Found in northern North America. **Season** September–October. **Not edible. Comment** The geranium smell is very distinct.

Cortinarius hemitrichus Fr. subgenus *Telamonia* **Cap** 2–5cm (3/4–2in) across, conical to hemispherical, then flattish with an umbo; cinnamon brown under the dense layer of tiny white hair scales; hygrophanous. **Gills** adnexed; pale brownish then cinnamon. **Stem** 30–60 × 3–8mm (1 1/4–2 1/4 × 1/8–5/16in), base may be swollen; woolly white all over with white, ringlike zones. **Flesh** pallid when dry, brown where wet, sometimes with a hint of violet when young. **Odor** none. **Taste** slight but distinct. **Spores** ellipsoid, finely roughened, 7–8 × 4–5μ, quotient 1.65. Deposit rusty brown. **Habitat** in mixed woods, especially with birch. Found in eastern and central North America. **Season** August–October. **Not edible. Comment** This mushroom was photographed in England.

Cortinarius rigidus Fr. sensu Kühner & Romagnesi subgenus *Telamonia* **Cap** 1.5–4cm (1/2–1 1/2in) across, conic with an acute umbo; hygrophanous, deep red-brown when wet, buffy brown when dry; smooth to fibrillose, the margin coated with silky white fibrils from the veil. **Gills** adnate; strong bright cinnamon color from the first. **Stem** 40–80 × 3–5mm (1 1/2–3 × 1/8–1/4in), equal; buffy color; fibrillose with distinct patches or bands of white veil remnants. **Flesh** reddish brown. **Odor** strong, of household geraniums. **Taste** distinct. **Spores** almond-shaped, finely roughened, 7.5–8.5 × 4.5–5.5μ, quotient 1.6. Deposit rusty brown. **Habitat** in coniferous woods. Occasional. Found in the Pacific Northwest and east to Colorado. **Season** September–October. **Not edible.**

Cortinarius alnetorum (Vel.) Moser subgenus *Telamonia* **Cap** 1–3cm (1/2–1 1/4in) across, rather flat but with a large, prominent umbo; cocoa brown or darker when wet, drying brownish fawn, umbo purplish black; covered in fine fibrils giving the cap a shiny appearance. **Gills** adnate, rather distant; dark cocoa brown from the beginning. **Stem** 25–65 × 2–4mm (1–2 1/2 × 3/32–3/16in), base sometimes swollen but not bulbous; dark brown to black with slight bands of white veil; fibrous, a little shiny. **Flesh** dark brown when wet, a little lighter on drying. **Odor** when crushed of geranium leaves. **Taste** slight. **Spores** ovoid, lightly roughened, 8.5–10 × 4.5–6μ, quotient 1.75. Deposit rusty brown. **Habitat** in alder bogs. Found in Washington and probably elsewhere. **Season** September–October. **Not edible. Comment** This is a European species not previously recorded in North America, but it fits perfectly and probably is much more common than this single record implies.

Cortinarius incisus Fr. subgenus *Telamonia* **Cap** 2–4cm (3/4–1 1/2in) across, convex-campanulate, with a distinct pointed umbo; hygrophanous, deep, dull blackish brown when wet, pallid grayish brown (snuff-colored) when dry; covered in fine silky fibrils, margin splitting in age. **Gills** adnexed; grayish ochre then more rusty. **Stem** 25–60 × 2–5mm (1–2 1/4 × 3/32–1/4in), equal; dull grayish brown, with remnants of white veil sometimes forming a belt. **Flesh** blackish brown when wet; dull grayish ochre when dry, more reddish in stem base. **Odor** none. **Taste** slight. **Spores** ellipsoid to pip-shaped, roughened, 7–8.7 × 4.6–5.5μ, quotient

(continued from page 164)
1–4cm (1/2–1 1/2in) across, conic, then flat with an umbo; hygrophanous, watery yellow-brown when wet, pale yellowish ochre when dry; smooth. **Gills** adnexed; pallid yellowish buff then rusty. **Stem** 60–100 × 2–5mm (2 1/4–4 × 3/32–1/4in), equal; creamy colored, a little darker near the base. **Flesh** yellowish. **Odor** slight, similar to iodine. **Taste** mild. **Spores** broadly ovate, rough, 6–7.5 × 4.25–5μ, quotient 1.46. Deposit rusty brown. **Habitat** under conifers. Occasional. Found in northern North America. **Season** September–October. **Not edible.**

Cortinarius paleaceus Fr. subgenus *Telamonia* **Cap** 1.5–4.5cm (1/2–1 3/4in), broadly conical; orangy brown in the center to brown nearer the edge under a layer of very fine white hair scales; hygrophanous. **Gills** adnate; pallid then milky coffee-colored. **Stem** 40–80 × 2–5mm (1 1/2–3 × 3/32–1/4in), equal or swollen a little near the base; covered in white veil, which tends to form into ringlike zones; the mycelium sometimes tends to be pallid violet. **Flesh** pallid, brownish. **Odor** strong, of household

1.55. Deposit rusty brown. **Habitat** in conifer woods, usually in ditches. Uncommon. Found in Colorado and New York, probably other states too. **Season** September–October. **Not edible. Comment** Often found growing in small clumps.

Cortinarius incognitus Ammirati & Smith subgenus *Dermocybe* **Cap** 1.5–6cm (½–2¼in) across; broadly conic, then flattish with a low umbo; yellowish brown to cinnamon; appressed fibrillose. **Gills** adnate to sinuate; amber yellow at first, then more rusty. **Stem** 20–60 × 2–8mm (¾–2¼ × ³⁄₃₂–⁵⁄₁₆in), equal or tapering downward; pallid yellowish, base rusty brown or darker; surface fibrillose. **Flesh** yellowish at first, then brown. **Odor** normal. **Taste** slight. **Spores** ellipsoid or ovate, rough, 7–9 × 4–5μ, quotient 1.8. Deposit rusty brown. **Habitat** in conifer woods, especially pine. Found in Michigan and New Mexico and probably other areas. **Season** July–November. **Not edible.**

Cortinarius tubarius Ammirati & Smith subgenus *Dermocybe* **Cap** 2–5cm (¾–2in) across, broadly campanulate becoming flattish with an umbo; yellowish olive then more olive-brown; covered in appressed fibrils. **Gills** adnexed; light yellow to yellowish green at first, eventually yellowish brown. **Stem** 50–100 × 4–7mm (2–4 × ³⁄₁₆–¼in), equal; dull buffy yellow at first, then more brownish; not covered in veil fibers. **Flesh** dull buffy olive, more brownish at the stem base. **Odor** slight. **Taste** slight. **Spores** ellipsoid to ovoid, roughened, 8.75–10.5 × 5–6μ, quotient 1.75. Deposit rusty brown. **Habitat** in mossy conifer woods. Occasional. Found in Washington and probably other similar habitats. **Season** September–October. **Not edible. Comment** KOH causes an olive-fuscous reaction on the cap surface and an olive reaction on the flesh.

Cortinarius huronensis var. *olivaceus* Ammirati & Smith subgenus *Dermocybe* **Cap** 2–4cm (¾–1½in) across, broadly conical, then flat with a low umbo; buffy olive in color; fibrillose. **Gills** adnexed; olive-green when young, then more brownish yellow. **Stem** 60–120 × 6–10mm (2¼–4¾ × ¼–½in), tall, slender, slightly thicker near the base; pallid olive to ochre; fibrillose, the base covered in whitish tomentum, which clings to the sphagnum moss. **Flesh** yellowish or olivaceous brown in the stem base. **Odor** not distinctive. **Taste** radishlike. **Spores** ovate, inequilateral, roughened, rather variable, 8.7–11(12.8) × 5–6(6.5)μ, quotient 1.8. Deposit rusty brown. **Habitat** in sphagnum bogs. Found in New York and Michigan and probably many other areas if sphagnum present. **Season** August–November. **Not edible. Comment** As the bogs dry out a little in late summer, this mushroom is sometimes found in profusion.

Cortinarius huronensis Ammirati & Smith subgenus *Dermocybe* **Cap** 2–4cm (¾–1½in) across, convex, umbonate; olivaceous, more brown in the center and eventually all over. **Gills** adnate to sinuate; yellow at first, eventually rusty yellow. **Stem** 30–75 × 3–6mm (1¼–3 × ⅛–¼in), equal or slightly swollen at the base; yellowish with reddish tints near the base, in age more olivaceous with brown base. **Flesh** yellowish. **Odor** radishy. **Taste** of radish. **Spores** ellipsoid, warty, 8.5–10 × 5–5.5μ, quotient 1.75. Deposit rusty brown. Clamp connections present. **Habitat** in mosses under conifers. Uncommon. I know only of records for Michigan and Washington, but it is probably more widespread. **Season** September. **Not edible. Comment** KOH causes a reddish-brown reaction on the cap.

Cortinarius incognitus ½ life-size

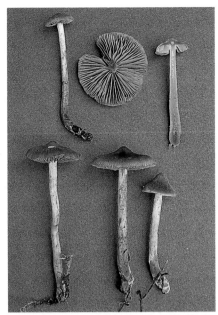

Cortinarius tubarius ½ life-size

Cortinarius huronensis var. *olivaceus* ⅓ life-size

Cortinarius huronensis ½ life-size

Cortinarius pseudotubarius Robar, Grund, & Harrison subgenus *Dermocybe* **Cap** 1.5–4cm (½–1½in) across, broadly conical or convex, with a low, broad umbo; brownish or reddish ochre, or deep golden; appressed fibrillose. **Gills** adnexed; rich, deep yellow. **Stem** 40–90 × 3–8mm (1½–3½ × ⅛–⁵⁄₁₆in), tall, equal or swollen toward the base; ochre or reddish ochre, especially near the base, which has a layer of red-brown fibrils. **Flesh** pallid ochre, becoming darker, more red-brown in age, especially near the stem base. **Odor** not distinctive. **Taste** a touch radishy. **Spores** ellipsoid, warty, 7.5–9 × 4.9–6μ, quotient 1.5. Deposit rusty brown. **Habitat** in sphagnum bogs or with other mosses in conifer or mixed woodlands. Found in New York, Nova Scotia, and probably other northern areas. **Season** August–October. **Not edible. Comment** The spores are smaller than those of *Cortinarius tubarius* (above), which it closely resembles.

Cortinarius pseudotubarius ⅓ life-size

Cortinarius sanguineus ¾ life-size

Cortinarius semisanguineus ½ life-size

Cortinarius cinnabarinus ½ life-size

Rozites caperata ½ life-size

Cortinarius malicorius ½ life-size

Cortinarius sanguineus (Fr.) S. F. Gray subgenus *Dermocybe* **Cap** 1.5–4cm (½–1½in) across, deep maroon-red; dry, fibrillose to slightly scaly. **Gills** adnexed with a slight decurrent tooth; deep maroon-red. **Stem** 40–80×3–8mm (1½–3×⅛–⁵⁄₁₆in), equal or slightly widening downward; dull red, lighter than the cap when young but becoming darker with age; cortina dark red. **Flesh** dull maroon. **Odor** slight, a little radishy. **Taste** mild, a touch of radish. **Spores** almond-shaped, lightly roughened, 7.3–8.9×4.5–5.2µ, quotient 1.65. Deposit rusty brown. **Habitat** under conifers. Rather rare. Found in northern North America. **Season** September–October. **Not edible.**

Cortinarius semisanguineus (Fr.) Gillet subgenus *Dermocybe*. **Cap** 2–6cm (¾–2¼in) across, convex then flattened, slightly umbonate; ochre to olive-buff or reddish; dry, minutely fibrillose-scaly. **Gills** adnate, narrow, crowded; deep cinnabar to blood red. **Stem** 20–100×4–12mm (¾–4×³⁄₁₆–½in); ochre to olivaceous, paler at

Phaeocollybia christinae ¾ life-size

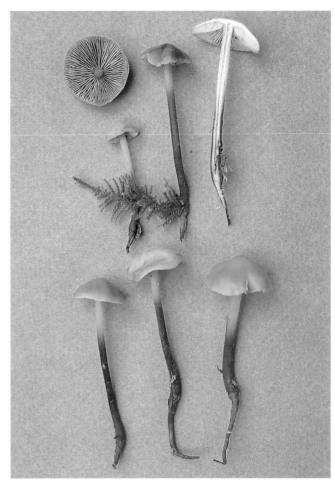

Phaeocollybia similis ⅔ life-size

apex, with sparse olive fibers below; base often with pinkish to reddish down. **Flesh** ochraceous. **Odor** slight, of radish. **Taste** slightly bitter. **Spores** ellipsoid, roughened, 6–8 × 4–5μ, quotient 1.55. Deposit rust-brown. **Habitat** in mixed conifer-deciduous woods. Frequent. Found throughout much of North America. **Season** August–October. **Edibility suspect** — best avoided.

Cortinarius cinnabarinus Fr. subgenus *Dermocybe* **Cap** 2–6cm (¾–2¼in) across, conic, then plane with a slight umbo; cinnabar red, fading with age; matted with dry, shining, innate fibrils. **Gills** adnate; cinnabar red then dark rusty red. **Stem** 20–50 × 3–8mm (¾–2 × ⅛–⁵⁄₁₆in), normally equal but base sometimes swollen; dull cinnabar red, fading; silky. **Flesh** pallid with reddish tints. **Odor** slight. **Taste** mild. **Spores** ellipsoid, very rough, 8.5–10 × 4–6μ, quotient 1.85. Deposit rusty brown. **Habitat** in oak and beech woods. Uncommon. Found in northeastern North America, south to North Carolina and west to Wisconsin. **Season** August. **Not edible. Comment** My specimens are not as completely cinnabar red as this species normally is. It is probable that American collections will be found to differ from the Friesian species and will be given another name, *Cortinarius heslerii* Ammirati & Smith ined.

Cortinarius malicorius Fr. syn. *Cortinarius croceofolius* Pk. subgenus *Dermocybe* **Cap** 2.5–5cm (1–2in) across, hemispherical then broadly convex; brownish cinnamon, margin more yellow; fibrillose to minutely scaly. **Gills** adnexed; bright orange at first, then bright cinnamon orange. **Stem** 20–40 × 4–7mm

(¾–1½ × ³⁄₁₆–¼in), equal or slightly swollen at the base, soon hollow; bright yellow often with reddish area near the base; cortina yellow. **Flesh** yellow. **Odor** slight. **Taste** slight. **Spores** ovoid, rough, 6–7.5 × 4–5μ, quotient 1.5. Deposit rusty brown. **Habitat** in coniferous woods. Uncommon. Found in northeastern North America. **Season** August–September. **Not edible.**

Rozites caperata (Fr.) Karsten **Cap** 5–15cm (2–6in) across, convex with flattened center, then soon expanded; pale ochraceous to tawny brown, with a hoary coating of fine white veil at center; dry. **Gills** adnate, broad; cream then pale tawny yellow. **Stem** 60–120 × 10–20mm (2¼–4¾ × ½–¾in), often wormy and fragile; pale ochre, with a white, fragile annulus in the middle. **Flesh** soft; white. **Odor** pleasant. **Taste** pleasant. **Spores** ellipsoid, warty, 12–14 × 7–9μ. Deposit rust-brown. **Habitat** in coniferous woods, often under blueberries or huckleberries (*Vaccinium*). Found in northern and eastern North America. **Season** July–September. **Edible** — good.

Phaeocollybia christinae (Fr.) Heim syn. *Phaeocollybia lateraria* Smith **Cap** 1–5cm (½–2in) across, sharply to bluntly conical with a pointed umbo; bright orange-red to cinnamon-foxy; moist, somewhat sticky, shining. **Gills** slightly adnexed, crowded, narrow; light to dark rust-yellow. **Stem** 40–120 × 3–6mm (1½–4¾ × ⅛–¼in), long, slender, rooting; pale reddish yellow at the top, darker wine-brown toward the base; cartilaginous, smooth. **Flesh** firm, brittle; same color as cap. **Odor** strong, plantlike. **Taste** rather tart. **Spores** almond-shaped, roughened, 8.7–9.8 × 4.7–5.3μ. Deposit

rust-brown. Cheilocystidia club-shaped. **Habitat** on poor soil in wet conifer woods. Found in Maine and reported from the Pacific Northwest. **Season** August–November. **Not edible. Comment** The material photographed was found in Maine. The material found in the Pacific Northwest is quoted as having much smaller spores, 5–6 × 3.5–3.8μ, making it very different from the European species described by European authors.

Phaeocollybia similis Smith **Cap** 1.5–3.5cm (½–1¼in) across, obtusely cone-shaped expanding to convex and then becoming flatter with a slight umbo; amber yellow to cinnamon or rusty red; smooth, moist, naked. **Gills** nearly free, crowded, moderately broad; dark, dull ochraceous tawny in maturity. **Stem** 60–100 × 1–3mm (2¼–4 × ¹⁄₁₆–⅛in), tapering downward into a long rootlike base; pale yellow-brown at the top, becoming very dark and almost black at the base; smooth, polished. **Flesh** thin, firm; yellowy buff. **Odor** strong, sweet to nauseating. **Taste** bitter. **Spores** ellipsoid to pip-shaped, warty, 7.5–8.7 × 5.4–6.2μ. Deposit rusty brown. **Habitat** in groups in virgin forest. Rare. Found in the Pacific Northwest. **Season** September–October. **Not edible. Comment** My spores are a little smaller than those described by Alexander H. Smith.

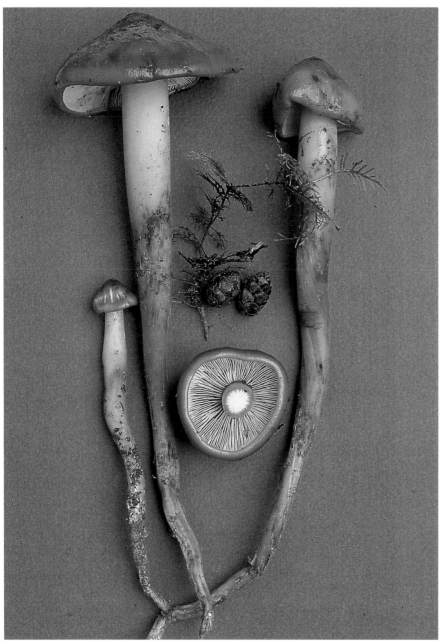

Phaeocollybia kauffmanii ½ life-size

Phaeocollybia kauffmanii Smith **Cap** 8–25cm
(3–10in) across, obtusely conic expanding to
broadly umbonate or nearly flat, with an
inrolled margin that never fully expands;
orangy brown or rusty red fading to reddish
orange; smooth, very slimy becoming merely
sticky, subhygrophanous. **Gills** free or slightly
adnexed, crowded, narrow becoming broad;
dirty cream becoming dirty orangish brown.
Stem 20–40 × 15–35mm (¾–1½ × ½–1¼in),
stuffed, tapering downward to a long rootlike
base; flesh-pink above, rusty, purplish brown
below, becoming darker in age; smooth with a
thick, cartilaginous rind. **Flesh** thick, firm,
with a cartilaginous rind near the surface,
moist; red-brown. **Odor** mealy when flesh is
crushed. **Taste** slightly mealy. **Spores**
ellipsoid, almond-shaped, roughened, 8–8.5 ×
5–5.5μ. Deposit rusty brown. **Habitat**
scattered or in groups under conifers. Found in
the Pacific Northwest. **Season** August–
October. **Not edible. Comment** This is a
remarkable mushroom because of its large size
and the depth to which the stem may extend.

Phaeocollybia attenuata Smith **Cap** 1.5–5cm
(½–2in) across, conical becoming convex then
almost flat with a low umbo, margin inrolled
at first, becoming wavy in age; amber to brown
when moist, fading to yellowish buff or pale
yellow; smooth, polished, hygrophanous. **Gills**
adnexed or nearly free, close, narrow; pallid to
dingy reddish buff, becoming pale tan with
darker, rust-brown spots. **Stem** 100–120 ×
3–10mm (4–4¾ × ⅛–½in), hollow, with a
long rootlike base; pale pinky-brown all over,
becoming darker rusty brown or blackish in
age; very cartilaginous, smooth, polished.
Flesh thin but cartilaginous; buff to watery
brown. **Odor** strongly radishy. **Taste** slowly
becoming very unpleasant. **Spores** ellipsoid to
almond-shaped, 7.5–9.6 × 5–6.3μ. Deposit rust-
brown. **Habitat** in groups under Sitka spruce
and redwood. Found in the Pacific Northwest
and California. **Season** September–November.
Not edible.

Phaeocollybia fallax Smith **Cap** 1–5cm (½–2in)
across, conical expanding to flat with a distinct
pointed umbo, margin inrolled when young,

becoming faintly striate when older; olive or
olive-buff fading to light yellow-green; smooth,
slimy when wet, hygrophanous. **Gills** free or
slightly adnexed, close, narrow; violet to violet-
gray becoming rust-brown. **Stem** 70–120 ×
3–10mm (2¾–4¾ × ⅛–½in), hollow, tapering
to a long rootlike base; faint bluish gray at the
top, brownish below from minute, fine hairs;
smooth, cartilaginous. **Flesh** firm, brittle, thick
at the disc, thin at the margin; olive-green
fading to olive-buff. **Odor** radishy. **Taste** mild.
Spores ovoid with a porelike beak at the tip,
warty, 8–9 × 5–6μ. Deposit dark rust. **Habitat**
scattered in large groups under Douglas fir,
Sitka spruce, or redwood. Found in the Pacific
Northwest and California. **Season** September–
November. **Not edible. Comment** Finding this
fungus in Washington was one of the highlights
of my years of American collecting.

Paxillus involutus (Fr.) Fr. **Cap** 5–15cm (2–6in)
across, broadly convex with an inrolled
margin, becoming flatter then somewhat
sunken in the center; ochraceous or dingy
brown with an olive tinge, becoming more
rusty brown then finally hazel or snuff brown;
sticky at the center when wet, then dry and
downy becoming smooth. **Gills** decurrent,
crowded; pale ochre then sienna, bruising
pinky-brown to chestnut. **Stem** 20–80 ×
5–40mm (¾–3 × ¼–1½in), solid, sometimes
slightly off-center; same color as cap, staining
chestnut in age or when bruised; firm, dry,
smooth. **Flesh** thick, firm; pale ochre in cap,
fulvous in stem base, darkening on cutting.
Odor mushroomy. **Taste** acidic. **Spores**
ellipsoid, smooth, 8–10 × 5–6μ. Deposit sienna.
Habitat scattered or in dense groups in woods,
parks, lawns, and on bog edges. Common.
Found widely distributed in North America.
Season July–November. **Deadly poisonous.**
Although sometimes eaten in Europe, it is now
known to cause hemolysis and kidney failure.

Paxillus atrotomentosus (Fr.) Fr. **Cap** 4–18cm
(1½–7in) across, convex becoming flatter,
depressed in the center with an inrolled
margin; snuff brown or sepia with sienna
patches; dry, downy. **Gills** generally decurrent,
crowded, often forked or veined near stem; tan
to dingy yellow. **Stem** 20–90 × 10–40mm
(¾–3½ × ½–1½in), sometimes lateral, solid,
rooting; top paler or more yellowish than rest
of stem, which is covered in fine olivaceous-buff
down that becomes coarser, more velvety and
darker brown in age. **Flesh** thick, firm to
tough; cream, ochre, or buff in stem. **Odor** not
distinctive. **Taste** not distinctive. **Spores**
ellipsoid, smooth, 5–6.5 × 3–4.5μ. Deposit
sienna. **Habitat** singly or in groups or clumps
on conifer stumps. Sometimes frequent. Found
in northern North America and southward
along coastal mountain regions. **Season**
July–October. **Not edible.**

Phaeocollybia attenuata ½ life-size

Phaeocollybia fallax ½ life-size

Paxillus involutus ⅔ life-size

Paxillus atrotomentosus ⅓ life-size

Gymnopilus luteus ½ life-size

Europe this mushroom would be considered a cortinarius rather than a gymnopilus.

Gymnopilus punctifolius (Pk.) Singer **Cap** 2.5–10cm (1–4in) across, convex becoming flatter with an inrolled margin; buttons pinky-brown lilac then dull greenish or bluish green, blue and yellow mixed; dry, slightly hairy or scaly around the disc, becoming smooth in age. **Gills** adnate to sinuate, close to subdistant, broad; olive-yellow when young, becoming dotted with yellow or rusty-red stains in age, particularly at the edges. **Stem** 100–150 × 5–10mm (4–6 × ¼–½in), stuffed then hollow; same color as cap, staining brownish yellow or olive-ochre within; wavy, lined. **No Veil. Flesh** thin at the edge; greenish yellow. **Odor** pleasant, not distinctive. **Taste** very bitter. **Spores** ovoid or ellipsoid, warty or roughened, 4–5.5 × 3.5–4μ. Deposit brownish rust. Clamp connections present. **Habitat** singly or scattered on debris, rich humus, and coniferous wood. Frequent. Found in the Pacific Northwest and California. **Season** August–December. **Not edible.**

Gymnopilus aeruginosus (Pk.) Singer **Cap** 2–5cm (¾–2in) across, convex; dull bluish gray-green or variegated with pink or red patches, becoming warm pinkish buff then drab brown when dried; dry then covered with small tawny or blackish scales and patches. **Gills** adnexed to adnate, crowded, broadish, numerous; creamy buff to pale yellowish orange. **Stem** 40–120 × 10–20mm (1½–4¾ × ½–¾in), solid, becoming more or less hollow; similar color as cap; smooth or minutely hairy, sometimes lined with a hairy base. **Veil** fibrillose, yellowish, often leaving a slight zone at top of stem. **Flesh** whitish tinged greenish or bluish green, becoming yellowish or pinkish brown when dry. **Odor** mild. **Taste** bitter. **Spores** ellipsoid, minutely hairy, 6–8.5 × 4–4.5μ. Deposit dark reddish orange or brownish rust. Pleurocystidia rare; clamp connections present. **Habitat** in tufts on logs or stumps on hardwoods and conifers. Found widely distributed throughout North America. **Season** May–November. **Not edible. Comment** Occasionally the cap can reach 23cm (9in) across. My photograph does not show the bluish gray-green colors very distinctly.

Gymnopilus ventricosus (Earle) Hesler **Cap** 6–10cm (2¼–4in) across, convex, obtuse, with an even margin sometimes hanging with bits of veil remnants; orange-yellow to reddish brown, often with a lighter disc; covered with minute yellow hairs, sometimes becoming scaly or almost smooth. **Gills** subsinuate, crowded, rather broad; pallid, becoming cinnamon in age. **Stem** 140–180 × 20–30mm (5½–7 × ¾–1¼in), solid, swollen in the middle; pale brownish with dense white hairs at the top and fine yellow hairs below; rooting and covered with a white mycelium at the base. **Veil** forms a thick, persistent jagged ring at top of stem. **Flesh** pale yellow. **Odor** none. **Taste** bitter. **Spores** ellipsoid or ovoid, warty, 7.5–9 × 4–5.5μ. Deposit rusty brown. Clamp connections present. **Habitat** in groups or dense tufts at the base of living pine. Found in the Pacific Northwest and California. **Season** September–December. **Not edible.**

Gymnopilus luteus (Pk.) Hesler **Cap** 5–10cm (2–4in) across, convex, with an incurved margin; saffron yellow to buff-yellow, often a little darker at the center; dry, silky or minutely downy, or sometimes with minute woolly scales toward the middle. **Gills** adnexed, thin, close, moderately narrow; pale yellow, becoming dark rust-red with age. **Stem** 40–75 × 6–16mm (1½–2¾ × ¼–½in), solid; same color as cap, becoming rusty yellowish when handled; minutely hairy. **Veil** forms a hairy or submembranous ring. **Flesh** firm; pale yellow. **Odor** pleasant. **Taste** bitter. **Spores** ellipsoid, minutely warty, 6–9 × 4–5μ. Deposit brownish rust. No caulocystidia; clamp connections present. **Habitat** on decaying wood, stumps, logs, and trees. Found in eastern North America. **Season** August–September. **Not edible. Comment** Often mistaken for *Gymnopilus junonius* (Fr.) P. D. Orton.

Gymnopilus terrestris Hesler **Cap** 3–7cm (1¼–2¾in) across, convex with an umbo, becoming flatter and sometimes depressed in age; amber brown, rusty red, or tawny, somewhat paler when drying; smooth, moist or sometimes slightly sticky, and with a translucently lined edge. **Gills** adnate to sinuate, close, broad to medium broad; dirty yellowish tawny, rusty brown, or rust-red. **Stem** 60–120 × 5–10mm (2¼–4¾ × ¼–½in), hollow; whitish or rusty brown to rust-red; smooth. **No veil. Flesh** fragile, disc thick, margin thin, watery; same color as cap fading to warm buff. **Odor** mild. **Taste** mild. **Spores** ellipsoid to subglobose, warty, 5.5–7 × 4.5–5.5μ. Deposit brownish rust. No pleurocystidia or cheilocystidia; clamp connections present. **Habitat** in soil or humus under conifers. Found in Michigan, Colorado, and the Pacific Northwest. **Season** June–October. **Not edible. Comment** In

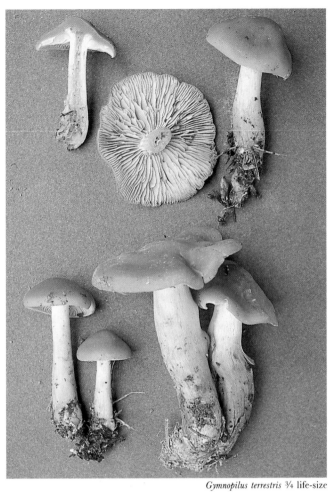

Gymnopilus terrestris ¾ life-size

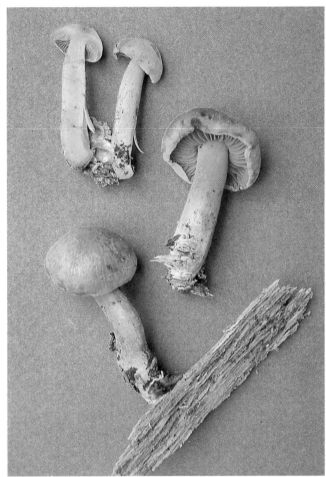

Gymnopilus punctifolius life-size

Gymnopilus aeruginosus ⅓ life-size

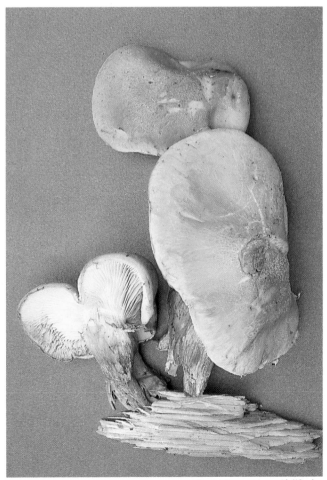

Gymnopilus ventricosus ⅓ life-size

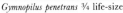

Gymnopilus penetrans ¾ life-size   *Gymnopilus humicola* nearly life-size

Gymnopilus penetrans (Fr. ex Fr.) Murr. syn.
Flammula penetrans (Fr. ex Fr.) Quél. **Cap**
2–5cm (¾–2in) across, bell-shaped to convex
then flattened, often with a wavy margin;
chrome yellow to golden then tawny and
fading yellowish in age; smooth. **Gills** adnate,
close, moderately broad; gold or yellowish
white becoming tawny-spotted. **Stem**
40–60 × 4–7mm (1½–2¼ × ³⁄₁₆–¼in),
sometimes enlarged toward base; yellowish;
base whitish with downy hairs. **Veil** white,
fibrous; leaves no ring. **Flesh** whitish. **Odor**
none or mild. **Taste** bitter. **Spores** ellipsoid,
warted, 7–9 × 4–5.5µ. Deposit orange-brown.
Cheilocystidia present; no pileocystidia; clamp
connections present. **Habitat** singly or in small
tufts on hardwoods and conifers. Found
throughout North America. **Season**
July–October. **Not edible.**

Gymnopilus humicola Harding ex Singer **Cap**
2–5cm (¾–2in) across, convex-flattened with
wavy inrolled margin; reddish brown, auburn,
tawny orange to russet; smooth, glabrous to
finely squamulose, cracking when dry. **Gills**
adnate-decurrent, broad; bright yellow then
orange-brown from the spores, staining deep
rust-red where bruised. **Stem**
25–100 × 3–10mm (1–4 × ⅛–½in), equal; pale
orange-buff to tawny orange, staining darker
where handled; fibrillose, base with cream-
colored tomentum. **No veil. Flesh** pale buff-
yellow. **Odor** pleasant, slightly aromatic. **Taste**
bitter. **Spores** warty, dextrinoid, 7–8.5(9) ×
4–5µ. Deposit bright rust-brown.
Pleurocystidia and cheilocystidia both present,
capitate. **Habitat** usually on soil in humus or
in flower beds mulched with wood chips.
Found widely distributed in North America

west of the Mississippi River. **Season**
June–December. **Not edible.**

Pholiota limonella (Pk.) Sacc. **Cap** 2.5–5cm
(1–2in) across, convex or nearly flat,
sometimes with an umbo; lemon yellow when
fresh, scattered with hairy, tawny red, suberect
scales; sticky. **Gills** sinuate-adnate, close,
narrow; whitish becoming rust-red. **Stem**
30–70 × 3–5mm (1¼–2¾ × ⅛–¼in); pallid or
yellowish with scattered recurved yellow scales;
smooth above the ring. **Veil** forms a cottony
evanescent yellow ring. **Flesh** thin; yellow.
Odor mild. **Taste** mild. **Spores** ovoid to
ellipsoid, smooth, distinct pore at apex,
6–7.5 × 4–5µ. Deposit rusty brown.
Pleurocystidia and cheilocystidia present;
caulocystidia rare. **Habitat** on fallen beech
trunks. Found in New York and Idaho. **Season**
September. **Not edible.**

Pholiota aurivella (Fr.) Kummer **Cap** 4–15cm
(1½–6in) across, bell-shaped to convex with a
broad umbo; ochre-orange to tawny; sticky to
slimy with large flattened spotlike scales, which
may disappear or become somewhat sticky
when wet. **Gills** adnate, close, moderately
broad; pale yellowish becoming tawny brown.
Stem 50–80 × 5–15mm (2–3 × ¼–½in), dry,
solid, central or off-center; yellowish to yellow-
brown; dry and cottony above the ring, hairy
and with downcurving scales toward the base.
Veil partial veil leaves evanescent ring or zone
on upper stalk; white. **Flesh** firm; yellow.
Odor sweet. **Taste** slight. **Spores** ellipsoid,
smooth, with pore at apex, 7–9.5 × 4.5–6µ.
Deposit rusty brownish. Caulocystidia absent;
pleurocystidia present. **Habitat** in clusters on
living trunks and logs of hardwoods and
conifers. Found in North America except the

Southeast. **Season** June–November. **Not
edible.**

Pholiota flavida (Fr.) Singer **Cap** 3–7cm
(1¼–2¾in) across, convex expanding to
almost flat, with an incurved margin with some
faint veil remnants; yellow to dingy, watery
yellow-ochre or tawny; thinly sticky and
smooth. **Gills** adnate to adnexed, close, narrow
to moderately broad, edges even; pallid,
becoming pale rusty brown in age. **Stem**
50–110 × 5–15mm (2–4½ × ¼–½in), solid,
slightly tapering to the base; pallid above fine
hairy zone of evanescent yellowish veil, dark
rust-brown from base upward; grooved and
finely hairy in lower part. **Flesh** thick, firm,
yellowish. **Odor** faintly fragrant. **Taste** mild.
Spores oval to subellipsoid, smooth, distinct
pore at apex, 7–9 × 4–5µ. Deposit cigar brown.
No pleurocystidia; cheilocystidia versiform and
caulocystidia similar. **Habitat** in large clusters
on logs and stumps and at the base of
coniferous and hardwood trees. Found in
Maine and the Pacific Northwest. **Season**
August–November. **Not edible.**

Pholiota flammans (Fr.) Kummer **Cap** 2–10cm
(¾–4in) across, convex then expanded with an
incurved margin; tawny yellow covered in
recurved lemon to sulphur-yellow fibrous
scales; sticky. **Gills** adnate, close, moderately
broad; pale yellow, darkening to rusty yellow
with age. **Stem** 40–120 × 5–10mm
(1½–4¾ × ¼–½in), sometimes enlarged at
base; bright yellow with concolorous cottony
ring near the top; densely covered in
concolorous scales below ring, silky above.
Veil partial veil, yellow, leaving evanescent
zone of fibers on upper stalk. **Flesh** firm,
pliant, thick; pale yellow. **Odor** mild. **Taste**

(continued on page 176)

Pholiota limonella ⅔ life-size

Pholiota aurivella ⅓ life-size

Pholiota flavida ½ life-size

Pholiota flammans ¾ life-size

Pholiota astragalina ½ life-size

Pholiota albocrenulata ⅓ life-size

Pholiota mutabilis ⅓ life-size

Pholiota malicola ⅓ life-size

Pholiota multifolia ½ life-size

Pholiota highlandensis ½ life-size

(continued from page 174)

mild. **Spores** oblong to ellipsoid, smooth, 4–5 × 2.5–3μ. Deposit rusty. Pleurocystidia lanceolate with pointed apex. **Habitat** singly or in small clusters on conifer logs and stumps. Found in northeastern North America, south to Tennessee, and in the Pacific Northwest. **Season** August–October. **Not edible**.

Pholiota astragalina (Fr.) Singer **Cap** 2–5cm (¾–2in) across, cone-shaped with an obtuse umbo, becoming flatly bell-shaped with a spreading or upturning or wavy margin; pinky-orange or apricot with a paler margin, fading in age and discoloring blackish in places; smooth, sticky or slimy when wet, pallid, hairy veil remnants on the margin. **Gills** sharply and deeply adnexed, appearing free, close, moderately broad; orange-yellow, but discoloring where bruised. **Stem** 50–90 × 4–7mm (2–3½ × ³⁄₁₆–¼in), hollow; pale yellow to drab orange at the base, brownish where handled; fibrous. **Veil** yellow. **Flesh** pliant, watery; orange to yellowish orange. **Odor** distinct. **Taste** bitter. **Spores** oval to ellipsoid, smooth, with a minute germ pore, 5.5–6.5 × 3.5–4μ. Deposit brown. **Habitat** singly or in small groups on logs, stumps, and rotting coniferous wood. Found in northeastern North America, the Pacific Northwest, and California. **Season** August–October. **Not edible**.

Pholiota albocrenulata (Pk.) Sacc. **Cap** 3–9cm (1¼–3½in) across, conic to convex, becoming flatter with an obtuse umbo; orange-brown to red-brown; glutinous to sticky, dotted with brown scaly patches of veil remnants, which sometimes also hang from the margin. **Gills** adnate to subdecurrent with a decurrent tooth, close, very broad; whitish becoming grayish then rusty umber; edges uneven, beaded with drops of white liquid. **Stem** 30–110 × 5–15mm (1¼–4½ × ¼–½in), stuffed becoming hollow; pale and grayish above, dark brown below; firm, fibrous, with rough brown scales up to the ring, covered with a bloom at the top. **Veil** brownish. **Flesh** thick; pallid. **Odor** not distinctive. **Taste** slight. **Spores** ellipsoid not equilateral, smooth, pore at the apex, thickened wall, 10–16 × 5.5–7.5μ. Deposit brown. No pleurocystidia; abundant cheilocystidia. **Habitat** singly or in small groups on stumps, logs, and trunks of hardwood trees, especially maple and elm. Frequent. Found in southern and eastern North America. **Season** July–October. **Not edible**.

Pholiota mutabilis (Schaeff. ex Fr.) Kummer syn. *Kuehneromyces mutabilis* (Schaeff. ex Fr.) Singer & Smith **Cap** 3–6cm (1¼–2¼in) across, convex to broadly umbonate; bright orange-cinnamon when moist, drying pale ochre from the center, and often appearing distinctly two-colored; smooth. **Gills** adnate; pallid to dull cinnamon. **Stem** 30–80 × 5–10mm (1¼–3 × ¼–½in); white above, ochre-tan below, and blackish brown at base; scaly-floccose below ring. **Veil** whitish. **Flesh** whitish brown. **Odor** not distinctive. **Taste** not distinctive. **Spores** ovate, with germ pore, 6–7.5 × 4–5μ. Deposit deep ochre-brown. **Habitat** in dense clusters on old logs and stumps of deciduous trees. Found widely distributed throughout North America. **Season** September–November. **Edible** and good, but beware of the deadly *Galerina autumnalis* (p. 186) and *Galerina marginata* (Fr.) Kühner, which lack stem scales and have rough spores.

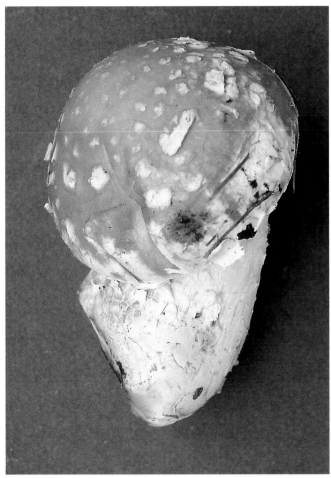

Pholiota destruens life-size

Pholiota squarrosoides ½ life-size

Pholiota malicola (Kauffman) Smith **Cap** 3–6cm (1¼–2¼in) across, obtusely cone-shaped, becoming flatter in age with an umbo; wax yellow or straw yellow, with a greenish tinge, then sulphur yellow with an opaque margin; sticky becoming dry and shining, with thin brown hairy patches of veil remnants when young. **Gills** adnate to sinuate, close to crowded; yellow becoming reddish cinnamon, edges white-fringed. **Stem** 40–120 × 4–12mm (1½–4¾ × 3/16–½in), solid; same color as cap or whitish at top, darker, more tawny below; hairy patches of pallid or yellowish veil remnants below the ring, becoming tawny like the base; evanescent submembranous or hairy ring, sometimes poorly formed. **Flesh** thick on disc, thin on margin, pliant; whitish. **Odor** not distinctive or slightly alkaline. **Taste** not distinctive or slightly alkaline. **Spores** ellipsoid to ovoid, smooth, small germ pore, 8.5–11 × 4.5–5.5μ. Deposit rust-red. No pleurocystidia; cheilocystidia and caulocystidia present. **Habitat** in dense clusters on stumps, debris, buried wood, or at the base of conifers and hardwoods. Found in northeastern North America. **Season** July–November. **Not edible**.

Pholiota multifolia (Pk.) Smith & Hesler syn. *Gymnopilus multifolius* (Pk.) Murr. **Cap** 5–8cm (2–3in) across, convex with an incurved margin, becoming flatter with a broad umbo; brilliant yellow becoming more tawny or orangish, sometimes paler on the margin; smooth, dry, and sometimes faintly hairy from veil remnants. **Gills** adnexed, crowded, narrow, with notched edges; similar color to cap or paler, becoming spotted or stained rusty brown or orange-brown when bruised. **Stem** 30–70 × 4–10mm (1¼–2¾ × 3/16–½in), solid,

tough, sometimes slightly enlarged at the base; yellow; woolly or hairy becoming nearly smooth. **Veil** yellowish, hairy; soon disappearing. **Flesh** fairly thick in big specimens; pale to bright yellow, becoming dirty rusty brown where bruised. **Odor** slight. **Taste** bitter. **Spores** ovoid to ellipsoid, smooth, 6.5–9 × 4.5–5μ. Deposit brownish rust. No caulocystidia; clamp connections present. **Habitat** on logs, sawdust, and decaying wood of deciduous trees. Found in eastern North America. **Season** June–October. **Not edible**.

Pholiota highlandensis (Pk.) Smith & Hesler **Cap** 2–5cm (¾–2in) across, broadly convex becoming flatter and somewhat depressed, sometimes with a low umbo; yellowy orange to cinnamon reddish brown with a paler margin, fading to ochraceous-buff colors; smooth except for veil remnants on the margin, hygrophanous. **Gills** adnate, close, broad, edges even or eroded; pallid or pale yellowish becoming cinnamon brown. **Stem** 20–40 × 3–6mm (¾–1½ × 1/8–¼in); top portion whitish to yellowish becoming dingy brown, lower portion pallid then brownish (darker than the top), with patches of pale yellow or buff veil remnants. **Flesh** thin; yellow. **Odor** not distinctive. **Taste** slightly disagreeable or none. **Spores** ellipsoid to oval, smooth, distinct pore at apex, 6–8 × 4–4.5μ. Deposit cinnamon brown. **Habitat** on burned-over soil or charred wood. Found in many parts of North America, though apparently not in the Northeast. **Season** April–November. **Not edible**.

Pholiota destruens (Brond.) Gillet **Cap** 8–20cm (3–8in) across, convex when young, expanding to more broadly convex in age, with a margin shaggy from veil remains; whitish, creamy, or

ochre, sometimes gradually darkening on the disc to nut brown or dark yellow-brown, with dingy white or buff, woolly scales or patches of veil remnants. **Gills** adnate to sinuate, close, broad, edges even; white when young, becoming deep rusty cinnamon from the spores. **Stem** 50–150 × 10–30mm (2–6 × ½–1¼in), enlarged to 70mm (2¾in) at the base, hard, solid; white then brownish in age, particularly in the lower section; numerous thick white cottony patches up to the evanescent, cottony ring; silky at the top. **Flesh** thick, firm; white. **Odor** not distinctive or mildly fungusy. **Taste** slightly disagreeable but hardly distinctive. **Spores** ellipsoid to oval, smooth, with pore at tip, 7–9.5 × 4–5.5μ. Deposit cinnamon brown. No pleurocystidia; caulocystidia abundant. **Habitat** singly or in clusters on logs and dead wood, particularly poplar, cottonwood, and aspen. Found in central and northern North America and New Mexico. **Season** September–November. **Edible**.

Pholiota squarrosoides (Pk.) Sacc. **Cap** 2.5–10cm (1–4in) across, obtusely convex with an umbo becoming flat; whitish becoming cinnamon with downcurved tawny scales scattered near the margin, clustered over the disc; sticky beneath the dry scales, veil remnants often hanging from the margin. **Gills** adnate becoming sharply adnexed, close to crowded, moderately broad; whitish, changing slowly to rust-brown as spores mature. **Stem** 50–150 × 5–15mm (2–6 × ¼–½in), stuffed or solid; whitish at the top, pale buff below the ring, with coarse downcurving ochre-tawny scales, sometimes staining rusty brown near the base; silky above the pale, fibrous, often

(continued on next page)

Pholiota bakerensis ¾ life-size

Pholiota lenta ½ life-size

(continued from preceding page)

evanescent ring. **Flesh** thick, pliant; whitish.
Odor not distinctive. **Taste** not distinctive.
Spores broadly ellipsoid, smooth, pore at apex
not evident, 4–5.5 × 3–3.5μ. Deposit brown.
Pleurocystidia abundant. **Habitat** singly or in
dense clusters on deciduous trees such as
beech, birch, maple, and alder. Common.
Found widely distributed throughout much of
North America. **Season** August–October.
Edible.

Pholiota bakerensis Smith & Hesler **Cap** 2–5cm
(¾–2in) across, broadly convex when young,
then flatter with disc shallowly depressed in
age, with an arched margin; yellowish tawny
brown on the disc, paler, more fawn-colored
toward the margin; sticky, opaque, with one or
more rows of appressed hairy scales glued to
the surface. **Gills** adnate, short decurrent by a
tooth, close, edges sawlike; tawny olive. **Stem**
30–50 × 3–5mm (1¼–2 × ⅛–¼in), solid or
hollowed by worms; clay color over lower part,
surface pallid from thin coating of pale buff
hairs; an evanescent hairy zone toward the top.
Flesh thin, pliant; pale, watery brown. **Odor**
faintly fragrant. **Taste** mild. **Spores** ellipsoid,
smooth, tiny pore at apex, 7–9 × 4–5μ. Deposit
cigar brown. Pleurocystidia abundant. **Habitat**
scattered on conifer sticks. Found in
Washington State. **Season** September. **Not
edible**.

Pholiota lenta (Fr.) Singer **Cap** 3–8cm (1¼–3in)
across, convex-hemispherical, becoming more
expanded in age; whitish to pinkish buff or

smoky gray, with a slightly darker disc; sticky
to slimy, with scattered white hairy scales of
veil remnants. **Gills** adnate or with a decurrent
tooth, close, narrow to medium-broad, edges
even to fringed; white becoming grayish brown.
Stem 30–100 × 4–12mm (1¼–4 × ³⁄₁₆–½in),
solid or spongy, sub-bulbous; white above,
brownish below; finely hairy. **Veil** copious,
cortinate; white; leaves an evanescent ring.
Flesh firm; white. **Odor** slight, radishy. **Taste**
mild. **Spores** ellipsoid to oblong, smooth, tiny
pore at apex, 5.5–7 × 3.5–4.5μ. Deposit cigar
brown. Pleurocystidia abundant. **Habitat** on
humus debris in mixed woods. Found in
eastern North America and California. **Season**
July–December. **Not edible**.

Pholiota decorata (Murr.) Smith & Hesler
Cap 3–8cm (1¼–3in) across, flatly convex
with a hairy fringed margin, becoming flatter
with an umbo and spreading margin; disc dark
brownish fading to fawn, the margin pale
brown to light fawn or dark cream; slimy to
sticky, with rows of concentric hairy scales
which often disappear before maturity; smooth
and appearing streaked with fine hairs beneath
the gluten in age. **Gills** adnate to sinuate,
close, moderately broad, edges even; white to
creamy yellowish. **Stem** 40–90 × 3–10mm
(1½–3½ × ⅛–½in), solid becoming hollow;
pallid to pale greenish yellow with dingy
pinkish brown flaky scales or patches on the
lower portion, top silky; sub-bulbous base with
brownish cortex. **Flesh** thick, watery to soft
and pliant; white, becoming yellowish in age.
Odor none or faintly fragrant. **Taste** mild.
Spores oval to ellipsoid, smooth, tiny pore at
apex, 6–7.5 × 3.5–4.5μ. Deposit cigar brown.
Pleurocystidia thick- or thin-walled. **Habitat**

singly or scattered on fallen conifer branches
and debris. Found from the Pacific Northwest
east to Idaho and Colorado, and in California.
Season July–November. **Not edible**.

Pholiota scamba (Fr.) Moser **Cap** 1.5–3cm
(½–1¼in) across, broadly convex becoming
flatter, sometimes with a low umbo; pallid to
pinkish cinnamon with similar color hairs
hanging from the margin; sticky but soon dry,
silky and glistening with minute hairs. **Gills**
adnate with a slight tooth, close, medium-
broad, edges even; pale yellow becoming pale
olive-brownish. **Stem** 15–30 × 1–3mm
(½–1¼ × ¹⁄₁₆–⅛in), solid, often curved; top
portion pale, clear yellow, bottom portion clay-
brown; minutely woolly or hairy and scaly
below, with stiff, rough hairs on the base.
Flesh very soft and watery, cartilaginous;
yellowish. **Odor** faintly fragrant. **Taste** mild.
Spores ovoid to ellipsoid, smooth, germ pore
in apex, 7–9 × 4.5–5.5μ. Deposit cigar brown.
Habitat in groups or dense clusters on mossy
conifer logs and debris. Found in northeastern
and northwestern North America. **Season**
June–October. **Not edible**.

Ground Pholiota *Pholiota terrestris* Overholts
Cap 2–10cm (¾–4in), obtusely conical to
convex, becoming flatter with a slight umbo;
sepia to cinnamon brown to dark gray-brown;
slimy beneath the numerous hairy scales, with
veil remnants hanging from the margin. **Gills**
adnate, crowded, narrow, edges slightly
uneven; pale becoming grayish brown to
brown. **Stem** 30–100 × 5–10mm
(1¼–4 × ¼–½in), solid but soon hollowed;
grayish staining yellow or brownish at base or
around worm holes, covered with dark brown
downcurved scales; finely hairy above the ring,

Pholiota decorata ⅔ life-size

Pholiota scamba ⅔ life-size

Hebeloma sacchariolens ⅓ life-size

Ground Pholiota *Pholiota terrestris* ⅓ life-size

Pholiota myosotis ½ life-size

Hebeloma sterlingii ¾ life-size

sheathlike below. **Flesh** rather thick, pliant, and tending to be rough; watery buff to brown. **Odor** mild. **Taste** mild. **Spores** ellipsoid, smooth, small pore at tip, 4.5–6.5 × 3.5–4.5μ. Deposit brown. Cheilocystidia numerous. **Habitat** on the ground in dense clusters in woods, lawns, along roadsides, and more rarely on buried wood. Found widely distributed in western North America eastward to Michigan. **Season** June–January. **Edible. Comment** This species appears to be terrestrial, but it has also been found growing on buried wood to which its mycelium may be attached, so its name is rather deceptive.

Pholiota myosotis (Fr.) Singer **Cap** 1.5–3cm (½–1¼in) across, broadly conic to convex becoming flatter; evenly olive-green to olive-bronze, fading to olive-buff; smooth except for occasional veil remnants on margin, sticky to slimy, opaque when young, becoming striate. **Gills** adnate to shallowly adnexed, subdistant, broad, edges minutely white-fringed and eroded; dingy white to olive-brown. **Stem** 100–150 × 2–5mm (4–6 × ³⁄₃₂–¼in), rooting,

hollow, very rigid and fragile; very pale yellow above, covered with patches of white or olivaceous veil remnants up to the ring; a dense bloom at the top. **Flesh** pliant; olivaceous. **Odor** not distinctive. **Taste** not distinctive. **Spores** almond-shaped, smooth, 14–17 × 7–9μ. Deposit dull rusty brown. Pleurocystidia present; cheilocystidia abundant; caulocystidia merely hyphal ends. **Habitat** in groups in boggy soil or on bog edges. Often common. Found in northern North America. **Season** August–September. **Not edible**.

Hebeloma sacchariolens Quél. **Cap** 2–8cm (¾–3in) across, soon flattened; white to pale cream; dry to viscid when wet. **Gills** crowded, narrow; pallid then milky-coffee color. **Stem** 40–100 × 4–10mm (1½–4 × ³⁄₁₆–½in), equal, soft; pallid; fibrous. **Flesh** soft; white. **Odor** strong, penetrating, very sweet and aromatic, like burnt sugar, orange blossom, or fruit candy. **Spores** lemon-shaped, warty, 12–17 × 7–9μ. Deposit pale brown. **Habitat** gregarious under pine. Found in New Jersey.

Season September–October. **Edibility not known** — best avoided. **Comment** Unmistakable because of its remarkable odor.

Hebeloma sterlingii (Pk.) Murr. **Cap** 1.5–2.5cm (½–1in) across, convex becoming flatter, margin incurved and with a few hanging veil fragments at first; gray or clay-colored with a brown disc; smooth and slightly sticky in the middle when moist. **Gills** adnexed, crowded, thin; pale to cinnamon. **Stem** 25–35 × 2–4mm (1–1¼ × ³⁄₃₂–³⁄₁₆in), solid, sometimes slightly enlarged at the base; whitish vertically streaked with brown and browner downward, brick red inside; woolly-hairy. **Veil** hairy or weblike partial veil leaves an evanescent ring. **Flesh** reddish in stem. **Odor** mealy. **Taste** mealy. **Spores** ellipsoid, smooth, 10.5–11.5 × 6–7μ. Deposit rusty clay. **Habitat** singly or in groups under spruce trees. Found in New Jersey and Washington. **Season** September–October. **Not edible**.

Poison Pie *Hebeloma crustuliniforme* ⅓ life-size

Hebeloma illicitum ⅔ life-size

Hebeloma edurum ½ life-size

Poison Pie *Hebeloma crustuliniforme* (Bull. ex St. Amans) Quél. **Cap** 3–11cm (1¼–4½in) across, convex then expanded and often obtusely umbonate, the margin remaining inrolled for a long time and often lobed; buff to pale ochre-tan, darker at the center; slightly greasy to sticky when moist. **Gills** adnate or notched, crowded; pale clay-brown exuding watery droplets in moist conditions, spotted when dry. **Stem** 40–130 × 5–20mm (1½–5 × ¼–¾in), solid with an enlarged base; whitish or tinged with cap color; top powdery or finely flaky, center finely hairy, base with white mycelium. **Veil** not showing when cap is extended. **Flesh** thick; white. **Odor** strongly of radish. **Taste** bitter. **Spores** almond-shaped, smooth or finely warted, 9–13 × 5.5–7.5μ. Deposit rust-brown. Cheilocystidia club-shaped, thin-walled, hyaline. **Habitat** singly or in groups on the ground on wood edges, on lawns, or under trees. Common. Found widely distributed in North America. **Season** September–November (through May in California). **Poisonous.**

Hebeloma illicitum (Pk.) Sacc. **Cap** 2.5–3.5cm (1–1¼in) across, broadly convex becoming expanded; hygrophanous, dark chestnut brown when moist, paler when dry, especially on the margin, disc remaining darker; smooth. **Gills** adnexed, crowded, broad; pale brown. **Stem** 35–50 × 4mm (1¼–2 × 3⁄16in), firm, hollow; cream or similar color to cap but much paler; lined at the top, white mycelium on the base. **Veil** white, slight. **Flesh** firm. **Spores** subellipsoid, 8.7–10 × 4.7–5.7μ. Deposit rusty clay. **Habitat** on decaying logs, twigs, branches in woods. Found in northeastern North America, west to Missouri. Not common. **Season** September–October. **Not edible.**

Hebeloma edurum Métrod syn. *Hebeloma sinuosum* (Fr.) Quél. **Cap** 3–10cm (1¼–4in) across, convex with inrolled margin, the edge wrinkled-furrowed when young; pale cream-ochre to hazel brown at center, to pale pinkish; dry. **Gills** sinuate; whitish then dull milky coffee. **Stem** 30–100 × 10–20mm (1¼–4 × ½–¾in), base swollen, slightly rooting radicant, difficult to dig up intact; whitish to pale clay; fibrous. **Veil** present. **Flesh** white. **Odor** faint, like cocoa, or absent. **Taste** mild to slightly radishy. **Spores** almond-shaped,

warty, 9–12 × 5–6.5μ. Deposit pale brown. **Habitat** in circles around pine. Found in New Jersey. **Season** October–November. **Not edible.**

INOCYBE *These are small, dull-colored mushrooms with conical or umbonate caps. Note presence or absence of a bulb at the stem base. Note the cap texture and the smell. Note any discoloration by bruising. A microscope will be needed to identify all but a few species. The spores and cystidia are very distinctive. Do not eat members of this genus.*

Inocybe cookei Bres. **Cap** 2–4.5cm (¾–1¾in) across broadly conic, then flat with a large umbo, eventually wavy or splitting at the edge; ochre to straw yellowish; smooth with fine silky fibrils. **Gills** adnexed; white at first, then pale ochre-cinnamon. **Stem** 25–60 × 4–8mm (1–2¼ × 3⁄16–5⁄16in); whitish flushing ochre, with a distinct marginate bulb. **Flesh** white to pale yellowish. **Odor** slight. **Spores** bean-shaped, 7.5–10.5 × 4.5–6μ. Deposit snuff brown. Pleurocystidia lacking. **Habitat** in mixed woods. Uncommon. Found in northeastern North America, south to North Carolina and west to Michigan. **Season** August–October. **Not edible.**

Inocybe calamistrata (Fr.) Gillet **Cap** 1–4cm (½–1½in) across, bell-shaped or broadly conical; brownish or dingy umber-colored; fibrous to very scaly in texture. **Gills** adnexed to adnate; pale buff at first, then dull orangy or with greenish-umber tints. **Stem** 30–90 × 2–8mm (1¼–3½ × ⅛–¼in); similar in color to the cap, except the base, which is suffused with a dull blue-green stain; surface fibrous to very scaly. **Flesh** pallid at first, then bruising reddish to dingy brown, except the base of the stem, which has blue-green tones. **Odor** odd, a mixture of pine resin and fish or of green corn. **Spores** bean-shaped, smooth, 10–12 × 5–6μ. Deposit snuff brown. Pleurocystidia lacking. **Habitat** in coniferous or mixed woods. Frequent. Found over most of North America north from North Carolina in the east and

Oregon in the west. **Season** July–October. **Not edible. Comment** The tall, scaly appearance together with the blue-green base of the stem are distinctive.

Inocybe fastigiata (Schaeff. ex Fr.) Quél. **Cap** 2–8cm (¾–3in) across, broadly conical, then flat with a distinct umbo, edge cracking with age; straw yellow to yellow-ochre, umbo more reddish orange; covered in radial silky fibers. **Gills** adnexed; pallid creamy gray, later dull grayish. **Stem** 30–90 × 4–12mm (1¼–3½ × 3⁄16–½in); white or pale straw-colored; no bulb. **Flesh** white. **Odor** slight, of meal. **Spores** bean-shaped, smooth, 9–13 × 5–7μ. Deposit snuff brown. Pleurocystidia lacking. **Habitat** in deciduous or mixed woods. Fairly common. Found throughout North America. **Season** June–October. **Poisonous. Comment** One of the best-known species, with its large straw-yellow cap splitting distinctly in age.

Inocybe fastigiata var. *microsperma* Bres. **Cap** 3–6cm (1¼–2¼in) across, broadly conical; yellow-ochre; with parallel fibrils, splitting at the edge. **Gills** adnexed; whitish becoming pale brown. **Stem** 50–80 × 6–10mm (2–3 × 1¼–½in), equal, with a small basal bulb; whitish, then more ochre in age; fibrillose. **Flesh** whitish. **Odor** spermatic. **Spores** ovoid, smooth, 8–10 × 5–6μ. Deposit snuff brown. Pleurocystidia absent. **Habitat** under broad-leaved trees. Found in eastern North America at least as far west as Michigan. **Season** August–September. **Not edible. Comment** This mushroom differs from *Inocybe fastigiata* (above) in that it has smaller spores, a small bulb at the stem base, and a spermatic odor.

Inocybe fastigiata var. *umbrinella* (Bres.) Heim **Cap** 1.5–4cm (½–1½in) across, conical, then flat with a distinct umbo; fibrillose brown, but soon radially splitting, especially when dry, except for the umbo, which remains brown and scaly. **Gills** adnexed; pallid at first, finally snuff brownish. **Stem** 15–35 × 2–3mm (½–1¼ × 1⁄16–⅛in), not bulbous; pallid whitish to browny gray. **Flesh** pallid. **Odor** slight. **Spores** ellipsoid, smooth, 10–13(16) × 5.2–7(8.5)μ. Deposit snuff brown. Pleurocystidia absent. In 2 percent KOH slight yellow tones were seen in the hyphae. **Habitat**

Inocybe cookei ⅓ life-size

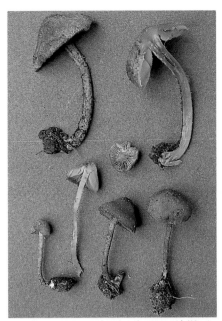

Inocybe calamistrata ½ life-size

Inocybe fastigiata ⅓ life-size

in mixed woods. Found in Mississippi and no doubt other southern states. **Season** June–August. **Not edible. Comment** *Inocybe fastigiella* Atkinson is very similar to this species and is also southern, but the spores are much smaller, not exceeding 10 × 5.5μ.

Inocybe calamistrata var. *mucidiolens* Grund & Stuntz **Cap** 2.5–5cm (1–2in) across, convex; pallid reddish brown, reddening where bruised; strongly fibrillose, at times squarrose (scaly). **Gills** adnexed to adnate; pallid at first, then more red brownish. **Stem** 40–90 × 4–12mm (1½–3½ × ³⁄₁₆–½in; colored like the cap, bruising reddish brown, especially on the scales, the base sometimes swollen and discoloring blue-green; coarsely fibrillose, the fibrils sometimes turning up into scales. **Flesh** pallid, but discoloring reddish brown on exposure; stem base usually shows blue-green color. **Odor** very strong, a mixture of pleasant flowery scent and ripe pears or green corn. **Spores** large, bean-shaped, smooth, 8.5–11.5 × 4.5–6μ. Deposit snuff brown. Pleurocystidia absent. **Habitat** in conifer woods. Rather frequent in Washington and also found in Nova Scotia, but will presumably turn up elsewhere as it becomes better known. **Season** August–October. **Not edible. Comment** It brings to mind the European *Inocybe bongardii* (Weinm.) Quél., which is also a large mushroom with a strong smell but entirely lacks the blue coloring in the stem base.

Inocybe rimosoides Pk. **Cap** 2–4cm (¾–1½in) across, broadly conical, then flat with a prominent umbo; golden or reddish ochre; fibrillose with long parallel fibers that tend to cause the cap to crack from the edge (rimose). **Gills** adnexed; pallid buff at first, then darker. **Stem** 25–55 × 3–7mm (1–2¼ × ⅛–¼in), base equal or swollen to sub-bulbous; whitish or with touches of ochre; pruinose at the apex. **Flesh** whitish. **Odor** faint, a touch spermatic. **Spores** ovoid to bean-shaped, smooth, 8–10 × 5–6μ. Deposit snuff brown. Pleurocystidia absent. **Habitat** in mixed woods. Uncommon. Found in eastern North America from North Carolina to Nova Scotia and west to Michigan. **Season** August. **Not edible.**

Inocybe calamistrata var. *mucidiolens* ⅓ life-size

Inocybe fastigiata var. *microsperma* ⅓ life-size

Inocybe rimosoides ⅔ life-size

Inocybe fastigiata var. *umbrinella* ⅔ life-size

Inocybe caesariata almost life-size

Inocybe subochracea ¾ life-size

Inocybe caesariata (Fr.) Karsten **Cap** 1.5–5cm
(½–2in) across, broadly convex, eventually
flattish; yellow-ochre, darker in the center,
covered in tiny brownish ochre scales. **Gills**
adnate-seceding; pallid ochre then darker.
Stem 15–40×2–6mm (½–1½ × ³⁄₃₂–¼in);
pallid olive; fibrillose to scaly. **Flesh** pallid
tinged ochraceous. **Odor** slight. **Spores** bean-
shaped to ovoid, smooth, 10–12μ. Deposit
snuff brown. Pleurocystidia absent. **Habitat** on
the ground and in grass on woodland tracks,
mainly with broad-leaved trees. Occasional.
Found in northern North America. **Season**
July–October. **Not edible**.

Inocybe subochracea (Pk.) Earle **Cap** 2.5–5cm
(1–2in) across, subconic then flat; yellow-
ochre; fibrillose to scaly. **Gills** adnexed; honey
yellow then more tawny olive. **Stem** 30–60×
3–6mm (1¼–2¼ × ⅛–¼in); whitish at first,
then rapidly ochre, ochre-tawny. **Flesh** white
to slightly yellowish. **Odor** strong. **Spores** long
almond-shaped, smooth, 8.7–10.9×4.6–5.8μ.
Deposit snuff brown. Pleurocystidia plentiful,
colored yellow in 3 percent KOH, very long
and narrow, thick-walled, 70–90×12–17μ.
Habitat in mixed woods. Fairly common.
Found over most of North America. **Season**
August–October. **Not edible. Comment** The
long yellow cystidia are most distinctive.

Inocybe atripes Atkinson **Cap** 1.5–3.5cm
(½–1¼in) across, broadly conic; tawny olive to
brown; fibrillose to scaly. **Gills** adnexed; pallid
then tawny. **Stem** 30–60×3–8mm (1¼–2¼ ×
⅛–⁵⁄₁₆in), equal or a little swollen at the base;
pallid at the apex, then brown darkening from
the base upward; pruinose and slightly
fibrillose. **Flesh** white. **Odor** spermatic.
Spores ovoid, smooth, 8–9×5–5.5μ. Deposit

snuff brown. Pleurocystidia clavate, thick-
walled, encrusted, yellowish in 3 percent
KOH, 40–45×13–15μ. **Habitat** on the soil
under mixed trees. Occasional. Found in
northeastern North America, west to
Wyoming. **Season** July–October. **Not edible.**
Comment My specimens do not show the
blackening stem base well.

Inocybe pyriodora (Pers. ex Fr.) Kummer **Cap**
3–5.5cm (1¼–2¼in) across, broadly conical at
first, then flatter with a distinct umbo; when
young, whitish with brown scales developing in
the center, gradually becoming much more
red-brown with age or when damaged;
fibrillose-scaly. **Gills** sinuate-adnexed; white
when young, then dingy reddish ochre. **Stem**
40–70×4–9mm (1½–2¾ × ³⁄₁₆–⅜in); pallid
whitish at first, discoloring reddish from the
base up; fibrillose. **Flesh** white, discoloring red
when damaged. **Odor** strong, of overripe
pears. **Spores** ovoid, smooth, 8.5–11×5–6μ.
Deposit snuff brown. Pleurocystidia walls of
medium thickness, encrusted, 45–70×10–20μ.
Habitat in broad-leaved or coniferous woods.
Uncommon. Found in northern North
America. **Season** August–September. **Not
edible**.

Inocybe lacera (Fr.) Kummer **Cap** 1–4cm
(½–1½in) across, convex at first, then flat with
an umbo; dark brown; densely fibrillose to
fibrillose-scaly. **Gills** adnate; grayish brown,
edges whitish. **Stem** 10–30×2–5mm
(½–1¼ × ³⁄₃₂–¼in), equal or with base a little
swollen; brown in color like the cap; fibrillose.
Flesh pallid brown. **Odor** none. **Spores** long
bean-shaped, smooth, 14–16.7×5–6μ. Deposit
snuff brown. Pleurocystidia cylindrical to
ventricose, thin-walled with encrustations up to

70μ long, showing slightly yellowish in 3
percent KOH. **Habitat** in conifer woods.
Frequent. Found all over North America.
Season August–September. **Not edible.**
Comment The long, bean-shaped spores are
the most distinctive characteristic.

Inocybe lacera (Fr.) Kummer var. *heterosperma*
Grund & Stuntz **Cap** 1–3cm (½–1¼in),
broadly conical, margin for a long time
rounded; dark brown; the surface very matted
and fibrous, sometimes arranged into scalelike
patches. **Gills** adnate; pallid brown, bruising
reddish brown then darker. **Stem** 20–35×
4–7mm (¾–1¼ × ³⁄₁₆–¼in), not bulbous; pallid
bruising brown, eventually dark brown,
especially at the base; very fibrous. **Flesh**
pallid, bruising a little brownish. **Odor** slight,
a touch spermatic. **Spores** very variable in size
and shape, from ovoid to very long almond-
shaped, smooth, 8–13.5×5.5–7.5μ. Deposit
snuff brown. Pleurocystidia ventricose,
transparent, thick-walled, some encrusted,
around 65×17μ. **Habitat** in mixed woods, in
the open on the roadside. Found in
Washington and Nova Scotia. **Season**
May–October. **Not edible. Comment** My
collection agreed quite well with that of Grund
& Stuntz except that mine bruised more.

Inocybe griseo-lilacina Lange **Cap** 1.5–3.5cm
(½–1¼in) across; flatly bell-shaped, eventually
flat with an umbo; pinkish buff with hints of
violet, more reddish on the umbo; densely
fibrillose. **Gills** adnexed; at first pallid with a
hint of violet, then becoming dingy buff. **Stem**
25–40×3–6mm (1–1½ × ⅛–¼in), base a little
swollen and truncate but not bulbous; pallid
violet; the surface appressed, base fibrillose.
Flesh pallid with touches of violet, especially

at the stem apex. **Odor** distinct, a little spermatic. **Spores** bean-shaped to almond-shaped, smooth, $8-10 \times 4.5-6\mu$. Deposit snuff brown. Pleurocystidia ventricose, transparent, medium-walled, encrusted, about $60 \times 20\mu$. **Habitat** under conifers and deciduous trees. Uncommon. Found on the West Coast and in Colorado and Michigan, occasionally in the Northeast. **Season** June–October. **Not edible. Comment** The pale cap color together with the violet stem is distinctive.

Inocybe eutheles (Berk. & Br.) Sacc. **Cap** 3–7cm (1¼–2¾in) across, conical, eventually flat with or without an umbo; creamy-straw to reddish-ochre color; the fibers break up and group together to form distinct scales, especially in the center. **Gills** adnexed; pallid then more cigar brown. **Stem** $50-100 \times 4-8$mm (2–4 × ³⁄₁₆–⁵⁄₁₆in), the base swollen but not bulbous; pallid ochre-creamish like the cap; pruinose above. **Flesh** whitish. **Odor** spermatic. **Spores** bean-shaped, smooth, $8-10 \times 5-8.5\mu$. Deposit snuff brown. Pleurocystidia $52-75 \times 11-15\mu$. **Habitat** under conifers. Found in California. (This species occurs in broad-leaved woods in Europe.) **Season** December in California, but if found elsewhere probably autumnal. **Not edible. Comment** This mushroom was photographed in England.

Inocybe geophylla (Sow. ex Fr.) Kummer **Cap** 1.5–3.5cm (½–1¼in) across; conical, soon expanding to flat with a distinct umbo; white or whitish; smooth and silky. **Gills** adnexed, crowded; cream at first, then darkening to clay. **Stem** $10-60 \times 3-6$mm (½–2¼ × ¹⁄₈–¼in), usually without a bulb; white; silky; **Flesh** white, unchanging. **Odor** varies from slight to spermatic or scented. **Spores** ellipsoid to bean-shaped, smooth, $8-10 \times 4.5-6\mu$. Deposit snuff brown. Pleurocystidia fusoid, medium thick-walled, encrusted. **Habitat** in coniferous or deciduous woods. Frequent. Found throughout North America. **Season** June–November. **Poisonous. Comment** The pure white cap and stem in conjunction with the clay-colored gills are distinct characteristics, but because the fibers on the cap are extremely fine this mushroom could sometimes be mistaken for another genus. Note: one of my specimens has a small bulb at the base of the stem, as sometimes occurs.

Inocybe atripes ½ life-size

Inocybe pyriodora ⅓ life-size

Inocybe lacera ¾ life-size

Inocybe lacera var. *heterosperma* ½ life-size

Inocybe geophylla ⅔ life-size

Inocybe griseo-lilacina ½ life-size

Inocybe eutheles ½ life-size

Inocybe subcarpta ⅓ life-size

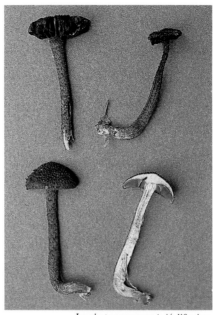

Inocybe taquamenonensis ½ life-size

Inocybe napipes ¾ life-size

Inocybe calospora ⅔ life-size

Inocybe grammata ½ life-size

Inocybe albodisca ½ life-size

Inocybe lanuginosa ¾ life-size

Inocybe subcarpta Kühner & Boursier **Cap**
3–7cm (1¼–2¾in) across, broadly conic,
showing an umbo as it flattens out; snuff
brown, lighter at the margin; surface strongly
fibrillose to scaly. **Gills** adnate; whitish at first,
brown in age. **Stem** 70–100×8–12mm
(2¾–4×⁵⁄₁₆–½in), equal without a bulb but
sometimes a little swollen, becoming hollow;
pale at first, becoming darker from the base
up, until it is colored as the cap; fibrous. **Flesh**
off-white to golden-colored. **Odor** slight.
Spores nodulose, 7–9.5×5–6.5μ. Deposit snuff
brown. Cystidia colorless, fusiform or cylindric,
not encrusted, 60–80×10–18μ. **Habitat** in
conifer woods. Rare or overlooked. Found in
the Pacific Northwest. **Season** September–
November. **Not edible. Comment** My spores
are a little shorter than those of the collection
by Smith.

Inocybe taquamenonensis Stuntz **Cap** 1.5–3.5cm
(½–1¼in) across, broadly convex; dark
purplish black; covered in blackish scales.
Gills adnate; violet or purplish at first,
eventually dark brown. **Stem** 30–75×3–7mm
(1¼–3×⅛–¼in); dark, purplish brown or
black as the cap; covered in dark scales. **Flesh**
purple or reddish colors, darkening after a few
hours. **Odor** slight, possibly a little radishy.
Spores nodulose, with very long, distinct
nodules, 5–8×4–5.5μ. Deposit snuff brown.
Cystidia thin-walled, cylindric or with a
swollen head like a bowling pin, without
crystals on the apex. **Habitat** in mixed woods
or hardwood forests. Probably frequent but
ignored. Found in northeastern North America
from Virginia northward and west as far as
Michigan. **Season** August–September. **Not
edible. Comment** At first I thought I had a
rare, scaly cortinarius; I discovered Smith said
that this mushroom reminded him of a very
small *Cortinarius violaceus* (p. 151).

Inocybe napipes Lange **Cap** 1.5–3cm (½–1¼in)
across, conical, then convex with an umbo;
hazel or snuff brown; silky-fibrillose, shining.
Gills adnexed; whitish when young, then pale
brown. **Stem** 40–60×4–8mm (1½–2¼×
³⁄₁₆–⁵⁄₁₆in), the base with a distinct bulb; pallid
at the apex, more brown below; fibrillose.

Flesh pallid, whitish. **Odor** strong, fruity or
rancid. **Spores** nodulose, with rather large
nodules, 7–9×5–6.5μ. Deposit snuff brown.
Pleurocystidia thinnish-walled, hyaline, some
encrusted, 40–55×13–19μ. **Habitat** in
coniferous woods. Uncommon. So far only
reported from eastern North America. **Season**
August–September. **Not edible.**

Inocybe calospora Quél. **Cap** 1–2.5cm (½–1in)
across, broadly conical, then flat with a distinct
umbo; dark brown drying a little lighter,
densely fibrillose, scaly. **Gills** adnate; pale
grayish brown at first, then more brown. **Stem**
20–60×1.5–3mm (¾–2¼×¹⁄₁₆–⅛in), with a
small whitish bulb; color varying from dark
brown to almost white; fibrillose. **Flesh**
brownish. **Odor** slight, pleasant. **Spores**
spheroid or broadly ellipsoid, with long distinct
spines giving the appearance of an ocean mine,
9.5–13.5×8.5–11μ. Deposit snuff brown.
Pleurocystidia uncommon, fusoid, only slightly
encrusted at the apex, walls thickish, 35–55×
10–20μ. **Habitat** in broad-leaved woods.
Uncommon in the north but more common in
southeastern and central North America from
Texas to Nova Scotia. **Season** July–August.
Not edible. Comment The spores are unique
under the microscope, and at first you might
even think you have a new genus.

Inocybe grammata Quél. **Cap** 2–4cm (¾–1½in)
across, bell-shaped to conical with a distinct
umbo; pallid straw color to brown, with a
superficial white layer in the center of the cap
often covering two-thirds of the surface; the
surface is covered in a distinct mass of fibers.
Gills adnexed; at first pallid, then ochre-
brownish. **Stem** 30–50×4–6mm
(1¼–2×³⁄₁₆–¼in), with a distinct marginate
bulb; paler in color than the cap, flesh-colored;
innately fibrillose but not scaly. **Flesh** whitish,
pallid. **Odor** strong, fruity, of ripe pears.
Spores very knobby, 9–10×6.5–7.5μ. Deposit
snuff brown. Pleurocystidia short and fat, club-
shaped, transparent, encrusted, rather thin-
walled, about 30×14μ. **Habitat** in mixed
woods and conifers. Rare. So far, I know of it
only from New York State. **Season**
August–September. **Not edible. Comment** I
have found no previous record of this species in
America, but I expect it has turned up in areas
apart from New York. It differs from *Inocybe
albodisca* (below) in smell and in having short,
fat cystidia.

Inocybe albodisca Pk. **Cap** 1.5–3.5cm (½–1¼in)
across, convex to broadly conic; pale brownish
flesh-colored or pallid grayish brown, except at
the center, which has a persistent creamish-
white superficial layer; smooth then a little
fibrillose, silky. **Gills** adnexed; white becoming
dingy pinkish brown. **Stem** 30–50×3–5mm
(1¼–2×⅛–¼in), with a distinct marginate
bulb; pale flesh-colored, lighter at the base
when fresh, later pruinose. **Flesh** pallid or a
touch pinkish. **Odor** spermatic. **Spores**
nodulose, 6.5–8×4.5–6μ. Deposit snuff
brown. Pleurocystidia plentiful, colorless,
ventricose, thin-walled, encrusted,
45–70×13–20μ. **Habitat** on bare soil with
hemlock or in mixed woods. Quite common.
Found in eastern North America, west to
Missouri, and also on the northern West
Coast. **Season** August–October. **Not edible.**

Comment The superficial whitish layer on the
cap center is very distinctive.

Inocybe lanuginosa (Bull. ex Fr.) Kummer **Cap**
1.5–3cm (½–1¼in), at first hemispherical,
then flattish with an umbo; dark red-brown;
densely scaly even in button stage. **Gills**
adnexed; pallid clay-colored then cinnamon.
Stem 20–40×2.5–4mm (¾–1½×⅛–³⁄₁₆in);
similar in color to the cap but paler; strongly
fibrillose to scaly. **Flesh** brownish. **Odor**
unpleasant. **Spores** nodulose, rather varied
from almost stellate to ovoid, 7.7–10.2×
5.5–7μ. Deposit snuff brown. Pleurocystidia
few, rather narrow-walled, spindle-shaped, not
usually encrusted, showing a touch yellowish in
3 percent KOH. **Habitat** conifer woods.
Frequent. Found across northern North
America. **Season** July–September. **Not edible.**
Comment Very distinct with its prominent
dark brown pointed scales; probably often
thought to be *Inocybe hystrix* Karst. when
identified in the field.

Conocybe lactea (Lange) Métrod **Cap** 1–2.5cm
(½–1in) across, conical to narrowly bell-
shaped with the margin often flared; whitish to
yellowish cream; smooth. **Gills** nearly free,
close, narrow, whitish becoming cinnamon or
tawny. **Stem** 30–100×1–2mm (1¼–4×
¹⁄₁₆–³⁄₁₆in), very fragile, sometimes ending in a
rounded bulb; whitish with a powdery bloom.
Flesh thin. **Odor** not distinctive. **Taste** not
distinctive. **Spores** ellipsoid, with a germ pore,
smooth, 11.5–15×7–10μ. Deposit reddish.
Cheilocystidia spindle-shaped and thin-walled.
Habitat scattered or in dense groups on lawns
and grassy areas. Common. Found in northern
North America, California, Mexico, and the
Gulf Coast. **Season** May–September. **Edibility
not known.**

Conocybe tenera (Schaeff. ex Fr.) Kühner **Cap**
1–3cm (½–1¼in) across, bluntly conical at
first, expanding to bell-shaped; ochre-brown to
cinnamon, drying more yellowish; dry, with
fine lines, **Gills** adnate, crowded, narrow;
whitish at first, then cinnamon brown. **Stem**
50–90×4–7mm (2–3½×³⁄₁₆–¼in), straight,
becoming hollow, fragile; whitish flushed
brownish with cap color; surface has a finely
powdered appearance. **Flesh** thin, same color
as cap. **Odor** mushroomy. **Spores** ovoid to
ellipsoid, smooth, 9–14×5–8μ. Deposit bright
yellow or reddish brown. **Habitat** scattered to
numerous in grass in lawns, fields, and woods.
Found widely distributed in North America.
Season May–July, September. **Not edible.**

Conocybe filaris Fr. syn. *Pholiotina filaris*
(Fr.) Singer **Cap** 0.5–2.5cm (¼–1in) across,
conical to convex-umbonate; yellow-brown;
glabrous. **Gills** sinuate-adnate, crowded,
broad; pallid to rust-brown. **Stem**
10–40×1–1.5mm (½–1½×¹⁄₃₂–¹⁄₁₆in), equal;
yellow to yellow-orange; a distinct,
membranous ring on stem apex. **Odor** not
distinctive. **Taste** not distinctive. **Spores**
smooth, with germ pore at tip, 7.5–13×
3.5–6.5μ. Deposit cinnamon brown. **Habitat** in
grass or on wood chips. Found throughout
most of North America. **Season** July–October.
Deadly poisonous. Comment This mushroom
contains poisons similar to those of the
Destroying Angel, *Amanita virosa* (p. 29).

Conocybe lactea ½ life-size

Conocybe tenera ⅔ life-size

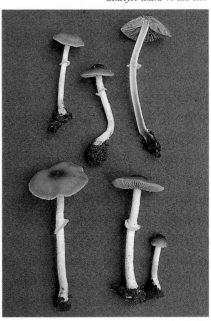
Conocybe filaris ⅔ life-size

Deadly Galerina *Galerina autumnalis* almost life-size

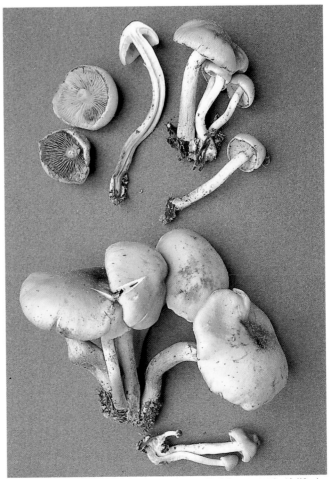

Hypholoma capnoides ½ life-size

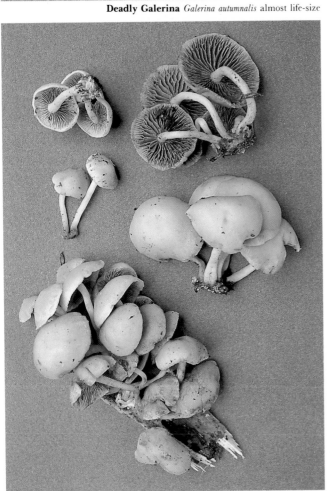

Hypholoma fasciculare ⅔ life-size

Hypholoma sublateritium ⅔ life-size

Ripartites tricholoma ¾ life-size

Deadly Galerina *Galerina autumnalis* (Pk.) Smith & Singer **Cap** 2.5–6.5cm (1–2½in) across, convex becoming flatter or with a slight umbo; dark brown to ochre-tawny, fading to yellowish or buff with disc sometimes remaining darker; smooth, sticky with translucent striate margin. **Gills** adnate or with a slight decurrent tooth, close, broad; yellowish becoming rust. **Stem** 30–90 × 3–10mm (1¼–3½ × ⅛–½in), hollow, sometimes enlarged toward base; whitish above, brownish to blackish base; vertically lined base and lower part covered in dense white mycelium; finely felty below evanescent ring on upper stalk with some white veil remnants. **Flesh** thick on disc, thinner at margin; watery brown fading to pale buff. **Odor** mealy when crushed. **Taste** mild to mealy. **Spores** ellipsoid, roughened, with a slight depression, 8–9 × 4.9–6μ. Deposit rust. Pleurocystidia abundant, 52μ long. **Habitat** scattered or clustered on hardwood and conifer logs and stumps. Common and sometimes abundant. Widely distributed throughout North America. **Season** May–October. **Deadly poisonous.**

Hypholoma capnoides (Fr. ex Fr.) Kummer syn. *Naematoloma capnoides* (Fr.) Karsten **Cap** 2–8cm (¾–3in) across, convex; pale yellow-ochre to cinnamon; smooth. **Gills** adnate, crowded, narrow; pallid then violaceous brown to gray-violet. **Stem** 50–80 × 3–8mm (2–3 × ⅛–⁵⁄₁₆in), equal; cream to pale yellow, turning rust-brown from the base up. **Flesh** pallid. **Odor** pleasant. **Taste** mild. **Spores** ellipsoid-ovate, with a distinct pore, 6–7 × 4–4.5μ. Deposit purple-brown. **Habitat** caespitose on fallen conifers. Common. Found throughout North America. **Season** August–November. **Edible** with caution.

Hypholoma fasciculare (Huds. ex Fr.) Kummer syn. *Naematoloma fasciculare* (Huds. ex Fr.) Karsten **Cap** 2–8cm (¾–3in) across, convex-flattened; bright yellow to ochre or even greenish yellow, center often brownish; smooth, dry. **Gills** adnate; lemon yellow then soon greenish, finally flushed purple-brown. **Stem** 50–120 × 3–10mm (2–4¾ × ⅛–½in); yellow with greenish flush, brown at base. **Flesh** yellow. **Odor** pleasant. **Taste** very bitter. **Spores** ovoid, with pore, 6.5–8 × 3.5–4μ. Deposit purple-brown. **Habitat** fasciculate (in dense clusters) on fallen hardwoods and conifers. Common. In most of North America. **Season** August–November. **Poisonous.**

Hypholoma sublateritium (Fr.) Quél. syn. *Naematoloma sublateritium* (Fr.) Karsten **Cap** 4–10cm (1½–4in) across, convex to expanded with inrolled margin; deep brick red, paler more ochre at margin; smooth, dry. **Gills** adnate; crowded, narrow; pallid to purple-gray when mature. **Stem** 50–100 × 5–15mm (2–4 × ¼–½in); whitish cream above, reddish yellow below, often staining slightly yellow; with slight ring zone at apex. **Flesh** firm; cream. **Odor** pleasant. **Taste** mild to bitter. **Spores** ellipsoid, with indistinct pore, 6–7 × 4–4.5μ. Deposit purple-brown. **Habitat** in clusters on hardwood logs and stumps. Found throughout North America. **Season** August–October. **Edible** with caution.

Ripartites tricholoma (Fr.) Karsten **Cap** 2–5cm (¾–2in) across, convex becoming flat then sunken in the center, with an incurved margin becoming radially lined with long, coarse hairs; white; tacky to dry with appressed silky hairs. **Gills** adnate, close, broad; white becoming dull pinkish cinnamon. **Stem** 25–50 × 3mm (1–2 × ⅛in), hollow, fragile; whitish to dingy ochre-brown; faint bloom to suedelike. **Flesh** very soft; pallid. **Odor** slight, not mealy. **Taste** mild. **Spores** ovoid to subglobose, warted, 4–5 × 3.5–4μ. Deposit pale brown. **Habitat** on humus or very decayed wood in forests. Not common. Found widely distributed in North America. **Season** August–November. **Edibility not known. Comment** This mushroom was photographed in England.

Tubaria furfuracea (Pers. ex Fr.) Gillet **Cap** 1–4cm (½–1½in) across, convex then flattened or centrally depressed; cinnamon to tan, drying pale buff; striate from margin inward when moist, slightly scurfy. **Gills** adnate to slightly decurrent, distant, broad; cinnamon. **Stem** 20–50 × 2–4mm (¾–2 × ³⁄₃₂–³⁄₁₆in), thin, fragile; similar color to cap; minutely hairy, silky at the top, base covered in white down. **Veil** hairy, white, evanescent partial veil. **Flesh** thin; brownish. **Odor** not distinctive. **Taste** not distinctive. **Spores** ellipsoid with rounded apex, smooth, 7.1–8.3 × 4.6–5.8μ. Deposit pale ochre. **Habitat** singly or in groups on soil, sticks, woody debris in woods and wastelands, and along paths. Common and abundant. Found widely distributed in North America. **Season** April –November (November–March in California). **Not edible.**

Galerina pallidispora Smith **Cap** 1–2.5cm (½–1in) across, flatly cone-shaped expanding to bell-shaped; russet becoming dingy yellowish tawny, fading to pale yellow or buff; smooth, moist, hygrophanous. **Gills** bluntly adnate, close, moderately broad; ochraceous tawny. **Stem** 30–40 × 2–3mm (1¼–1½ × ³⁄₃₂–⅛in), sometimes slightly enlarged downward; same color as cap; covered with a bloom above, smooth below, with mycelium around the base. **Spores** narrowly ovoid, smooth or minutely dotted, 9–13 × 5–6.5μ; pale in 3 percent KOH. Deposit ochre. No pleurocystidia; cheilocystidia often curved. **Habitat** scattered on hardwood logs, especially alder. Found in the Pacific Northwest. **Season** May–June. **Not edible.**

Hypholoma elongatipes Pk. **Cap** 1–2.5cm (½–1in) across, bell-shaped, conical; honey yellow to olive; margin slightly sulcate. **Gills** adnate; pallid then gray-brown. **Stem** 40–100 × 1–2mm (1½–4 × ¹⁄₁₆–⅛in), whitish, darker reddish at base. **Flesh** thin. **Taste** bitterish. **Spores** ellipsoid, 9–13 × 5.5–7μ. Deposit purple-brown. **Habitat** in moss in swamps and marshes. Found widely distributed throughout northern North America. **Season** August–October. **Not edible.**

Tubaria furfuracea ¾ life-size

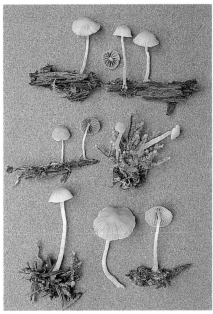

Galerina pallidispora ½ life-size

Hypholoma elongatipes ½ life-size

Hypholoma dispersum ½ life-size

Hypholoma udum ½ life-size

Hypholoma olivaceotinctum ¼ life-size

Hypholoma olivaceotinctum Kauffman **Cap** 1.5–3cm (½–1¼in) across, bell-shaped to flattened; dull olive-drab or greenish, to pinkish buff when dry. **Gills** adnate, close, thin; pale olive-buff. **Stem** 30–50 × 1.5–2.5mm (1¼–2 × ³⁄₃₂–⅛in), equal, brittle; bright reddish brown. **Flesh** thin. **Odor** not distinctive. **Taste** not distinctive. **Spores** ellipsoid, smooth, 9–12 × 4.5–5.5μ. Deposit purplish brown. **Habitat** on rich humus and debris in wet areas of conifer forests. Found in northern North America from the Pacific Northwest to New York. **Season** August–October. **Not edible**.

Hypholoma dispersum (Fr.) Quél. syn. *Naematoloma dispersum* (Fr.) Karsten **Cap** 1–4cm (½–1½in) across, conic-convex then expanded; tawny brown to orange-brown, paler, more ochre with age; dry, smooth. **Gills** adnate, crowded; pallid to olive-yellow then purplish brown. **Stem** 60–100 × 2–5mm (2¼–4 × ⅛–¼in), equal; deep reddish brown below, paler above; silky-fibrillose. **Flesh** thin. **Odor** pleasant. **Taste** mild. **Spores** ellipsoid, 7–10 × 4–5μ. Deposit purplish brown. **Habitat** usually single but often in troops, on fragments of conifer wood or wood-chip trails. Found in northern North America. **Season** August–November. **Not edible**.

Hypholoma udum (Pers. ex Fr.) Kühner syn. *Naematoloma udum* (Fr.) Karsten **Cap** 1–4cm (½–1½in) across, obtusely cone-shaped or convex, becoming flatter with an umbo; orangish brown to tawny, fading to more yellowish with olivaceous tints if water-soaked; smooth, sometimes a few sparse veil remnants on cap. **Gills** adnate, close, broad; yellowish at first, finally drying purple-brown. **Stem** 40–120 × 1–4mm (1½–4¾ × ¹⁄₁₆–³⁄₁₆in), fragile; rusty, paler above, browner below; smooth. **Flesh** thin. **Odor** mushroomy. **Taste** bitter. **Spores** elongated ellipsoid, 14–18 × 5–7μ. Deposit purplish brown. **Habitat** scattered or in groups in bogs. Very common. Widely distributed in northern North America. **Season** August–October. **Edibility not known.**

Black- or Purple-Brown-Spored Agarics
pp. 188–206. See Key D, p. 12

AGARICUS *The normal store-bought mushrooms are in this genus. They are usually divided into two sections, those tending to yellow when cut or bruised and those showing pink discoloration. Note smells. Avoid eating the ones that bruise yellow until you are certain they are edible.*

False Meadow Mushroom *Agaricus andrewii* Freeman **Cap** 2–6cm (¾–2¼in) across, convex then flattened, but with inrolled margin until fully mature; pure white, to cream when old; smooth, silky-fibrillose, margin of cap with floccose remnants of white veil. **Gills** free, crowded, broad; bright pink when young, then soon chocolate brown, and finally black. **Stem** 25–50 × 10–15mm (1–2 × ½in), equal to tapered at the base; white; fibrillose to woolly below the faint evanescent ring zone. **Flesh** firm; white. **Odor** very pleasant. **Taste** very pleasant. **Spores** broadly ellipsoid, 7–8 × 4–5μ. Deposit purplish brown. Marginal cystidia sparse, prominent and turnip-shaped to club-

shaped. **Habitat** As yet the exact distribution of this species is uncertain because of confusion with *Agaricus campestris* (below). However, it would appear to be widespread at least in the eastern United States as far south as North Carolina. **Season** late September–November. **Edible** and choice, it has doubtless been mistakenly collected many times as *Agaricus campestris*. **Comment** The more familiar *Agaricus campestris* lacks any marginal cystidia and may not be as common in America as is usually supposed. Apart from the microscopic differences, *Agaricus andrewii* would appear to differ hardly at all macroscopically, except that it seems to have a more consistently smooth and purer white cap than *Agaricus campestris*.

Meadow Mushroom *Agaricus campestris* Fr. **Cap** 3–8cm (1¼–3in) across, convex; white to dirty brown at center; smooth to woolly tomentose or even squamulose. **Gills** free; bright pink at first, then soon brown. **Stem** 25–60 × 10–20mm (1–2¼ × ½–¾in), equal to tapered at base; white; smooth above ring, often fibrillose below: ring thin, soon torn and often almost absent. **Flesh** white to slightly pinkish brown when bruised. **Odor** pleasant. **Taste** pleasant. **Spores** ovate, 6.5–7.5 × 4–5μ. Deposit deep brown. **Habitat** in meadows, fields, and grassy roadsides. Abundant. Found throughout North America. **Season** June–October. **Edible** — excellent. **Comment** There are some other, almost identical species best distinguished by their microscopic features: *Agaricus andrewii* (above), with large marginal cystidia on the gills (*campestris* has none), and *Agaricus solidipes*, with spores up to 10 × 6μ.

Agaricus altipes Møller **Cap** 4–7cm (1½–2¾in) across, convex to broadly flattened; white to slightly buff on disc; smooth to slightly tomentose-floccose. **Gills** free, crowded; bright rosy pink when young, brown when old. **Stem** 80–100 × 12–20mm (3–4 × ½–¾in), equal to slightly clavate; white, bruising slightly pinkish-buff color; fibrillose below; ring high on stem, white, thin, fragile, simple. **Flesh** firm; white bruising flesh-color. **Odor** mild. **Taste** mild. **Spores** ovate, 6.5–7.5 × 4.5–5.5μ. Deposit deep chocolate brown. **Habitat** in grass in mixed woods, mostly conifers. Found in the Pacific Northwest. **Season** September. **Edible. Comment** This collection agreed very well with the European *Agaricus altipes*, differing only in the disagreeable odor of the original description; the odor may have been missed if the flesh was not bruised, or perhaps the specimens were too young.

Agaricus porphyrocephalus Møller **Cap** 3–8cm (1¼–3in) across, convex with flattened disc; brown to purplish brown with small fibrillose scales. **Gills** free, crowded; dull pink then deep brown. **Stem** 30–50 × 10–20mm (1¼–2 × ¼–¾in), equal to tapered at base; white; fibrillose below ring; ring thin and soon almost absent. **Flesh** firm; white to slightly pinkish when cut. **Odor** not distinctive. **Taste** not distinctive. **Spores** ovoid, 5–7 × 3–4.5μ. Deposit deep brown. **Habitat** in lawns and meadows. Uncommon. Found in eastern North America. **Season** July–October. **Edible** — good.

False Meadow Mushroom *Agaricus andrewii* ½ life-size

Meadow Mushroom *Agaricus campestris* ½ life-size

Agaricus altipes ½ life-size

Agaricus porphyrocephalus ⅔ life-size

Agaricus bitorquis ⅓ life-size

Agaricus subrutilescens ⅓ life-size

Agaricus silvaticus ½ life-size

Agaricus porphyrizon ⅔ life-size

Agaricus bitorquis (Quél.) Sacc. syn. *Agaricus rodmanii* Pk. **Cap** 5–10cm (2–4in) across, convex-flattened with inrolled margin; white to dull ivory; very thick-fleshed, smooth, dry. **Gills** free, crowded, very narrow; pale pink then brown. **Stem** 50–100×15–25mm (2–4×½–1in), tapered at base; white; with a double ring stretched to form almost a volvalike structure at times. **Flesh** thick; white to pale flesh-colored when cut. **Odor** pleasant. **Taste** pleasant. **Spores** subglobose, 5–6.5×4–5μ. Deposit deep brown. Cheilocystidia present, numerous. **Habitat** especially along roadsides, paths, in parks and towns, often in bare soil or gravel, even pushing up asphalt. Found throughout most of North America. **Season** July–October. **Edible** — good.

Agaricus subrutilescens (Kauffman) Hot. & Stuntz **Cap** 5–13cm (2–5in) across, convex; whitish; fibrillose-striate, to slightly scaly; fibers deep vinaceous brown. **Gills** free, crowded, narrow; whitish pink then deep brown. **Stem** 60–150× 10–25mm (2¼–6×½–1in), clavate; whitish; smooth above the ring, with white to vinaceous veil remnants below, often in bands; ring large, thin, woolly below, high on stem. **Flesh** white. **Odor** pleasant. **Taste** pleasant. **Spores** ellipsoid, 4.5–6×3–4μ. Deposit deep brown. **Habitat** in mixed woods. Found on the West Coast. **Season** August–December. **Not edible.**

Agaricus silvaticus Schaeff. **Cap** 4–10cm (1½–4in) across, convex with a slightly flattened disc; reddish brown to umber on a paler ground; with distinct, scattered, pointed scales. **Gills** free; slightly pink then chocolate brown, edge paler; sterile. **Stem** 50–100×5–15mm (2–4×¼–½in), bulbous; white; squamulose to smooth; ring large, white, with underside slightly woolly-floccose. **Flesh** white bruising red on cutting. **Odor** pleasant, slightly aniselike. **Taste** pleasant. **Spores** ovoid 4.5–6×3–3.5μ. Deposit deep brown. **Habitat** under pine. Found in eastern North America. **Season** September–October. **Edible** — good.

Agaricus porphyrizon (Cke.) Orton **Cap** 4–10cm (1½–4in) across, convex; whitish with purplish-lilac adpressed scales and fibrils, darker at center, bruising dull yellow; smooth. **Gills** free, crowded; white to pinkish then brown. **Stem** 50–100×12–20mm (2–4× ½–¾in), clavate; white, staining yellow when touched; smooth; ring membranous, thin, simple. **Flesh** white staining yellowish in stem, especially at base. **Odor** pleasant, of almonds. **Taste** pleasant, of almonds. **Spores** ovoid, 4.5–5.6×3.2–3.8μ. Deposit deep brown. **Habitat** in mixed woodlands, mostly conifers. Found in Colorado. **Season** September. **Edible. Comment** This species does not satisfactorily fit any current American taxa but does agree very well with the European name given here, particularly in the very narrow spores.

The Prince *Agaricus augustus* Fr. **Cap** 10–20cm (4–8in) across, convex with flattened disc, with fine squamulose-fibrillose scales; the scales a deep tawny yellow over a pallid ground, bruising dark yellow-orange. **Gills** free; white then pale pink, finally deep brown. **Stem** 80–150×15–35mm (3–6×½–1¼in), equal to clavate; paler than the cap; smooth above the ring, densely squamulose-woolly below; ring large, membranous, with brownish veil remnants on the underside. **Flesh** firm; white.

Odor pleasant, of almonds. **Taste** pleasant, of almonds. **Spores** ovoid, 8–10(12)×5–6μ. Deposit deep brown. **Habitat** in open grassy areas, woodland clearings, road edges. Widely distributed throughout North America. **Season** July–October. **Edible** — good.

Agaricus diminutivus Pk. **Cap** 1–4cm (½–1½in) across, ovate then expanded-umbonate; pinky-buff, darker in the center, bruising yellowish; smooth to slightly fibrillose, dry; fibrils pinkish brown to purplish. **Gills** free; dull pink then brown. **Stem** 30–60×3–6mm (1¼–2¼× ⅛–¼in), equal, often with basal bulb; white bruising yellowish; smooth to slightly floccose-fibrillose below the ring; ring single, thin, white. **Flesh** thin; white. **Odor** pleasant. **Taste** pleasant. **Spores** broadly ellipsoid, 4.5–5.5×3.5–5μ. Deposit deep brown. **Habitat** scattered on soil and leaf litter. Found throughout northern North America. **Season** July–September. **Probably edible** but too small to eat.

The Prince *Agaricus augustus* ½ life-size

Agaricus diminutivus ½ life-size

Agaricus leucotrichus ⅓ life-size

Agaricus silvicola ⅔ life-size

Agaricus xanthoderma ¾ life-size

Agaricus pocillator ½ life-size

Agaricus leucotrichus Møller **Cap** 8–13cm (3–5in) across, ovate-campanulate, flattened at disc; white to pale straw yellow with age; silky, densely fibrillose-hairy with tiny erect, pointed scales. **Gills** free, crowded; white to pink then dark brown. **Stem** 80–120 × 15–25(30)mm (3–4¾ × ½–1[¼]in), equal to clavate-swollen; white to buff; smooth above the ring, floccose-tomentose below; ring white, thin, pendant, undersurface with small veil remnants. **Flesh** white, pinkish buff when old or bruised. **Odor** pleasant, almondlike. **Taste** pleasant, almondlike. **Spores** ellipsoid-ovoid, 6.5–7.5(8) × 4.5–5μ. Deposit deep brown. **Habitat** in spruce woods. Found in Colorado. **Season** September. **Edible. Comment** This group of specimens agrees very well with the European description, except that the gills are rather more pink than recorded. Pending a better American name being found for this fungus, the European name is applied here.

Agaricus silvicola (Vitt.) Pk. **Cap** 5–10cm (2–4in) across, convex to slightly flattened at center; white, soon flushed yellow-ochre with age or upon bruising; smooth. **Gills** free, crowded, narrow; pale grayish pink then deep brown. **Stem** 50–80 × 10–15mm (2–3 × ½in), equal with bulbous base; white; smooth; ring large, pendulous, with cogwheel-like veil remnants on underside, white. **Flesh** firm; white, bruising slightly yellowish. **Odor** pleasant, of aniseed. **Taste** pleasant, of aniseed. **Spores** ovoid 5–6 × 3–4μ. Deposit deep brown. **Habitat** in woods, usually coniferous. Found widely distributed throughout North America. **Season** August–October. **Edible — good.**

Agaricus xanthoderma Genevier **Cap** 5–15cm (2–6in) across, convex with flat disc, then soon expanded; white, often stained dirty gray-brown, when skin is scratched it turns bright chrome yellow; smooth to quite coarsely scaly in dry weather. **Gills** free; white then gray-pink, finally deep brown. **Stem** 50–120 × 10–25mm (2–4¾ × ½–1in), often slightly bulbous; white; ring large, feltlike, skirtlike. **Flesh** white, intense chrome yellow in stem base. **Odor** strong, inky or phenolic. **Taste** similar. **Spores** ovoid, 4.5–6 × 3–4.5μ. Deposit deep brown. **Habitat** on roadsides, in gardens, at edges of woods. Found in the Pacific Northwest. **Season** August–December. **Poisonous** Causes sweating, nausea, and severe stomach cramps, though some people are unaffected.

Agaricus pocillator Murr. **Cap** 3–10cm (1¼–4in) across, convex to flattened at disc; white with minute blackish-gray scales, darker at disc. **Gills** free; white to pinkish then dark brown. **Stem** 40–80 × 6–12mm (1½–3 × ¼–½in), equal with a distinct, often marginate, small bulb at base; white; smooth above and below ring; ring large, of two distinct layers, membranous. **Flesh** white, yellowing slightly in stem base. **Odor** not distinctive. **Taste** not distinctive. **Spores** ellipsoid, 4.5–6 × 3–3.8μ. Deposit deep brown. **Habitat** often in large numbers in mixed woods. Found in southeastern and eastern North America. **Season** July–August. **Not edible**.

Horse Mushroom *Agaricus arvensis* Fr. **Cap** 8–20cm (3–8in), convex with flattened disc, especially when young; white to dull brassy yellow with age or when bruised; smooth to distinctly floccose-scaly, sometimes quite coarsely in dry weather. **Gills** free, crowded; white to grayish (not pink), then dark brown. **Stem** 60–150 × 10–25mm (2¼–6 × ½–1in), equal to clavate; white; smooth to slightly floccose-scaly below the ring; ring large, thick, pendant, upper side smooth, underside with

cogwheel-like veil remnants. **Flesh** firm; white, when cut turning slightly buff to yellowish. **Odor** pleasant, almondlike. **Taste** pleasant, almondlike. **Spores** ovate, 7–9 × 4.5–6μ. Deposit deep brown. **Habitat** in fields and meadows. Found throughout North America. **Season** June–October. **Edible.**

Agaricus micromegethus Pk. **Cap** 2.5–6cm (1–2¼in) across, convex with flattened disc; cream with darker innate fibrils, fibers dull yellowish brown to brown, staining stronger yellow when bruised. **Gills** free; white to grayish then dark brown. **Stem** 25–50 × 6–10mm (1–2 × ¼–½in), equal; white staining yellowish; smooth above the ring, slightly fibrillose below; ring thin, single, often vanishing, white. **Flesh** white, discoloring yellowish. **Odor** pleasant. **Taste** pleasant. **Spores** ovoid to ellipsoid, 4.5–5.5 × 3.5–4μ. Deposit dark brown. **Habitat** in open pastures and meadows. Frequent. Found east of the Great Plains. **Season** July–September. **Edible. Comment** The collection illustrated was found in mixed woodlands in Oregon.

Horse Mushroom *Agaricus arvensis* ⅓ life-size

Agaricus micromegethus ⅓ life-size

Agaricus praeclaresquamosus ⅓ life-size

Agaricus placomyces ⅓ life-size

Agaricus praeclaresquamosus Freeman syn. *Agaricus meleagris* of American authors **Cap** 5–20cm (2–8in) across, convex with flattened disc; with gray to gray-brown or blackish flattened scales on a white background; dry. **Gills** free, crowded; white to grayish then deep brown. **Stem** 80–150 × 10–30mm (3–6 × ½–1¼in), equal to clavate; white, often discoloring reddish brown; smooth; ring white, thick, feltlike, membranous, very persistent. **Flesh** firm; white, bruising bright yellow in the extreme base of the stem, finally reddish brown. **Odor** unpleasant, phenolic, inklike, especially when flesh is crushed or cooked. **Taste** similar. **Spores** ovoid, 4–6.5 × 3–3.5(4)μ. Deposit deep brown. **Habitat** under mixed woods, along roads and paths. Frequent. Found throughout western North America. **Season** September–December. **Not edible** — poisonous to many. **Comment** The name *meleagris* cannot be used for this since another fungus — formerly placed in *Agaricus* — was given this name earlier.

Agaricus placomyces Pk. **Cap** 5–12cm (2–4¾in) across, hemispherical to flat; sooty brown in the center, breaking up into tiny scales on a whitish ground. **Gills** free, crowded; pallid at first, then flesh-pink, finally blackish. **Stem** 60–120 × 10–12mm (2¼–4¾ × ½in), equal with a swollen bulbous base; white, bruising lemon yellow on the stem base and the rings when fresh; ring pendant, white, rather thick. **Flesh** bright lemon yellow in the stem base when cut, otherwise pallid brownish after a few minutes. **Odor** like carbolic acid or ink. **Taste** also inky. **Spores** ovoid, 5–7 × 3.5–4.6μ.

Deposit deep brown. **Habitat** in broad-leaved woods. Found throughout North America. **Season** July–November. **Not edible** — poisonous to some people. **Comment** This mushroom has been placed in synonymy with *Agaricus meleagris*, now called *Agaricus praeclaresquamosus* (above), but Freeman has pointed out that the spores differ in size and has therefore retained it as a separate taxon.

Agrocybe praecox (Fr.) Fayod **Cap** 3–10cm (1¼–4in) across, convex to broadly umbonate; pale cream to buff or clay; smooth, soft, sometimes cracking when dry. **Gills** adnate, crowded; pallid then clay-brown. **Stem** 30–100 × 4–10mm (1¼–4 × 3/16–½in), equal; pale buff below ring, with white rhizomorphs at base; ring membranous, high, white, often torn. **Flesh** thin; white. **Odor** farinaceous-mealy. **Taste** farinaceous-mealy. **Spores** truncate, 8–11 × 5–6μ. Deposit dark brown. **Habitat** in wood chips, humus, and grass, in fields and woodlands. Common. Found throughout most of North America. **Season** May–July. **Not edible**.

Agrocybe dura (Fr.) Singer **Cap** 4–10cm (1½–4in) across, convex then flattened; white to pale buff on disc; smooth, soft, dry, often cracking in dry weather. **Gills** adnate; pale brown to umber. **Stem** 40–100 × 5–15mm (1½–4 × ¼–½in), equal, firm; white; smooth to fibrillose; ring soon vanishing or left as tatters on cap margin. **Flesh** white. **Odor** not distinctive. **Taste** a little bitter, unpleasant. **Spores** truncate, 10–14 × 6.5–8μ. Deposit dull brown. **Habitat** in fields, pastures, and

wastelands, shrub borders. Found in northern North America. **Season** May–July in hot, wet weather. **Not edible**.

Agrocybe semiorbicularis (Bull. ex St. Amans) Fayod syn. *Naucoria semiorbicularis* (Bull. ex St. Amans) Quél. **Cap** 1–2cm (½–¾in) across, hemispherical to flattened-convex; ochraceous to tan; smooth, slightly tacky. **Gills** adnate, very broad; pale ochraceous at first, becoming dark cinnamon with age. **Stem** 25–40 × 2–3mm (1–1½ × 3/32–⅛in) across, twisted and slightly bulbous at the base; pallid flushed with the cap color; dry, slightly scurfy. **Odor** mealy. **Spores** ovoid, 11–13 × 7.5–8μ. Deposit cigar brown. **Habitat** scattered or in groups in grass or sand or along roadsides. Frequent. Found widely distributed in North America. **Season** July–September. **Not edible**.

Agrocybe acericola (Pk.) Singer **Cap** 3–10cm (1¼–4in) across, convex then flattened; dark yellow-brown then pale buff when dry; smooth, dry, hygrophanous. **Gills** adnate or with decurrent tooth, crowded; pale gray-brown. **Stem** 50–100 × 5–10mm (2–4 × ¼–½in); pallid buff or whitish, with white threads at base; ring large, flaring, white. **Flesh** thin; white. **Odor** farinaceous. **Taste** farinaceous. **Spores** truncate, smooth, 8–10.5 × 5–6.5μ. Deposit dull rust-brown. **Habitat** on decaying logs and woody debris in hardwoods. Common. Found throughout North America. **Season** July–September. **Not edible**.

Agrocybe praecox ⅓ life-size

Agrocybe dura ½ life-size

Agrocybe semiorbicularis ⅔ life-size

Agrocybe acericola ⅔ life-size

Agrocybe erebia ½ life-size

Agrocybe pediades ⅓ life-size

Stropharia hardii ⅓ life-size

Stropharia squamosa ⅓ life-size

Stropharia albonitens ⅔ life-size

Agrocybe erebia (Fr.) Kühner **Cap** 3–6cm
(1¼–2¼in) across, convex-flattened with broad
umbo; deep reddish brown to umber when
moist, paler dull clay-brown when dry. **Gills**
adnate to often decurrent; pallid then deep
umber brown. **Stem** 60–80 × 6–12mm
(2¼–3 × ¼–½in), equal; whitish, then
darkening brown from the base upward; ring
prominent when young but soon torn and
easily lost, whitish and grooved above. **Flesh**
pale brown. **Odor** not distinctive. **Taste** not
distinctive. **Spores** ellipsoid, 11–15.5 × 5–6.5μ.
Deposit dull brown. **Habitat** on soil in damp
woods. Found widespread in northern North
America. **Season** July–September. **Not edible.**

Agrocybe pediades (Fr.) Fayod **Cap** 1–3cm
(½–1¼in) across, convex; ochre-buff to darker
brown when wet; smooth. **Gills** adnate,
crowded; cream then rust-brown. **Stem**
20–50 × 2–3mm (¾–2 × ³⁄₃₂–⅛in); pale buff;
fibrillose. **Odor** not distinctive. **Taste** not
distinctive. **Spores** truncate, 9–13 × 6.6–7.5μ.
Deposit dark brown. **Habitat** grasslands,
pastures. Common. Found throughout North
America. **Season** May–June. **Not edible.**

Stropharia hardii Atkinson **Cap** 3–10cm
(1¼–4in) across, convex-flattened; yellow-
ochre; slightly viscid, then soon dry. **Gills**
adnate, crowded; pale gray-buff then purple-
brown. **Stem** 50–80 × 5–15mm (2–3 × ¼–½in),
equal; pale yellowish; slightly scaly to smooth,
with white rhizomorphs at base. **Flesh** soft;
white. **Odor** not distinctive. **Taste** not
distinctive. **Spores** ellipsoid, with germ pore at
tip, 5–9 × 3–5μ. Deposit purple-brown.
Habitat in deciduous woods in leaf litter.
Found in eastern North America, west to Ohio.
Season July–September. **Edibility not
known.**

Stropharia squamosa (Fr.) Quél. **Cap** 3–5cm
(1¼–3in) across, convex-campanulate; dull
yellow-ochre to tawny, with paler, faint scales
at margin; viscid. **Gills** adnate, crowded; pallid
then purple-brown. **Stem** 60–120 × 3–10mm
(2¼–4¾ × ⅛–½in), long, rigid; brownish;
scaly below the small ring at apex. **Flesh** thin;
whitish. **Odor** not distinctive. **Taste** not
distinctive. **Spores** ellipsoid, with pore at tip,

12–14 × 6–7.5μ. Deposit purple-brown.
Habitat on decayed wood chips in mixed
woodlands. Found throughout northern North
America. **Season** August–October. **Not
edible.**

Stropharia albonitens (Fr.) Karsten **Cap** 2–6cm
(¾–2¼in) across, convex, often umbonate;
white to cream with yellowish center; smooth,
viscid. **Gills** sinuate; pale violet-gray-brown.
Stem 40–80 × 3–5mm (1½–3 × ⅛–¼in); white,
yellow-floccose below the apical ring zone.
Flesh thin; white. **Odor** not distinctive. **Taste**
not distinctive. **Spores** ellipsoid, 8–9 × 4–5μ.
Deposit purple-black. **Habitat** in grassy
meadows. Found in the Pacific Northwest.
Season July–November. **Not edible.**

Red-cap Psilocybe *Psilocybe squamosa* (Pers. ex
Fr.) Orton var. *thrausta* (Schultz ex Kalchb.)
Lange syn. *Stropharia squamosa* var. *thrausta*
(Schultz ex Kachlb.) Lange **Cap** 2.5–7cm
(1–2¾in) across, convex to obtuse, then
campanulate-flattened, with central umbo;
intense reddish orange to brick red; viscid,
smooth, but with numerous small white
evanescent scales at margin. **Gills** adnate,
rather crowded to almost distant, broad;
whitish to gray then almost black with pale
margin. **Stem** 60–120 × 3–8mm (2¼–4¾ ×
⅛–⁵⁄₁₆in), long, slender, densely scaly in lower
half; brownish orange to red like cap, whitish
above the distinct membranous cottony ring,
which is soon stained purple-black from spores.
Flesh thin; pale cream. **Odor** not distinctive.
Taste not distinctive. **Spores** ellipsoid, with
germ pore distinctly eccentric, 12–14 × 6–7.5μ.
Deposit purple brown. **Habitat** in small
scattered clusters on wood chips and twigs.
Rather rare. Found over most of North
America. **Season** August–October. **Not
edible. Comment** Frequently confused with
the much duller colored, ochre-brown *Psilocybe
squamosa* (Pers. ex Fr.) Orton, of which it is
often regarded as a variety; but the germ pore
in that species is centrally placed. Both species
possess chrysocystidia, which removes them
from the genus *Stropharia*, where they were
formerly placed.

Red-cap Psilocybe *Psilocybe squamosa* var. *thrausta* life-size

Stropharia cyanea ½ life-size

Stropharia aeruginosa ½ life-size

Stropharia hornemannii ⅓ life-size

Stropharia ambigua ½ life-size

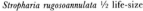
Stropharia rugosoannulata ½ life-size

Stropharia rugosoannulata (white form) ½ life-size

Stropharia cyanea (Bolton ex Secr.) Tuomikoski **Cap** 3–8cm (1¼–3in) across, convex then obtuse-umbonate; bluish green to yellow-green, soon discolored to pale straw yellow with only hint of green; viscid, with fine, evanescent white veil flakes at margin. **Gills** emarginate; pale vinaceous buff to tobacco brown when old. **Stem** 40–100×3–12mm (1½–3×⅛–½in); colored as cap; with fugacious ring zone. **Flesh** pale blue-green. **Odor** not distinctive. **Taste** not distinctive. **Spores** ellipsoid, 7–9×4.5–5μ. Deposit umber-brown. **Habitat** in grass and leaf litter. Found in Washington State, probably widespread in the Northwest. **Season** August–October. **Not edible. Comment** This species has long been confused with *Stropharia aeruginosa* (below), which is more strongly colored blue-green and has a copious white veil and white then darker, purple-brown gills.

Stropharia aeruginosa (Curt. ex Fr.) Quél **Cap** 2–8cm (¾–3in) across, convex to bell-shaped, then flattened and slightly umbonate, the margin hung with veil remnants; blue to blue-green fading to yellowish; slimy with gluten, flecked with white scales. **Gills** adnate, close, broad; pallid then purple-brown, often with a white edge. **Stem** 40–100×5–15mm (1½–4× ¼–½in); whitish to blue or greenish blue; top smooth, covered in small whitish scales below; whitish partial veil leaves a spreading membranous ring on the upper stem. **Flesh** soft; whitish blue. **Odor** not distinctive. **Taste** not distinctive. **Spores** ellipsoid, smooth, with

a pore at the tip, 7–10×5μ. Deposit brownish purple. Chrysocystidia present on the gills. **Habitat** singly or in small groups in rich soil, humus, wood debris, or grass, in woodland. Found widely distributed in many parts of North America. **Season** August–October (December in southern California). **Poisonous. Comment** The very similar *Stropharia cyanea* (above) is a grassland or garden species.

Stropharia hornemannii (Fr.) Lundell **Cap** 6–15cm (2¼–4in) across, broadly convex-umbonate; dull reddish brown or purple-brown, with white veil remnants at margin; very viscid when wet. **Gills** adnate, crowded; pallid then purple-brown. **Stem** 60–120× 10–20mm (2¼–4¾×½–¾in); white; strongly fibrillose-scaly below the prominent ring. **Flesh** white. **Odor** a little unpleasant. **Taste** a little unpleasant. **Spores** ellipsoid, with germ pore, 10–14×5.5–7μ. Deposit purple-brown. **Habitat** on rotting conifer logs. Found in northern North America. **Season** August–November. **Not edible** — possibly poisonous.

Stropharia ambigua (Pk.) Zeller **Cap** 5–15cm (2–6in) across, convex to broadly umbonate; pale yellow to bright ochre; smooth, but with veil remnants at margin. **Gills** adnate, crowded; pale gray then soon purplish brown. **Stem** 75–150×10–20mm (3–6×½–¾in); white with fibrillose-scaly covering below faint ring zone; with many white rhizomorphs at base. **Flesh** white. **Odor** not distinctive. **Taste** not distinctive. **Spores** ellipsoid, smooth, with germ pore at tip, 11–14×6–7.5μ. Deposit

purple-brown. **Habitat** in conifer woods. Often numerous. Found in the Pacific Northwest and on the California coast. **Season** August–November. **Edibility not known.**

Stropharia rugosoannulata Farlow ex Murr. **Cap** 5–20cm (2–8in) across, convex-flattened to umbonate; deep purplish red to dull brown or even grayish or white with age; smooth, not viscid. **Gills** adnate, crowded; pallid then gray and finally purple-brown. **Stem** 100–180× 10–25mm (4–7×½–1in), equal to clavate; white; smooth; ring large, prominent, deeply wrinkled or segmented below, very thick, white. **Flesh** firm; white. **Odor** pleasant. **Taste** pleasant. **Spores** ellipsoid, with germ pore, 10–13×7.5–9μ. Deposit purple-brown. **Habitat** on wood chips and bark mulch and around flower beds. Very common. Found widely distributed in northern North America. **Season** June–October. **Edible** — delicious. **Comment** An almost pure white form is not infrequent; also a closely related (probably undescribed) yellow species with *viscid* cap may be found at the same time.

Psilocybe atrobrunnea ½ life-size

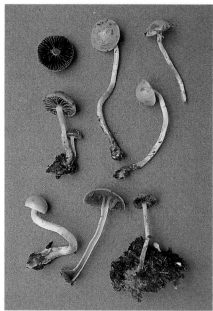

Psilocybe merdaria ½ life-size

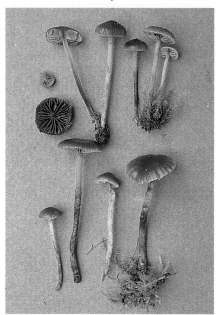

Psilocybe montana ⅔ life-size

Liberty Cap *Psilocybe semilanceata* ¾ life-size

Psilocybe cubensis ½ life-size

Psilocybe merdaria (Fr.) Ricken syn. *Stropharia merdaria* (Fr.) Quél. **Cap** 1–3cm (½–1¼) across, convex to slightly umbonate; dull yellow-orange to yellow-brown; smooth, viscid. **Gills** distant, broad; pale yellowish then purple-brown. **Stem** 20–40×1–3mm (¾–1½× ¹⁄₁₆–⅛in), equal; yellow; dry, with veil leaving a ring zone at apex. **Flesh** pallid becoming brownish in stem. **Odor** not distinctive. **Taste** not distinctive. **Spores** broadly ellipsoid, 14–17×7–8μ. Deposit purple-brown. **Habitat** on horse dung. Uncommon. Found throughout most of North America. **Season** April–October. **Not edible**.

Psilocybe atrobrunnea (Lasch) Gillet **Cap** 1.5–5cm (½–2in) across, conic-campanulate or convex; dark reddish brown, becoming blackish brown with age; pale tan when faded, margin striate when moist, smooth, viscid. **Gills** adnate, crowded, broad; pale cinnamon buff, then dark violaceous brown when mature. **Stem** 60–120×2–4mm (2¼–4¾×³⁄₃₂–³⁄₁₆in), equal, flexuous, fibrous; covered with pale fibrils, becoming darker brown from the base up with age. **Flesh** thin; pallid. **Odor** slight, mealy. **Taste** mealy. **Spores** ellipsoid, smooth, 9–12.5×5–7μ. Deposit deep purple brown. **Habitat** gregarious in swamps and bogs. Frequent. Found in northeastern North America. **Season** August–October. **Not edible**.

Psilocybe montana (Fr.) Quél. **Cap** 0.5–2.5cm (¼–1in) across, convex to umbonate; deep reddish brown, drying paler; smooth, slightly tacky. **Gills** adnate, distant; reddish brown. **Stem** 25–50×1.5mm (1–2×¹⁄₁₆in); dark brown; smooth. **Odor** not distinctive. **Taste** not distinctive. **Spores** ellipsoid, with germ pore at tip, 5.5–8×4–5μ. Deposit purple-brown. **Habitat** in moss in mountainous areas. Found in western North America. **Season** July–September. **Not edible**.

Liberty Cap *Psilocybe semilanceata* (Fr. ex Secr.) Kummer **Cap** 1–2.5cm (½–1in) across, sharply conical, hardly expanding; pale buff to whitish ochre, often olive to grayish when wet or old; smooth, slightly viscid. **Gills** adnate, narrow, crowded; gray then brown. **Stem** 50–100×1–2mm (2–4×¹⁄₁₆–³⁄₃₂in); pale cream, bruising blue, especially at the base or all over with aging. **Flesh** pallid. **Odor** not distinctive. **Taste** slightly unpleasant, grassy. **Spores** ellipsoid to lemon-shaped, 11–13.5× 7–8μ. Deposit purple-brown. **Habitat** in fields and pastures, often gregarious. Common in the Pacific Northwest. Found widely distributed in North America. **Season** August–November. **Edible** but not recommended. Hallucinogenic and widely collected for this purpose, but should be regarded as mildly poisonous; also, there is the possibility of misidentification.

Psilocybe cubensis (Earle) Singer **Cap** 1.5–9cm (½–3½in) across, broadly conical to bell-shaped, becoming convex and flatter with an umbo, the margin sometimes with hanging veil remnants; white with a yellowish or brownish center, becoming entirely yellowish buff to yellowish brown, bruising and aging bluish; sticky when moist then dry, smooth or some small, whitish veil remnants when young. **Gills** adnate, close, narrow; gray becoming deep purple-gray to almost black, edges whitish. **Stem** 40–150×4–15mm (1½–6×³⁄₁₆–½in), often enlarged toward the base; white or yellowish, bruising blue or bluish green; smooth, grooved at the top; membranous partial veil leaves a persistent white ring on the upper stalk which is blackened by the falling spores. **Flesh** firm; white, bruising bluish green. **Spores** ellipsoid, smooth, with a distinct pore at the tip, $11–17 \times 7–12\mu$. Deposit purple-brown. **Habitat** singly or in groups on horse dung or cow manure in cattle pastures. Common. Found on the Gulf Coast. **Season** nearly all year. **Edibility suspect** — hallucinogenic, possibly poisonous.

Psathyrella hirta Pk. syn. *Psathyrella coprobia* (Lange) Smith **Cap** 1–2.5cm (½–1in) across, convex; dark brown; covered with white veil flakes. **Gills** adnate, crowded; pallid then purple-brown. **Stem** $25–50 \times 2–3$mm ($1–2 \times \frac{1}{16}–\frac{1}{8}$in); white with fine squamules. **Flesh** pallid. **Odor** pleasant. **Taste** pleasant. **Spores** ellipsoid, $10–13 \times 5.5–6.5\mu$. Deposit purple-brown. Cystidia swollen with long throat. **Habitat** on animal dung. Found widely distributed throughout North America. **Season** June–October. **Not edible**.

Psathyrella hydrophila (Fr.) Maire **Cap** 2–5cm (¾–2in) across, conic-convex to flattened, margin with white appendiculate veil fragments; hygrophanous, deep brown becoming pale tan when dry; glabrous. **Gills** adnexed, crowded, narrow; buff to reddish brown. **Stem** $30–100 \times 2–6$mm ($1¼–4 \times \frac{1}{16}–\frac{1}{4}$in); pallid. **Flesh** thin; whitish. **Odor** not distinctive. **Taste** bitter. **Spores** ellipsoid, $4.5–6 \times 3–3.5\mu$. Deposit purplish brown. Cystidia thin-walled, fusiform. **Habitat** caespitose on hardwoods. Found throughout North America. **Season** June–October. **Edible** but bitter and not worthwhile.

Psathyrella gossypina (Bull. ex Fr.) Pearson & Dennis **Cap** 1–2.5cm (½–1in) across, hemispherical to bell-shaped, expanding slightly, margin hung with cottony white veil fragments; chestnut brown drying ochre; partly covered with veil remnants, striate. **Gills** adnexed, crowded, narrow; pallid becoming purple-gray. **Stem** $20–50 \times 3–6$mm (¾–2 × ⅛–¼in), slightly enlarged toward the base; white, cream below; minutely cottony. **Flesh** yellow in the stem when young. **Spores** ellipsoid, $6–8 \times 3.5–4\mu$. Deposit black-brown. Cystidia lance-shaped. **Habitat** on the ground often on fire sites. Rare but probably underrecorded. Found in the Pacific Northwest. **Season** September–October. **Not edible**.

Psathyrella cernua (Vahl ex Fr.) Moser **Cap** 2–4cm (¾–1½in) across, broadly convex; dark, watery brown; margin finely striate, almost white when dry. **Gills** adnexed, close; pallid then dull purple-brown. **Stem** $30–50 \times 2–4$mm ($1¼–2 \times \frac{1}{16}–\frac{3}{16}$in); white. **Flesh** pallid. **Odor** not distinctive. **Taste** not distinctive. **Spores** ellipsoid, $7–8 \times 4–5\mu$. Deposit purple-brown. Cystidia thick-walled. **Habitat** on fallen trees, especially poplars. Rather rare. Distribution uncertain. **Season** July–October. **Edibility not known**.

Psathyrella ochracea (Romagnesi) Kits van Waveren **Cap** 0.8–2cm (⁵⁄₁₆–¾in) across, conical-campanulate; ochre to reddish yellow, drying paler; wrinkled-veined. **Gills** adnate, distant; dull sooty brown. **Stem** $20–40 \times 1–2$mm (¾–1½ × $\frac{1}{16}$–⅛in); yellowish to ochraceous. **Flesh** watery pallid. **Spores** ellipsoid, $12.5–15 \times 6.5–7\mu$. Deposit purplish brown. Pleurocystidia sharp, awl-like. **Habitat** on soil. Found in the Pacific Northwest and possibly elsewhere. **Season** July–October. **Not edible**.

Psathyrella candolleana (Fr.) Maire **Cap** 3–10cm (1¼–4in) across, conic to expanded-umbonate; bright honey color when moist, then soon dull buff to ivory when dry; glabrous. **Gills** adnate to seceding, crowded, narrow; white to grayish

(continued on page 203)

Psathyrella hirta ¾ life-size

Psathyrella hydrophila ⅓ life-size

Psathyrella gossypina ½ life-size

Psathyrella cernua ⅓ life-size

Psathyrella ochracea ½ life-size

Psathyrella candolleana ¼ life-size

Psathyrella velutina life-size

Psathyrella conissans ⅔ life-size

Psathyrella caputmedusae ½ life-size

Psathyrella delineata ½ life-size

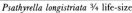

Psathyrella longistriata ¾ life-size

Psathyrella gracilis ⅔ life-size

(continued from page 201)

brown or purplish. **Stem** 40–100 × 3–8mm (1½–4 × ⅛–¼in), very fragile; white; with slight remains of fibrillose veil. **Flesh** white. **Odor** pleasant. **Taste** pleasant. **Spores** ellipsoid to ovoid, smooth, 7–10 × 4–5µ. Deposit purplish brown. Pleurocystidia absent; cheilocystidia thin-walled, cylindric. **Habitat** in clusters, usually on hardwood debris or stumps. Common. Found throughout North America. **Season** April–July. **Edible**.

Psathyrella velutina (Fr.) Singer syn. *Lacrymaria velutina* (Pers. ex Fr.) Singer **Cap** 5–10cm (2–4in) across, convex-flattened; tawny ochraceous to rust-brown; densely fibrillose, margin with veil remnants. **Gills** adnate, edges often beaded with moisture; deep rust-brown to almost blackish brown. **Stem** 50–150 × 4–20mm (2–6 × ¼–¾in), soon hollow; colored like cap below fine veil zone at apex. **Flesh** ochre to brownish. **Odor** pleasant. **Taste** pleasant. **Spores** ellipsoid, minutely warty, with a prominent germ pore, 9–12 × 6–7µ. Deposit blackish brown. **Habitat** on rich soil in woodlands and gardens. Common. Found throughout most of North America. **Season** August–November. **Edible** but not recommended; similar species are known to be poisonous.

Psathyrella conissans (Pk.) Smith **Cap** 2–5cm (¾–2in) across, convex with an inrolled margin, expanding to become almost flat; dark reddish brown fading to buff-brown or pinkish-tinged; moist, smooth, sometimes wrinkled

with the margin radially lined. **Gills** adnate, crowded, narrow; brown becoming pinky-reddish brown. **Stem** 25–50 × 3–5mm (1–2 × ⅛–¼in), rigid, hollow; whitish tinged pink; finely hairy to scurfy. **Flesh** white. **Spores** oblong to ellipsoid, smooth, with an indistinct pore at the tip, 6.5–8 × 3.5–5µ. Deposit pinkish red. **Habitat** in clusters at the base of trees and stumps. Found in northeastern North America, south to Tennessee and west to Michigan. **Season** August–October. **Not edible**.

Psathyrella caputmedusae (Fr.) Konrad & Maublanc **Cap** 4–5cm (1½–2in) across, campanulate; dull date brown when moist, with small fibrillose scales that have darker tips. **Gills** adnate, crowded, narrow; white then pinkish gray to brown. **Stem** 50–100 × 7–8mm (2–4 × ¼–⁵⁄₁₆in); white; fibrillose-scaly below a membranous ring. **Flesh** white. **Odor** pleasant. **Taste** pleasant. **Spores** ovoid to ellipsoid, 9–12 × 4.5–6µ. Deposit purplish brown. Pleurocystidia obtuse to rounded. **Habitat** on conifer wood. Rather rare. Found in the Pacific Northwest. **Season** September–November. **Edibility not known**.

Psathyrella delineata (Pk.) Smith **Cap** 3–10cm (1¼–4in) across, convex-flattened, margin slightly appendiculate; dark rusty brown; surface often rugulose (wrinkled), with fine fibrillose coating. **Gills** adnate, crowded, broad; deep brown. **Stem** 50–100 × 5–15mm (2–4 × ¼–½in); white; fibrillose. **Flesh** white. **Odor** not distinctive. **Taste** not distinctive. **Spores** slightly bean-shaped, 6.5–9 × 4.5–5.5µ.

Deposit purplish brown. Pleurocystidia with fingerlike apical projection. **Habitat** on woody debris of hardwoods. Found throughout eastern North America. **Season** July–September. **Edibility not known**.

Psathyrella longistriata (Murr.) Smith **Cap** 3–10cm (1¼–4in), convex-umbonate; vinaceous brown to gray-brown; with thin coating of veil fibers. **Gills** adnate seceding, crowded, broad; buff to dull brown. **Stem** 40–100 × 4–10mm (1½–4 × ³⁄₁₆–½in); white; with veil fragments below membranous, striate ring. **Flesh** white. **Odor** not distinctive. **Taste** not distinctive. **Spores** ellipsoid, slightly truncate, 7–9 × 4–5µ. Deposit purplish brown. Pleurocystidia ventricose with acute apex. **Habitat** on leaf litter and needle duff in mixed woods. Found in the Pacific Northwest. **Season** August–December. **Not edible**.

Psathyrella gracilis (Fr.) Quél. **Cap** 2–5cm (¾–2in) across, campanulate; yellow-ochre to buff-brown, pinkish when dry; glabrous. **Gills** adnate, crowded, broad; pallid then pinkish brown. **Stem** 60–120 × 2–3mm (2¼–4¾ × ¹⁄₁₆–⅛in); white; pruinose at first. **Flesh** whitish. **Spores** ellipsoid, 11–14 × 6.5–8µ. Deposit purplish brown. Pleurocystidia with an acute apex, 54–75 × 10–16µ. **Habitat** scattered on soil. Found throughout North America. **Season** August–October. **Edibility not known**.

Shaggy Mane *Coprinus comatus* ⅔ life-size

Coprinus disseminatus ¾ life-size

Coprinus lagopides ⅓ life-size

Shaggy Mane *Coprinus comatus* (Fr.) S. F. Gray
Cap 3–7cm (1¼–2¾in) across when
expanded, more or less a tall ovoid when
young, becoming more cylindrical as it
expands; white and very shaggy-scaly, often
with a pale brownish "skullcap" at apex;
margin of the cap dissolves away and
progresses steadily upward until the entire cap
has liquified away, including the gills. **Gills**
free, crowded, very narrow; white becoming
black and inky from the margin upward. **Stem**
60–120 × 10–20mm (2¼–4¾ × ½–¾in), very
tall, straight, with a slightly bulbous base,
hollow in center; white; smooth, with a ring of
veil tissue left lower down on the stem. **Flesh**
soft, fibrous; white. **Odor** (when young)
pleasant. **Taste** similar. **Spores** ellipsoid,
smooth, with germ pore at apex,
(12)13–17(18) × 7–9μ. Deposit black. **Habitat**
often in large numbers on roadsides, lawns,
and other urban sites, especially where the soil
has been disturbed. Found throughout North
America. **Season** sometimes in the spring but
usually July–November. **Edible** and delicious
when young.

Coprinus disseminatus (Pers. ex Fr.) S. F. Gray
syn. *Psathyrella disseminata* (Pers. ex Fr.) Quél.
Cap 0.5–1.5cm (¼–½in) high, ovoid at first,
expanding to convex or bell-shaped; pale buff
with buff or honey-buff center; deeply grooved,
minutely scruffy. **Gills** attached, nearly distant,
broad; white then amber to black, but not inky
or deliquescing. **Stem** 15–40 × 1–3mm
(½–1½ × 1/16–⅛in), hollow, fragile; white with
a buff tinge near the base, which is covered in
white down; smooth to minutely hairy. **Flesh**
fragile. **Odor** none. **Spores** ellipsoid to
almond-shaped, smooth, 7–9.5 × 4–5μ. Deposit
dark brown or blackish. Dermatocystidia thin-
walled, blunt, cylindrical, with a swollen base,
75–100 × 20–30μ. **Habitat** in large groups
(sometimes hundreds) on stumps and debris of
deciduous wood and on lawns and grassy
areas. Found widely distributed in eastern
North America and California. **Season**
May–October (November–March in southern
California). **Edible** but not worthwhile.

Coprinus lagopides Karsten **Cap** 1.5–5cm
(½–2in) high, cylindrical to conical, expanding
with margin up to 5cm (2in) broad and often
torn or split; white to gray at first, covered
with conspicuous whitish to grayish felty scales
from the veil remnants; becoming smooth,
striate from the margin inward. **Gills** attached
to top of stem, crowded, broad; white to dark
pinky-brown then black. **Stem** 30–110 ×
3–12mm (1¼–4½ × ⅛–½in), slightly enlarging
toward the base; white; densely felty white,
becoming smooth with the base white and
woolly. **Odor** none. **Spores** ellipsoid to
subglobose, 6–9 × 5–7μ. Deposit violaceous
black. **Habitat** scattered or in groups on
charred wood or burned soil. Sometimes
abundant. Found in the Great Lakes area and
the Pacific Northwest. **Season** September–
October. **Edibility not known. Comment**
Superficially similar to *Coprinus lagopus* (Fr.) Fr.
but distinguished by habitat and the scales on
the cap.

Tippler's Bane *Coprinus atramentarius* (Bull. ex
Fr.) Fr. **Cap** 3–7cm (1¼–2¾in) high, ovoid at
first, then broadly conical when expanded,
with the margin irregularly puckered at first,
then becoming split; gray to gray-brown; dry,
smooth or silky with minute scales or veil
remnants, especially near the center. **Gills** free,

crowded, broad; white then lavender-gray then inky black and soon deliquescing. **Stem** 70–170 × 9–20mm (2¾–6½ × ⅜–¾in), hollow; whitish; dry, silky-fibrous; fibrous white partial veil leaving ring zone near base. **Odor** faint and pleasant or none. **Spores** ellipsoid, smooth, with pore at tip, 7–11 × 4–6μ. Deposit black. **Habitat** usually in clusters on the ground near rotting or buried wood or in grass. Found widely distributed throughout North America. **Season** May–September (November–April in California). **Edible** but dangerous because it causes alarming symptoms (nausea, palpitations) when taken in conjunction with alcohol; indeed, it has been given to alcoholics to cause these symptoms and eventually cure their habit. **Comment** Good black drawing ink used to be made from the deliquesced caps by boiling the black "ink" with a little water and cloves.

Mica Cap *Coprinus micaceus* (Bull. ex Fr.) Fr. **Cap** 1.5–5cm (½–2in) high, ovoid expanding to bell-shaped, with a split or sometimes rolled-back margin that is lined and grooved almost to the center; tawny becoming cinnamon toward the center; covered with white powdery granules from the veil, especially when young, becoming smooth. **Gills** attached, close, moderately broad; white becoming date brown then black. **Stem** 25–85 × 2–5mm (1–3¼ × ³⁄₃₂–¼in), hollow, fragile; white, discoloring buff in lower part; smooth or slightly felty. **Odor** none. **Spores** ellipsoid, smooth, with pore at tip, 7–10 × 4.5–6μ. Deposit date brown to blackish. **Habitat** in dense tufts around stumps or on bruised wood. Very common. Found widely distributed throughout North America. **Season** April–October (all year in southern California). **Edible**.

Coprinus quadrifidus Pk. **Cap** 2–7cm (¾–2¾in) across at base, 4–7cm (1½–2¾in) high, egg-shaped expanding to bell-shaped; whitish at first from veil remnants, then becoming dingy buff to gray or grayish brown; felty at first, breaking up into loose cottony-flaky patches or warts. **Gills** free, crowded, broad; white becoming purplish, then blackish or inky. **Stem** 40–120 × 5–10mm (1½–4¾ × ¼–½in), hollow; white with brownish rhizomorphs connected to the base; felty, scaly; white partial veil leaving evanescent ring. **Taste** unpleasant. **Spores** ellipsoid, smooth, with pore at tip, 7.5–10 × 4–5μ. Deposit blackish. **Habitat** growing in large clusters on hardwoods, especially elm and ash. Found in northeastern North America west to the Great Lakes. **Season** June–August. **Edibility suspect** — may cause stomach upset in some people.

Coprinus silvaticus Pk. **Cap** 1–3cm (½–1¼in) high, ovoid expanding to conical-convex, with a deeply striate margin grooved to the center and often split; cream-buff with darker sienna or cinnamon center; apparently smooth but actually minutely hairy. **Gills** free, crowded; white then gray-umber becoming black. **Stem** 40–85 × 3–6mm (1½–3¼ × ⅛–¼in), rather fragile; white discoloring pale buff, especially in the lower section; silky-lined with top finely woolly and lower part becoming smooth. **Odor** none. **Spores** almond-shaped, ornamented with low warts and ridges, 11–15 × 8–10μ. Deposit black. Dermatocystidia 90–180 × 16–25μ. **Habitat** on soil attached to burned wood. Rare. Found in the Pacific Northwest. **Season** September–October. **Edibility not known.**

Panaeolus retirugis Fr. **Cap** 3–6cm (1¼–2¼in) across, bell-shaped; pale tan-brown, but often white when old and dry; glabrous, surface often pitted or wrinkled. **Gills** adnate, crowded, broad; purplish gray at first, then blackish brown. **Stem** 90–150 × 3–7mm (3½–6 × ⅛–¼in), hollow; whitish to pale gray; densely pruinose for most of its length; beaded above with droplets when fresh. **Flesh** thin; brown. **Odor** not distinctive. **Taste** slightly foul. **Spores** lemon-shaped, with an apical pore (11)12–16(18) × 8–11μ. Deposit blackish. **Habitat** on dung or on heavily manured soil. Found widespread throughout North America. **Season** April–August. **Not edible.**

Tippler's Bane *Coprinus atramentarius* ⅓ life-size

Mica Cap *Coprinus micaceus* ⅓ life-size

Coprinus quadrifidus ½ life-size

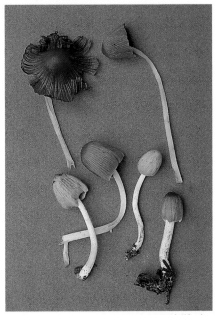

Coprinus silvaticus ½ life-size

Panaeolus retirugis ⅓ life-size

Panaeolus semiovatus ½ life-size

Angel Wings *Pleurocybella porrigens* ⅓ life-size

Oyster Mushroom *Pleurotus ostreatus* ⅓ life-size

Panaeolus semiovatus (Fr.) Lundell & Nannfeldt **Cap** 5–8cm (2–3in) across, deeply campanulate to obtusely conic; pale tan to gray-buff; smooth, viscid, sometimes cracking. **Gills** free, crowded, narrow; pale gray then mottled black. **Stem** 50–150×6–12mm (2–6×¼–½in), rigid, slightly swollen at base; whitish; small ring around center. **Flesh** thin, pallid. **Odor** not distinctive. **Taste** not distinctive. **Spores** pip-shaped, 15–20×8–11μ. Deposit black. **Habitat** on horse dung. Found widely distributed throughout North America. **Season** May–October. **Not edible**.

Lateral-Stemmed Agarics pp. 206–212
These mushrooms have lateral rather than central stems and are found growing on wood.

Angel Wings *Pleurocybella porrigens* (Pers. ex Fr.) Singer syn. *Pleurotellus porrigens* (Pers. ex Fr.) Kühner & Romagnesi and *Pleurotus porrigens* (Pers. ex Fr.) Kummer **Fruit body** a stemless cap laterally attached to wood. **Cap** 4–10cm (1½–4in) across, fanlike to elongately ear-shaped; pure white; smooth. **Gills** decurrent, crowded, narrow; white. **Flesh** white. **Odor** pleasant. **Taste** pleasant. **Spores** rounded to short ellipsoid, 6–7×5–6μ. Deposit white. **Habitat** on decaying conifer logs. Found across most of North America. **Season** September–October. **Edible** — good.

Oyster Mushroom *Pleurotus ostreatus* (Jacq. ex Fr.) Kummer **Cap** 6–14cm (2¼–5½in) across, shell-shaped, convex at first, then flattening or slightly depressed; often wavy and lobed at the margin or splitting; rather variable in color, but deep bluish gray to gray-brown when young, buff-brown with age. **Gills** slightly decurrent on the short stem, crowded, narrow; white to pale cream-ochre. **Stem** 20–30×10–20mm (¾–1¼×½–¾in), eccentric to lateral, or absent; white; with a woolly base. **Flesh** white. **Odor** rather pleasant. **Taste** rather pleasant. **Spores** subcylindric, 7.5–11×3–4μ. Deposit lilac. **Habitat** in large clusters on stumps, logs, and trunks of deciduous trees. Common. Found throughout North America. **Season** usually late, October–January, often through the winter whenever the weather is mild. **Edible** — delicious.

Jack O'Lantern *Omphalotus illudens* (Schw.) Bigelow **Cap** 5–20cm (2–8in) across, convex then soon flattened and then funnel-shaped with incurved margin; a brilliant and intense yellow-orange in color; smooth. **Gills** decurrent, crowded; bright yellow-orange. **Stem** 50–200×10–20mm (2–8×½–¾in), tapered at base, solid; colored as cap but darkening at base; smooth. **Flesh** firm; pale orange. **Odor** not distinctive. **Taste** not distinctive. **Spores** globose, 3.5–5×3.5–5μ. Deposit pale cream. **Habitat** often in enormous clusters at base of stumps or on buried roots (the latter is very common in gardens and lawns) of oaks and some other deciduous trees. Common. Found throughout much of North America, particularly the eastern United States. **Season** July–September but sometimes to November. **Poisonous** but usually not fatal, typically causing gastric upset for some hours or even days. **Comment** When fresh the gills of this species glow a bright greenish yellow in the dark. Based upon cultural evidence, this may be the same as *Omphalotus olearius* (DC ex Fr.) Singer of southern Europe, which name would then take precedence. On the West Coast the species *Omphalotus olivascens* Bigelow, Miller & Thiers is found, which differs in its duller brownish-orange to olivaceous cap and larger spores. It is also poisonous.

Pleurotus elongatipes Pk. **Cap** 2–4cm (¾–1½in) across, convex becoming flatter; creamy or tinged pinkish with watery spots over the disc; moist, smooth, hairless. **Gills** adnexed or notched, subdistant, broad; creamy buff to pinkish buff. **Stem** 40–220 × 15–25mm (1½–8½ × ½–1in), sometimes gradually tapering toward the base and often curved or bent, becoming soft and hollow; white; smooth, hairless except at the base. **Flesh** quite thick and firm; white to creamy or pinkish buff. **Spores** subglobose, smooth, 4–5.5 × 3.5–4.5μ. Deposit buff-colored. **Habitat** singly or in dense clusters at the base of living hardwood trees or on decaying logs and stumps. Rare. Found in northern North America. **Season** August–October. **Edible**.

Pleurotus dryinus (Pers. ex Fr.) Kummer **Cap** 5–15cm (2–6in) across, convex then slowly expanding, margin inrolled; white to cream; surface dry, felty-hairy to slightly scaly. **Gills** decurrent, crowded, narrow, often cross-veined on the stem; white. **Stem** 50–100 × 10–30mm (2–4 × ⅜–1¼in), lateral to just off-center; white; felty, with a slight membranous ring at apex when young, soon vanishing or leaving fragments on cap margin. **Flesh** firm; white. **Odor** pleasant. **Taste** pleasant. **Spores** cylindrical, 9–12 × 3.5–4μ. Deposit white. **Habitat** on deciduous timber. Found throughout most of northern North America. **Season** July–October. **Edible.**

Hypsizygus tessulatus (Bull. ex Fr.) Singer **Cap** 5–15cm (2–6in) across, convex becoming flatter and rather sunken; white to buff-yellow, creamy tan or crust brown in the center; moist, smooth, minutely hairy, becoming cracked with scaly patches. **Gills** adnexed to sinuate, close to subdistant, broad; whitish becoming cream. **Stem** 40–110 × 10–30mm (1½–4½ × ½–1¼in), solid, off-center, enlarged toward the base; white; dry, smooth, sometimes hairy. **Flesh** thick, firm; white. **Odor** mushroomy. **Taste** mild. **Spores** globose, smooth, 5–7 × 5–7μ. Deposit white to buff. **Habitat** singly or scattered on old hardwood trees, especially elm, often quite high up. Frequent. Found in northern North America. **Season** September–December. **Edible** but tough. **Comment** Formerly known as *Pleurotus ulmarius* sensu American authors.

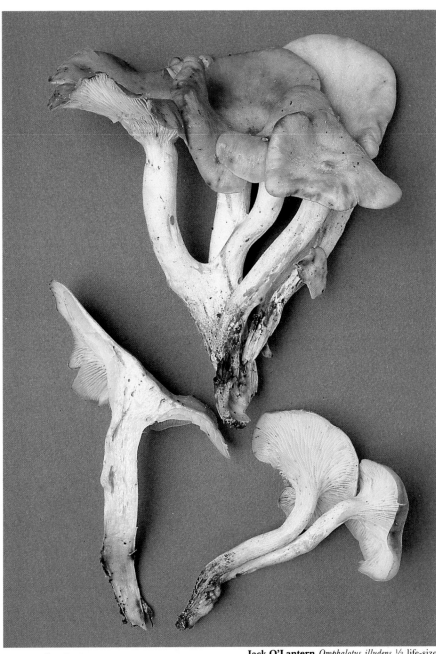

Jack O'Lantern *Omphalotus illudens* ⅓ life-size

Pleurotus elongatipes ⅓ life-size

Pleurotus dryinus ¼ life-size

Hypsizygus tessulatus ¼ life-size

Panus strigosus nearly ¼ life-size

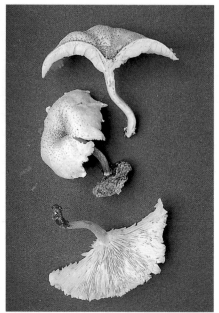

Lentinus tigrinus ⅓ life-size

Lentinus crinitis ⅔ life-size

Bear Lentinus *Lentinellus ursinus* ⅓ life-size

Lentinellus omphalodes ⅓ life-size

Lentinellus cochleatus ⅓ life-size

Panus strigosus Berk. & Curt. **Cap** 10–40cm (4–16in) across, fan-shaped to broadly convex, becoming flatter or slightly sunken; whitish to buff, creamy, or yellowish; dry, coarsely hairy all over. **Gills** decurrent, close to subdistant, broad; whitish tinged brownish or pale mauve, yellowing in age. **Stem** 20–150×10–40mm (¾–6×½–1½in), generally lateral or off-center, solid, sometimes thicker below; white to buff, yellowish in age; tough, coarsely hairy toward the base. **Flesh** thick, quite tough; white or yellowish. **Taste** mild. **Spores** oblong, smooth, nonamyloid, 11–13×3.5–5μ. Deposit white. **Habitat** singly or in small clusters of 3–4 in the wounds of living hardwoods, especially maple and yellow birch. Rare. Found widely distributed in eastern and western North America. **Season** August–October. **Said to be edible** but too tough and rare to be worthwhile.

Lentinus tigrinus (Fr.) Fr. **Cap** 1.5–10cm (½–4in) across, convex to convex-depressed to funnel-shaped, with a somewhat wavy, ragged margin; at first grayish brown to black, with white to buff only showing in maturity; dry, covered in a dense coating of dark brown or blackish hairs or scales, which become sparser in age. **Gills** decurrent, crowded, narrow; white to yellowish white with ragged edges. **Stem** 20–60×4–10mm (¾–2¼×³⁄₁₆–½in), central to eccentric, tapering downward and often bent; grayish yellowy fawn, lighter at the top, darker at the base; hairy to scaly; creamy partial veil leaves a slight ring or zone toward the top of the stem which may disappear in age, or veil may remain intact, covering the gills. **Flesh** thin, tough, fibrous; white. **Odor** mild or none. **Taste** not distinctive. **Spores** narrowly cylindric, smooth, nonamyloid, 6–9.5×2.5–3.5μ. Deposit white. **Habitat** singly or more commonly in groups or clusters on water-soaked hardwood logs or stumps. Sometimes common. Found widely distributed in North America east of the Rockies, but more abundant southward. **Season** May–September. **Edibility not known.**

Lentinus crinitis (Fr.) Fr. **Cap** 1–4cm (½–1½in) across, funnel-shaped or with a deeply depressed center and a somewhat wavy margin; yellowish brown to dark reddish brown; covered with dense, stiff brown hairs. **Gills** decurrent, very crowded, narrow; whitish to slightly yellowish. **Stem** 10–40×2–6mm (½–1½×³⁄₃₂–¼in), slightly expanded at base and apex, sometimes a little twisted; similar color to cap or paler; white cottony-scurfy with a few tiny dark hairs at the base. **Flesh** thin, tough; whitish. **Spores** narrow, cylindric, 5.5–7×1.8–2.7μ. Deposit white. **Habitat** scattered or in groups on dead wood. Frequent. Found along the Gulf of Mexico. **Season** July–August. **Not edible** although Brazilian Indians are reported to eat it boiled.

Bear Lentinus *Lentinellus ursinus* (Fr.) Kühner **Fruit body** a fan- or kidney-shaped cap without a stem. **Cap** 3–11cm (1¼–4½in) wide, 2–5cm (¾–2in) from front to back; yellow-brown to reddish brown; densely covered in fine stiff hairs or pubescence. **Gills** broad, close, with edges minutely serrated or raggedly toothed; pale pinkish brown to reddish brown. **Flesh** thin, pale brown. **Odor** not distinctive. **Taste** extremely acrid. **Spores** nearly round, ornamented with strongly amyloid spines, 2.5–3.5×2–3μ. Deposit white. **Habitat** often gregarious, growing laterally from hardwood logs, particularly oak, but also coniferous timber. Found widely distributed in North America. **Season** July–October. **Not edible.** **Comment** The similar *Lentinellus montanus*

The Train Wrecker *Lentinus lepideus* ½ life-size

Lentinus detonsus ¾ life-size

(Sow: Fr.) Kuhn & Maire grows near western snowbanks in spring, while *Lentinellus vulpinus* (Sow: Fr.) Kuhn & Maire differs in its paler whitish pubescence, radially ribbed cap, and stalklike base.

Lentinellus omphalodes (Fr.) Karsten syn. *Lentinellus bisus* Quél. **Cap** 1.5–5cm (½–2in) across, convex to nearly flat or depressed to umbilicate; beige with a fleshy or brownish tone, darkening to brown; smooth, moist. **Gills** adnate, nearly distant, broad, distinctly saw-edged; whitish then creamy pinkish brown. **Stem** 4–35 × 1–3mm (³⁄₁₆–1¼ × ¹⁄₁₆–⅛in), central or eccentric; beige-brown to reddish brown; dry, usually with longitudinal furrows. **Flesh** pallid. **Odor** sharp. **Taste** peppery. **Spores** ellipsoid, short, with amyloid spines, 4.5–6.5 × 3.5–5µ. Deposit buff. **Habitat** singly or in small groups on the ground or on buried wood or on coniferous or deciduous wood debris. Quite common. Found widely distributed in North America. **Season** August–November. **Edibility not known.**

Lentinellus cochleatus (Pers. ex Fr.) Karsten **Cap** 2–5cm (1–2in) across, irregularly funnel-shaped, margin incurved becoming lobed and wavy; flesh-color to reddish brown; smooth, moist, brittle with scattered hairs. **Gills** decurrent, close, broad; often torn with toothed edges; whitish to pale flesh-colored. **Stem** 20–60 × 5–15mm (¾–2¼ × ¼–½in), central or lateral, often rooting, fused at the base in dense clusters; reddish brown darkening toward the base; deeply furrowed. **Flesh** tough, pinkish. **Odor** of anise. **Taste** distinctive. **Spores**

subglobose, ornamented with spines, amyloid, 4.5–5 × 3.5–4µ. **Habitat** in clusters or tufts on logs, stumps, and decayed wood of deciduous trees. Frequent in the East, rare in the West. Found in eastern North America and rarely in western North America. **Season** July–September. **Not edible**.

The Train Wrecker *Lentinus lepideus* (Fr. ex Fr.) Fr. **Cap** 5–12cm (2–4¾in) across, convex to nearly flat, with an incurved margin that becomes straight in age; whitish to buff with cinnamon-brown scales; shiny, dry and scaly. **Gills** adnexed, close, broad, with toothed edges; whitish to buff bruising brownish. **Stem** 30–100 × 10–15mm (1¼–4 × ½in), solid, either narrow at the base or bulbous; white, to reddish brown in age; minutely hairy above ring, scaly below; partial veil forms membranous whitish ring on the upper stalk. **Flesh** white, aging or bruising dirty yellowish. **Odor** fragrant, like anise. **Taste** rather disagreeable. **Spores** almost cylindrical, smooth, nonamyloid, 9–12 × 4–5µ. Deposit white. **Habitat** growing singly, scattered, or in clusters on logs, stumps, fence posts, railroad ties, decaying coniferous wood, and occasionally hardwoods. Common. Found widely distributed throughout much of North America. **Season** May–September (later in California). **Edible — good. Comment** The stem of this mushroom is extremely tough and woody. Even larger forms of this mushroom can be found in the Southwest; I found a specimen 28cm (11in) across in northern Arizona.

Lentinus detonsus Fr. **Cap** 3–7cm (1¼–2¾in) across, broadly cone-shaped becoming flatter and somewhat wavy, the margin incurved at first, then expanding; pale orangish yellow or yellowish buff, more orangish in the center; smooth, moist. **Gills** adnexed, very crowded; whitish cream bruising reddish or brownish. **Stem** 15–30 × 4–6mm (½–1¼ × ³⁄₁₆–¼in), eccentric, hollow; similar color to cap or paler, darker at the base. **Flesh** tough. **Odor** very strong, garlicky. **Taste** faintly of garlic. **Spores** smooth, nonamyloid, 4.5–6 × 2–3µ. **Habitat** scattered or in groups on the trunks or branches of dead or living hardwoods. Found along the Gulf of Mexico. **Season** July–August. **Edibility not known. Comment** This fungus dries quickly in hot, dry weather but revives as soon as it is rewetted. This species is quite common in the South and has been described by Earle not as a lentinus but as a lentinula. Pegler in his world monograph places it in synonymy with *Lentinula boryana* (Berk. & Montagne) Pegler.

Crepidotus mollis ½ life-size

Crepidotus herbarum ⅔ life-size

Crepidotus crocophyllus ½ life-size

Crepidotus applanatus ⅓ life-size

Luminescent Panellus *Panellus stipticus* ½ life-size

Tectella patellaris ⅔ life-size

Schizophyllum commune ½ life-size

Crepidotus mollis (Fr.) Stde. **Cap** 1–8cm (½–3in) across, shell- or kidney-shaped bracket; pale brown to olive-brown; smooth to slightly scaly. **Gills** radiating from base of cap, crowded, thin; white then brown. **No stem. Flesh** thin, gelatinous; pale brown. **Odor** not distinctive. **Taste** mild. **Spores** broadly ellipsoid, smooth, 7–11×4.5–6.5μ. Deposit brown. **Habitat** in clusters on dead deciduous wood. Found throughout most of North America. **Season** July–October. **Not edible**.

Crepidotus herbarum (Pk.) Sacc. **Cap** 0.5–2cm (¼–¾in) across, kidney-shaped or almost round; white; minutely hairy or downy. **Gills** radiating from a naked lateral or eccentric point, subdistant, narrow; white becoming dingy yellow or brownish. **Stem** none or rudimentary. **Flesh** very thin; white. **Spores** lance-shaped to ellipsoid, smooth, 6–7.5×3–4μ. Deposit pale yellowish brown. **Habitat** scattered or in groups on dead stems, twigs, branches, and debris of shrubs and hardwoods. Common. Found widely distributed in northern North America, south to New York, and in California. **Season** September–October, possibly earlier. **Edibility not known**.

Crepidotus crocophyllus (Berk.) Sacc. **Cap** 0.5–1.5cm (¼–½in) across, convex to kidney-shaped or fan-shaped; ochraceous brown; covered in minute appressed scales. **Gills** rather broad, crowded, rounded behind; yellow to ochre-orange. **No stem. Flesh** soft, thin. **Spores** subglobose, slightly roughened, pale ochre-yellow, 5.7–6×5.7–6μ. Deposit dull brown to cinnamon brown. **Habitat** on dead or decaying hardwood logs and stumps. Found widely distributed in North America. **Season** May possibly through to September. **Edibility not known**.

Crepidotus applanatus (Pers. ex Pers.) Kummer **Cap** 1–4cm (½–1½in) across, shell-shaped to petal-like; white to pale buff; smooth, margin striate when wet. **Gills** radiating from where the cap joins the wood on which it grows, crowded; white to pale brown. **No stem. Flesh** thin; white. **Odor** not distinctive. **Taste** mild. **Spores** globose, minutely spiny, 4–5.5×4–5.5μ. Deposit brown. **Habitat** in clusters on dead deciduous wood. Common. Found throughout most of North America. **Season** July–October. **Not edible**.

Panellus serotinus ½ life-size

Phyllotopsis nidulans ½ life-size

Luminescent Panellus *Panellus stipticus* (Fr.) Karsten **Cap** 5–25mm (¼–1in) across, tongue- to kidney-shaped; pale buff to tan; dry, minutely hairy to rough-scaly. **Gills** crowded, narrow; pinkish buff. **Stem** rudimentary, 10–20 × 3–10mm (½–¾ × ⅛–½in), off-center; colored as cap; densely hairy. **Flesh** thin, tough; pale buff. **Taste** very acrid-astringent. **Spores** oblong, smooth, 3–4 × 2μ. Deposit white. **Habitat** densely gregarious on logs and stumps of hardwoods. Common. Found throughout North America. **Season** May–December. **Not edible. Comment** Its specific name alludes to its traditional reputation for stanching blood. Its common name refers to the way the gills glow greenish white in the dark.

Tectella patellaris (Fr.) Murr. **Cap** 1–1.5cm (⅜–½in) across, convex with inrolled margin; dull brown; slightly viscid. **Gills** arising from point of attachment of the cap edge, close, narrow; pale brown, covered at first with a pale whitish-buff veil. **Stem** (if present) 1–3mm (¹⁄₁₆–⅛in), laterally attached. **Spores** sausage- shaped, smooth, amyloid, 3–4 × 1–1.5μ. Deposit white. **Habitat** in groups on fallen branches and logs of deciduous trees. Found throughout most of North America. **Season** July–October. **Not edible.**

Schizophyllum commune Fr. **Fruit body** fan- shaped or lobed, laterally attached to the substrate on which it grows; upper surface whitish, densely hairy; lower surface gill-like, the "gills" split down their long axis with their sides recurved. **Cap** 1–5cm (½–2in) across. **Odor** pleasant. **Taste** pleasant. **Spores**

cylindric, 3–4 × 1–1.5μ. Deposit white. **Habitat** in clusters on dead wood of hardwoods. Found throughout North America. **Season** all year. **Not edible.**

Panellus serotinus (Fr.) Kühner **Fruit body** 3–7(15)cm (1¼–2¾[6]in) across, kidney- shaped to fan-shaped, laterally attached to wood, stemless; upper surface subtomentose to glabrous; olive-green or yellowish, often with a flush of violet. **Gills** narrow, crowded; ochre- yellow to slightly violet on edges. **Spores** curved cylindrical, usually amyloid when dry, 4–6 × 1.5μ. Deposit white. **Habitat** on fallen logs of both hardwoods and conifers. Found throughout much of North America. **Season** August–October. **Not edible.**

Phyllotopsis nidulans (Fr.) Singer **Fruit body** a laterally attached, bracketlike cap without a stem. **Cap** 3–8cm (1¼–3in) across, circular to kidney-shaped, margin inrolled when young; bright yellow-orange when young, then tawny buff; densely hairy surface. **Gills** narrow, rather crowded; bright orange-yellow. **Flesh** pale orange-buff. **Odor** sharp, very unpleasant. **Taste** sharp, very unpleasant. **Spores** sausagelike, smooth, 6–8 × 3–4μ. Deposit pinkish. **Habitat** on fallen timber, often in overlapping clusters. Common. Found throughout most of North America. **Season** August–October. **Not edible.**

Gomphidius smithii ½ life-size

Gomphidius smithii Miller **Cap** 2.5–7.5cm (1–3in) across, convex-flattened to depressed; grayish to vinaceous gray; smooth, viscid when wet. **Gills** decurrent, broad, almost waxy; white then pale gray and finally blackish gray. **Stem** 30–75 × 5–15mm (1¼–3 × ¼–½in), tapered at base; white to dull grayish, with no yellow at base; dry above veil zone, viscid below. **Flesh** white. **Odor** pleasant. **Taste** pleasant. **Spores** long ellipsoid, 15–20 × 4.5–7μ. Deposit blackish. **Habitat** under Douglas fir. Found in coastal California. **Season** August–November. **Edible.**

Gomphidius subroseus life-size

Chroogomphus tomentosus ½ life-size

Scaly Vase Chanterelle *Gomphus floccosus* ⅓ life-size

Gomphus clavatus ⅔ life-size

Gomphidius oregonensis ⅓ life-size

GOMPHIDIACEAE *The mushrooms in this small group are all found under conifers. They have rather primitive thick gills that run down the stem (decurrent).*

Gomphidius subroseus Kauffman **Cap** 4–6cm (1½–2¼in) across, convex; pink to reddish salmon; smooth, viscid when wet. **Gills** decurrent, subdistant, thick; white then smoky gray. **Stem** 35–100×6–20mm (1¼–4×¼–¾in), equal; white and smooth above the glutinous ring, ivory to yellow at base; viscid from veil covering. **Flesh** pallid, yellow in base. **Odor** mild. **Taste** mild. **Spores** tapering at ends, 15–21×4.5–7μ. Deposit black. **Habitat** under conifers. Found in the Rocky Mountains, Pacific Northwest, and rarely in eastern North America. **Season** June–September (December on the Pacific Coast). **Edible** but poor.

Chroogomphus tomentosus (Murr.) Miller **Cap** 2.5–8cm (1–3in) across; bluntly conical to almost flattened-umbonate; bright ochre-tawny; dry with fibrillose-tomentose surface. **Gills** decurrent, distant, thick; ochre-yellow when young, then purplish. **Stem** 40–150×5–25mm (1½–6×¼–1in), tapered to spindle-shaped; colored like cap; thin, hairy veil leaving slight traces on the upper stem. **Flesh** firm, pale ochre-orange. **Odor** mild. **Taste** mild. **Spores** ellipsoid, smooth, 15–25×6–8μ. Deposit blackish gray. **Habitat** in moss under conifers, especially Douglas fir and hemlock. Found in the Pacific Northwest. **Season** August–October. **Edible.**

Gomphidius oregonensis Pk. **Cap** 2.5–15cm (1–6in) across, broadly convex; ochraceous salmon, turning blackish brown when old; smooth, viscid when wet. **Gills** decurrent, distant, broad; white then smoky gray-brown. **Stem** 40–100×10–25mm (1½–4×½–1in); white above, with yellow flush at base; a white, fibrillose veil above, covered with a secondary glutinous veil; several fruit bodies often clumped together with their stems fused at base. **Flesh** solid; white, yellow at base of stem. **Odor** mild. **Taste** mild. **Spores** spindle-shaped to narrowly ellipsoid, tapered at ends, 10–13×4–6μ. Deposit smoky black. **Habitat** normally emerging from a large fleshy mass buried deep in the soil, usually under conifers. Found along the Pacific Coast. **Season** August–October. **Edible** but poor.

Chroogomphus vinicolor (Pk.) Miller **Cap** 2–8cm (¾–3in) across, convex with a sharp to obtusely pointed umbo; pale orange-ochre to reddish brown or wine color; viscid when wet but soon dry, smooth. **Gills** decurrent, thick; pale ochre, then blackish when mature. **Stem** 50–100×5–15mm (2–4×¼–½in), tapered at base; pale ochre to vinaceous; with slight ring zone at apex. **Flesh** ochre-orange. **Odor** pleasant. **Taste** pleasant. **Spores** subfusiform, 17–23×4.5–7.5μ. Deposit blackish. **Habitat** in groups under pine. Found in eastern North America. **Season** July–September. **Edible.**

Scaly Vase Chanterelle *Gomphus floccosus* (Schw.) Singer syn. *Cantharellus floccosus* Schw. **Fruit body** up to 20cm (8in) high by 1–3cm (½–1¼in) across at base; cap and stem forming at first a small cylindrical fruit body but soon expanding to form a deeply funnel-shaped mushroom, hollow almost to base. **Cap** 3–15cm (1¼–6in) across; yellow-orange, ochre, or tawny; surface may be smooth to fibrillose or even coarsely scaly. **Fertile undersurface** broad, low ridges or wrinkles arranged longitudinally, covering almost the entire outer surface; buffy ochre to slightly vinaceous or brownish where bruised. **Stem** white at base, pale cream to buff above, becoming yellowish with age and bruising brownish; smooth. **Flesh** firm; whitish. **Spores** cylindrical to ellipsoid, 11.5–20×6–10μ; surface roughened, with ornamentation of coarse warts and ridges up to 0.5μ high. Deposit dull ochre. **Habitat** often in rings, in mixed woods. Found over most of North America with the exception (perhaps) of the southwestern United States. **Season** June–September. **Edible** but not recommended; contains indigestible acids which are often sour. **Comment** *Gomphus bonarii* (Morse) Singer can look very similar in some of its forms, although it is usually a more reddish hue with paler, whitish hymenium. Both species vary in their scaliness, and microscopic examination is often necessary in difficult cases; the spores of *Gomphus bonarii* are less warty.

Gomphus clavatus (Fr.) S. F. Gray **Fruit body** 2.5–10cm (1–4in) wide, up to 15cm (6in) high, compressed and partially fused, the cap flat with a sunken center and wavy margin; violet becoming yellowish buff; smooth, moist then dry, felty becoming scaly on the disc. **Fertile undersurface** shallow, wrinkled, sometimes with folds or pits; violet when young, becoming duller and more brownish in age. **Stem** 10–50× 10–20mm (½–2×½–¾in), very short, often curved, sometimes fused with adjacent stems; buff to pale lilac; smooth to minutely hairy. **Flesh** solid; whitish to pale pink. **Odor** none or faintly earthy. **Taste** mushroomy. **Spores** ellipsoid to narrowly ovoid, warty, 10.3–15.5×4.3–7μ. Deposit ochre to dark olive-buff. **Habitat** growing singly or in overlapping clusters or arcs or circles of up to 40 fruiting bodies, under conifers. Common in the Pacific Northwest, rarer in the East. Found in northern California and northern North America. **Season** August–October. **Edibility questionable** — some people get severe gastric upsets from it, others find it excellent.

CHANTERELLES *These mushrooms have primitive foldlike or even absent gills.*

Black Trumpet *Craterellus fallax* Smith **Fruit body** a trumpet-shaped mushroom with the opening recurved back, margin often wavy, irregular; inner surface dry, finely scaly, gray to dark blackish brown when wet. **Fertile undersurface** smooth to irregularly veined; pale brown to gray with a whitish bloom when young, soon developing a flush of salmon pink with age. **Flesh** thin, brittle; gray-brown. **Odor** fragrant, reminiscent of apricots. **Taste** similar. **Spores** broadly ellipsoid, smooth, 10–20×7–11.5μ. Deposit ochre-buff to pale orange. **Habitat** under mixed deciduous trees. Common and abundant. Found throughout North America. **Season** July–October. **Edible** — delicious. **Comment** The very similar but much rarer *Craterellus cornucopiodes* (L. ex Fr.) Pers. has a white spore print and does not develop the salmon-colored flush to the fruit body.

Chroogomphus vinicolor ½ life-size

Black Trumpet *Craterellus fallax* ⅓ life-size

Chanterelle *Cantharellus cibarius* ¾ life-size   **White Chanterelle** *Cantharellus subalbidus* ⅓ life-size

Chanterelle *Cantharellus cibarius* Fr. **Cap** 3–15 cm (1¼–6in) across, convex then becoming flattened to slightly depressed at center, margin inrolled and wavy; bright egg yellow to pale yellow-orange; surface smooth to minutely tomentose. **Fertile undersurface** of cap without true gills; instead with shallow thick-edged wrinkles and veins, often crossveined, quite distant, descending stem; pale yellow to orange. **Stem** 20–60 × 5–25mm (¾–2¼ × ¼–1in), usually tapered below; color and surface similar to cap. **Flesh** solid, firm; white to pale yellowish. **Odor** pleasant, of apricot. **Taste** mild to slightly peppery. **Spores** ellipsoid, smooth, 8–11 × 4–6μ. Deposit pale ochre or yellow. **Habitat** in both deciduous and coniferous woods. Common and often abundant. Found throughout North America. **Season** July–August in the eastern states, September–November in the Northwest, and November–February in California. **Edible** — delicious, but be sure of your identification. **Comment** One of the most famous edible species in the world, it comes in a large number of different forms, some of which are considered good species in their own right by some mycologists. Compare carefully with the poisonous Jack O'Lantern, *Omphalotus illudens* (p. 206).

Smooth Chanterelle *Cantharellus lateritius* (Berk.) Singer syn. *Craterellus cantharellus* Schw. **Cap** 3–10cm (1¼–4in) across, convex then flattened and often depressed at center, margin inrolled and wavy or lobed; pale yellow-orange to orange; smooth to slightly tomentose.

Fertile undersurface of cap without any gills; instead there is a smooth to very slightly wrinkled or veined surface, sometimes crossveined; pale orange-yellow to pinkish. **Stem** 25–100 × 5–25mm (1–4 × ¼–1in), thick, tapered toward base, often curved and off-center; orange-yellow. **Flesh** solid but becoming hollow in stem; white. **Odor** fragrant, fruity, of apricot. **Taste** pleasant. **Spores** ellipsoid, smooth, 7.5–12.5 × 4.5–6.5μ. Deposit pinkish yellow. **Habitat** under oak, especially along pathsides. Common to abundant. Found in northeastern North America. **Season** July–October. **Edible**.

Cantharellus minor Pk. **Cap** 0.5–3cm (¼–1¼in) across, convex with an inrolled margin, then flat to depressed or funnel-shaped, thin, wavy at the margin; yellow to pale orange; smooth. **Fertile undersurface** of cap with very narrow, blunt ridges and wrinkles, often crossveined and descending stem; pale yellow-orange. **Stem** 15–50 × 3–10mm (½–2 × ⅛–½in); yellow-orange; smooth. **Flesh** soft; pale yellow. **Odor** pleasant. **Taste** pleasant. **Spores** ellipsoid, smooth, 6–11.5 × 4–6.5μ. Deposit pale yellow. **Habitat** on damp, mossy soils in deciduous woods. Frequent. Found in eastern North America. **Season** July–September. **Edible**.

Cantharellus tubaeformis Fr. **Cap** 2–8cm (¾–3in) across, convex then soon flat and depressed, funnel-shaped in center, margin inrolled, wavy; deep yellow to yellow-brown, paler with age. **Gills** decurrent, narrow, blunt, and irregularly branched and veinlike; yellowish to gray-violet.

Stem 25–80 × 4–10mm (1–3 × 3/16–½in), hollow, often flattened or grooved; yellow to dull yellow-orange. **Flesh** pallid yellow. **Odor** pleasant. **Taste** pleasant. **Spores** ellipsoid, smooth, 8–12 × 6–10μ. Deposit white. **Habitat** often in large troops in wet, mossy bogs. Found throughout northern North America. **Season** July–October. **Edible** — good. **Comment** The very similar *Cantharellus infundibuliformis* Fr. has a darker cap, duller stem, and cream to yellowish spores.

White Chanterelle *Cantharellus subalbidus* Smith & Morse **Cap** 5–13cm (2–5in) across, flat to broadly depressed with a somewhat wavy margin; whitish bruising yellowy or orange or orange-brown; smooth or slightly scaly in age. **Gills** close, often forked, crossveined, distant ridges descending stalk; whitish. **Stem** 20–60 × 10–30mm (¾–2¼ × ½–1¼in), stout; white discoloring brownish; dry, smooth. **Flesh** thick, firm; white. **Spores** ellipsoid, smooth, 7–9 × 5–5.5μ. Deposit white. **Habitat** scattered or in groups on the ground under mixed conifers or mixed oaks. Common. Found in the Pacific Northwest. **Season** September–November. **Edible** — excellent, and much sought after in western North America.

Cantharellus minor ⅓ life-size

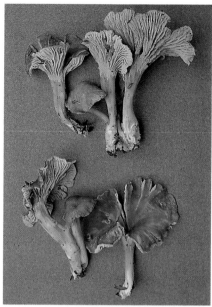

Cantharellus tubaeformis ⅓ life-size

Red Chanterelle *Cantharellus cinnabarinus*
Schw. **Cap** 1–5cm (½–2in) across, nearly flat
to slightly funnel-shaped; margin inrolled when
young, irregular, lobed or scalloped; brilliant
cinnabar red, then pinkish orange with age,
finally nearly pallid. **Gills** decurrent, narrow,
irregularly branched and veinlike, blunt-edged;
pink. **Stem** 20–50 × 3–9mm (¾–2 × ⅛–⅜in);
concolorous with cap. **Flesh** solid; white. **Odor**
pleasant. **Taste** pleasant. **Spores** ellipsoid,
smooth, 6–11 × 4–6μ. Deposit pinkish cream.
Habitat in open areas at edges of mixed
woods, often in large numbers. Common.
Found widespread throughout North America.
Season June–September (later in the West).
Edible.

Cantharellus luteocomus Bigelow **Cap**
0.5–2.5cm(¼–1in) across, convex and
shallowly depressed with an inrolled margin,
becoming flatter or more vase-shaped with a
wavy, scalloped margin; orange-yellow; moist,
smooth. **Fertile undersurface** running down
the stem; orange-yellow to pinky-brownish with
a whitish bloom in older specimens; smooth or
wrinkled. **Stem** 15–30 × 3–6mm (½–1¼ ×
⅛–¼in), stuffed; orange-yellow; smooth. **Flesh**
thin, soft; orange-yellow. **Odor** not distinctive.
Taste not distinctive. **Spores** ellipsoid, smooth,
nonamyloid, 10–13 × 6–8.5μ. **Habitat** in
groups or dense clusters on damp, mossy
ground under birch and mixed woods. Found
in Vermont and New York. **Season**
July–September. **Edibility not known**.

Cantharellus xanthopus (Pers.) Duby syn.
Cantharellus lutescens sensu Fr. **Cap**
1–6cm(½–2¼in) across, convex becoming
flatter with a crimped, wavy margin and a
sunken center, later becoming vase-shaped
with a raised margin; orange-yellow to
brownish yellow; small coarse brownish hairs
or scales give whole surface a brownish tinge.
Fertile undersurface descending stalk; dingy
yellowish brown to buff; smooth to slightly
veined or wrinkled. **Stem** 20–50 × 1–15mm
(¾–2 × 1/16–½in), stuffed becoming hollow,
often curved or compressed; orange; smooth
and slightly hairy at base. **Flesh** very thin;
pale buff to orangish. **Odor** not distinctive.
Taste not distinctive. **Spores** ellipsoid, smooth,
nonamyloid, 9–11 × 6–7.5μ. Deposit pale
orange-buff. **Habitat** in groups or clusters on
damp, mossy wood in low, wet mixed woods.
Found in eastern North America. **Season**
July–September. **Edible**.

Smooth Chanterelle *Cantharellus lateritius*
⅓ life-size

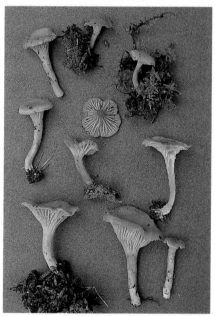

Red Chanterelle *Cantharellus cinnabarinus*
½ life-size

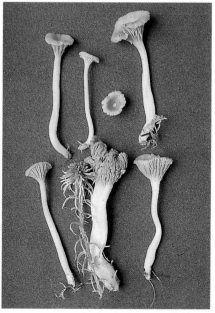

Cantharellus luteocomus ½ life-size

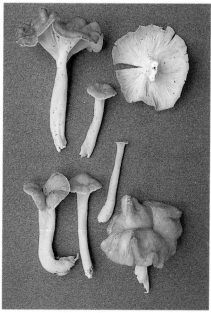

Cantharellus xanthopus ⅔ life-size

Craterellus cinereus var. *multiplex* ⅓ life-size

Cantharellus ignicolor ½ life-size

Gilled Bolete *Phylloporus rhodoxanthus*
subsp. *rhodoxanthus* ⅓ life-size

Craterellus cinereus var. *multiplex* Smith **Fruit body** 1–4cm (½–1½in) across, basically funnel-shaped becoming deeply perforated; sooty black; smooth becoming scaly. **Fertile undersurface** folded at first, then becoming more gill-like. **Stem** 20–40×3–8mm (¾–1½×⅛–5⁄16in) individually, central, compound; grayish; smooth. **Flesh** thin, tough. **Taste** very mildly disagreeable. **Spores** ellipsoid, 8–11×5–6μ. **Habitat** in compound clusters on the ground in mixed woods. Found in the Great Lakes area. **Season** August–September. **Edibility not known.**

Cantharellus ignicolor Petersen **Cap** 1–5cm (½–2in), convex with a slight depression and an inrolled margin, becoming flat with a deep depression and a decurved to wavy margin; apricot orange to yellow-orange, becoming somewhat dingy in age; smooth to rough or uneven. **Fertile undersurface** descending stem, narrow, distant, forked ridges with crossveins; orange-yellow, becoming wine-buff or violet-tinged when spores are mature. **Stem** 20–60×2–15mm (¾–2¼×3⁄32–½in), compressed, stuffed becoming hollow; dingy orange becoming paler. **Flesh** thin; concolorous with cap. **Odor** none or very slightly fragrant. **Taste** none. **Spores** broadly ellipsoid, smooth, nonamyloid, 9–13×6–9μ. Deposit ochre-salmon. **Habitat** scattered, in groups, or in dense clusters on the ground under deciduous or coniferous trees. Found in eastern North America, south to Georgia and west to Michigan. **Season** July–September. **Edibility not known.**

Boleti pp. 216–252. See Key E, p. 12
This section contains mushrooms with caps and stems, but the gills of the agarics are replaced by tubes that can easily be separated from the cap flesh. There are a great many genera in this section, principally: BOLETUS *(pp. 216–237) The stems are usually short and fat, not viscous, and without scales but often with a network (reticulum).* LECCINUM *(pp. 239–241) Caps are dry, stems normally long with small woolly scales.* TYLOPILUS *(pp. 241–245) The spores of this genus are pinkish, and they will tend to discolor the mature tube pores pinkish.* SUILLUS *(pp. 245–251) They generally have viscous caps, and often the stems are viscous too. Some have a distinct ring (annulus), and many have spots (puncti) on the stem.*

Gilled Bolete *Phylloporus rhodoxanthus* (Schw.) Bres. **Cap** 3–12cm (1¼–4¾in) across, broadly convex; bright reddish brown to yellowish olive; dry, dull velvety to subtomentose. **Tubes** (pseudogills) decurrent, subdistant, broad, thick, often forked and with crossveins, often almost porelike near stem; bright golden yellow, bruising blue; easily separable from cap. **Stem** 40–100×5–15mm (1½–4×¼–½in), equal to swollen; reddish to rust-yellow. **Flesh** pallid to yellowish. **Taste** not distinctive. **Odor** not distinctive. **Spores** narrowly ellipsoid, 9–12×3.5–5μ. Deposit olivaceous. **Habitat** under mixed conifers and hardwoods. Quite common. Found widely distributed throughout North America. **Season** July–October.

Edible — quite good. **Comment** Two forms have been given subspecific status, subspecies *rhodoxanthus* and subspecies *albomycelinus*.

Phylloporus foliiporus (Murr.) Phillips & Kibby comb. nov. Basionym *Phylloporus rhodoxanthus* ssp. *foliiporus* (Murr.) Singer **Cap** 2–14cm (¾–5½in) across, convex then soon flattened, margin incurved; maroon to dull brown, fading to ochre-brown, tan with age; dry, subtomentose. **Tubes** (pseudogills) decurrent, distant, broad, often forked with transverse veins, especially at their bases; pale whitish to yellowish, bruising strongly deep blue, finally deep brownish. **Stem** 16–55×5–20mm (½–2×¼–¾in), tapered below; concolorous with cap, olivaceous below. **Flesh** soft; yellowish-white, blueing when cut. **Odor** pleasant. **Taste** pleasant. **Spores** subfusiform-ellipsoid, smooth, 11–15×4.5–5.8μ. Deposit yellow-brown to slightly olivaceous. **Habitat** on sandy soil under mixed conifers and oak. Often abundant. Found in Florida and the New Jersey pine barrens. **Season** August–September. **Edible. Comment** Because the characters by which it differs from *Phylloporus rhodoxanthus* (above) — the blueing flesh and longer spores — are exactly those characters traditionally used in the differentiation of species in the Boletaceae, it is felt that specific status is warranted for this distinctive fungus. Intermediates have not been noted between the two and their range of distribution overlaps, further evidence of their being two separate species.

Red-cracked Bolete *Boletus chrysenteron* Bull. ex St. Amans **Cap** 3–8cm (1¼–3in) across, broadly convex; deep olive-brown to reddish brown, soon cracking to reveal reddish flesh in cracks; dry, velvety. **Tubes** sulphur yellow bruising greenish. **Pores** large, angular; concolorous with tubes and bruising greenish. **Stem** 40–80×10–15mm (1½–3×½in), yellow at apex, becoming carmine red over most of stem surface, more olivaceous below. **Flesh** cream or lemon yellow in cap, brown to reddish buff in stem, usually pale red just below cap cuticle, blueing slightly above tubes and in stem base. **Odor** not distinctive. **Taste** not distinctive. **Spores** subfusiform, (9)12–15×3.5–5μ. Deposit olivaceous brown. **Habitat** with deciduous trees, especially oak. Common, but rarer in the West. Found widely distributed in North America. **Season** July–October. **Edible.**

Boletus subtomentosus Fr. **Cap** 5–15cm (2–6in) across, soon flattened; bright olive-yellow to citrine, duller yellow-brown to reddish brown with age or when wet; dry, subtomentose-velvety. **Tubes** pale yellow. **Pores** rather large; yellow to yellow-olive. **Stem** 40–100×10–25mm (1½–4×½–1in), equal; dull yellow to buff; often with surface floccose-scabrous to slightly reticulate above. **Flesh** pallid. **Odor** pleasant. **Taste** pleasant. **Spores** subfusiform, 10–13×3.5–5μ. Deposit olive-brown. **Habitat** usually under conifers. Frequent. Found throughout northern North America. **Season** July–October. **Edible.**

Red-cracked Bolete *Boletus chrysenteron* ½ life-size

Boletus subtomentosus ⅔ life-size

Phylloporus rhodoxanthus subsp. *albomycelinus* ¾ life-size

Phylloporus foliiporus ½ life-size

Boletus mirabilis ⅓ life-size

Boletus parasiticus ¾ life-size

Boletus projectellus ⅓ life-size

Boletus mirabilis Murr. **Cap** 5–15cm (2–6in) across, convex-flattened, margin inrolled; deep reddish brown, liver-colored; moist to soon dry, woolly or even squamulose. **Tubes** depressed around stem; yellowish. **Pores** olive-yellow when mature. **Stem** 80–150 × 35–50mm (3–6 × 1¼–2in); deep brown; smooth with reticulum at apex. **Flesh** firm; white, pinkish in stem. **Odor** pleasant. **Taste** pleasant. **Spores** ellipsoid, 19–24 × 7–9μ. Deposit olive-brown. **Habitat** on or near logs of fir, hemlock, or western red cedar. Found in the Pacific Northwest. **Season** September–December. **Edible** — good.

Boletus parasiticus Fr. **Cap** 2–8cm (¾–3in) across; tawny yellow to olive-yellow or ochre; dry to slightly viscid when wet. **Tubes** rather deep, up to 1cm (½in), depressed around stem; concolorous with pores. **Pores** angular; ochre-yellow to tawny, often stained reddish orange. **Stem** 30–60 × 5–15mm (1¼–2¼ × ¼–⅝in), tapered at base, always attached to an earthball (*Scleroderma citrinum*); same color as cap; dry. **Flesh** pale yellow, not changing color when cut. **Odor** not distinctive. **Taste** not distinctive. **Spores** inequilateral fusiform, 12–18.5 × 3.5–5μ. Deposit deep olive. **Habitat** in damp hardwoods, only on the common earthball, particularly in wet years. Found in eastern North America. **Season** August–October. **Edible** but not recommended.

Boletus projectellus Murr. syn. *Boletellus projectellus* (Murr.) Singer **Cap** 4–15cm (1½–6in) across, convex; dark cinnamon to yellow-brown or reddish bay with age; dry, subtomentose, squamulose, and slightly cracking. **Tubes** pale creamy olive. **Pores** rather large (1–2mm [1/16–3/32in]), pale cream to yellowish olive. **Stem** 60–110 × 10–30mm (2¼–4½ × ½–1¼in), equal to slightly clavate; reddish buff to vinaceous buff with deep, coarse reticulum from top to bottom. **Flesh** firm; pallid to pale vinaceous. **Odor** pleasant. **Taste** acidic. **Spores** long ovoid, 18–33 × 7.5–10μ. Deposit olive. **Habitat** under pine. Locally common. Found throughout the eastern United States, west to Michigan. **Season** August–September. **Edible.**

Boletus zelleri Murr. **Cap** 5–12cm (2–4¾in) across, convex; blackish to fuscous brown with olive tints, cuticle sometimes cracking to reveal reddish flesh beneath; smooth or quite rugulose-wrinkled, often velvety. **Tubes** yellow. **Pores** yellow, blueing when bruised. **Stem** 50–100 × 7–20mm (2–4 × ¼–¾in) equal; yellow overlaid with a reddish flush, becoming dull olive with age; dry, pruinose-punctate. **Flesh** yellow, slightly blueing when bruised. **Odor** pleasant. **Taste** pleasant. **Spores** spindle-shaped to ellipsoid, 12–15 × 4–5.5μ. Deposit olive-brown. **Habitat** under mixed trees, especially conifers. Found in the Pacific Northwest. **Season** August–November. **Edible.**

Spotted Bolete *Boletus affinis* Pk. **Cap** 5–10cm (2–4in) across, convex; color very variable, reddish brown, vinaceous brown to yellow-brown, often spotted and blotched with pallid, yellowish spots in the var. *maculosus;* dry, often with white bloom. **Tubes** sunken around stem; white to yellowish. **Pores** off-white to pale buff with age, bruising olivaceous. **Stem** 50–120 × 10–20mm (2–4¾ × ½–¾in), cylindrical to clavate; pale fawn, reddish brown in midportion, white at base, in the var. *maculosus* sometimes completely pallid; dry, smooth to very slightly reticulate. **Flesh** white. **Odor** pleasant. **Taste** pleasant. **Spores** smooth, ellipsoid, (9)12–16 × 3–4(5)μ. Deposit yellow-brown. **Habitat** abundant at times under deciduous trees, especially beech. Found

(*continued on page 221*)

Boletus zelleri ⅔ life-size

Boletus zelleri (wrinkled form) ⅔ life-size

Spotted Bolete *Boletus affinis* var. *maculosus* ½ life-size

Spotted Bolete *Boletus affinis* ⅔ life-size

Gold-pored Bolete *Boletus auriporus* ½ life-size

Boeltus spadiceus ⅔ life-size

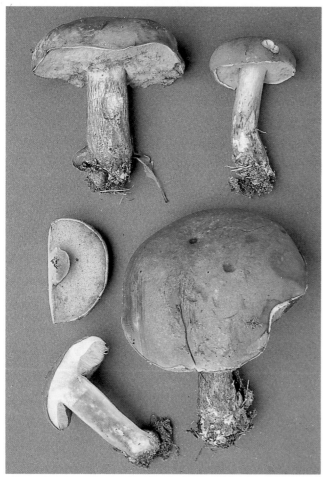

Boletus miniato-pallescens ½ life-size

Boletus fraternus ¾ life-size

(*continued from page 218*)
widespread in eastern North America, from eastern Canada to Florida. **Season** June–October. **Edible. Comment** The young buttons are dark brown and the mature specimens often light tan.

Gold-pored Bolete *Boletus auriporus* Pk. **Cap** 3–7cm (1¼–2¾in) across, convex then flattened; dull yellow-brown to reddish or pinkish brown; dry but soon viscid, tacky when wet or when handled for any length of time, with innate fibrils. **Tubes** brilliant chrome yellow, unchanging. **Pores** 2–3 per mm; concolorous with tubes, unchanging. **Stem** 30–60×10–15mm (1¼–2¼×½in), often swollen at center, spindle-shaped; pale yellow to pinkish brown, darker stains developing where handled; viscid when wet, smooth. **Flesh** pale yellow, unchanging when cut. **Odor** none. **Taste** slightly acidic. **Spores** smooth, subellipsoid, 8–11×3.5–4.5µ. Deposit olive-brown. Pleurocystidia in tubes of remarkable size, 38–70(100)×9–16µ, filled with golden sap. **Habitat** under mixed deciduous trees. Uncommon to rare. Found in eastern North America. **Season** July–August. **Edible. Comment** Easily confused with other bright-pored boletes, such as *Boletus illudens* (below), but that species does not develop the noticeably sticky cap of this fungus, however wet it becomes.

Boletus spadiceus Fr. **Cap** 5–10cm (2–4in) across; deep olive to buff or yellow-brown to date brown when wet, often cracked when old and dry, cracks pallid; subtomentose-velvety. **Tubes** dull yellow. **Pores** rather large; bright yellow-ochre. **Stem** 50–100×10–25mm (2–4×½–1in), equal; dull yellowish to olive, slightly pinkish brown below; dry, often with coarse reticulate ridges on upper half. **Flesh** pale yellow. **Odor** pleasant. **Taste** pleasant. **Spores** subfusiform, 11–14×4.5–5.5µ. Deposit olive-brown. **Habitat** in mixed woods and along trackways and banks. Fairly common. Found in the Pacific Northwest and in northern and eastern North America. **Season** July–October. **Edible. Comment** Dilute ammonia (NH₄OH) placed on cap surface gives a green or blue-green coloration that quickly fades.

Boletus miniato-pallescens Smith & Thiers **Cap** 8–20cm (3–8in) across, convex to plane; brick red fading to apricot buff or orange-yellow; smooth, glabrous to minutely fibrillose, dry, soon with surface cracked. **Tubes** adnate to subdecurrent; bright yellow. **Pores** very small (1–2 per mm); chrome yellow to wax yellow, often flushed orange-red with age, bruising greenish blue. **Stem** 60–140×10–40mm (2¼–5½×½–1½in), tapered below or equal; bright yellow above, flushed orange to brick red below; strongly pruinose when young, more or less persistently. **Flesh** pale yellow, turning blue when cut. **Odor** pleasant. **Taste** pleasant. **Spores** subfusiform, (11)12–16(17)×4–5µ. Deposit olive-brown. **Habitat** usually under oak. Probably quite common. Found in eastern North America, west to Michigan. **Season** July–September. **Edibility not known.**

Boletus fraternus Pk. **Cap** 2–5cm (¾–2in) across, convex, soon flattened; deep red then pinkish, often with a cracked surface showing yellow flesh beneath, dry, subtomentose to minutely scabrous. **Tubes** bright yellow. **Pores** small; yellow quickly bruising blue. **Stem** 20–50×

5–12mm (¾–2×¼–½in), equal; colored as cap but with pallid base, color fading to brown rather quickly. **Flesh** soft, often buggy; pale yellow, soon blue when bruised. **Odor** pleasant. **Taste** pleasant. **Spores** fusiform, 12–15×6–8µ. Deposit olive-brown. **Habitat** often gregarious in grass in mixed woods or gardens. Found in eastern North America. **Season** July–September. **Edible** but poor quality.

Boletus illudens Pk. **Cap** 3–9cm (1¼–3½in) across, convex to flattened; pale pinkish buff to cinnamon, brighter, more lemon yellow at margin; dry, velvety, then moist but not viscid. **Tubes** adnate-decurrent; honey yellow to olivaceous. **Pores** large, angular; lemon yellow, then brownish where bruised or old. **Stem** 30–90×6–12mm (1¼–3½×¼–½in), tapered below; pale brownish above becoming yellowish to mustard yellow at base; usually with coarse ridges and wrinkles above, but not truly reticulate. **Flesh** pallid, mustard yellow in base and below stem cortex. **Odor** not distinctive. **Taste** not distinctive. **Spores** subfusiform, 10–14×4–5µ. Deposit olive-brown. **Habitat** in oak or mixed woods. Often common. Found in northeastern North America. **Season** July–September. **Edible. Comment** A drop of ammonia on the cap cuticle produces a deep green reaction; the similar *Boletus subtomentosus* (p. 217) and *Boletus nancyae* Smith & Thiers turn purple-brown with ammonia.

Boletus rubellus Krombh. **Cap** 3–8cm (1¼–3in) across, broadly convex then flattened; scarlet red when young, becoming dull olivaceous red with age, margin often yellowish; dry and velvety, finally glabrous, and often areolate. **Tubes** dull yellow. **Pores** lemon yellow then greenish yellow, bruising blue. **Stem** 40–80×4–8mm (1½–3×³⁄₁₆–⁵⁄₁₆in), equal; bright yellow at apex, shading to bright rose red or scarlet below, with yellow basal mycelium. **Flesh** yellowish, staining blue when cut. **Odor** pleasant. **Taste** slightly soapy. **Spores** subellipsoid, 10–13×4–5µ. Deposit olive-brown. **Habitat** often gregarious in grassy woodlands, especially oak. Found in the northeastern United States. **Season** July–September. **Edible** but often maggoty. **Comment** This is one of a complex of very closely related species, often separable only with microscopic examination.

Boletus campestris Smith & Thiers **Cap** 3–5cm (1¼–2in) across, broadly convex; deep rose red to blood when young, paler, browner with age; dry, velvety, often cracking. **Tubes** 6–8mm (¼–⁵⁄₁₆in) deep, depressed at the stem, bright lemon yellow then greenish yellow, bruising blue. **Pores** 1–2 per mm, round to angular; same color as tubes. **Stem** 40–50×5–10mm (1½–2×¼–½in), equal; yellow at apex, reddish below, base with bright yellow tomentum. **Flesh** bright yellow bruising blue-green. **Odor** pleasant. **Taste** pleasant. **Spores** inequilateral long ellipsoid, 11–14×4.5–6µ. Deposit olive-brown. **Habitat** gregarious in short grass in lawns and woodland edges. Found widely distributed in eastern North America. **Season** June–September. **Edible.**

Boletus illudens ⅓ life-size

Boletus rubellus ⅓ life-size

Boletus campestris ½ life-size

Two-colored Bolete *Bolete bicolor* ½ life-size

Boletus pseudosensibilis ½ life-size

Pallid Bolete *Boletus pallidus* ⅓ life-size

Boletus badius ½ life-size

Two-colored Bolete *Boletus bicolor* Pk. **Cap**
5–15cm (2–6in) across, convex then flattened;
deep rose red to pinkish, fading with age, paler
toward margin; dry, subtomentose, then soon
smooth with age, often cracking in dry
weather. **Tubes** yellow. **Pores** 1–2 per mm;
bright yellow, blue when bruised. **Stem**
50–100 × 10–30mm (2–4 × ½–1¼in), equal to
slightly clavate below; colored as cap for lower
two-thirds, yellow above, slowly bruising blue;
smooth, dull, dry. **Flesh** firm; pale yellow,
slowly bruising blue. **Odor** not distinctive.
Taste not distinctive. **Spores** ellipsoid,
8–11 × 3.5–4.5(5)μ. Deposit olive-brown.
Habitat in oak woods. Common. Found
widespread in eastern North America. **Season**
July–October. **Edible** — good, but see
Comment. **Comment** The very similar *Boletus
sensibilis* (p. 235), which has been reported as
poisonous, differs in its brick-red cap and its
instant color change to blue when cut. *Boletus
miniato-olivaceus* Frost also has a red cap, but it
has a mostly yellow stem.

Boletus pseudosensibilis Smith & Thiers **Cap**
6–15cm (2¼–6in) across, broadly convex;
brick red to ferruginous, fading to yellow-
brown or cinnamon; unpolished, dry, glabrous,
cracking when dry. **Tubes** shallow,
subdecurrent down stem; yellow. **Pores**
minute; bright yellow, instantly deep blue
when bruised. **Stem** 80–160 × 15–30mm
(3–6¼ × ½–1¼in), equal to slightly flared at
apex; pale yellow flushed pinkish to darker red
below; smooth. **Flesh** solid; bright yellow,
instantly blue when cut. **Odor** mild, pleasant.
Taste mild, pleasant. **Spores** subfusiform,
9–12 × 3–4μ. Deposit olive-brown. **Habitat** in
mixed deciduous woods, especially oak. Often
abundant. One of the commonest summer
boletes in the eastern United States, especially
New Jersey, occurring north and west to
Michigan. **Season** June–September. **Edible**
but not recommended because of risk of
confusion. **Comment** Dilute ammonia
(NH₄OH) applied to cap turns blue-green.

Pallid Bolete *Boletus pallidus* Pk. **Cap** 4–15cm
(1½–6in) across, broadly convex; pale cream-
buff to pale leather-colored with age; dull, dry
when young, can be slightly viscid when wet.
Tubes adnate to stem or slightly depressed;
pale yellow then olive. **Pores** pallid yellowish,
not changing color on bruising. **Stem**
50–120 × 8–30mm (2–4¾ × ⁵⁄₁₆–1¼in), equal
or clavate; pallid to yellowish at apex and often
flushed reddish at base; smooth. **Flesh** dirty
white. **Odor** not distinctive. **Taste** not
distinctive. **Spores** subfusiform, 9–15 × 3–4.5μ.
Deposit olive-brown. **Habitat** often gregarious,
even caespitose, under oak on sandy soil.
Common. Found in eastern North America
Season July–September. **Edible.**

Boletus badius Fr. **Cap** 3–10cm (1¼–4in) across,
convex then flattened; brick red to chestnut or
bay brown; minutely tomentose to smooth or
slightly viscid when wet. **Tubes** pale yellow to
olive. **Pores** small; olive-yellow, bruising blue-
gray. **Stem** 40–90 × 10–20mm
(1½–3½ × ½–¾in), cylindrical; colored as cap
but often with slight rose tints, yellowish at
base; somewhat pruinose when fresh. **Flesh**
firm; pallid, slightly pink under cap cuticle,
staining weak vinaceous to blue when cut.
Odor pleasant. **Taste** pleasant. **Spores**
ellipsoid, 10–14 × 4–5μ. Deposit olive-brown.
Habitat in mixed woods, often on decaying
tree stumps. Rather common. Found from
eastern Canada down to North Carolina, west
to Minnesota. **Season** July–November.
Edible — good.

Boletus lignicola Kallenbach **Cap** 5–20cm
(2–8in) across, convex with inrolled margin;
reddish brown to yellow-brown or rust;
subtomentose, floccose at first, then smooth.
Tubes decurrent on stem; bright yellow. **Pores**
bright yellow, bruising blue-green. **Stem**
30–80 × 5–25mm (1¼–3 × ¼–1in), often
eccentric, tapered below; rust-yellow to brown;
dry, pulverulent. **Flesh** firm; pale lemon
yellow. **Odor** faint, aromatic. **Taste** pleasant.
Spores ellipsoid, 6.5–9 × 2.8–3.8μ. Deposit
olive. **Habitat** always on stumps or trunks of
conifers, exceptionally on sawdust; often
associated with the polypore *Phaeolus schweinitzii*
(p. 261). Rare. Found widely distributed in
eastern North America. **Season** July–
September. **Edible.**

Boletus albisulphureus Murr. **Cap** 5–8cm (2–3in)
across, convex; milk-white; smooth, soft. **Tubes**
bright sulphur yellow. **Pores** sulphur yellow.
Stem 50–100 × 15–25mm (2–4 × ½–1in); white
with a yellow reticulum on upper half of stem.
Flesh very soft; white to pale sulphur yellow.
Odor none. **Taste** pleasant. **Spores** sausage-
shaped, smooth, 12 × 3.5μ. Deposit pale
brown. **Habitat** in mixed woods. Found in
Florida, Mississippi, North Carolina, and other
southern and southeastern states. **Season**
July–August. **Edibility not known.**

Boletus stramineum (Murr.) H. V. Smith &
Smith **Cap** 4.5–8.5cm (1¾–3¼in) across,
convex with an incurved margin at first,
becoming expanded and somewhat knobby or
irregularly shaped in age; whitish becoming
straw-colored, then grayish or brownish;
smooth or with a very slight bloom on the
margin when young, shining when dry,
cracked. **Tubes** 3–10mm (⅛–½in) long,
adnate and sometimes with a slightly decurrent
tooth; white becoming dirty cream to
brownish. **Pores** small, round to angular;
similar color to tubes or slightly bruising
brownish. **Stem** 30–60 × 10–35mm
(1¼–2¼ × ½–1¼in), solid, somewhat swollen
in the middle; pure white to whitish with white
mycelium at the base; smooth. **Flesh** thick,
firm becoming softer; white. **Odor** faintly fruity
or none. **Taste** mild. **Spores** cylindrical,
smooth, thin-walled, 10.5–14.5 × 2.5–3.5μ.
Deposit brownish. **Habitat** in lawns, gardens,
open places, and woods under pine and oak.
Found in southeastern North America. **Season**
July–September. **Edibility not known.**

Boletus lignicola ⅔ life-size

Boletus albisulphureus ½ life-size

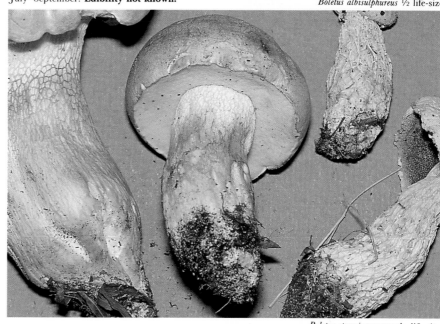

Boletus stramineum nearly life-size

Boletus pulverulentus ⅓ life-size

Boletus longicurvipes ¾ life-size

Boletus piperatus life-size

Boletus subglabripes ½ life-size

Boletus hortonii life-size

Boletus rubricitrinus ⅔ life-size

Boletus pulverulentus Opatowski **Cap** 4–10cm (1½–4in) across, broadly convex; deep yellow-brown to blackish brown, sometimes with reddish hues; subtomentose to dull, dry, or glabrous, tacky when moist. **Tubes** yellow, but instantly deep blue when cut. **Pores** large and angular; lemon yellow, instantly deep blue when touched. **Stem** 40–80 × 10–30mm (1½–3 × ½–1¼in), equal to tapering below; bright yellow-orange on apex, reddish brown below, turns instantly blue-black when handled; surface pruinose. **Flesh** soft; yellow then deep blue to almost black when cut. **Odor** not distinctive. **Taste** not distinctive. **Spores** subfusiform, 11–14(15) × 4.5–6μ. Deposit olive-brown. **Habitat** in grassy oak woods and in garden lawns, particularly on slopes and banks. Often common. Found throughout northeastern North America. **Season** July–August. **Edible. Comment** One of the most easily identifiable boletes, with its instant and very deep blue color change of all parts. Ammonia on the cap cuticle gives a fleeting green coloration.

Boletus longicurvipes Snell & Smith **Cap** 2–6cm (¾–2¼in) across, convex; reddish orange to dull ochre; glabrous, viscid-tacky, with separable pellicle, often wrinkled-reticulate. **Tubes** pale yellow. **Pores** small; yellow then greenish. **Stem** 50–100 × 6–15mm (2–4 × ¼–½in), long, slender, and often curved; pale pinkish brown, dull red with age; surface scabrous-scurfy. **Flesh** soft; white to pale yellow. **Odor** mild. **Taste** mild. **Spores** narrowly subfusiform, 13–17 × 4–5μ. Deposit olive-brown. **Habitat** Northeastern North America, west to Michigan, south to New Jersey. **Season** August–September. **Edible.**

Boletus piperatus Bull. ex Fr. **Cap** 2–10cm (¾–4in) across, convex then soon flattened; tawny, yellow-brown, or rust; dry or slightly viscid when wet, smooth. **Tubes** adnate; cinnamon red. **Pores** angular; bright cinnamon red. **Stem** 20–100 × 4–12mm (¾–4 × ³⁄₁₆–½in); ochre-brown, base bright yellow. **Flesh** bright yellow. **Odor** pleasant. **Taste** very peppery. **Spores** fusiform, smooth, 8–12 × 4–5μ. Deposit dull brown. **Habitat** under birch or pine. Found in most of North America. **Season** July–October (September–February on the West Coast). **Not edible.**

Boletus subglabripes Pk. syn. *Leccinum subglabripes* (Pk.) Singer **Cap** 4–10cm (1½–4in) across, convex then expanding to almost plane; light brown to rich cinnamon, yellow-brown, or reddish brown; dry, glabrous to slightly viscid when wet. **Tubes** deeply depressed around stem; lemon yellow to olive-yellow. **Pores** yellow to amber yellow, not changing on injury. **Stem** 50–100 × 10–20mm (2–4 × ½–¾in), even and tapered at the base; pale to bright yellow, occasionally staining reddish at base; entire surface covered with scurfy, scabrous squamules (never reticulate), dry, often with distinct white mycelial remains at base. **Flesh** pale to bright lemon yellow, sometimes faintly blue on cutting. **Odor** not distinctive. **Taste** mild to slightly acidic. **Spores** subfusiform, smooth, (11)12–14(17) × 3–3.5(5)μ. Deposit pale olive-brown. **Habitat** often gregarious under mixed deciduous trees, sometimes under spruce. Found in eastern and particularly northern North America. **Season** June–September. **Edible** — good, but soon very soft. **Comment** Placed by some authors in the genus *Leccinum*,

but it does not have the darkening squamules on the stem typical of that genus.

Boletus hortonii Smith & Thiers syn. *Boletus subglabripes* var. *corrugis* Pk. **Cap** 4–12cm (1½–4¾in) across, convex-flattened; pale reddish tan; dry, extremely wrinkled-rugulose to pitted. **Tubes** yellow, sometimes weakly staining blue. **Pores** very small; yellow, sometimes bruising weakly blue. **Stem** 60–100 × 10–20mm (2¼–4 × ½–¾in), equal to clavate; pale yellow to tan or reddish; smooth. **Flesh** firm; almost white. **Odor** pleasant. **Taste** pleasant. **Spores** subfusiform, 12–15 × 3.5–4.5μ. Deposit olive-yellow. **Habitat** in mixed deciduous woods. Rather rare. Found in eastern North America, west to Michigan. **Season** July–September. **Edible. Comment** Distinguished from the superficially similar *Leccinum rugosiceps* (p. 240) by the smooth stem, more rugose cap, and flesh not turning red.

Boletus rubricitrinus (Murr.) Murr. **Cap** 4–15cm (1½–6in) across, broadly convex; brick red to orange-cinnamon or madder brown with age; subtomentose. **Tubes** lemon yellow. **Pores** small; bright yellow, blueing when touched. **Stem** 50–150 × 15–25mm (2–6 × ½–1in); mustard yellow, the surface covered with red squamules or floccules. **Flesh** yellow, deep red in base, blueing when cut. **Odor** mild. **Taste** mild. **Spores** subfusiform, 12.5–18.8 × 4.5–7.5μ. Deposit olive-brown. **Habitat** under mixed oak and pine. Common in Florida, rare farther north (as far as southern New Jersey). **Season** May–November. **Edibility not known.**

Boletus calopus ⅓ life-size

Boletus rubripes ⅓ life-size

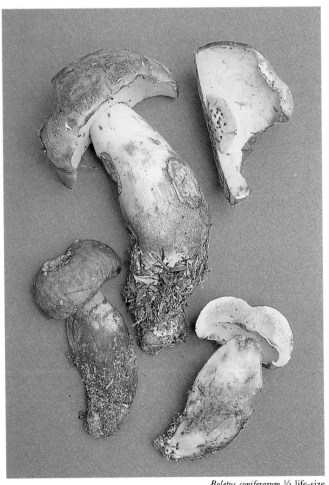

Boletus coniferarum ⅓ life-size

Boletus inedulis ⅓ life-size

Boletus pseudopeckii ½ life-size

Boletus speciosus ½ life-size

Boletus calopus Fr. **Cap** 10–25cm (4–10in) across, broadly convex to flattened; dull buffy tan to olive-brown or yellow-brown; dry, subtomentose, often cracked-areolate. **Tubes** pale yellow. **Pores** small; pale yellow, bright blue when bruised. **Stem** 100–150 × 30–70mm (4–6 × 1¼–2¾in), fat, bulbous, distinctly reticulate over upper half; yellow to slightly orange-yellow, reticulum whitish, base pinkish. **Flesh** solid; yellow then whitish, very quickly sky blue when cut. **Odor** none. **Taste** bitter. **Spores** subfusiform, smooth, 13–19 × 5–6μ. Deposit olive-brown. **Habitat** in conifer woods. Rather common. Found in the Pacific Northwest and Michigan. **Season** August–October. **Not edible.**

Boletus rubripes Thiers **Cap** 9–16cm (3½–6¼in) across, convex-flattened; pale buff to dull tawny brown, darker where handled; dry, dull, tomentose. **Tubes** up to 1.5cm (½in) deep; primrose yellow. **Pores** small, angular; bright yellow, blue where bruised. **Stem** 80–130 × 20–40mm (3–5 × ¾–1½in), slightly clavate; yellow at apex, pinkish red to deep red at base; dry, smooth. **Flesh** firm, solid; yellow-buff, blue when cut. **Odor** unpleasant. **Taste** bitter. **Spores** spindle-shaped to ellipsoid, 12.5–17.6 × 4–5μ. Deposit olive-brown. **Habitat** in mixed woods, especially oak and pine. Found in western North America. **Season** August–December. **Not edible.**

Boletus coniferarum Dick & Snell **Cap** 10–30cm (4–12in) across; dull olive-brown to olive-gray; dry, subtomentose, cracking with age. **Tubes** yellow. **Pores** yellow bruising blue. **Stem** 100–150 × 30–60mm (4–6 × 1¼–2¼in), bulbous to equal; pale yellow to olive-yellow, blue where handled. **Flesh** thick, solid; pale

yellow. **Odor** pleasant. **Taste** bitter. **Spores** subfusiform, 11–14 × 3.5–5μ. Deposit olive-brown. **Habitat** in conifer forests. Uncommon. Found in the Pacific Northwest. **Season** August–October. **Not edible. Comment** This species closely resembles *Boletus calopus* (above) but differs in lack of red tints on the stem and is less reticulate.

Boletus inedulis Murr. **Cap** 4–10cm (1½–4in) across, convex to flattened; pale, whitish at first, then buff to tan; dry, subtomentose when young, often conspicuously areolate when old. **Tubes** pale greenish yellow, turning blue when cut. **Pores** small (1.5–2 per mm); pale yellow, blue on bruising. **Stem** 60–100 × 10–25mm (2¼–4 × ½–1in), equal; yellow overall with a pink flush over base; surface reticulate over upper half, very fine and often almost smooth. **Flesh** firm; yellowish then pale blue when cut. **Odor** pleasant. **Taste** bitter. **Spores** subfusiform, 9–12 × 3.3–4.5μ. Deposit olive-brown. **Habitat** in oak and hickory woods. Found in northeastern North America, west to Michigan and south to New York. **Season** July–September. **Not edible. Comment** It is often mistaken for *Boletus calopus* (above), from which it differs in its slender stature, finer reticulum, and small spores.

Boletus pseudopeckii Smith & Thiers **Cap** 4–10cm (1½–4in) across, convex then flattened; rosy to brick red but soon with gray-brown overtones; dry to finely tomentose. **Tubes** slightly depressed around stem, 10mm (½in) deep; pale to bright yellow, staining blue when injured. **Pores** 2–3 per mm; yellow bruising blue. **Stem** 40–120 × 10–30mm (1½–4¾ × ½–1¼in), cylindric to clavate, sometimes deep in soil; yellow overall, but with reddish pink or

purple tones at base and apex; dry, very finely reticulate over upper half. **Flesh** thick, solid; pale yellow, staining bluish when cut; FeSO₄ bleaches away the blue color. **Odor** not distinctive. **Taste** not distinctive. **Spores** fusoid, smooth, 10–14 × 3.5–4.5μ. Deposit olive-brown. **Habitat** usually under beech. Rather uncommon but often misidentified. Found in upper northeastern North America. **Season** August–September. **Edibility not known. Comment** Often recorded in error as *Boletus speciosus* Frost or *Boletus peckii* Frost in Pk. and looking very much like the European *Boletus appendiculatus* Schaeff. ex Fr. *Boletus peckii* has a distinctly bitter taste, while *Boletus speciosus* (below) is more uniformly red, especially when mature, and the flesh has a grayish reaction with FeSO₄.

Boletus speciosus Frost **Cap** 6–15cm (2¼–6in) across, rounded to flattened; bright red, rose red, to dull vinaceous with age; dry, smooth, fibrillose with age. **Tubes** bright yellow to olive-yellow with age, bruising blue. **Pores** small (2 per mm); bright yellow bruising blue. **Stem** 50–130 × 15–40mm (2–5 × ½–1½in), equal to clavate; pale lemon yellow, reddish at base; reticulate over upper half, sometimes overall. **Flesh** solid, thick; pale yellow, quickly blue when cut, often chrome yellow in base of stem; grayish with FeSO₄. **Odor** pleasant. **Taste** pleasant, not distinctive. **Spores** subfusiform, smooth, 11–15 × 3–4μ. Deposit olive-brown. **Habitat** usually solitary in mixed woods. Rather rare. Found throughout northeastern North America. **Season** July–September. **Edible. Comment** See also the very similar *Boletus pseudopeckii* (above).

Ornate-stalked Bolete *Boletus ornatipes* ⅔ life-size

Boletus ornatipes (older specimens) ½ life-size

Boletus impolitus ¼ life-size

Boletus griseus ⅓ life-size

Ornate-stalked Bolete *Boletus ornatipes* Pk.
Cap 4–20cm (1½–8in) across; with a whitish
bloom when young, then gray to yellowish or
olive, sometimes strongly yellow; dry and dull
to slightly tomentose, slightly viscid when wet.
Tubes lemon yellow to tawny. **Pores** small;
lemon yellow bruising orange-brown. **Stem**
80–150×15–30mm (3–6×½–1¼in), cylindric
to slightly clavate, usually rather long; chrome
yellow throughout, bruising orange-brown;
surface with a prominent network, or
reticulum, of raised ridges. **Flesh** chrome
yellow. **Odor** none. **Taste** slightly bitterish.
Spores ellipsoid, smooth, subfusiform,
9–13×3–4μ. Deposit olive-brown to yellow-
brown. **Habitat** solitary or often clustered on
path sides, woodland edges, and clearings
under deciduous trees, usually beech or oak.
Common. Found in northeastern North
America. **Season** July–September.
Edible — quite good; although some authors
report bitterness in the flesh, this collection
was mild. **Comment** When young the stem is
usually a brilliant yellow, but with age it may
become white.

Boletus impolitus Fr. **Cap** 5–12cm (2–4¾in)
across; pale buff to ochre or cinnamon,
sometimes olivaceous; minutely tomentose then
smooth. **Tubes** lemon yellow. **Pores** small;
lemon chrome, not changing when bruised.
Stem 60–100×30–50mm (2¼–4×1¼–2in),
swollen, tapered at base; pale lemon yellow
above, darker, straw yellow to ochre, with
some reddish patches below. **Flesh** solid, firm;
pale yellowish. **Odor** pleasant to slightly
iodoform in stem base. **Taste** mild. **Spores**
subfusiform, 10–14×4.5–5.5μ. Deposit olive-
brown. **Habitat** in mixed woods under
hemlock. Rare. Found in New York. **Season**
August. **Edible.**

Boletus griseus Frost apud Pk. **Cap** 5–15cm
(2–6in) across, broadly convex; pale gray with
darker fibrils, with age slightly ochraceous;
dry, tomentose-felty, sometimes slightly
fibrillose-scaly. **Tubes** pale grayish then
brownish. **Pores** pallid to slightly brown where
bruised. **Stem** 40–110×10–30mm
(1½–4×½–1¼in), equal to tapered below,
solid; pallid above, yellow at base and then
soon overall; with strong reticulum overall,
concolorous with stem then slightly brown to
blackish. **Flesh** pallid to greenish yellow below,
yellow throughout with age, not changing color
when bruised. **Odor** not distinctive. **Taste** not
distinctive. **Spores** subfusiform, 9–12(13)×
3.5–4μ. Deposit deep olive-buff. **Habitat**
scattered in open deciduous woodland,
especially oak. Quite common. Found in
northeastern North America. **Season** July–
September. **Edible.**

Frost's Bolete *Boletus frostii* Russ. apud Frost
Cap 5–15cm (2–6in) across, convex then flat;
deep blood red with a white bloom at first,
soon disappearing, margin of cap with a very
narrow yellow zone; smooth, quite viscid at
first, then tacky to dry. **Tubes** sunken around
stem; yellow to yellow-green, bruising blue.
Pores very fine; intense deep blood red to
purple, with white bloom when young, fading
to orange-red when old. **Stem** 40–120×
15–25mm (1½–4¾×½–1in), equal to slightly
clavate; colored as cap but with prominent
raised network over entire surface, the ridges of
the network yellow overlying the blood-red
background. **Flesh** yellow, instantly turning
blue when cut. **Odor** pleasant. **Taste** pleasant.
Spores fusiform, smoooth, 11–15×4–5μ.
Deposit olive-brown. **Habitat** in oak woods.
Often locally common. Found throughout

eastern North America from the Great Lakes
region to Florida. **Season** July–September.
Edible but not recommended.

Austroboletus subflavidus (Murr.) Wolf syn.
Porphyrellus subflavidus (Murr.) Singer **Cap**
4–11cm (1½–4½in) across, convex; pale
cream, yellowish buff to clay, with a pink tint
when old. **Tubes** pale grayish white with a
vinaceous tint. **Pores** rather wide; concolorous
with tubes. **Stem** 45–145×7–30mm (1¾–5¾×
¼–1¼in), long, tapering; pallid yellowish
white; with remarkable raised reticulation of
3–4mm (⅛–³⁄₁₆in) deep flaps of tissue almost
like wrinkled gills. **Flesh** pure white, in base of
stem usually rich yellow, not changing on cutting. **Odor** slight, fruity. **Taste** rather bitter.
Spores ellipsoid, 14.5–18(20)×6.5–8.3(8.8)μ,
with a slightly crenulate outline, with thin cylindric spines. Deposit reddish brown. **Habitat**
on sandy soil under oak. Very southern in distribution, Florida and in the south New Jersey
pine barrens. **Season** June–October. **Edibility
not known** but probably too bitter.

Frost's Bolete Boletus frostii ⅔ life-size

Austroboletus subflavidus ⅔ life-size

Boletus separans ⅓ life-size

Boletus flammans ⅓ life-size

Boletus subluridellus ⅓ life-size

Red-speckled Bolete *Boletus morrisii* ⅓ life-size

Boletus variipes ½ life-size

Boletus vermiculosus ¾ life-size

Boletus separans Pk. **Cap** 6–15cm (2¼–6in) across, broadly convex; dark, dull red to liver brown or bay; dry to somewhat tacky, often slightly wrinkled-rugulose. **Tubes** white to yellowish. **Pores** small; pallid then pale yellow, not bruising. **Stem** 60–150 × 10–25mm (2¼–6 × ½–1in), even to clavate; usually paler than cap, or concolorous; covered with fine network overall. **Flesh** firm; white, reddish near cuticle. **Odor** pleasant. **Taste** pleasant. **Spores** subfusiform, 12.5–16 × 3.5–4.5μ. Deposit olive-brown. **Habitat** in mixed woods. Uncommon. Found in northeastern North America. **Season** August–October. **Edible** — good. **Comment** Ten percent KOH on cap or stem surface turns bright green.

Boletus flammans Dick & Snell **Cap** 4–12cm (1½–4¾in) across, convex, sometimes irregular; deep red to red-brown, becoming deep rosy red to brick red with age, bruising blue; dry, subtomentose, viscid when wet. **Tubes** depressed around stem, 8–12mm (⁵⁄₁₆–½in) deep; pale yellow. **Pores** small; bright red to carmine, blue when bruised. **Stem** 65–80 × 10–15mm (2½–3 × ½ in), equal; with bright red reticulations on upper half, brownish red to yellowish below; smooth or longitudinally ridged below. **Flesh** pale yellow, rapidly blue when cut. **Odor** not distinctive. **Taste** not distinctive. **Spores** subfusiform, 10–13 × 3.5–5μ. Deposit olive-brown. **Habitat** under conifers. Found from Nova Scotia to Pennsylvania. **Season** July–September. **Edibility not known** — not advised.

Boletus subluridellus Smith & Thiers **Cap** 5–10cm (2–4in) across, broadly convex; from

deep blood red to vermilion or orange-red, dull brown when old, instantly deep blue when touched; velvety-tomentose. **Tubes** yellow bruising blue. **Pores** minute; deep carmine red, instantly bruising blue. **Stem** 40–90 × 15–25mm (1½–3½ × ½–1in), equal; yellow ground color overlaid with red pruina, especially at base; extreme base with pale yellow tomentum; surface bruising deep blue. **Flesh** lemon yellow, blue when cut. **Odor** pleasant. **Taste** pleasant. **Spores** subfusoid, smooth, 10.8–13(15) × (3.8)4–5.5μ. Deposit olive. **Habitat** under oak. Frequent. Found in northeastern North America. **Season** July–September. **Comment** It lacks the red hairs at stem base found in the similar *Boletus subvelutipes* Pk. **Edibility** suspect, best avoided.

Red-speckled Bolete *Boletus morrisii* Pk. **Cap** 3–10cm (1¼–4in) across, broadly convex; deep smoky brown to olivaceous, becoming reddish brown at center, with orange-yellow margin; dry, finely pulverulent, then smooth. **Tubes** usually deeply depressed around stem; yellow to ochre, reddish where bruised. **Pores** small; orange to brick red. **Stem** 40–80 × 8–15mm (1½–3 × ⁵⁄₁₆–½in), equal to slightly swollen; bright yellow with very distinct and quite widely separate bright red squamules or dots nearly to apex. **Flesh** yellow with discolored areas of vinaceous or dark purple, especially in stem. **Odor** not distinctive. **Taste** not distinctive. **Spores** ellipsoid-subfusiform, 10–15(16) × 3.5–5.5(6.5)μ. Deposit olivaceous. **Habitat** gregarious or even subcaespitose in deciduous woods. Rather rare. Found from Massachusetts to northern Georgia, not known from western

North America. **Season** July–September. **Edibility not known.**

Boletus variipes Pk. **Cap** 6–10cm (2¼–4in) across, soon nearly flat; grayish tan to smoky buff; smooth, dry, often cracked when old. **Tubes** up to 3cm (1¼in) deep; white to pale yellow. **Pores** white to greenish yellow when mature. **Stem** 80–150 × 10–35mm (3–6 × ½–1¼in), equal to clavate; concolorous with cap or paler; surface with faint reticulations over apex, sometimes overall. **Flesh** firm; white. **Odor** mild, pleasant. **Taste** mild, pleasant. **Spores** subfusiform, 12–16(18) × 3.5–5.5μ. Deposit olive-brown. **Habitat** often in large numbers in mixed woodlands, especially beech and oak. Found throughout northern and eastern North America. **Season** July–September. **Edible** — good.

Boletus vermiculosus Pk. **Cap** 5–10cm (2–4in) across, convex-flattened; deep brown to bay, reddish cinnamon, paler with age; subtomentose to deeply velvety, slightly tacky when wet. **Tubes** dull greenish yellow. **Pores** reddish brown, often very dark, bruising blue-black. **Stem** 40–80 × 10–25mm (1½–3 × ½–1in), equal, firm; surface a dull yellow covered overall with dark reddish-brown minute floccose squamules, bruising blue-black where handled. **Flesh** firm; yellow, then instantly blue when cut. **Odor** mild. **Taste** mild. **Spores** subfusiform, 10–13.5 × 4–5μ. Deposit olive-brown. **Habitat** in beech woods. Rather uncommon (although much misidentified). Found in northern and eastern North America. **Season** July–August. **Edibility not known.**

Boletus edulis var. *aurantio-ruber* ½ life-size

Cep *Boletus edulis* ⅓ life-size

Cep *Boletus edulis* Bull. ex Fr. **Cap** 8–20cm
(3–8in) across, convex to nearly flat; very
variable in color, but in type form yellow-
brown to orange-brown or bright tawny;
smooth, glabrous, slightly viscid when wet,
often wrinkled-rugose. **Tubes** up to 25mm
(1in) deep; white to yellowish. **Pores** small;
pallid then greenish yellow when old. **Stem**
100–180×20–50mm (4–7×¾–2in), swollen-
clavate; pallid to pale brown or pinkish brown,
with a very fine white network over upper half
or even completely. **Flesh** firm, thick; white,
unchanging when bruised. **Odor** pleasant.
Taste pleasant. **Spores** subfusiform, 14–17(19)×
4–6.5µ. Deposit olive-brown. **Habitat** in mixed
woods, especially conifers. Found in the Pacific
Northwest and northern North America, rarer
in the East. **Season** July–October.
Edible — delicious, one of the best-known
edible fungi. **Comment** Var. *aurantio-ruber* Dick
& Snell differs in its ferruginous-red cap, and
pores staining yellow-olive when bruised.

Boletus barrowsii Smith **Cap** 6–25cm (2¼–10in)
across, convex becoming broadly convex to
flat; whitish, creamy with a pinkish tinge, to
buff; smooth or suedelike. **Tubes** up to 25mm
(1in) deep; white to yellowish. **Pores** small;
white or pallid when young, becoming yellow
to greenish yellow in age. **Stem** 60–200×
20–60mm (2¼–8×¾–2¼in), slightly enlarged
below in young specimens, firm, solid; whitish
to buff, sometimes with brownish stains;
reticulate, particularly over the upper section.
Flesh thick; white, unchanging or blueing very
slightly near the tubes. **Odor** pleasant. **Taste**
pleasant. **Spores** spindle-shaped to ellipsoid,
smooth, 13–15×4–5µ. Deposit dark olive-
brown. **Habitat** singly, scattered, or in groups
under coniferous and deciduous trees.
Common. Found in southwestern North
America, especially abundant in New Mexico
and Arizona. **Season** August–January.
Edible — excellent.

Boletus barrowsii ½ life-size

Boletus auripes ⅓ life-size

Boletus russellii ½ life-size

Boletus aereus Fr. **Cap** 8–20cm (3–8in) across, convex-expanded; deep blackish brown when young, to sepia or date brown when older, finally tan-brown, often with irregular decolored areas; dry, slightly roughened to subtomentose. **Tubes** white then yellowish. **Pores** white to pale greenish yellow. **Stem** 80–150 × 20–50mm (3–6 × ¼–2in), equal to clavate-swollen; pale to quite dark brown with brown reticulum. **Flesh** firm; white. **Odor** pleasant. **Taste** pleasant. **Spores** spindle-shaped, smooth, 12–15 × 4–5μ. Deposit olive-brown. **Habitat** under oak. Uncommon. Found in western North America. **Season** October–November. **Edible** — delicious.

Boletus russellii (Frost) Gilbert **Cap** 3–10cm (1¼–4in) across, convex; buffy brown to reddish brown; dry, subtomentose, then very cracked-areolate. **Tubes** yellowish olive. **Pores** large and angular; olive-yellow. **Stem** 100–180 × 10–25mm (4–7 × ½–1in), equal; deep reddish brown; entirely and strongly lacerated-reticulate. **Flesh** firm; yellow, not changing when cut. **Odor** pleasant. **Taste** pleasant. **Spores** fusiform, 15–20 × 7–11μ, with longitudinal striations and grooves. Deposit olive-brown. **Habitat** in oak woods. Locally common. Found in eastern North America, west to Michigan. **Season** July–September. **Edible.**

Boletus smithii Thiers **Cap** 10–16cm (4–6¼in) across, broadly convex; olive to olive-yellow flushed in part or entirely rose red, especially with age; dry. **Tubes** yellow. **Pores** small; bright yellow bruising blue. **Stem** 100–160 × 10–35mm (4–6¼ × ½–1¼in), equal to swollen;

yellowish flushed rose red, especially in upper half. **Flesh** firm; pale yellow, slightly blue when bruised. **Odor** pleasant. **Taste** pleasant. **Spores** subfusiform, 14–19 × 4–6μ. Deposit olive-brown. **Habitat** often in large numbers under conifers. Found in the Pacific Northwest and Idaho. **Season** July–September. **Edible.**

Brick-cap Bolete *Boletus sensibilis* Pk. **Cap** 6–15cm (2¼–6in) across, convex; brick red to sienna, becoming dull cinnamon with age; dry and unpolished. **Tubes** 1–1.5cm (½in) long; bright yellow. **Pores** 1–2 per mm; bright yellow, instantly blue where bruised, then slowly reddish. **Stem** 80–120 × 10–30mm (3–4¾ × ½–1¼in), equal to clavate; bright yellow with base a dull red; apex very slightly reticulate. **Flesh** bright yellow, then instantly blue when cut (see Comment). **Odor** mild. **Taste** mild. **Spores** subfusiform, 10–13 × 3.5–4.5μ. Deposit olive-brown. **Habitat** gregarious in mixed deciduous woods on sandy soils. Found in eastern North America. **Season** July–September. **Edibility uncertain** — to be avoided. **Comment** The similar *Boletus miniato-olivaceus* Frost turns blue very slowly; and *Boletus bicolor* (p. 223), also blueing strongly, is a brighter rose red on cap and stem.

Boletus auripes Pk. **Cap** 6–15cm (2¼–6in) across, convex to flattened; yellow-brown to gold when young, soon uniformly brown when old; dry, subtomentose to pruinose. **Tubes** bright yellow. **Pores** soon free of stem; bright yellow, unchanging. **Stem** 60–120 × 10–45mm (2¼–4¾ × ½–1¾in), bulbous to equal; bright golden yellow; surface finely reticulate on upper half. **Flesh** firm; yellow. **Odor** pleasant.

Taste mild. **Spores** ellipsoid to subfusiform, 9.5–15 × 3.5–5μ. Deposit ochre-brown. **Habitat** in mixed hardwood trees. Rather uncommon. Found in eastern North America. **Season** August–September. **Edible. Comment** The cap of this species changes color markedly from young to old and is strongly reminiscent of *Xanthoconium* (= *Boletus*) *affine* (Pk.) Singer, with which it might be better grouped.

Pulveroboletus ravenelii (Berk. & Curt.) Murr. **Cap** 2–10cm (¾–4in) across, obtuse to convex, becoming flatter with an incurved margin; bright sulphur yellow, becoming orange-red to brownish red on the disc; dry and powdery becoming felty-hairy. **Tubes** 5–8mm (¼–⁵⁄₁₆in) deep, adnate; lemon yellow to olive-tinted, bruising greenish blue to brown or black. **Pores** 2–3 per mm, smooth, round to angular; bright yellow bruising greenish-blue. **Stem** 40–150 × 4–15mm (1½–6 × ³⁄₁₆–½in), sometimes irregular or tapering toward the top; at first covered with yellow powder, then brilliant yellow with yellowish or white strands around the base; dry to moist; universal veil leaves a faint ring on the stem. **Flesh** white to yellow bruising pale blue then dull brown or yellowish. **Odor** very faint. **Taste** rather sour. **Spores** ellipsoid, smooth, 8–10.5 × 4–5μ. Deposit smoky olive. **Habitat** singly or scattered in pine and hardwood forests. Found in eastern North America, west to Michigan and Texas, reported in California. **Season** July–September. **Edible.**

Boletus piedmontensis Grund & Smith **Cap** 5–18cm (2–7in) across, convex becoming flatter; pinkish cinnamon to pinkish olive when

Boletus smithii ¼ life-size

Brick-cap Bolete *Boletus sensibilis* ¼ life-size

Pulveroboletus ravenelii ¾ life-size

older, sometimes grayish or pallid when young; dry, with a faint bloom, becoming smoother in age. **Tubes** 6–10mm (¼–½in) deep, moderated, depressed to depressed-decurrent; greenish yellow, blueing. **Pores** 2–4 per mm, subangular and round; deep red to orange-red with occasional patches of dark olive-brown in age, blueing when bruised. **Stem** 50–100 × 25–50mm (2–4 × 1–2in), bulbous when young, equal in age; buff-yellow overlaid with rose red; reticulate at apex only, base with dull grayish tomentum. **Flesh** thick, firm; white with yellow patches, blue when cut. **Odor** pleasant. **Taste** mild to slightly bitter. **Spores** oblong to narrowly ellipsoid, 9–12.5 × 4–4.5μ. Deposit light olive-green. **Habitat** under mixed hardwoods. Found in North Carolina. **Season** August–September. **Not edible.**

Boletus aereus ¼ life-size

Boletus piedmontensis ½ life-size

Boletus caespitosus ½ life-size

Gyroporus cyanescens ⅓ life-size

Boletus caespitosus Pk. **Cap** 2–8cm (¾–3in) across, convex to flattened; ochre-brown to reddish brown or pinkish; dry to distinctly viscid when wet. **Tubes** bright gold. **Pores** chrome yellow to golden yellow, unchanging. **Stem** 25–80 × 5–15mm (1–3 × ¼–½in), usually swollen-attenuate at base, narrowed at apex and often caespitose; pale to yellow to pinkish buff below, brown on handling; dry to slightly viscid. **Flesh** pallid buff. **Odor** when crushed, strong unpleasant, like earthballs (scleroderma species). **Taste** mild. **Spores** ovoid to ellipsoid or fusiform, 8–11 × 3.5–5μ. Deposit olive-ochre. **Habitat** usually in dense clusters in mixed hardwoods, especially along stream and river edges. Found in northeastern North America, south to North Carolina. **Season** July–September. **Edible. Comment** This species is usually confused with *Boletus auriporus* (p. 220), from which it differs in its smaller spores and the odor of the crushed flesh. The latter feature does not appear to have been noted by any previous author but has been confirmed on numerous collections by several different persons.

Boletus satanas Lenz sensu American authors **Cap** 7–30cm (2¾–12in) across, convex, becoming nearly flat in age; pallid then grayish to olive-buff or pinkish, particularly toward the edge. **Tubes** yellow to greenish. **Pores** deep red, becoming pink or orange or yellowish in age, bruising blue-black. **Stem** 60–150 × 20–60mm (2¼–6 × ¾–2¼in), with an enormous basal bulb up to 15cm (6in) broad, solid, firm; color similar to cap at the top, pink or reddish pink at base, particularly when young, fading in age; finely but distinctly reticulate toward the top. **Flesh** white to yellow, blueing when bruised. **Odor** unpleasant. **Taste** unpleasant. **Spores** spindle-shaped to ellipsoid, smooth, 11–15 × 4–6μ. Deposit brown to olive-brown. **Habitat** singly or in groups under oak, frequently at pasture edges. Sometimes abundant. Found in California. **Season** September–December. **Poisonous. Comment** *Boletus satanas* Lenz as described in Europe is much less pink overall, and the network on the stem is red.

Boletinellus merulioides (Schw.) Murr. **Cap** 5–15cm (2–6in) across, soon flattened and then depressed, margin incurved, often wavy, irregular in shape; dull yellow-brown or tan; dry, subtomentose to polished when old. **Tubes** decurrent, 3–5mm (⅛–¼in) deep, with a radiating pattern; light yellow changing to dark olive, reddish brown when bruised. **Pores** large, compound, with shallower pores within pores. **Stem** 25–50 × 5–10mm (1–2 × ¼–½in), often positioned off-center; concolorous with cap. **Flesh** pallid to yellowish, sometimes stains pale blue-green when cut. **Odor** mild. **Taste** mild. **Spores** ovate, smooth, 7–10 × 6–7.5μ. Deposit olive-brown. **Habitat** often in large numbers on moist ground under ash trees. Common. Found widespread throughout northern and eastern North America. **Season** June–September. **Edible** but not very good.

Gyroporus cyanescens (Fr.) Quél. **Cap** 4–12cm (1½–4¾in) across, broadly convex becoming almost flat or shallowly depressed, with a thin, incurved margin often splitting in age; pale ochre to buff, bruising completely blue when old; dry, pitted or wrinkled, with minute flattened fibers. **Tubes** deeply depressed around the stem; white then pale yellow flushed greenish with age; instantly bruising indigo blue. **Pores** small (2 per mm), round; instantly bruising indigo blue. **Stem** 40–100 × 10–25mm (1½–4 × ½–1in), often thicker toward the base, hollow and easily broken when removed from the ground; pale straw to buff, instantly bruising blue; dry, downy, but becoming smoother toward the top and in age. **Flesh** firm, thick, brittle; pale, but instantly turning blue when cut or handled. **Odor** not distinctive. **Taste** not distinctive. **Spores** ellipsoid, smooth, 8–10 × 5–6μ. Deposit pale yellow. **Habitat** singly, scattered, in groups, or in dense tufts, along roads or railways or on exposed or sandy soil among hardwoods. Frequent and sometimes abundant. Found in eastern North America. **Season** July–September. **Edible — good.

Gyroporus castaneus (Fr.) Quél. **Cap** 3–10cm (1¼–4in) across, broadly convex becoming flatter or shallowly depressed, with the margin often split or flaring in age; color variable, orange to reddish to cinnamon or tawny; dry and slightly downy at first, becoming smoother. **Tubes** free and deeply sunken around the stem; white becoming yellowish. **Pores** 1–2 per mm, round to angular. **Stem** 30–90 × 6–15mm (1¼–3½ × ¼–½in), becoming hollow, fragile, pinched at the base and sometimes bulbous; same color as the cap but often paler toward the top; dry, smooth. **Flesh** thick, fragile; white. **Odor** not distinctive. **Taste** mild. **Spores** ellipsoid, smooth, 8–12.6 × 5–6μ. Deposit pale lemon yellow. **Habitat** singly, scattered, or in groups, on the ground in oak woods and mixed conifer and hardwood forests. Sometimes quite common. Found in eastern North America and on the West Coast. **Season** July–October. **Edible** — excellent.

Gyroporus purpurinus (Snell) Singer **Cap** 1–9cm (½–3½in) across, broadly convex becoming flatter and often somewhat sunken; dark vinaceous red or burgundy; dry, uneven, minutely velvety. **Tubes** adnate, depressed becoming deeply depressed; white slowly aging yellow. **Pores** 1–3 per mm; white becoming yellow. **Stem** 30–60 × 3–8mm (1¼–2¼ × ⅛–⅝₁₆in), stuffed but becoming hollow; same color as cap or browner, with top often pallid; somewhat roughened or velvety. **Flesh** fragile; white. **Odor** slight. **Taste** mild. **Spores** ovoid to ellipsoid, smooth, 8–11 × 5–5.6μ. Deposit bright yellow. **Habitat** under oak and other hardwoods. Found in eastern North America. **Season** August–September. **Edible.**

Gyroporus castaneus ¼ life-size

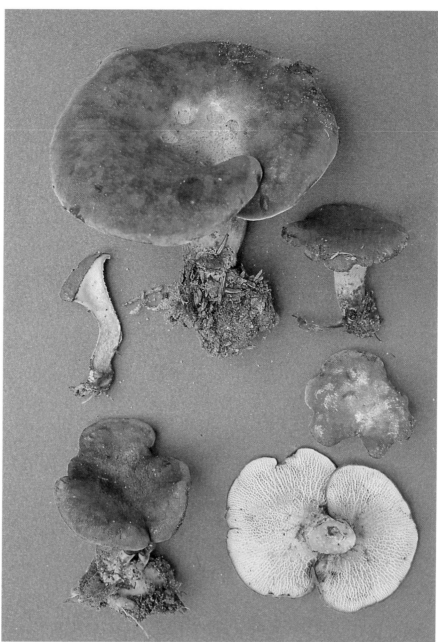

Boletinellus merulioides ½ life-size

Gyroporus purpurinus ¾ life-size

Boletus satanas ⅓ life-size

Black-stemmed Leccinum *Leccinum atrostipitatum* ⅓ life-size

Leccinum scabrum ½ life-size

Leccinum insigne ½ life-size

Leccinum holopus var. *americanum* ¾ life-size

Black-stemmed Leccinum *Leccinum atrostipitatum* Smith, Thiers, & Watling **Cap** 6–15cm (2¼–6in) across, convex with overhanging marginal flaps of tissue; pale orange-buff to apricot orange, then dingy pinkish tan often with grayish fibrils; dry, woolly-fibrillose to smooth and viscid in wet weather. **Tubes** pale olivaceous when young, then deeper brown. **Pores** minute; pallid olivaceous then olive-brown, bruising darker. **Stem** 80–150×20–50mm (3–6×¾–2in), massive, swollen clavate, solid, firm; white but with a dense black covering of squamules over entire stem even when in the button stage. **Flesh** white then staining pinkish gray to violaceous brown, bluish green near cortex. **Odor** pleasant. **Taste** pleasant. **Spores** subfusoid, smooth, 13–17×4–5µ. Deposit dull yellow-brown. **Habitat** under birch. Occasional. Found in northern and eastern North America. **Season** July–October. **Edible** — good.

Leccinum scabrum (Fr.) S. F. Gray **Cap** 4–10cm (1½–4in) across, cushion-shaped to broadly convex with an even margin; gray-brown to dull yellowish brown, flushed with olive tints in age. **Tubes** 8–15mm (⁵⁄₁₆–½in) deep, deeply depressed around the stem; pallid, slowly becoming wood brown. **Pores** small; same color as sides and sometimes bruising yellow. **Stem** 70–120×7–15mm (2¾–4¾×¼–½in), enlarging downward, solid; background color pallid with dark brown or blackish pits or scales, and staining red or blue in the lower section if wet. **Flesh** white, sometimes bruising slightly brownish if cut. **Odor** mild. **Taste** mild. **Spores** subfusiform, smooth, 15–19× 5–7µ. Deposit snuff brown. **Habitat** under birch. Frequent. Found throughout North America. **Season** July–October. **Edible** but not worthwhile.

Leccinum insigne Smith, Thiers, & Watling **Cap** 4–16cm (1½–6¼in) across, round to broadly convex becoming flatter; color varies from bright orange to reddish brown or brown, often paler and duller in age; dry, smooth or minutely hairy and sometimes scaly, then pitted in age. **Tubes** adnate or decurrent; whitish then olive-gray bruising pinkish brown. **Pores** whitish bruising yellow to olive-brown. **Stem** 60–150×10–25mm (2¼–6×½–1in), solid, swollen at the base; whitish when young, but covered with numerous small projecting scabers which turn from reddish brown to blackish in age, base often bruises blue; tough, fibrous. **Flesh** thick, soft; white turning violet-gray or dingy brown when cut. **Spores** spindle-shaped to ellipsoid, smooth, 11–18×4–6µ. Deposit brownish or yellowish-brown. **Habitat** scattered or in groups under aspen or birch in woods and along wood edges. Often common and sometimes abundant. Found widely distributed in North America. **Season** June–September. **Edible** — good.

Red-cap Bolete *Leccinum aurantiacum* ½ life-size

Leccinum holopus Smith & Thiers var. *americanum* **Cap** 5–12cm (2–4¾in) across, broadly convex; white then soon pallid buff to olivaceous or pinkish buff at disc, sometimes more greenish; glabrous, dry, then slightly viscid with age. **Tubes** deeply depressed around stem; white then brownish. **Pores** small; white staining yellowish when bruised. **Stem** 60–120× 10–15mm (2¼–4¾×½in), long, cylindric; pallid, whitish below a darker blackish-brown ornamentation of fine dots and squamules, slowly staining reddish to gray. **Flesh** firm and white when young, soon soft and cut surface staining reddish with grayish streaks. **Odor** not distinctive. **Taste** not distinctive. **Spores** subfusiform, 14–19×5–6µ. Deposit cinnamon brown. **Habitat** in wet, mossy bogs. Common. Found widespread in northeastern North America. **Season** August–October. **Edible**. **Comment** This variant differs from the typical form of *Leccinum holopus* in the cut flesh staining clearly pinkish.

Red-cap Bolete *Leccinum aurantiacum* (Fr.) S. F. Gray **Cap** 5–15cm (2–6in) across, convex; bright orange-brown to reddish orange, often with pallid areas where covered by leaves; dry, woolly-fibrillose, margin with sterile band of tissue up to 1cm (½in) deep hanging in irregular segments. **Tubes** 1–2cm (½–¾in) long; olive-buff then darker brownish. **Pores** small; pallid when young, then pallid olivaceous. **Stem** 80–160×20–30mm (3–6¼×¾–1¼), narrowly clavate, solid, fibrous, very firm; entirely pallid at first with finely woolly-scabrous surface, soon discoloring brown, finally blackish, especially in basal half, apex remaining pallid. **Flesh** white then rapidly staining grayish vinaceous, then fuscous, and in base of stem often bright blue mixed with reddish brown. **Odor** pleasant. **Taste** pleasant. **Spores** subfusoid, 13–16(18)×3.8–4.5(5)µ. Deposit deep yellow-brown. **Habitat** under aspen and pine. Rather common. Found in northern and eastern North America. **Season** July–October. **Edible** — good.

Leccinum rugosiceps ⅓ life-size

Leccinum oxydabile ½ life-size

Leccinum albellum ½ life-size

Tylopilus chromapes ½ life-size

Leccinum rugosiceps (Pk.) Singer **Cap** 5–15cm (2–6in) across, convex; yellow-ochre to orange-yellow or yellow-brown when old; dry or moist, not viscid, surface pitted, rugulose-wrinkled, often becoming cracked-areolate with maturity. **Tubes** depressed or free around stem; yellow. **Pores** small; bright yellow. **Stem** 80–100 × 20–30mm (3–4 × ¾–1¼in), equal to clavate, fleshy, firm; pale yellow to pallid overall; surface floccose, punctate with a resinous feel, floccules darkening to blackish with age. **Flesh** thick, firm; pale yellow, turning reddish when cut. **Odor** pleasant. **Taste** pleasant. **Spores** fusiform, (14)16–21 × 5–5.5μ. Deposit olive-brown. **Habitat** often gregarious in grassy oak woods. Found throughout northeastern North America. **Season** July–September. **Edible** — good.

Leccinum oxydabile (Singer) Singer **Cap** 2.5–3cm (1–1¼in) across, convex with an even margin; white at first, soon becoming yellowish then crust brown in age; slightly felty becoming smooth and dry. **Tubes** up to 10mm (½in) deep, depressed to almost free; white becoming cream-buff when cut. **Pores** minute; white staining dark yellowish when bruised. **Stem** 60–70 × 8–13mm (2¼–2¾ × ⁵⁄₁₆–½in), tapering slightly at the top; pallid slowly staining dull pink, base blue where handled; fine white ornamentation darkens downward and dries yellowish on the basal section. **Flesh** white staining pinky-buff. **Odor** mild. **Taste** mild. **Spores** bluntly fusiform, smooth, 15–21 × 5–6.5μ. Deposit ochraceous snuff brown. **Habitat** along roadsides and logging roads in hardwood forests, under birch. Rare. Found in the Great Lakes region and Pacific Northwest. **Season** July–August. **Edible** but not worthwhile.

Leccinum albellum (Pk.) Singer **Cap** 3–6cm (1¼–2¼in) across, broadly convex becoming flatter, sometimes somewhat depressed in the center in age; whitish to pale buff or tan, sometimes olive-brown or yellowy; smooth, occasionally velvety at first, sometimes pitted, then often with scaly patches in age. **Tubes** deeply depressed; whitish at first. **Pores** up to 1mm (¹⁄₁₆in) broad; whitish, unchanging. **Stem** 50–80 × 7–11mm (2–3 × ¼–½in), sometimes slightly twisted; white to pale olive-buff, becoming rough with dark ornamentation. **Flesh** thick; white, unchanging. **Spores** ellipsoid to subfusiform, 15–19.5 × 4–6μ. Deposit olive-brown. **Habitat** under hardwoods, especially oak. Quite common. Found in southern and eastern North America. **Season** July–October. **Edibility not known.**

Tylopilus chromapes (Frost) Smith & Thiers syn. *Leccinum chromapes* (Frost) Singer **Cap** 3–15cm (1¼–6in) across, convex becoming flatter, with margin often flared and irregular in age; rose-pink, fading in age. **Tubes** 5–12mm (¼–½in) deep, depressed to nearly free; white, yellowish, finally pinkish or brownish. **Pores** 2–3 per mm, round to angular; white when young. **Stem** 40–150 × 10–25mm (1½–6 × ½–1in), solid, usually pinched off at the base; whitish above to paler pink then deep chrome yellow at base; firm, dry, dotted with pink or reddish scales. **Flesh** soft; white or pink-tinted. **Odor** not distinctive. **Taste** very slightly acid. **Spores** ellipsoid, smooth, 11–17 × 4–5.5μ. Deposit rosy brown. **Habitat** singly or scattered on the ground under hardwoods or conifers. Found in northeastern North America, south to Georgia. **Season** June–August. **Edible** — good.

Austroboletus gracilis (Pk.) Wolfe syn. *Porphyrellus gracilis* (Pk.) Singer **Cap** 3–10cm (1¼–4in) across, convex to broadly convex; reddish chestnut brown to cinnamon brown; dry, granulose becoming cracked. **Tubes** up to 2 cm (¾in) deep, deeply depressed around the stalk, uneven; white to flesh-colored then pinkish brown. **Pores** 1–2 per mm; white to pinkish brown. **Stem** 60–150 × 4–10mm (2¼–6 × ³⁄₁₆–½in), long, solid, slender, often curved; same color as cap or paler cinnamon tan, white within, base white; longitudinally lined, with a bloom or finely granulose. **Flesh** white or tinged reddish near cuticle. **Odor** not distinctive. **Taste** mild. **Spores** ellipsoid, often punctate, 11–17 × 5–7μ. Deposit dark reddish brown. **Habitat** singly or scattered on the ground in woods under aspen, oak, pine, and hemlock. Found in northeastern North America, south to Georgia. **Season** June–October. **Edibility not known** — probably good.

Tylopilus porphyrosporus (Fr.) Smith & Thiers **Cap** 4–8cm (1½–3in) across, broadly convex with an even margin; dark dull brown or olive-brown becoming drab, with a distinct ochraceous tinge when dry; dry, suedelike, sometimes becoming smooth in age. **Tubes** about 10mm (½in) long, nearly free, sharply depressed around the stem; dark dull cocoa brown, wood brown when the sides are cut. **Pores** same color as sides when young, staining dark chocolate brown when bruised. **Stem** 60–150 × 8–20mm (2¼–6 × ⁵⁄₁₆–¾in), tapering toward the top, slightly bulbous at the base; whitish beneath, but brownish like cap on top from velvety or minutely powdered surface, staining dark brown where handled or in age. **Flesh** firm, quite tough; white slowly tinged grayish brown or reddish brown when cut. **Odor** slightly disagreeable. **Taste** slightly mealy. **Spores** subellipsoid, smooth, 13–17 × 6–8μ. Deposit chocolate gray or deep, dull pinkish brown. **Habitat** singly along trails or in mixed deciduous woods under hemlock, spruce, and balsam fir. Found in eastern North America, south to South Carolina and west to Michigan and Mississippi. **Season** August. **Edible.**

Tylopilus pseudoscaber (Secr.) Smith & Thiers syn. *Porphyrellus pseudoscaber* (Secr.) Singer **Cap** 5–15cm (2–6in) across, convex, becoming flatter in age; very dark earth brown to olive-brown or vinaceous brown; dry, minutely hairy to fibrous. **Tubes** 1.5–2cm (½–¾in), subdecurrent becoming deeply depressed with age; pale grayish brown, bruising green-blue. **Pores** small, round; dark yellow-brown. **Stem** 40–120 × 10–30mm (1½–4¾ × ½–1¼in), solid-stuffed, becoming larger toward the base; brown outside, white within, often blackening with age at base; dry, minutely felty, sometimes longitudinally and irregularly lined. **Flesh** white, bruising bright blue then reddish to dull brown. **Spores** ellipsoid, smooth, 12–18 × 6–7.5μ. Deposit reddish-brown. **Habitat** scattered or in groups on the ground or along old roads in hardwood and mixed forests. Quite common. Found in eastern North America, the Pacific Northwest, and northern California. **Season** September–November. **Edible** — but not good.

Austroboletus gracilis ⅓ life-size

Tylopilus porphyrosporus ⅓ life-size

Tylopilus pseudoscaber ½ life-size

Lilac-brown Bolete *Tylopilus eximius* ½ life-size

Tylopilus alboater ½ life-size

Violet-gray Bolete *Tylopilus plumbeoviolaceus* ⅓ life-size

Tylopilus ballouii ½ life-size

Tylopilus griseocarneus ⅓ life-size

Tylopilus intermedius ⅓ life-size

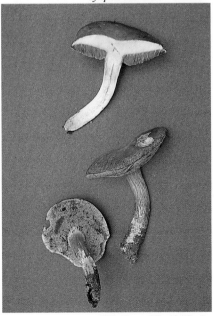

Tylopilus sordidus ⅓ life-size

Lilac-brown Bolete *Tylopilus eximius* (Pk.) Singer syn. *Leccinum eximium* (Pk.) Pomerleau **Cap** 5–20cm (2–8in) across, convex for a long time before expanding; pale dark chocolate brown with lilac or purplish overtones, with whitish bloom when young; viscid in wet weather, surface often slightly pitted. **Tubes** sunken around the stem or free; brownish purple. **Pores** very small; chocolate brown or purplish, paler with age, not changing or bruising. **Stem** 50–110 × 10–30mm (2–4½ × ½–1¼in), clavate or cylindrical; colored as cap but densely, minutely punctate with darker, purple-brown dots. **Flesh** thick; pale purple-brown, especially in stem apex, cream or yellowish in stem base, unchanging when cut. **Spores** ellipsoid, subfusiform, (11)11.5–15(17) × 3.5–4(6)μ. Deposit vinaceous brown. **Habitat** under mixed conifers. Common. Found in eastern North America from Georgia to Maine and west to Michigan. **Season** July–September. **Edible** with caution; a case of poisoning has been reported. **Comment** The spores of this collection were slightly narrower and the cap colors rather paler than usually reported, but it otherwise agrees with the original description.

Tylopilus alboater (Schw.) Murr. **Cap** 3–20cm (1¼–8in) across, convex, soon flattened; deep blackish brown with white bloom when young; dry, velvety. **Tubes** adnate, sunken around stem; pallid then pink. **Pores** small; pinkish. **Stem** 40–110 × 20–40mm (1½–4½ × ¾–1½in), equal; white to slightly pink. **Odor** pleasant. **Taste** mild. **Spores** fusiform, smooth, 8–11 × 3.5–5μ. Deposit pinkish. **Habitat** under deciduous trees, especially oak. Frequent. Found in eastern North America, west to Michigan and Texas. **Season** July–September. **Edible.**

Tylopilus griseocarneus Wolfe & Halling. **Cap** 4–11cm (1½–4¼in) across, convex; pale charcoal to dark brownish gray, sometimes bruising darker; dry, velvety, occasionally cracked. **Tubes** gray, staining orange finally fuscous where injured; depressed around stem. **Pores** black when young, gray in age staining grayish orange. **Stem** 45–85 × 10–35mm (1¾q–3¼ × ½–1¼in), thickest at the apex, tapering at the base; blackish brown or charcoal, lighter above, darker below; velvety with a distinct network on the upper half. **Flesh** solid, pallid gray to dark gray, staining strongly orange to orangy-red. **Odor** slight. **Taste** mild. **Spores** fusiform-ellipsoid, smooth, 7.8–14.3 × 3–5.2μ. Deposit pinkish brown. **Habitat** in sandy soil under oak and pine. Found in the New Jersey Pine Barrens and southern Louisiana. **Season** July–August. **Edibility** not known. **Comment** A new species published only in 1989. Easily distinguished from *T. alboater* (above) by the strong orange to red discoloration on cutting or damaging a fresh specimen.

Violet-gray Bolete *Tylopilus plumbeoviolaceus* (Snell & Dick) Singer **Cap** 8–20cm (3–8in) across, broadly convex; at first deep violet, but very quickly grayish lavender, finally dingy cinnamon, often with a white bloom; dry, unpolished. **Tubes** 1–2cm (½–¾in) deep; cream to dirty vinaceous. **Pores** small; pallid becoming pinkish tan. **Stem** 80–120 × 10–20mm (3–4¾ × ½–¾in), long, equal, rarely clavate; remaining a beautiful violet for some time after cap fades, often mottled darker violet; smooth to minutely reticulate at apex, paler at base. **Flesh** white, hardly discoloring or only very slight browning. **Odor** pleasant. **Taste** bitter. **Spores** ellipsoid, smooth, 10–13 × 3–4μ. Deposit pinkish brown. **Habitat** under oak, hickory, and aspen on sandy soil. Common and can be abundant in wet years. Quebec to Florida, west to Minnesota and Texas. **Season** June–September. **Not edible.**

Tylopilus ballouii (Pk.) Singer **Cap** 5–12cm (2–4½in) across, convex to flattened, often irregular in outline; bright fiery orange to reddish brown or cinnamon; dry, slightly roughened to subtomentose. **Tubes** white bruising brown. **Pores** cream to smoky brown where bruised. **Stem** 25–120 × 7–25mm (1–4¾ × ¼–1in), equal to swollen-clavate; pallid to pale yellow-orange, white at base; dry, smooth to subreticulate at apex. **Flesh** spongy; white, turning slightly brownish orange where eaten by insects. **Odor** pleasant. **Taste** mild. **Spores** ellipsoid, smooth, 8–10 × 4–5μ. Deposit vinaceous brown. **Habitat** under mixed deciduous trees, especially beech. In some years quite abundant. Found from New York down to North Carolina. **Season** July–September. **Edible.**

Tylopilus intermedius Smith & Thiers **Cap** 6–15cm (2¼–6in) across, broadly convex becoming almost flat, with an incurved margin that turns up in age; whitish when young, becoming buff then slowly pale tan, gradually staining brown where bruised; dry, uneven surface with a bloom, often wrinkled like parchment. **Tubes** 10–15mm (½in) deep, deeply depressed around stem or free; white becoming pinky-brown. **Pores** 1–2 per mm, round; white then dark pinky-brown. **Stem** 80–140 × 10–40mm (3–5½ × ½–1½in), solid, enlarged toward the base; white, staining dingy yellow-brown where handled, white within, base whitish; practically smooth. **Flesh** thick, hard; white discoloring brown around larval tunnels. **Odor** mushroomy. **Taste** very bitter. **Spores** narrowly boat-shaped, smooth, 10–15 × 3–5μ. Deposit fawn. **Habitat** in groups under oak. Rare, but abundant during seasons of heavy rainfall. Found in northeastern North America as far west as Michigan. **Season** August–September. **Not edible.**

Tylopilus sordidus (Frost) Smith & Thiers **Cap** 3–12cm (1¼–4¾in) across, convex becoming nearly flat, with an even margin that sometimes turns up in age; olive-brown; dry, velvety, but quickly becoming cracked into patches. **Tubes** 1–2cm (½–¾in) deep; graying to olive-gray becoming chocolate, bruising blue then brownish. **Pores** 1–2 per mm; graying to olive-gray, staining blue then chocolate. **Stem** 20–60 × 10–15mm (¾–2¼ × ½in), solid; surface similar color to cap with a copper blue zone at the top and whitish at the base, staining reddish or brownish when cut; brownish velvety bloom on the surface. **Flesh** thick, firm; yellowish white slowly staining pale bluish green then dingy pinkish gray. **Odor** mild but pungent and odd. **Taste** mild to slightly acidulous. **Spores** subellipsoid to inequilateral, smooth, 11–15 × 5–6.5μ. Deposit purple-brown to wood brown. Cheilocystidia clavate to vesiculose-pedicellate, up to 15μ thick. **Habitat** singly, in groups, or occasionally in clusters on sandy, open soil along streams and roads and in thin oak woods. Uncommon. Found in eastern North America and rarely in the Northwest. **Season** June–September. **Not edible.**

Bitter Bolete *Tylopilus felleus* ½ life-size

Tylopilus badiceps ½ life-size

Old Man of the Woods *Strobilomyces floccopus* ½ life-size

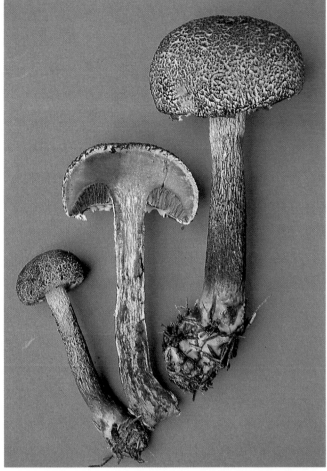

Confusing Bolete *Strobilomyces confusus* ½ life-size

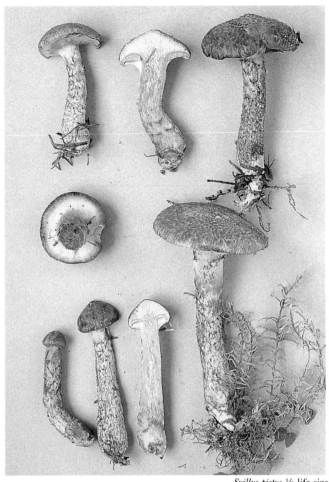

Suillus pictus ½ life-size

Suillus decipiens ⅔ life-size

Bitter Bolete *Tylopilus felleus* (Bull. ex Fr.) Karsten **Cap** 5–30cm (2–12in) across, convex then flattened, tan-brown to buff or ochre; slightly viscid when wet but soon dry, minutely tomentose. **Tubes** rather deep; white at first, then pinkish. **Pores** pale pinkish bruising darker. **Stem** 40–100 × 10–30mm (1½–4 × ½–1¼in), robust, equal or clavate; colored as the cap; covered with a raised mesh or network over most of the surface. **Flesh** firm; white. **Odor** pleasant. **Taste** very bitter. **Spores** fusiform, smooth, 11–15(17) × 3–5μ. Deposit vinaceous brown. **Habitat** in mixed woodlands. Common. Found from Maine to Florida and across to Texas. **Season** June–October. **Not edible.**

Tylopilus badiceps (Pk.) Smith & Thiers **Cap** 4–8cm (1½–3in) across, broadly convex, becoming somewhat centrally depressed when mature; dark brownish red or dark maroon; dry, velvety. **Tubes** adnate; white, becoming dingy with age. **Pores** minute; white, becoming dingy with age. **Stem** 40–50 × 15–30mm (1½–2 × ½–1¼in), solid, sometimes slightly swollen in the middle; brownish; smooth, radiating. **Flesh** white, when bruised brown spots slowly develop. **Odor** mild, sweet, suggestive of molasses. **Taste** similar. **Spores** narrowly ellipsoid, smooth, 8–10 × 3.5–4.5μ. Deposit purplish brown. **Habitat** on the ground in oak and mixed woods. Found in northeastern North America. **Season** August–September. **Edible** — good.

Old Man of the Woods *Strobilomyces floccopus* (Fr.) Karsten syn. *Boletus floccopus* Vahl ex. Fr. and *Boletus strobilaceus* Fr. **Cap** 5–12cm (2–4¾in) across; smoke gray to black, covered with large floccose-woolly scales or warts, with white flesh showing in cracks between scales; thick concolorous scales overhang cap margin. **Tubes** white to gray, coral then red upon bruising. **Pores** large, angular; similarly colored. **Stem** 80–120 × 10–25mm (3–4¾ × ½–1in), solid, slightly enlarged toward the base; white to grayish above ring zone; shaggy grayish black below. **Flesh** firm; white gradually turning vinaceous to coral, then brown or black on cutting. **Odor** not distinctive. **Taste** not distinctive. **Spores** subglobose, with reticulate ornamentation, 10–12 × 8.5–11μ. Deposit violaceous black. **Habitat** in mixed coniferous and hardwood forests. Very common. Found widespread from Nova Scotia to Florida, west to Michigan and Texas. **Season** July–October. **Edible** — good.

Confusing Bolete *Strobilomyces confusus* Singer **Cap** 3–10cm (1¼–4in) across; fuscous black to black; covered with acute, rigid, and erect spines or warts, which are denser toward the disk. **Tubes** white bruising coral to red. **Pores** large, angular; similarly colored. **Stem** 40–80 × 10–20mm (1½–3 × ½–¾in), solid, tapering downward; fuscous black to grayish; shaggy. **Flesh** white then bruising red to brownish black. **Odor** not distinctive. **Taste** not distinctive. **Spores** subglobose, spiny to warty with very incomplete network connecting spines, 10.5–12.5 × 9.5–10μ. Deposit violaceous black. **Habitat** in mixed coniferous and hardwood forests. Apparently widespread although exact distribution is not known; however, it appears to be somewhat more southern than *Strobilomyces floccopus* (above). Recorded from Massachusetts to Florida, west to Ohio and Tennessee. **Season** July–October.

Edible — quite good. **Comment** This species can be distinguished from *Strobilomyces floccopus* (above) in the field by the acute, erect warts or spines, which feel very firm if tapped gently with the fingers; in *floccopus* the warts or scales feel soft and woolly.

Suillus pictus (Pk.) Smith & Thiers **Cap** 3–12cm (1¼–4¾in) across, cone-shaped or convex becoming flat, with an incurved margin often hung with veil remnants; red to reddish yellow with coarse scales, often gray; surface dry and flaky, sometimes tacky in wet young specimens. **Tubes** adnate to decurrent; bright yellow. **Pores** large, angular; yellow bruising brownish. **Stem** 40–120 × 8–25mm (1½–4¾ × 5⁄16–1in), solid, sometimes wider at the bottom; yellow at the top above the ring, scaly, flaky, and patchy below, similar to the cap. **Veil** white with pink patches; delicate, fibrous, leaving a dull-colored ring on the stem. **Flesh** downy; yellow changing to dull pink or reddish if bruised. **Odor** mild. **Taste** mild. **Spores** ellipsoid, 8–12 × 3.5–5μ. Deposit olivaceous. **Habitat** scattered to gregarious under eastern white pine. Common. Found in northeastern North America. **Season** June–November. **Edible** — good. **Comment** It can often be found in large quantities in pine woods, where it is frequently the dominant species.

Suillus decipiens (Berk. & Curt.) O. Kuntze **Cap** 3.5–7cm (1¼–2¾in) across, convex becoming flatter; maize yellow or paler, with pinkish-cinnamon or pale tan appressed scales; sometimes appearing darker when wet. **Tubes** up to 5mm (¼in) long; honey yellow. **Pores** irregular, compound; same color as tubes.

(*continued on page 247*)

Suillus caerulescens ½ life-size

Larch Bolete *Suillus grevillei* ½ life-size

Suillus lakei ⅔ life-size

Hollow-stemmed Boletus *Suillus cavipes* ½ life-size

Suillus grevillei var. *clintonianus* ⅓ life-size

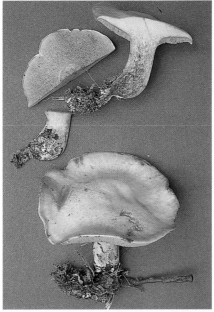

Suillus subaureus ⅓ life-size

(*continued from page 245*)
Stem 40–70 × 7–15mm (1½–2¾ × ¼–½in),
solid, tapering and hooked toward the base;
yellowish to cinnamon buff or cinnamon pink
below; cottony-scaly. **Veil** sheathlike, forms a
slight gray to whitish ring. **Flesh** straw yellow,
mostly unchanging, but becoming dull pinky-
tawny in places. **Odor** agreeable, mild. **Taste**
mild. **Spores** cylindrical to subellipsoid,
smooth, 9–12 × 3.5–5μ. Deposit ochraceous
brown. **Habitat** in dense groups on the ground,
in humus, in mixed pine-oak forests, and
sphagnum bogs. Frequent. Found in
southeastern North America, north to New
Jersey. **Season** June–September. **Edible.**

Suillus caerulescens Smith & Thiers **Cap** 6–14cm
(2¼–5½in) across, convex becoming flat; dull
reddish yellow, more yellow toward the
margin; small, scaly patches at first, then
smooth and slimy. **Tubes** adnate; yellow
bruising pinky-brown. **Pores** irregularly
angular; yellow. **Stem** 25–80 × 20–30mm
(1–3 × ¾–1¼in), webbed at the top above the
ring; yellow and smooth with some dots above,
red-brown with minute hairs below the ring,
base bruising blue when handled. **Veil** fibrous,
pallid; leaving bandlike ring on stalk. **Flesh**
pale yellow, sometimes flushed with pink,
bruises blue in stem. **Odor** mild or slightly
sour. **Taste** mild or slightly sour. **Spores**
ellipsoid, 8–11 × 3–5μ. Deposit dull cinnamon.
Habitat gregarious in humus under mixed
conifers, particularly Douglas fir. Common.
Found in the Pacific Northwest and California.
Season August–January. **Edible. Comment**
The blue bruising of the stem may take a few
minutes but is a distinctive feature.

Larch Bolete *Suillus grevillei* (Klotzch) Singer
Cap 5–15cm (2–6in) across, convex to nearly
flat; yellow to chrome or bright rusty red;
smooth, slimy, and sticky with pale lemon
gluten and removable pellicle. **Tubes** adnate to
subdecurrent; bright yellowish becoming
greenish and bruising pinky-brown. **Pores**
small, angular; lemon yellow bruising
rust. **Stem** 40–100 × 15–30mm
(1½–4 × ½–1¼in),

solid; pale yellow above the ring, flushed
cinnamon below, bruising brownish. **Veil**
white to yellowish; leaving cottony ring on
stem. **Flesh** pale yellow in cap, darker lemon
chrome in stem, bruises pinkish. **Odor** none or
faintly metallic. **Taste** mild or very slightly
bitter. **Spores** subfusiform-ellipsoid,
8–10 × 2.8–3.5μ. Deposit olive-brown to dull
cinnamon. **Habitat** gregarious or growing in
dense tufts under larch (tamarack). Common.
Found in northern and eastern North America.
Season September–November. **Edible** — good
after peeling and removing slime. **Comment**
The specimens in the photograph were found
in the West, where you usually find the red
color form; in the East the yellow color form is
generally found.

Suillus grevillei (Pk.) Singer var. *clintonianus*
Cap 5–15cm (2–6in) across, semicircular,
becoming broadly convex then flatter; color
variable from rich reddish brown to yellow;
smooth, glutinous, with a separable pellicle.
Tubes 10–15mm (½in) deep, adnexed
to depressed becoming subdecurrent; amber
yellow becoming olive-yellow then reddish
brown when bruised. **Pores** 1–2 per mm,
angular. **Stem** 40–100 × 10–30mm
(1½–4 × ½–1¼in), solid, slightly club-shaped;
yellow to pale yellow within and without,
surface soon developing chestnut patches. **Veil**
cottony partial veil, which may have a
gelatinous outer layer in wet weather, leaves a
distinct ring; dingy in age, staining pinky-
brown after handling. **Flesh** rather thick;
yellow becoming flesh-pink or salmon pink
when bruised. **Odor** none or slightly metallic.
Taste mild to slightly bitter. **Spores** ellipsoid,
smooth, 8–10 × 2.8–3.5μ. Deposit olive-brown.
Habitat in groups or dense clusters, often in
semicircles associated with larch. Common.
Found in northern North America. **Season**
September–October. **Edible. Comment**
Distinguished from *Suillus grevillei* (above) by
its much darker reddish-brown colors, although
some mycologists do not recognize it as a
separate variety.

Suillus lakei (Murr.) Smith & Thiers **Cap**
6–14cm (2¼–5½in) across, broadly convex to

flat, with an incurved margin that often has
veil remnants attached to it; yellow with dry
reddish-brown flakes; slimy and sticky beneath
the dry, scaly layer. **Tubes** adnate to
decurrent, shallow; dirty yellow. **Pores** large,
angular; bruising brownish. **Stem**
30–90 × 10–20mm (1¼–3½ × ½–¾in), solid,
enlarged toward base; yellow above veil, with
dry brownish scales on lower part; younger
specimens bruise slightly green or bluish. **Veil**
white to pale yellow; downy, leaving line or
partial ring near top of stem. **Flesh** yellow,
becoming pinky where exposed. **Odor** none.
Taste none. **Spores** ellipsoid, 8–11 × 3–4μ.
Deposit dull cinnamon. **Habitat** scattered to
gregarious in mixed coniferous woods under
Douglas fir. Common. Found in Rocky
Mountains, Pacific Northwest, and California.
Season July–August in high mountains,
September–November (January in California).
Edible — good. **Comment** The specimens
shown in the photograph are bruised from
handling. They were much lighter and fresher-
looking when collected.

Hollow-stemmed Boletus *Suillus cavipes*
(Opatowski) Smith & Thiers **Cap** 3–10cm
(1¼–4in) across, convex expanding to flat,
with an inrolled margin which sometimes has
veil remnants adhering to it; dull yellow to
rusty red-brown; dry, densely hairy, suedelike.
Tubes decurrent, short; pale yellow flushing
olivaceous. **Pores** large, angular; same color as
tubes. **Stem** 40–80 × 10–20mm (1½–3 ×
½–¾in), solid becoming distinctly hollow in
base; lemon yellow with an indistinct net above
the ring; darker like cap below and covered in
fine hairs. **Veil** white, netlike; leaving
remnants on margin and a distinct band on
stalk. **Flesh** white to yellowish and sometimes
pinkish in stem. **Odor** pleasant. **Taste**
pleasant. **Spores** subcylindric to fusiform,
7–10 × 3–4μ. Deposit olivaceous. **Habitat**
scattered to clustered under larch. Common.
Found in northern North America. **Season**
September–November. **Edible** — good.
Comment To be sure you have the correct
species cut a mature one in half and check that
the stem is hollow.

Suillus subaureus (Pk.) Snell **Cap** 3–12cm
(1¼–4¾in) across, convex with an inturned
margin, becoming flat with a fluted margin;
mustard yellow; surface sticky under tiny
patches of tomentum and flaky scales which
become red and spotlike. **Tubes** subdecurrent
to decurrent; ochraceous. **Pores** small, round
to angular; dirty yellow. **Stem** 40–80 ×
10–20mm (1½–3 × ½–¾in), solid, sticky when
young; yellow, staining dull brown when
bruised or handled; white mycelium at base.
Veil none except in the tiniest buttons. **Flesh**
up to 3cm (1¼in) thick; yellow staining
reddish brown. **Odor** slightly fragrant. **Taste**
mild, slightly acid. **Spores** ellipsoid to
subfusoid, 7–10 × 2.7–3.5μ. Deposit olive-
brown. **Habitat** scattered to gregarious under
white pine, aspen, and scrub oak. Sometimes
common. Found in eastern North America.
Season June–September. **Edible.**

Slippery Jack *Suillus luteus* ½ life-size

Suillus intermedius ½ life-size

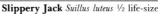

Slippery Jack *Suillus luteus* (Fr.) S. F. Gray. **Cap** 5–17cm (2–6½in) across, rounded becoming convex and clear flatter in age; chestnut to sepia; smooth, slimy and covered with brown gluten, shiny on drying. **Tubes** adnate to subdecurrent; lemon yellow to straw-colored. **Pores** round; yellow becoming brown-dotted. **Stem** 30–90×10–30mm (1¼–3½×½–1¼in), solid, equal or slightly tapering at base; pale straw-colored with darker dots above the ring, white below becoming darker pinky-brown. **Veil** finely webbed, shiny white; leaving a purplish sleevelike ring around the stem. **Flesh** white or pale yellow, often slightly pinky toward the stem. **Odor** not distinctive. **Taste** not distinctive. **Spores** subfusiform to elongate ellipsoid, 7–10×2.5–3.5μ. Deposit dull cinnamon. **Habitat** scattered to gregarious under conifers, especially Scotch pine. Common. Widespread in North America. **Season** September–December. **Edible** — good after peeling and removing slime.

Suillus intermedius (Smith & Thiers) Smith & Thiers syn. *Suillus acidus* var. *intermedius* Smith & Thiers **Cap** 5–10cm (2–4in) across, convex then expanded, slightly umbonate; tawny yellow, buff to ochre-brown; sticky or glutinous when wet, often appearing patchy or streaked from dried gluten. **Tubes** subdecurrent to adnate; pale yellow. **Pores** small; pale yellow or ochre, unchanging when bruised, often weeping droplets of clear fluid when young, yellowing with age. **Stem** 50–100×10–15mm (2–4×½in), often long in proportion to cap, slightly clavate; same color as cap, with

distinctly darker, glandular dots over most of the surface, darkening on handling to olive-brown. **Veil** leaves clear, gelatinous, pale buff ring or band of remnants at apex of stem. **Flesh** thick; pale cream to yellow, unchanging when cut. **Odor** none or slight. **Taste** of cap skin rather acid, of flesh mild. **Spores** subfusiform, smooth, 7–9(12)×2.5–3(4)μ. Deposit dull cinnamon. **Habitat** under pine, usually red pine (*Pinus resinosa*). Rather uncommon. Found in eastern North America extending north and west, but not beyond the Rockies. **Season** July–September. **Edible. Comment** The spores of this collection were rather shorter than in some published accounts, but it otherwise agreed well.

Suillus americanus (Pk.) Snell ex Slipp & Snell **Cap** 3–10cm (1¼–4in) across, broadly convex with a low umbo and incurved margin, which may have remnants of a yellowish, cottony veil hanging from it; bright yellow with red or brownish streaks and hairy patches; slimy and sticky. **Tubes** adnate to decurrent; yellow, staining reddish brown when bruised. **Pores** large, angular; yellow becoming darker with age. **Stem** 30–90×4–10mm (1¼–3½×³⁄₁₆–½in), often crooked, becoming hollow; lemon yellow with dots that bruise darker, as do other parts of stem if handled. **Veil** partial veil, not attached to stalk, leaves no ring on stem. **Flesh** mustard yellow, staining pinky-brown when bruised. **Odor** not distinctive. **Taste** not distinctive. **Spores** ellipsoid to subfusoid, 8–11×3–4μ. Deposit dull cinnamon. **Habitat** singly or gregarious under

eastern white pine. Very common. Found in northeastern North America. **Season** July–October. **Edible** — good. **Comment** Similar to *Suillus sibiricus* (p. 251), which has a ring and is found in the West.

Fragrant Suillus *Suillus punctipes* (Pk.) Singer syn. *Boletus punctipes* Pk. **Cap** 3–10cm (1¼–4in) across, convex then flattened; dull ochre or pale yellowish, browner with age; slightly tomentose when young, but soon sticky to glutinous, glabrous with age. **Tubes** adnate or slightly decurrent; grayish brown, then honey yellow with age. **Pores** round to angular when mature; brown at first, to honey or ochre-yellow, unchanging when bruised. **Stem** 40–100×10–15mm (1½–4×½in), cylindrical to clavate below; dull ochre to orange-ochre, discoloring brown, thickly covered with sticky, glandular, brown dots; stains brown and will also stain the hands ochre upon handling. **Flesh** soft; pale yellowish to ochre, bruises brown. **Odor** very distinctive, particularly when many fruit bodies are present or when flesh is cut; of bitter almonds according to some authors; the collection illustrated smelled of parsley. **Taste** mild. **Spores** subfusiform, smooth, (7)8–9.5(10)×2.5–3(4)μ. Deposit olive-brown. **Habitat** under spruce and balsam, often in sphagnum. Uncommon. Found in northeastern North America, west to Michigan. **Season** July–October. **Edible** but rather soft. **Comment** A rather rare species well characterized by its odor and the very prominent dots on the stem, which stain the hands.

Suillus umbonatus Dick & Snell **Cap** 3–6cm (1¼–2¼in) across, broadly convex with an acute umbo; olive tan, sometimes paler toward margin and darker on the umbo; slimy and sticky, streaked with rusty-brown gluten. **Tubes** adnate to subdecurrent, separating clearly; buff-colored. **Pores** subangular; pale yellow to greenish yellow, staining dirty pink when bruised. **Stem** 25–40×3–8mm (1–1½× ⅛–⁵⁄₁₆in), solid; pale yellow above the ring, whitish but streaked and blotched with brownish dots from below the ring to the base. **Veil** gelatinous becoming dirty pink; leaving a gelatinous ring. **Flesh** very soft; pale yellow, becoming tinged with pinkish buff when cut. **Odor** not distinctive. **Taste** slightly sour. **Spores** ellipsoid, 7–10×3–4.5μ. Deposit dull cinnamon. **Habitat** gregarious and sometimes growing in dense tufts among moss under pine. Often common. Found in the Rocky Mountains and Pacific Northwest. **Season** July–November. **Probably edible** but not recommended.

Suillus albidipes (Pk.) Singer **Cap** 4–10cm (1½–4in) across, convex, with a false veil of cottony whitish material along the margin, becoming patchy with age; whitish, becoming yellow to reddish buff in age; glutinous to sticky. **Tubes** adnate becoming adnexed; darkish yellow. **Pores** tiny, round; yellow, hardly staining when bruised. **Stem** 30–60×10–15mm (1¼–2¼×½in), solid; white, but yellowish toward top and brownish toward bottom portion, which also has dark dots in age. **No veil. Flesh** white, slowly becoming yellow. **Odor** not distinctive. **Taste** not distinctive. **Spores** oblong, 6.6–8.8× 2.5–3μ. Deposit dull cinnamon. **Habitat** scattered or abundant under pine. Found widespread in northern North America and sometimes very common in the Great Lakes area. **Season** August–November. **Edible.**

Suillus americanus ⅔ life-size

Fragrant Suillus *Suillus punctipes* ¼ life-size

Suillus umbonatus ⅓ life-size

Suillus albidipes ⅓ life-size

Granulated Boletus *Suillus granulatus* ¾ life-size

White Suillus *Suillus placidus* ½ life-size

Suillus pallidiceps ⅔ life-size

Suillus borealis ½ life-size

Suillus sibiricus ⅓ life-size

Suillus brevipes ½ life-size

Suillus albivelatus ½ life-size

Granulated Boletus *Suillus granulatus* (Fr.) Kuntze **Cap** 4–15cm (1½–6in) across, broadly convex; rusty brown to yellowish; sticky or glutinous when wet, shiny when dry, smooth. **Tubes** adnate; buff to pale yellow. **Pores** small; pale yellow becoming dingy, sometimes exuding pale milky droplets. **Stem** 35–80 × 8–25mm (1¼–3 × 5/16–1in), solid; lemon yellow flushed pink or cinnamon orange toward base; upper region covered in pinkish-tan dots that exude milky droplets. **No veil. Flesh** pale lemon yellow, lemon chrome in stem, paler in cap. **Odor** slight, pleasant. **Taste** slight, pleasant. **Spores** subfusiform-ellipsoid, 7–10 × 2.5–3.5μ. Deposit ochraceous sienna. **Habitat** scattered to gregarious under conifers, especially white pine. Common. Found widespread throughout North America. **Season** July–November. **Edible** — good. **Comment** The cap color may be rather variable, but the glandular dots are a characteristic feature.

White Suillus *Suillus placidus* (Bon.) Singer syn. *Boletus placidus* Bon. **Cap** 3–10cm (1¼–4in) across, convex then expanded and often slightly umbonate, finally flattened; white to ivory white at first, then discoloring yellowish or even olive when wet; surface sticky, appearing shiny when dry. **Tubes** slightly decurrent; white then soon yellowish to ochre. **Pores** white then yellowish to ochre, often exuding pinkish droplets when young. **Stem** 40–120 × 10–30mm (1½–4¾ × ½–1¼in), cylindrical; white to pinkish vinaceous, covered with dark pinky-brown glandular dots, surface yellowing with age. **No veil. Flesh** white then yellow, bruising pale pinky-brown when cut. **Odor** mild to acidulous. **Taste** mild. **Spores** ellipsoid-oblong, smooth, 7–9 × 2.5–4μ. Deposit dull cinnamon. **Habitat** under white pine. Often abundant. Found throughout the tree's range, except in the drier states. **Season** July–October. **Edible** — good.

Suillus pallidiceps Smith & Thiers **Cap** 3–8cm (1¼–3in) across, round then convex and flatter; whitish then becoming pale yellow and darkening to cinnamon; smooth, very sticky and glutinous, becoming somewhat patchy beneath the gluten in age. **Tubes** adnate to decurrent; yellow, darkening slowly with age. **Pores** small; pale yellow. **Stem** 10–20 × 12–16mm (½–¾ × ½–¾in), solid, bulbous at base; white, with edges of worm holes staining dirty yellow. **No veil. Flesh** white then yellow. **Odor** not distinctive. **Taste** not distinctive. **Spores** ellipsoid, 8–11 × 3.5–4.2μ. Deposit pale cinnamon. **Habitat** gregarious under pine. Rare. Found only in Idaho, as far as I know. **Season** July–October. **Edibility not known.**

Suillus borealis Smith, Thiers, & Muller **Cap** 4–12cm (1½–4¾in) across, convex; dark reddish brown to chocolate brown; with gelatinous purple-brown veil. **Tubes** 4–7mm (3/16–¼in). **Pores** about 2 per mm; pale, dull yellow. **Stem** 10–50 × 10–30mm (½–2 × ½–1¼in), white becoming yellowish, dotted with sticky glands. **Veil** dull lavender or purplish; gelatinous. **Flesh** white. **Odor** slight. **Taste** slight, pleasant. **Spores** fusoid, 7–8 × 2.8–3μ. Deposit dull cinnamon. **Habitat** gregarious or growing in dense tufts under western white pine. Common. Found in the northern Rocky Mountains. **Season** October–November. **Edible** — excellent.

Suillus sibiricus (Singer) Singer **Cap** 3–10cm (1¼–4in) across, obtusely convex becoming flat, with a margin that sometimes has dingy yellow veil fragments attached to it; bright yellow or darker, spotted with brown flakes, especially toward the edge; sticky to glutinous. **Tubes** adnate to decurrent; dark dingy yellow. **Pores** angular; dark dingy yellow, staining strong cinnamon when bruised. **Stem** 50–100 × 7–15mm (2–4 × ¼–½in), solid; yellow, darker dots all over, pink-brown at base, occasionally with a ring. **Veil** pale, dry, open; usually disappears early, leaving fragments on cap margin. **Flesh** pale olive-yellow, turning dull brown when bruised. **Odor** distinctive. **Taste** mildly bitter. **Spores** ellipsoid-fusoid, 8–11 × 3.8–4.2μ. Deposit dull cinnamon. **Habitat** solitary or gregarious under white pine. Sometimes very common. Found widespread in the Pacific Northwest. **Season** September–November. **Probably edible** but not recommended. **Comment** *Suillus americanus* (p. 248), which superficially looks very similar, never has a ring and is found in the East.

Suillus brevipes (Pk.) Kuntze **Cap** 5–10cm (2–4in) across, convex to nearly plane, with margin sometimes slightly lobed; dark red- or gray-brown becoming more red or yellowish brown; smooth and very slimy. **Tubes** adnate to decurrent; honey yellow, becoming more greenish in age. **Pores** small, round; pale yellow. **Stem** 20–50 × 10–20mm (¾–2 × ½–¾in), solid; white becoming pale yellowy, sometimes with faint dots. **No veil. Flesh** white becoming pale yellow. **Odor** not distinctive. **Taste** not distinctive. **Spores** ellipsoid, 7–10 × 2.5–3.2μ. Deposit cinnamon. **Habitat** scattered to gregarious in sandy soil under conifers. Common. Found widespread in southern, eastern, and western North America. **Season** June–November. **Edible** — good.

Suillus albivelatus Smith, Thiers, & Miller **Cap** 4–12cm (1½–4¾in) across, convex, sometimes with part of the margin intergrown with the stem; pale at first, becoming dull reddy or yellowy brown; at first covered with a white hairy veil, which leaves dry scales on the cap, particularly near the margin. **Tubes** 5–10mm (¼–½in) deep; pale ochre. **Pores** small; pale yellow-ochre. **Stem** 10–40 × 1.5–2.5mm (½–1½ × 1/16–1/8in), short and squat; pallid whitish; dryish, smooth. **Veil** white, viscid. **Flesh** white then lemon yellow, bruising reddish around worm holes. **Odor** not distinctive. **Taste** not distinctive. **Spores** fusoid, 7–8.5 × 2.8–3μ. Deposit cinnamon brown. **Habitat** gregarious and often growing in fairly dense tufts under white pine. Common in Idaho. Found in the Pacific Northwest. **Season** June–November. **Edible.**

Fuscoboletinus ochraceoroseus ¾ life-size

Found widely distributed in northern North America. **Season** July–October. **Edible** — good.

POLYPORES *and relatives have tubes which are part of the cap and cannot be separated from it as can the tubes of boletes. The first small group have central stems and grow on the ground. The vast majority of the rest grow on wood in the form of brackets or form crusts on wood.*

Albatrellus caeruleoporus (Pk.) Pouz. **Fruit body** annual. **Cap** up to 6cm (2¼in) wide, one or several growing from a branched base, circular, with an acute or rounded margin; indigo to blue-gray, becoming grayish brown or orange-brown in age; smooth to slightly rough and scaly. **Tubes** up to 3mm (⅛in) deep; indigo becoming reddish orange. **Pores** 2–3 per mm, angular; surface gray to blue, becoming grayish brown to bright reddish orange when dry. **Stem** up to 75 × 25mm (2¾ × 1in), central or off-center; indigo, discoloring with age; smooth to slightly pitted. **Flesh** up to 10mm (½in) thick, firm when dry; cream-colored to pale buff. **Odor** slight. **Taste** mild, pleasant. **Spores** ovoid to subglobose, smooth, 4–6 × 3–5μ. Deposit white. Hyphal structure monomitic. **Habitat** singly or gregariously on the ground in mixed hemlock and deciduous woods. Found in northeastern North America. **Season** September–October. **Edible.**

Sheep Polypore *Albatrellus ovinus* (Fr.) Murr. syn. *Polyporus ovinus* Fr. **Fruit body** annual. **Cap** 5–15cm (2–6in) across, usually single but sometimes several fused together, circular to irregular when fused, convex then depressed, dish-shaped; white to pale buff, tan; dry, smooth, or a little scaly with age. **Tubes** 1–2mm (¹⁄₁₆–³⁄₃₂in) deep, decurrent; white. **Pores** 2–4 per mm, angular; white to yellowish. **Stem** 20–75 × 10–30mm (¾–2¾ × ½–1¼in), slightly swollen, pointed at base, usually central; white bruising pinkish; smooth. **Flesh** 5–20mm (¼–¾in) thick, firm; white, dries yellowish. **Odor** pleasant, fungusy, aromatic. **Taste** mild, sometimes slightly bitter (see Comment). **Spores** subglobose-ellipsoid, 3–4.5 × 3–3.5μ. Deposit white. Hyphal structure monomitic. **Habitat** on the ground by conifers, especially at high elevations. Found throughout North America. **Season** December–June. **Edible. Comment** Similar are *Albatrellus confluens* (Fr.) Kotlaba & Pouz., which is darker, orange-hued, with a bitter flavor, and *Albatrellus subrubescens* (Murr.) Pouz., which bruises orange.

Albatrellus peckianus (Cke.) Niemela **Fruit body** annual. **Cap** up to 5cm (2in) across, one or several from a branched base, circular to kidney-shaped, depressed in the center becoming concave then funnel-shaped at maturity, with an acute, inrolled margin; yellowish, drying yellowy tan to pale buff, sometimes with faint concentric bands of color; smooth to minutely scurfy. **Tubes** up to 2mm (³⁄₃₂in) deep, decurrent; yellow to light buff. **Pores** 4–6 per mm, circular to angular; surface bright yellow, drying yellow to light buff. **Stem** up to 40 × 10mm (1½ × ½in), central or lateral; yellowish to light buff; minutely felty. **Flesh** up to 3mm (⅛in) thick; pale yellow to light buff. **Spores** ovoid to ellipsoid, smooth, 3.5–4.5 × 2.5–3μ. Deposit white. Hyphal structure monomitic; clamps present. **Habitat** singly to gregarious on the ground or on buried wood such as beech and linden. Found in northeastern North America. **Season** August–September. **Edibility not known. Comment** My specimens are rather dry and do not show the good yellow color.

Fuscoboletinus ochraceoroseus (Snell) Pomerleau & Smith **Cap** 8–25cm (3–10in) across, convex becoming broadly convex and slightly umbonate, with an incurved margin sometimes adorned with veil remnants; variable in color but generally lemon yellow along the margin and rose-pink toward the disc; dry, uneven, with a dense, whitish felt sometimes becoming scurfy. **Tubes** 5mm (¼in) deep, adnate to decurrent; yellow, ochre, or dingy brown. **Pores** elongated to angular, radially arranged. **Stem** 30–50 × 10–30mm (1¼–2 × ½–1¼in), solid, sometimes swollen at the base; yellowish, and often reddish or brownish at base; netlike pattern, unpolished or felty below ring. **Veil** thin, membranous, whitish to yellowish; leaving remnants on cap margin and evanescent ring. **Flesh** thick, soft; yellowish, with a pink zone under the cuticle, may slightly bruise bluish green. **Odor** acid. **Taste** slightly acrid. **Spores** ellipsoid, smooth, 7.5–9.5 × 2.5–3.2μ. Deposit reddish brown. **Habitat** scattered or in groups under western larch. Common. Found in the Pacific Northwest. **Season** August–October. **Edibility**

suspect — not advisable.

Fuscoboletinus aeruginascens (Secr.) Pomerleau & Smith syn. *Suillus aeruginascens* (Secr.) Snell **Cap** 3–12cm (1¼–4¾in) across, round becoming convex then almost flat; smoky gray to wood brown, becoming paler, yellowish or almost whitish, in age with dingy spots; slimy when moist, finely hairy to somewhat scaly, sometimes cracking when dry. **Tubes** 6–9mm (¼–⅜in) deep, adnate or subdecurrent; white or pale brown at first, becoming brownish gray to smoke gray, bruising bluish green. **Pores** round, becoming irregular and angular; same color as tubes. **Stem** 40–60 × 8–12mm (1½–2¼ × ⁵⁄₁₆–½in), solid, tapering slightly upward; whitish above the ring, smoke gray or brownish gray below; somewhat sticky with netlike ridges. **Veil** cottony, grayish to yellowish; leaving a flattened ring. **Flesh** firm at first, then soft; whitish to yellowish, turning bluish green where bruised or handled. **Odor** not distinctive. **Taste** mild, fruity. **Spores** ellipsoid to subfusiform, smooth, 8–12 × 3.5–5μ. Deposit pinkish brown. **Habitat** scattered or in groups under larch. Common.

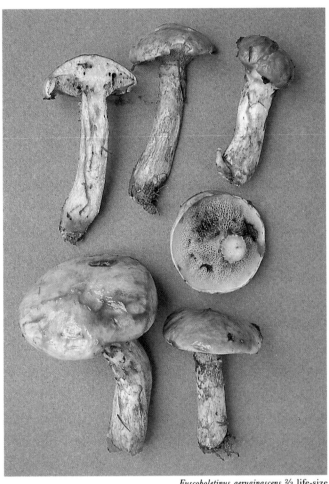

Fuscoboletinus aeruginascens ⅔ life-size

Albatrellus caeruleoporus ⅔ life-size

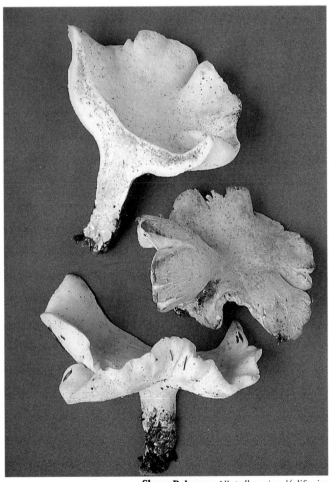

Sheep Polypore *Albatrellus ovinus* ⅓ life-size

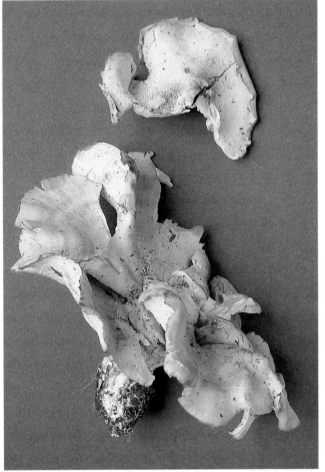

Albatrellus peckianus ½ life-size

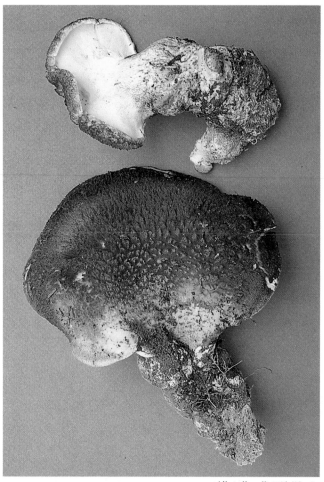

Albatrellus ellisii ¼ life-size

Bondarzewia montana ½ life-size

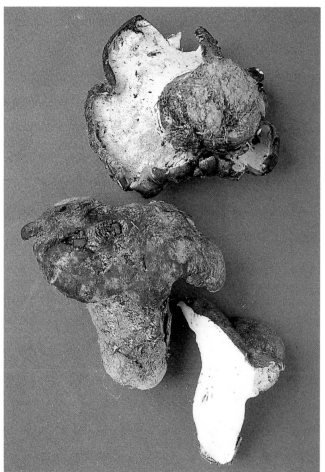

Boletopsis subsquamosa ½ life-size

Bone Polypore *Oligoporus obductus* ½ life-size

Albatrellus ellisii (Berk.) Pouz. **Cap** 8–25cm (3–10in) across, convex becoming flatter, circular, lobed or depressed; margin often wavy and inrolled at first; yellowish green or yellow-brown, with darker stains in age; dry, hairy, velvety, then matted forming coarse scales. **Tubes** 2–6mm (³⁄₃₂–¼in) deep, often decurrent; creamy to slightly greenish. **Pores** 1–2 per mm, circular to angular; white to cream, bruising greenish and becoming greenish or yellowish in age. **Stem** 30–120 × 20–60mm (1¼–4¾ × ¾–2¼in), off-center or lateral, solid, greenish yellow or similar to cap; tough, minutely hairy. **Flesh** thick, firm; white, sometimes bruising greenish slowly. **Spores** ellipsoid, smooth, 8–11 × 5–8μ. Deposit white. **Habitat** singly, scattered, or in groups or clusters in soil in coniferous woods. Not common but sometimes abundant. Found widely distributed in North America. **Season** August–October. **Said to be edible** when well cooked.

Boletopsis subsquamosa (Fr.) Kotlaba & Pouz. **Fruit body** annual. **Cap** up to 15cm (6in) across, 4cm (1½in) thick in center, circular to irregular outline, with thin wavy margin; bluish black to grayish brown tinged with olive; fleshy becoming soft or brittle and slightly wrinkled when dry. **Tubes** up to 8mm (⁵⁄₁₆in) deep; white to greenish white, paler than the flesh. **Pores** 1–3 per mm, angular, thin-walled, decurrent; surface white drying pale grayish. **Stem** up to 7 × 3cm (2¾ × 1¼in), central to lateral; gray to sordid olive-brown; smooth or with fine dark scales, fleshy becoming wrinkled when dry. **Flesh** up to 3cm (1¼in) thick; white when fresh but darkens when touched, becoming greenish gray when dry, often darker just above the tubes. **Odor** slight. **Taste** weak to bitterish when fresh, sweetish to spicy when dry. **Spores** angular to oval, with warty projections, 5–7 × 4–5μ. Deposit light or dark brown. Hyphal structure monomitic; clamps present. **Habitat** on the ground in deciduous or coniferous woods, especially pine. Found in eastern North America, the Pacific Northwest, and California. **Season** September–October. **Edible.** **Comment** Although the name suggests a boletus, this is, in fact, a polypore.

Bone Polypore *Oligoporus obductus* (Berk.) Gilbertson & Ryv. syn. *Osteina obducta* (Berk.) Donk **Fruit body** from one to several circular or spoon-shaped caps; up to 12 × 13 × 2cm (4¾ × 5 × ¾in); convex at first, then slightly depressed; white to grayish or buff; smooth. **Tubes** 1–3mm (¹⁄₁₆–⅛in) deep, slightly decurrent. **Pores** 3–5mm (⅛–¼in), minute, angular; white to yellowish. **Stem** 10–30 × 3–10mm (½–1¼ × ⅛–½in), usually off-center (stems may also fuse together as in the example shown); white to cream. **Flesh** 3–10mm (⅛–½in) thick; white. **Spores** cylindrical, smooth, 4.5–6.5 × 2–3μ. Deposit white. **Habitat** on dead conifers and birch. Found from northeastern North America across to the Pacific Northwest, California, and Colorado. **Season** July–November. **Not edible. Comment** The generic name *Osteina* means "bonelike" and refers to the extreme hardness of the dried fruit bodies.

Berkeley's Polypore *Bondarzewia berkeleyi* ⅓ life-size

Bondarzewia montana (Quél.) Singer **Fruit body** annual. **Cap** up to 11cm (4½in) across, 1cm (½in) thick, one or several on a branched stem, convex becoming flat and sunken; purplish brown or ochraceous brown; scurfy or finely felty becoming wrinkled. **Tubes** up to 2mm (³⁄₃₂in) deep, often decurrent on stem, continuous with flesh; cream-colored. **Pores** 1–3 per mm, angular; surface cream-colored. **Stem** up to 120 × 40mm (4¾ × 1½in), central or off-center; brown; velvety, rooting. **Flesh** up to 1cm (½in) thick, firm, hard; cream-colored. **Odor** pleasant, nutlike. **Spores** globose to subglobose, amyloid, ornamented with irregular short amyloid ridges, 6–8 × 5–7μ. Deposit white. Hyphal structure dimitic. **Habitat** on the ground or on buried wood under conifers — pine, spruce, fir, Douglas fir. Found in the Pacific Northwest and California. **Season** September–November. **Edible.**

Berkeley's Polypore *Bondarzewia berkeleyi* (Fr.) Bond. et Singer syn. *Polyporus berkeleyi* Fr. **Fruit body** annual. **Cap** up to 25cm (10in) across, 15cm (6in) wide, 3cm (1¼in) thick, one or several overlapping in large clusters, usually fan-shaped; tan to yellowish; smooth, finely felty or rough and pitted. **Tubes** up to 2cm (¾in) deep, decurrent, continuous with the flesh; pale buff. **Pores** 1–2 per mm, circular to angular; surface tan. **Stem** up to 8cm (3in) thick, lateral, usually branched, developing from an underground sclerotium; yellowish. **Flesh** up to 3cm (1¼in) thick, corky; pale buff. **Spores** globose to subglobose, ornamented with short irregularly arranged amyloid ridges, 7–9 × 6–8μ. Deposit ochraceous. Hyphal structure dimitic. **Habitat** growing from the base or stumps of hardwood and deciduous trees, particularly oak and chestnut. Found in eastern North America, west to Texas and Louisiana. **Season** July–October. **Edible.**

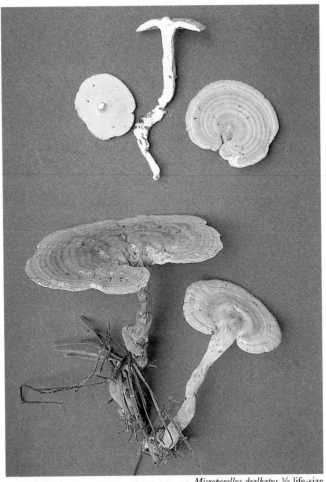

Microporellus dealbatus ½ life-size

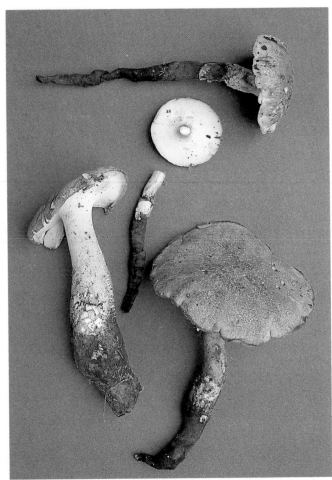

Polyporus radicatus ⅓ life-size

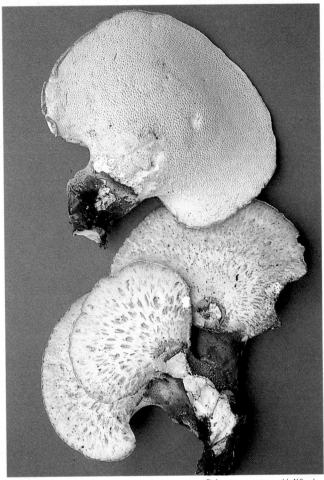

Polyporus squamosus ¼ life-size

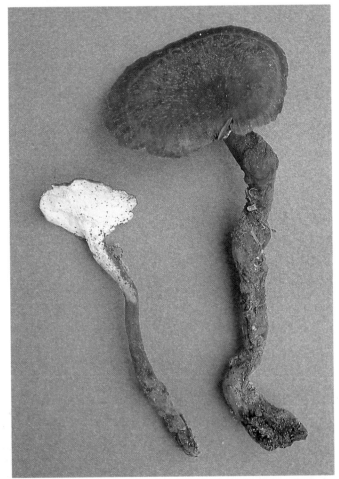

Polyporus melanopus ¾ life-size

Microporellus dealbatus (Berk. & Curt.) Murr. syn. *Polyporus dealbatus* Berk. & Curt. **Cap** 2–10cm (¾–4in) across, circular to kidney-shaped or fan-shaped; gray-brown to liver brown with numerous concentric zones of color; minutely velvety or scurfy-hairy. **Tubes** 1–4mm (¹⁄₁₆–³⁄₁₆in) deep. **Pores** 6–7 per mm, angular, entire; flesh-colored, drying darker. **Stem** 50–70 × 5–10mm (2–2¾ × ¼–½in), lateral or eccentric; same color as cap; drying hard and rigid. **Flesh** tough, drying very hard; white. **Spores** ellipsoid to broadly ellipsoid, smooth, 5–6 × 4–5μ. **Habitat** on the ground, probably attached to buried wood. Very common. Found in southeastern North America. **Season** June–September. **Not edible.**

Polyporus radicatus Schw. **Cap** 3.5–25cm (1¼–10in) across, circular, convex to sunken; yellowish brown to soot brown; dry, velvety to scurfy. **Tubes** 1–5mm (¹⁄₁₆–¼in) deep, decurrent. **Pores** 2–3 per mm, angular; whitish to yellowish. **Stem** 60–140 × 5–25mm (2¼–5½ × ¼–1in), central, with a long black rooting base; dingy yellow; scurfy to slightly scaly. **Flesh** white, dense. **Spores** ovoid to ellipsoid, smooth, 12–15 × 6–8μ. Deposit white. **Habitat** usually singly on the ground around stumps or attached to buried roots. Not common. Found in central and eastern North America. **Season** August–October. **Not edible.**

Polyporus squamosus Huds. ex Fr. **Cap** 5–30cm (2–12in) across, 0.5–5cm (¼–2in) thick, initially circular or fan-shaped; whitish to ochraceous cream; surface covered in concentric dark brown hairy scales. **Tubes** 2–8mm (³⁄₃₂–⁵⁄₁₆in) deep, decurrent; white to creamy. **Pores** irregular and angular; whitish to ochraceous cream. **Stem** 10–50 × 10–40cm (½–2 × ½–1½in), stubby, lateral or occasionally off-center; blackish toward the base. **Flesh** 1–3cm (½–1¼in) thick, succulent when fresh, drying corky; white. **Odor** strongly of meal. **Taste** mild and pleasant when young. **Spores** oblong ellipsoid, smooth, 10–16 × 4–6μ. Deposit white. **Habitat** singly or in overlapping clusters on the wounds of dead trees. Found in eastern Canada and North America south to the Carolinas and west as far as the Midwest. **Season** May–November. **Edible. Comment** This parasite causes intensive white rot in trees.

Polyporus melanopus Fr. **Cap** 3–10cm (1¼–4in) across, convex with an even margin, then funnel-shaped or flattish with a wavy margin which is deflexed in older specimens; bay brown or purplish brown with pale yellowish lines; smooth and velvety, then very wrinkled, especially in the middle. **Tubes** up to 3mm (⅛in) deep, decurrent in specimens with a lateral stem; cream to straw-colored. **Pores** 4–7 per mm, round; whitish to cream then pale straw. **Stem** 10–50 × 2–8mm (½–2 × ³⁄₃₂–⁵⁄₁₆in), central or lateral; dark brown then black, white inside; velvety becoming

longitudinally wrinkled. **Flesh** about 1mm (¹⁄₁₆in) thick; white to cream. **Spores** cylindrical to oblong ellipsoid, nonamyloid, 6–8 × 3–4μ. **Habitat** on the ground or on buried wood. Found in northern North America, especially in the far north. **Season** July–September. **Not edible.**

Polyporus alveolaris (DC ex Fr.) Bond. & Singer syn. *Favolus alveolaris* (DC ex Fr.) Quél. **Fruit body** annual; often with a short lateral stem. **Cap** 1–9cm (½–3½in) across, semicircular to ovoid; pallid reddish yellow; scaly at first. **Pores** rather large (0.5–3mm across), hexagonal; white or dull creamy; decurrent down the stem. **Tubes** up to 5mm (¼in) deep; whitish tan. **Stem** lateral, short or absent, about 10mm (½in) long; colored like the pores. **Flesh** white. **Spores** cylindrical, smooth, 9–11 × 3–3.5μ. Deposit white. **Habitat** on dead wood of deciduous trees. Found throughout most of North America. **Season** May–August, sometimes later. **Edible** but tough.

Polyporus brumalis Fr. **Fruit body** annual. **Cap** 1.5–10cm (½–4in) across, circular, convex or depressed with an inrolled margin; yellow-brown to reddish brown or blackish brown; dry, densely hairy when young, becoming almost smooth. **Tubes** 1–3mm (¹⁄₁₆–⅛in) deep, slightly decurrent. **Pores** 2–3 per mm, circular to angular; whitish. **Stem** 20–60 × 1–5mm (¾–2¼ × ¹⁄₁₆–¼in), central or off-center; grayish or brownish; minutely hairy or smooth. **Flesh** 1–2mm (¹⁄₁₆–³⁄₃₂in) thick; white. **Spores** cylindrical to sausage-shaped, smooth, 5–7 × 1.5–2.5μ. Deposit white. **Habitat** on dead hardwoods, especially birch. Common. Found in eastern North America, west to the Great Plains, and occasionally in the Pacific Northwest. **Season** June–October. **Not edible.**

Polyporus varius (Pers. ex Fr.) **Cap** 3–10cm (1¼–4in) across, funnel-shaped or irregularly kidney-shaped, depressed above the point of attachment to the stem, wavy and often lobed at the margin; ochre-brown with fine radial lines, becoming tobacco brown in age; dry, nearly smooth. **Tubes** 1–3mm (¹⁄₁₆–⅛in) deep, decurrent down the stem; white to cream. **Pores** 4–7mm, circular; white becoming ochraceous brown to light reddish brown. **Stem** 5–70 × 1–15mm (¼–2¾ × ¹⁄₁₆–½in), lateral or off-center; tan with basal part brown-black; smooth or finely hairy. **Flesh** 1–5cm (½–2in) thick, tough, leathery; white when fresh, drying corky and cream-colored. **Odor** faint and mushroomy. **Taste** slightly bitter. **Spores** ellipsoid to fusiform, smooth, 7–10 × 2–3.5μ. Deposit white. **Habitat** singly or several together on dead or dying deciduous trees. Sometimes quite common. Found widely distributed in North America. **Season** June–November (later in California). **Not edible. Comment** *Polyporus elegans* Fr. has often been separated from this fungus but now has been synonymized.

Polyporus alveolaris ⅓ life-size

Polyporus brumalis ½ life-size

Polyporus varius ¼ life-size

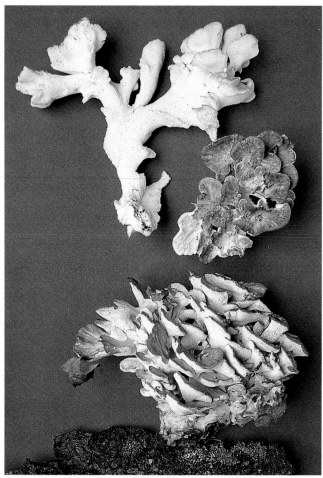

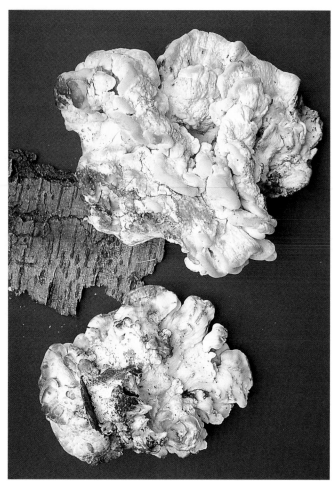

Hen of the Woods *Grifola frondosa* ⅓ life-size

Meripilus giganteus ¼ life-size

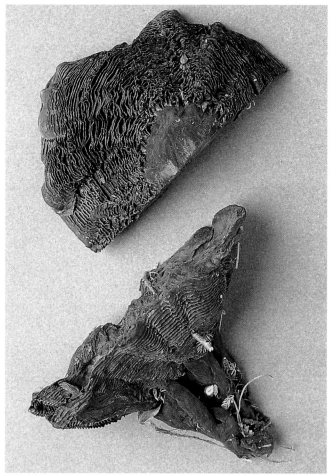

Coltricia perennis ⅔ life-size

Coltricia montagnei (previously var. *greenii*) ½ life-size

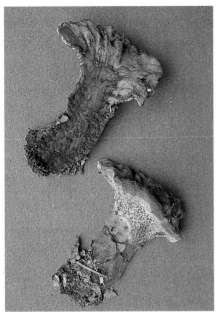

Coltricia montagnei (previously var. *montagnei*)
½ life-size

Coltricia cinnamomea ¾ life-size

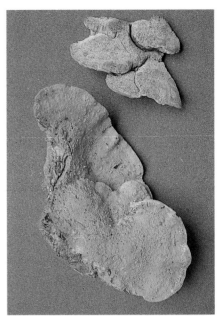

Hapalopilus nidulans ⅓ life-size

Hen of the Woods *Grifola frondosa* (Fr.) S. F. Gray **Fruit body** annual; a large clustered mass of flattened, fan- or tongue-shaped caps joined by short stems to a central base; 15–40cm (6–16in) across. **Cap** 2–8cm (¾–3in) across, overlapping each other, flat, thin (2–7mm [³⁄₃₂–¼in]), dry; pale grayish to gray-brown above; smooth to finely roughened. **Tubes** 2–3mm (³⁄₃₂–⅛in) deep, decurrent onto stem; white. **Pores** 1–3 per mm, angular; white to pale yellowish. **Stems** laterally attached to rear of caps; white; very tough, many branched from the central core. **Flesh** tough; white. **Odor** rather sharp, pungent. **Taste** pleasant when young, soon acid. **Spores** ellipsoid, smooth, 5–7 × 3.5–5μ. Deposit white. Hyphal structure dimitic; clamps present. **Habitat** on ground at base of deciduous trees, especially oak. Frequent. Found widely distributed in eastern North America, rarer elsewhere. **Season** September–November. **Edible** — delicious if the softer caps are well cooked. **Comment** This easily recognized species can often reach a weight of 5–10 pounds and a span of 25–50cm (10–20in).

Meripilus giganteus (Pers. ex Fr.) Karsten **Fruit body** 50–80cm (20–32in) across, rosettelike, consisting of numerous flattened fan-shaped caps around common base. **Cap** 5–25cm (2–10in) across, 1–2cm (½–¾in) thick, spoon-shaped with a thin, sharp margin; grayish to dull yellow, becoming smoky and dark with concentric zones of lighter and darker brown; radially grooved and covered in very fine brown scales. **Tubes** 1–5mm (¹⁄₁₆–¼in) deep; whitish bruising blackish. **Pores** small, often late in forming; cream. **Stem** very short and thick; ochre; smooth or fibrous. **Flesh** soft, fibrous; white. **Odor** pleasant. **Taste** slightly sour. **Spores** broadly ovoid to subglobose, smooth, hyaline, 5.5–6.5 × 4.5–5μ. Deposit white. **Habitat** at the base of deciduous trees or stumps, especially oak and beech. Quite common. Found in southeastern North America from Massachusetts to Missouri. **Season** July–September. **Edible.**

Coltricia perennis (Fr.) Murr. **Fruit body** annual. **Cap** up to 10cm (4in) across, circular but often blending into adjacent specimens when growing in groups; thin, wavy margin; pale cinnamon to deep brown then grayish in age, with dense, concentric bands of color; tough and leathery, becoming brittle and hard when dry; velvety with different tomentum from one color zone to another, reflecting different growth conditions. **Tubes** up to 3mm (⅛in) deep; cinnamon to rusty brown. **Pores** 2–4 per mm, angular, thin-walled, slightly decurrent; surface golden brown to cinnamon or dark brown in age. **Stem** 15–35 × 2–10mm (½–1¼ × ³⁄₃₂–½in), central; dark brown. **Flesh** 1–2mm (¹⁄₁₆–³⁄₃₂in) thick, dense; rusty brown, paler toward the cap. **Spores** ellipsoid, smooth, 6–9 × 3.5–5μ. Deposit pale yellowish brown. Hyphal structure monomitic with two types of generative hyphae. **Habitat** on the ground on paths, roadsides, clearings, edges of fires, in coniferous woods. Quite common. Found widely distributed throughout North America in conifer zones. **Season** August–November. **Not edible.**

Coltricia montagnei (Fr.) Murr. **Fruit body** annual. **Cap** up to 12cm (4¾in) across, 1–2cm (½–¾in) thick, circular or irregular, depressed toward the stem, with a wavy margin; cinnamon to deep reddish, rusty brown when older, sometimes with uneven furrowed bands of color and a paler margin in growing specimens; velvety or finely felty becoming hairy, warted, or scaly, particularly toward the center. **Tubes** up to 4mm (³⁄₁₆in) deep, rarely 8mm (⁵⁄₁₆in) near the stem. **Pores** 1–3mm, angular, often expanded and radially elongated toward the stem; in some specimens the pores join together to form pseudogills with 1–3mm (¹⁄₁₆–⅛in) between the pseudogills; pore surface cinnamon to rusty brown. **Stem** 10–40 × 5–10mm (½–1½ × ¼–½in), central or lateral, expanding toward the pore surface; cinnamon to deep rusty brown; felty to warted with smooth patches in age. **Flesh** up to 2cm (¾in) thick at center; cinnamon to rusty brown; upper part soft then corky, lower part distinctly denser. **Spores** ellipsoid, smooth, 9–14 × 5.5–7.5μ. Deposit pale brown. Hyphal structure monomitic. **Habitat** on the ground, often on footpaths and clay banks in hardwood forests. Found in eastern North America and in Oregon. **Season** July–October. **Not edible** — too tough. **Comment** In the past, this fungus has been split into two varieties. Those with the concentric pseudogills were known as

Coltricia montagnei var. *greenii* Fr., which was also known as *Cyclomyces greenii* Berk., and the poroid variety, known as *Coltricia montagnei* Fr. var. *montagnei*.

Coltricia cinnamomea (Pers.) Murr. **Fruit body** annual. **Cap** up to 5cm (2in) across, single but often joined to other fruit bodies, circular, flat to funnel-shaped, with a thin lined or slightly fringed margin; brown to deep reddish brown with concentric bands of color; dry, velvety, shiny. **Tubes** up to 2mm (³⁄₃₂in) deep, 1mm (¹⁄₁₆in) thick, narrow, pliant and fibrous; rusty to reddish brown. **Pores** 2–4 per mm, thin-walled, angular; surface reddish brown. **Stem** up to 40 × 6mm (1¾ × 1¼in), central, expanded toward base; yellowy brownish red to deep reddish brown; finely velvety. **Flesh** thin, reddish brown. **Spores** oblong to ellipsoid, smooth, 6–10 × 4.5–7μ. Hyphal structure monomitic. **Habitat** on the ground in dense masses and along paths in deciduous woods. Found in forest regions in many parts of eastern and western North America. **Season** June–November. **Edibility not known.**

Hapalopilus nidulans (Fr.) Karsten **Fruit body** annual; no stem. **Bracket** up to 10cm (4in) across, 4cm (1½in) thick at the base, convex, fan-shaped, with an acute margin; upper surface cinnamon to yellowish brown; soft and watery when fresh, becoming light and somewhat brittle; finely felty then smooth. **Tubes** up to 10mm (½in) deep; yellowish brown or white. **Pores** 2–4 per mm, thin-walled and angular; surface ochraceous to cinnamon brown, often with a few large cracks in bigger specimens. **No stem. Flesh** up to 4cm (1½in) deep; light cinnamon and distinctly lighter toward the cap; soft, fibrous, and quite brittle. **Odor** sweetish. **Spores** ellipsoid to cylindrical, smooth, nonamyloid, 3.5–5 × 2–2.5μ. Deposit white. Hyphal structure monomitic; clamps present. **Habitat** on dead deciduous wood, and on pine in the Southwest. Found commonly in eastern and southwestern North America but rarely elsewhere in the West. **Season** June–November. **Not edible. Comment** This species is easily recognized by the vivid violet reaction that occurs on all parts of the fruit body when KOH is applied.

Sulphur Shelf or **Chicken of the Woods Mushroom** *Laetiporus sulphureus* ⅓ life-size

Pycnoporellus alboluteus ⅓ life-size

the Pacific Northwest, and in California.
Season May–November. **Edible** — excellent.
Comment This is the most popular edible
polypore. Fresh young specimens are delicious
when cooked, but older specimens can be
slightly indigestible, so they are best avoided.
There is a form of this fungus which has a
white pore surface, and some authors recognize
this as *Laetiporus sulphureus* var. *semialbinus*.

Pycnoporellus alboluteus (Ellis & Ev.) Kotlaba &
Pouz. **Fruit body** annual; spongelike to
cushion-shaped, 5–100cm (2–40in) across, with
pores splitting into long spinelike teeth; bright
orange-ochre to paler whitish yellow with age.
Tubes 2cm (¾in) deep; same color as flesh.
Pores 1–3cm (½–1¼in) across. **No stem.**
Flesh thin; orange, turns red in KOH. **Spores**
cylindrical, 6–9 × 2.5–4μ. Deposit white.
Habitat on undersides of conifer logs. Found in
the Pacific Northwest. **Season** July–October.
Edibility not known.

Dye Polypore *Phaeolus schweinitzii* (Fr.) Pat.
syn. *Polyporus schweinitzii* Fr. **Fruit body**
sometimes an irregular cushion, but usually
subcircular, 10–30cm (½–1¼in) across, with a
short, thick stalk; deep sulphur yellow
becoming rust- or dark brown, finally blackish
when old; soft and spongy, upper surface
concave, rough and hairy or woolly, concentrically grooved at first. **Tubes** 3–6mm
(⅛–¼in) deep, decurrent; concolorous with the
upper surface. **Pores** 0.3–2.5mm (¹⁄₃₂–⅛in)
across, circular to angular; yellow then olive,
finally brown to maroon, glistening in the light.
Stem brown, very short, thick, merging into
cap and with the tubes decurrent down it.
Flesh rust-brown; fibrous, soft then fragile and
light when dry. **Spores** ovate-ellipsoid,
5.5–7.5 × 3.5–4μ. Deposit white to yellowish.
Hyphal structure monomitic; generative
hyphae lacking clamps. **Habitat** parasitic on
conifers, usually growing from the roots. Found
over most of North America. **Season**
June–November. **Not edible.**

Pycnoporus cinnabarinus (Fr.) Karsten **Fruit body**
3–10cm (1¼–4in) across, semicircular to
kidney-shaped; bright cinnabar red to orange;
upper surface roughened, wrinkled. **Tubes**
1–6mm (¹⁄₁₆–¼in) deep; cinnabar red. **Pores**
fine, angular; concolorous with cap. **No stem.**
Flesh tough, leathery; black with KOH.
Spores sausagelike, smooth, 5–6 × 2–2.5μ.
Deposit white. **Habitat** on fallen timber,
mainly oak. Found over much of North
America. **Season** all year. **Not edible.**

Beefsteak Fungus or **Ox Tongue** *Fistulina
hepatica* Schaeff. ex Fr. **Bracket** 8–25cm
(3–10in) across, 2–6cm (¾–2¼in) thick,
usually single, tongue-shaped or semicircular;
upper surface pinkish to orange-red and finally
purple-brown; rough with rudimentary pores,
especially toward the margin; moist or tacky.
Tubes up to 15mm (½in) deep; arising free,
but adhering in maturity; whitish or yellowish.
Pores 3 per mm, circular; whitish at first,
bruising reddish brown. **Stem** none or
rudimentary; short, thick, blood red. **Flesh**
thick, succulent; mottled, dark flesh-pink with
lighter veining, with bloodlike sap; reminiscent
of raw meat. **Odor** pleasant. **Taste** sourish.
Spores ovoid, smooth, 4.5–6 × 3–4μ. Deposit
pinkish salmon. **Habitat** singly or sometimes
several in a cluster on the base of living oaks
or chestnuts, also dead hardwood stumps.
Frequent; common in the East. Found
particularly in eastern North America. **Season**
July–October. **Edible** — good. **Comment**
Infected oak timber has a much richer, darker
color and is much sought after by furniture
makers.

Sulphur Shelf or **Chicken of the Woods
Mushroom** *Laetiporus sulphureus* (Bull. ex Fr.)
Murr. **Fruit body** annual. **Bracket** up to 40cm
(16in) across; flat, fan-shaped or irregularly
semicircular, thick and fleshy, often with an
undulating rounded margin; lemon yellow to
orange, fading to straw or pale brown and
bleaching white in old, deteriorating
specimens; minutely felty to smooth, radially
furrowed. **Tubes** up to 4mm (³⁄₁₆in) deep,
distinct; sulphur yellow drying pale buff. **Pores**
3–4 per mm, angular; surface sulphur yellow,
fading to pale tan on drying; thin dissepiments
that soon lacerate. **No stem** or with a lateral
stemlike attachment. **Flesh** up to 2cm (¾in)
thick; white; succulent and exuding a yellowish
juice when squeezed, drying crumbly or
chalky. **Odor** fungusy. **Taste** pleasant. **Spores**
ovoid to ellipsoid, smooth, 5–8 × 4–5μ. Deposit
white. Hyphal structure dimitic. **Habitat** singly
or in large overlapping clusters on stumps,
trunks, and logs, of dead or living deciduous
and coniferous trees. Found in many areas of
North America from the Midwest eastward, in

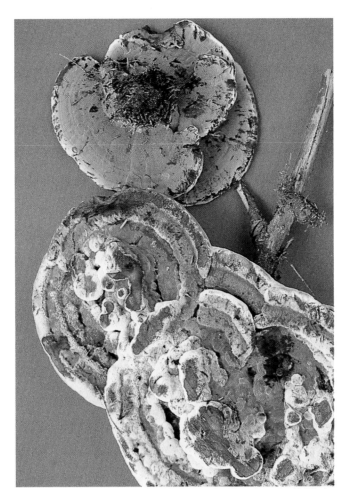

Laetiporus sulphureus (var. *semialbinus*) ⅓ life-size

Dye Polypore *Phaeolus schweinitzii* ⅓ life-size

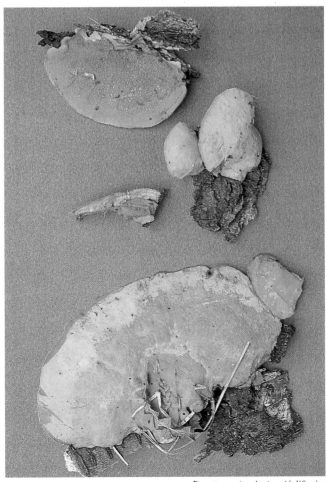

Pycnoporus cinnabarinus ½ life-size

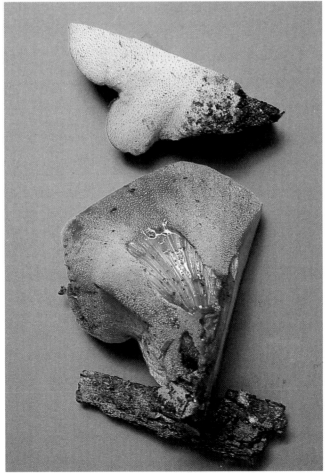

Beefsteak Fungus or **Ox Tongue** *Fistulina hepatica* ½ life-size

Pine Conk *Phellinus pini* ¾ life-size

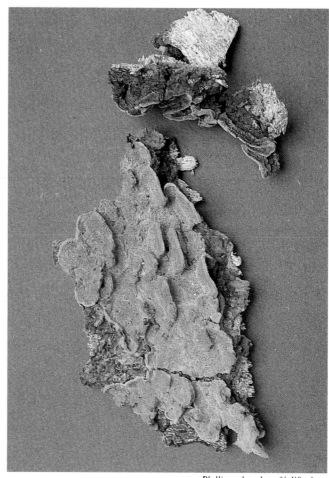

Phellinus chrysoloma ¾ life-size

Pine Conk *Phellinus pini* (Fr.) Ames **Bracket** 2–20cm (¾–8in) across, 1–15cm (½–6in) thick; hoof-shaped, fan-shaped, or shelflike; tawny to dark reddish brown or brownish black in age, with the margin often brighter; hard, crusty, rough or cracked, minutely hairy, generally curved. **Tubes** up to 6mm (¼in) deep. **Pores** circular to angular; dingy yellow-tawny. **Stem** minute or none. **Flesh** tough; tawny to tan or ochre. **Spores** globose or subglobose, smooth, 4–6 × 3.5–5μ. Deposit brown. **Habitat** singly or in rows on living or recently dead coniferous trunks. Common. Widely distributed in North America. **Season** perennial. **Not edible. Comment** A very destructive fungus that attacks the heartwood of living trees, resulting in "conk rot" causing more timber loss than any other fungus.

Phellinus chrysoloma (Fr.) Donk **Bracket** 1–8cm (½–3in) across, thin, flat, and crustlike, with a sharp margin; orangy brown to russet brown with a paler margin, becoming dark brown to blackish; concentric rings and hairy ridges. **Tubes** up to 5mm (¼in) deep, accumulating layers annually. **Pores** 2–5mm (³⁄₃₂–¼in), angular or elongated; bright ochre-tawny. **No stem. Flesh** 1–3mm (¹⁄₁₆–⅛in) thick; tawny or dingy yellow; separated from cap hair by black line. **Spores** subglobose, smooth, 4.5–5.5 × 4–5μ. Deposit light brown. Setae present. **Habitat** in dense overlapping clusters or partly fused rows on decaying or living conifer trunks. Found in most of North America. **Season** sometimes perennial. **Not edible.**

Hemlock Varnish Shelf *Ganoderma tsugae* Murr. **Fruit body** annual. **Cap** 5–25cm (2–10in) across, kidney- or fan-shaped; reddish

to maroon or brownish, margin often white or yellow; surface smooth to wrinkled with a shiny, lacquered appearance. **Tubes** up to 1.5cm (½in) deep; pale purplish brown. **Pores** 5–6 per mm, circular to angular; surface cream-colored bruising yellowish or brownish. **Stem** up to 30–150 × 15–30mm (1¼–6 × ½–1¼in), usually lateral; reddish brown to mahogany or almost black; surface has a highly varnished crust. **Flesh** up to 5cm (2in) thick, upper part soft and spongy, lower part corky when dried; creamy-colored to buff, with a darker layer next to tubes. **Spores** ellipsoid, blunt at one end, with thick double wall, 9–11 × 6–8μ. Deposit rust-brown. Hyphal structure dimitic; clamps present. **Habitat** singly or in clusters on dead or dying coniferous wood, especially hemlock and spruce. Found in eastern North America, California, and Arizona. **Season** May–November. **Comment** The similar *Ganoderma lucidum* (Curt. ex Fr.) Karsten grows on deciduous trees. *Ganoderma curtisii* (Berk.) Murr., which occurs in the South, is probably not a distinct species but a form of *Ganoderma lucidum*. It has a distinct, usually central stem, and the cap is pallid or bright ochre rather than red-brown.

Ganoderma oregonense Murr. **Fruit body** annual; no stem or very small lateral stemlike attachment. **Bracket** up to 100cm (40in) across, 40cm (16in) wide, 20cm (8in) thick, semicircular or fan-shaped; upper surface yellowy brown to dark reddish brown or mahogany, with concentric bands of color; quickly develops a shiny, waxy crust that cracks extensively on older specimens. **Tubes**

up to 3cm (1¼in) deep; pale purplish brown. **Pores** 2–3 per mm, circular to angular; surface cream-colored bruising brown or purplish brown. **Flesh** up to 15cm (6in) thick, soft-fibrous; cream-colored to pale buff. **Spores** ellipsoid, blunt at one end, with a thick double wall, 13–17 × 8–10μ. Deposit rusty brown. Hyphal structure trimitic; clamps present. **Habitat** singly or occasionally overlapping on dead conifers or conifer stumps. Found in the Pacific Northwest. **Season** all year. **Not edible. Comment** Gilbertson has remarked that *Ganoderma oregonense* may be found to be a large form of *Ganoderma tsugae* (above) when further work is carried out.

Ischnoderma resinosum (Fr.) Karsten **Fruit body** annual; no stem, broadly attached, or with a small stemlike attachment. **Bracket** up to 15cm (6in) across, 12cm (4¾in) wide, 3cm (1¼in) thick at base, semicircular, with a thick rounded or lobed margin; at first fleshy and exuding resin, then hard and brittle; upper surface ochre to rusty brown to blackish; finely felty and fairly smooth, becoming concentrically zoned and ridged and wrinkled with a glossy black resinous crust. **Tubes** up to 10mm (½in) deep; brownish. **Pores** 4–6 per mm, angular to round; surface creamy white, darker when touched, later pale brown. **Flesh** up to 10mm (½in) thick at base, soft becoming harder; whitish drying yellowish brown to pale cinnamon. **Spores** cylindrical, smooth, 5–7 × 1.5–2μ. Deposit white. Hyphal structure dimitic; clamps present. **Habitat** singly or occasionally overlapping on logs and stumps of hardwoods. Found throughout North America. **Season** September–October. **Not edible.**

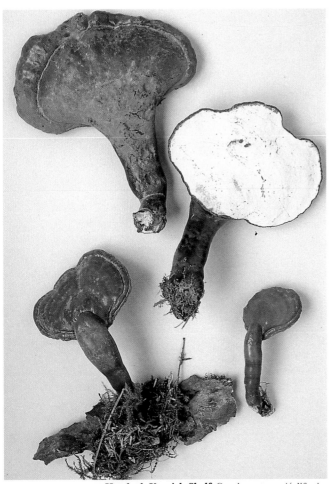

Hemlock Varnish Shelf *Ganoderma tsugae* ¼ life-size

Ganoderma oregonense ⅓ life-size

Ganoderma curtisii ½ life-size

Ischnoderma resinosum ⅓ life-size

Hoof Fungus or **Tinder Fungus** *Fomes fomentarius* ½ life-size

Phellinus igniarius ⅔ life-size

Fomitopsis cajanderi ½ life-size

Fomitopsis pinicola ⅓ life-size

Hoof Fungus or **Tinder Fungus** *Fomes fomentarius* (L. ex Fr.) Kickx **Fruit body** perennial; no stem. **Bracket** up to 15cm (6in) across, hoof-shaped; gray to gray-brown or gray-black with dark zones, light brown near margin; tough, woody, concentrically grooved, minutely velvety with a hard, smooth crust. **Tubes** 0.5–6cm (¼–2¼in) deep; very thick, not distinctly layered; light brown, becoming stuffed with white mycelium. **Pores** 4–5 per mm, circular; surface pale brown, darkening when handled. **Flesh** up to 1cm (½in) thick, tough, fibrous; yellowish brown. **Spores** cylindrical, smooth, 12–18×4–7μ. Deposit lemon yellow. Hyphal structure trimitic. **Habitat** on dead or living hardwood trees, including birch, beech, maple, and poplar. Found throughout northern North America. **Season** all year. **Not edible. Comment** In Europe, before matches were invented, this fungus was ground to a powder and used in tinderboxes.

Phellinus igniarius (Fr.) Quél. syn. *Fomes igniarius* (L. ex Fr.) Gillet **Bracket** 5–25cm (2–10in) across, 1–5cm (½–2in) wide, 2–12cm (¾–4¾in) thick, hoof-shaped to convex or semicircular, and flat with a rounded margin; rusty brown when young, later gray, and finally black; concentrically ridged, becoming smooth then cracking, margin stays velvety for a long time. **Tubes** 3–5mm (⅛–¼in) deep in each annual layer; rusty brown. **Pores** 4–6 per mm, circular; rusty cinnamon to grayish brown. **No stem. Flesh** hard, woody; rusty brown. **Odor** mushroomy. **Taste** sour or bitter. **Spores** more or less globose, smooth, thick-walled, 5–7×4.5–6μ. Deposit whitish. Setae present. **Habitat** singly or in small groups on hardwood trunks, especially birch or aspen. Sometimes common. Found widely distributed in North America. **Season** all year. **Not edible. Comment** This parasitic fungus can be very destructive, causing intensive white rot that reduces its host to a spongy mass.

Fomitopsis cajanderi (Karsten) Kotlaba & Pouz. **Fruit body** perennial; no stem. **Bracket** up to 20cm (8in) across, 10cm (4in) wide, 7cm (2¾in) thick, flat to convex, with a sharp margin; light pinky-beige becoming darker pinkish brown or gray to blackish; finely felty becoming hairy or smooth. **Tubes** 1–3mm (1/16–⅛in) deep per season; pale pinky-brown; layers stratified, up to 2cm (¾in) thick. **Pores** 4–5 per mm, circular to angular; surface rose-colored. **Flesh** up to 1cm (½in) thick, corky; rosy pink to light pinkish brown. **Spores** cylindrical, curved, smooth, 5–7×1.5–2μ. Deposit whitish. Hyphal structure dimitic; clamps present. **Habitat** singly or overlapping on dead conifer wood; rarely on hardwoods. Found throughout North America. **Season** all year. **Edibility not known.**

Fomitopsis pinicola (Swartz ex Fr.) Karsten **Fruit body** perennial; no stem. Up to 38cm (15in) across, 20cm (8in) wide, 15cm (6in) thick, convex to hoof-shaped, with a thickened, rounded margin; upper surface with a sticky reddish-brown resinous crust, then grayish to

brown or black; hard, woody, smooth or glossy-looking. **Tubes** up to 6mm (¼in) deep per season; cream to buff. **Pores** 5–6 per mm, circular; surface cream-colored. **Flesh** up to 12cm (4¾in) thick, corky, hard, woody; cream to buff, sometimes zoned. **Spores** cylindrical-ellipsoid, smooth, 6–9×3.5–4.5μ. Deposit whitish. Hyphal structure trimitic; clamps present. **Habitat** on dead conifer stumps and logs and occasionally on living trees. Found throughout most of North America except the South from Texas eastward. **Season** all year. **Not edible. Comment** The most commonly collected polypore in North America. The cap colors are rather variable.

Phellinus ferruginosus (Fr.) Pat. **Fruit body** 4–15cm (1½–6in) across, 0.5–1.5cm (¼–½in) thick, growing closely attached to the wood, spreading widely and irregularly with little swellings along the margin; cinnamon to rusty brown; leathery and flexible. **Tubes** 1–3mm (1/16–⅛in) deep in each layer, up to 10mm (½in) in total; tawny or umber-brown. **Pores** 5–6 per mm, round and entire on horizontal parts, split elsewhere; surface even, with little swellings or slightly wavy; tawny to umber-brown with yellowish-brown margin in growing specimens. **No stem. Flesh** irregular, thickness up to 5mm (¼in), loose, cottony; cinnamon to rusty brown. **Spores** broadly ellipsoid, nonamyloid, 4–5×3–3.5μ. **Habitat** on dead hardwoods. Found in northern North America. **Season** annual to perennial. **Not edible.**

Phellinus nigricans (Fr.) Karsten **Bracket** 4–14cm (1½–5½in) across, 2–9cm (¾–3½in) wide, 2–6cm (¾–2¼in) thick at base, semicircular to elongated, triangular in section; zones of whitish, gray, or light cinnamon and concentrically grooved black zones; with a distinct crust, except on the margin, old parts deeply cracked, base often mossy. **Tubes** up to 5cm (2in) deep in layers; cinnamon to rusty brown, or filled with white mycelium in old parts. **Pores** 5–6 per mm, round; cinnamon to deep rusty brown or gray when weathered. **No stem. Flesh** up to 10mm (½in) thick, dense; rusty brown. **Spores** globose to subglobose, smooth, nonamyloid, 6–7×5.5–6.5μ. **Habitat** on deciduous wood, especially beech. Found in northern North America. **Season** perennial. **Not edible.**

Inonotus obliquus (Fr.) Pilát **Fruit body** sterile conk 25–40cm (10–16in) across; black; deeply cracked, very hard and brittle when dry. Fertile portion 5mm (¼in) thick, crustlike, thin; dark brown. **Tubes** 3–10mm (⅛–½in) deep, brittle, usually split in front. **Pores** 6–8 per mm, circular; whitish becoming dark brown. **No stem. Flesh** corky, faintly zoned; bright yellowish brown. **Spores** broadly ellipsoid to ovoid, smooth, 9–10×5.5–6.5μ. Setae present. **Habitat** beneath the bark or outer layers of wood on living, dead, standing, or fallen trees, erupting into conspicuous black conks, generally on birch, elm, and alder. Found in northern North America. **Season** all year. **Not edible.**

Phellinus ferruginosus ⅓ life-size

Phellinus nigricans ⅓ life-size

Inonotus obliquus ½ life-size

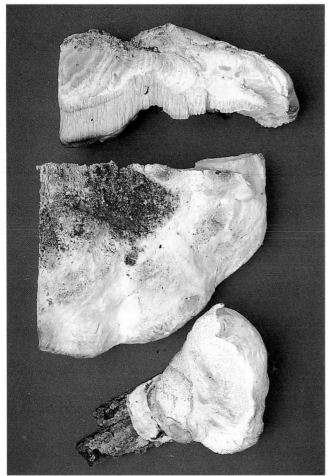

Tyromyces fissilis ⅓ life-size

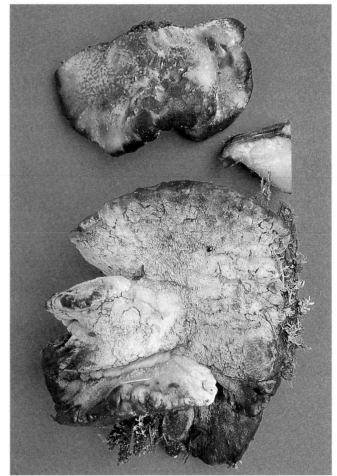

Oligoporus fragilis ½ life-size

Oxyporus populinus ½ life-size

Inonotus radiatus ¾ life-size

Tyromyces fissilis (Berk. & Curt.) Donk **Bracket** 6–30cm (2¼–12in) across, 4–10cm (1½–4in) wide, 1–7cm (½–2¾in) thick, semicircular; whitish, discoloring or drying yellowish then reddish; almost hairy. **Tubes** 4–10mm (³⁄₁₆–½in) deep. **Pores** 1–3 per mm, angular; white or pale flesh-pink, bruising yellowish then deep reddish brown and becoming waxy on drying. **No stem. Flesh** fibrous, tough, zoned, drying hard; white to pale flesh-colored. **Odor** strong, cheeselike. **Taste** bitter or astringent. **Spores** ovoid or ellipsoid, smooth, 3.5–4.5×3–3.5μ. **Habitat** growing on the wounds of living hardwoods. Found in eastern North America, west to Ohio. **Season** July–August. **Not edible.**

Oligoporus fragilis (Fr.) Gilbertson & Ryv. syn. *Tyromyces fragilis* (Fr.) Donk **Bracket** 2–10cm (¾–4in) across, semicircular, fan-shaped, or elongated in outline; white becoming reddish or pinkish brown in age, bruising yellow then reddish when handled; finely hairy, becoming matted in age. **Tubes** 2–8mm (³⁄₃₂–⁵⁄₁₆in) deep. **Pores** 2–4 per mm, angular to irregular; white bruising yellowish then rusty red. **No stem. Flesh** soft when fresh; white then reddish, bruising yellowish. **Odor** strong. **Spores** cylindrical to sausage-shaped, smooth, 4–5×1–2μ. Deposit whitish. **Habitat** singly or in overlapping groups or fused clusters on decaying conifer wood. Quite common. Found widely distributed in northeastern North America, south to North Carolina, west to Colorado, and in California. **Season** September–October (later in the West). **Not edible.**

White Cheese Polypore *Tyromyces chioneus* (Fr.) Karsten syn. *Tyromyces albellus* Pk. **Bracket** 2–12cm (¾–4¾in) across, 2–8cm (¾–3in) wide, 1–3.5cm (½–1¼in) thick; fan-shaped to subcircular, succulent; upper surface white or buff then yellowing; suedelike. **Tubes** 3–7mm (⅛–¼in) deep; white, yellowing with age. **Pores** 3–4 per mm, circular to elongate; whitish or yellowish. **No stem. Flesh** soft and watery at first, drying hard and chalky; white. **Odor** fragrant when fresh. **Taste** nasty, soapy. **Spores** subcylindrical to sausage-shaped, smooth, 4–6×1.5–2μ. Deposit white. **Habitat** singly or in groups on dead hardwoods and occasionally on conifers. Common. Found widely distributed in North America. **Season** July–November (later in California). **Not edible.**

Oligoporus caesius (Schrad.: Fr.) Gilbertson & Ryv., syn. *Tyromyces caesius* (Schrad. ex Fr.) Murr. **Bracket** 1–8cm (½–3in) across, 1–4cm (½–1½in) wide, 0.3–1cm (⅛–½in) thick, convex to flat, semicircular, with a sharp margin; upper surface whitish, becoming blue-gray with age or hardening; covered in fine long hairs. **Tubes** 2–8mm (³⁄₃₂–⁵⁄₁₆in) deep; white, later gray-blue. **Pores** 2–4 per mm, circular or angular to elongated; white, becoming grayish or gray-blue with age. **No stem. Flesh** soft, spongy, watery when fresh; white, gray or yellowish in age. **Odor** often fragrant. **Taste** soapy. **Spores** sausage-shaped, smooth, amyloid, 4–5×0.7–1μ. Deposit pale

blue or whitish with a gray-blue tinge. **Habitat** singly or in overlapping groups on dead wood. Frequent. Found throughout North America. **Season** August–November. **Not edible.**

Oxyporus populinus (Schum. ex Fr.) Donk syn. *Fomes connatus* (Weinm.) Gillet **Bracket** 3–6cm (1¼–2¼in) across, 2–3cm (¾–1¼in) wide, 1–4cm (½–1½in) thick, semicircular to elongated in tiers; upper surface whitish gray to pale gray-buff, often with an ochraceous tint, frequently tinged green from algae or moss; uneven, finely and densely hairy becoming smooth to warty, often overgrown with moss. **Tubes** 2–5mm (³⁄₃₂×¼in) deep, up to 4cm (1½in) deep in old specimens, distinct new layer each season; whitish at first, then straw yellow. **Pores** 4–7 per mm, circular or slightly angular; whitish. **No stem. Flesh** watery, firm; white to ochre. **Odor** slightly fungusy. **Spores** subglobose, smooth, 3.5–4.5×2.5–4μ. Deposit white. **Habitat** on trunks or in wounds of hardwood trees, especially living maple. Frequent. Found in eastern North America, south to Georgia and east to Missouri. **Season** all year. **Not edible.**

Inonotus radiatus (Sow. ex Fr.) Karsten **Fruit body** annual; no stem but with a stalklike attachment. **Bracket** up to 5cm (2in) across, 3cm (1¼in) wide, 1.5cm (½in) thick, woody with a thin margin; upper surface yellowy brown, then rusty becoming blackish with age, with concentric bands of color; finely felty, then smooth and radially wrinkled. **Tubes** up to 7mm (¼in) deep; whitish within, tube layer darker brown. **Pores** 2–5 per mm, angular; surface pale yellowish brown, becoming darker brown with age, glancing silvery in the light; with hairy dissepiments that become thin and lacerate. **Flesh** up to 1cm (½in) thick; yellowish brown to reddish brown, lustrous. **Odor** faint and sweet. **Taste** bitter. **Spores** ellipsoid to ovoid, smooth, 5–6.5×3–4.5μ. Deposit yellowish. Hyphal structure monomitic. Setae in tubes. **Habitat** singly or in overlapping clusters on a large number of hardwood trees. Found in eastern North America, south to the southern Appalachians, and in the Pacific Northwest. **Season** all year. **Not edible.**

Inonotus tomentosus (Fr.) Teng **Fruit body** annual. **Bracket** up to 11cm (4½in) across, circular to fan-shaped, sometimes lobed, with a central depression; yellowish brown to wood brown, sometimes with faint concentric bands of color; felty. **Tubes** up to 3mm (⅛in) deep, decurrent; white stuffed and paler than flesh. **Pores** 2–4 per mm, angular; surface pale buff becoming darker brown; with thick entire dissepiments that become thin and lacerate with age. **Stem** 35×15mm (1¼×½in), central or lateral; dark brown. **Flesh** up to 4mm (³⁄₁₆in) thick, soft, spongy upper layer, firm fibrous lower layer; yellowish brown. **Spores** ellipsoid, smooth, 5–6×3–4μ. Hyphal structure monomitic. **Habitat** singly or several branching from a common base on coniferous wood. Found throughout most of North America except the Midwest. **Season** all year. **Not edible.**

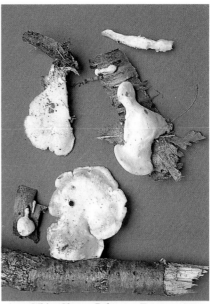

White Cheese Polypore *Tyromyces chioneus* ⅓ life-size

Oligoporus caesius ½ life-size

Inonotus tomentosus ⅓ life-size

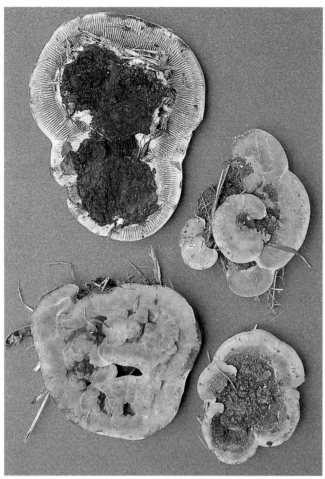

Gloeophyllum sepiarium ⅔ life-s ze

Daedalea quercina ⅓ life-size

Blushing Bracket *Daedaleopsis confragosa* ⅓ life-size

Lenzites betulina ½ life-size

Birch Polypore *Piptoporus betulinus* ⅓ life-size

Spongipellis pachyodon ¼ life-size

Trichaptum biforme ⅓ life-size

Gloeophyllum sepiarium (Fr.) Karsten **Fruit body** annual to perennial; no stem. **Bracket** up to 12cm (4¾in) across, 7cm (2¾in) wide, 0.5–1cm (¼–½in) thick at base of cap, fan-shaped with a sharp, slightly wavy margin, corky; upper surface bright yellowish brown, then darker reddish brown, and finally grayish to black; softly hairy at first, then bristly, coarsely concentrically ridged and radially wrinkled, reflecting different stages of growth. **Pores** 15–20 per cm, densely and radially arranged, often fusing together irregularly, giving a gill-like appearance; light golden brown becoming umber-brown. **Flesh** up to 5mm (¼in) thick; dark brown. **Spores** cylindrical, smooth, 9–13 × 3–5μ. Deposit white. Hyphal structure trimitic; clamps present. **Habitat** often overlapping in clusters from a common base on dead coniferous wood and often on processed boards. Found widely distributed throughout North America. **Season** June–November. **Not edible.**

Daedalea quercina Fr. **Fruit body** perennial; no stem. **Bracket** up to 20cm (8in) across, 15cm (6in) wide, 8cm (3in) thick, single or occasionally in shelved groups, semicircular, upper surface flat to slightly convex with a smooth, thick margin; creamy or yellow-brown tinged with gray becoming black; corky and woody, smooth or finely velvety, some specimens with scattered warts. **Tubes** up to 40mm (1½in) deep; ochraceous cream on the inner walls. **Pores** 1–4mm (¹⁄₁₆–³⁄₁₆in) wide, walls 1–3mm (¹⁄₁₆–⅛in) thick, large, irregular and mazelike, often elongate resembling gills; surface ochraceous. **Flesh** up to 1cm (½in) thick; ochraceous to pale wood-colored with very faint annual zones. **Odor** faintly acrid or mushroomy. **Spores** ellipsoid to cylindrical, smooth, 5.5–6 × 2.5–3.5μ. Deposit white. Hyphal structure trimitic. **Habitat** on dead oak and other deciduous trees. Found in eastern North America and occasionally in Oregon and California. **Season** all year. **Not edible.**

Blushing Bracket *Daedaleopsis confragosa* (Bolton ex Fr.) Schroet. **Fruit body** annual; no stem. **Bracket** up to 12cm (4¾in) across, single or tiered, convex to flat, with a thin, acute margin; buff to light brown, with concentric bands of color; hairy to smooth.

Tubes up to 10mm (½in) deep, continuous with flesh and same color. **Pores** up to 1mm (¹⁄₁₆in) across, variable, circular or elongated and slotlike, mazelike, or gill-like; surface whitish, readily bruising pink to red-brown when fresh. **Flesh** up to 20mm (¾in) thick, firm, corky; pale buff to brown. **Odor** none. **Taste** slightly bitter. **Spores** cylindrical, slightly curved, smooth, 9–11 × 2–2.5μ. Deposit white. Hyphal structure trimitic. **Habitat** on dead wood or in the wounds of living trees in hardwood forests. Common throughout eastern North America and found in the Northwest and Southwest, but infrequent in western North America. **Season** June–December, but can be found all year as fruit bodies persist for several years. **Not edible.**

Lenzites betulina (Fr.) Fr. **Fruit body** annual; no stem; broadly attached or with a small stemlike attachment. **Bracket** up to 8cm (3in) across, 5cm (2in) wide, 2cm (¾in) thick, flat, semicircular or fan-shaped, with an even or lobed margin; upper surface white, cream, grayish, or brownish, older specimens often have a green tinge because of algae growing in the fine hairs; tough and leathery with an uneven surface, concentrically grooved, zoned and hairy. **Gills** forked and fused together in places; white then cream to yellowish brown; undulating or flexuous. **Flesh** 1–2mm (¹⁄₁₆–³⁄₃₂in) thick, thin, fibrous; white, lighter than the gills. **Spores** subcylindrical, slightly curved, smooth, 5–6 × 2–3μ. Deposit white. Hyphal structure trimitic; clamps present. **Habitat** singly or in overlapping groups on hardwoods and coniferous wood. Common. Found in midwestern and eastern North America, the Pacific Northwest, and California, but extremely rare elsewhere. **Season** July–November. **Not edible.**

Birch Polypore *Piptoporus betulinus* (Bull. ex Fr.) Karsten syn. *Polyporus betulinus* Bull. ex Fr. **Bracket** 5–25cm (2–10in) across, 2–6cm (¾–2¼in) thick, subglobose at first, expanding to hoof-shaped with a thick and rounded margin; whitish when young, darkening to fleshy gray-brown with age; tough, smooth, with a thin separable skin. **Tubes** 1–5mm

(¹⁄₁₆–¼in) deep; white when fresh, becoming grayish brown. **Pores** 3–4 per mm, circular; white at first, later pale gray-brown. **Stem** rudimentary, stubby, or none. **Flesh** up to 3cm (1¼in) thick, rubbery; white. **Odor** strong and pleasant. **Taste** slightly bitter. **Spores** cylindrical to bean-shaped, smooth, 3–6 × 1.5–2μ. Deposit white. **Habitat** singly or in groups on dead or living birch. Common. Found in northern North America. **Season** all year, annual. **Not edible. Comment** This species has been used in a variety of ways; as tinder, as a razor strop, as an anesthetic, and for mounting insect specimens.

Spongipellis pachyodon (Pers.) Kotlaba & Pouz. **Fruit body** 5–20cm (2–8in) across, crustlike or spreading on surface of logs, often many smaller patches or caps fused together into sheets. If fruit body is projecting, caplike, then upper surface is finely tomentose, cream-colored; lower surface with toothlike or flattened-spiny projections up to 12mm (½in) long, white to cream color. **Flesh** tough, corky; white. **Odor** mild. **Taste** mild. **Spores** oval, smooth, 5.5–7.5 × 5–6μ. Deposit white. **Habitat** on fallen logs of maple, oak, and beech. Found widely distributed throughout North America. **Season** July–October. **Not edible.**

Trichaptum biforme (Fr. in Klotzch) Ryv. **Bracket** 1–8cm (½–3in) across, semicircular, fan-shaped, flat; color variable in concentric zones, ochre to dark brown, white to grayish, brownish or black, violet margins; hairy becoming smooth. **Tubes** 1–10mm (¹⁄₁₆–½in) deep. **Pores** 2–5 per mm, angular, becoming toothlike; white to brownish with mauve tinge and mauve along the margin. **No stem. Flesh** 0.5–1.5cm (¼–½in) thick; white to yellow. **Spores** cylindrical, smooth, 5–6.5 × 2–2.5μ. Deposit white. **Habitat** numerous, single, or overlapping caps on dead stumps of trees of deciduous wood, reducing them to sawdust. Very common. Found widely distributed throughout North America. **Season** May–December, but often persisting all year. **Not edible.**

Climacocystis borealis ½ life-size

False Turkey-tail *Stereum ostrea* ⅔ life-size

Climacocystis borealis (Fr.) Kotlaba & Pouz.
Fruit body annual; no stem. **Bracket** 15cm
(6in) across, 8cm (3in) wide, 4cm (1½in)
thick, fan-shaped to broadly stalkless, flat and
semicircular, often overlapping; white to cream
or straw-colored when fresh, becoming darker
when dry; soft and watery when fresh, dry,
brittle and often with radial lines when dry;
felty to hairy, becoming partly smooth and
partly covered in stiff hairs when dry. **Tubes**
up to 5mm (¼in) deep; same color as pores.
Pores 1–2 per mm, thin-walled, angular; white
to cream or light straw. **Flesh** duplex, with a
lower dense layer up to 2cm (¾in) thick;
whitish. **Taste** mild. **Spores** broadly ellipsoid,
smooth, thin-walled, 4.5–6.5 × 3–4.5μ. Hyphal
structure monomitic. **Habitat** on dead or living
conifers and rarely on deciduous wood. Often
abundant. Found widely distributed in
coniferous forests in North America, excluding
the southern pine region. **Season** August–
November. **Not edible.**

False Turkey-tail *Stereum ostrea* (Blume &
Nees ex Fr.) Fr. **Cap** 1–10cm (½–4in) across,
irregular, semicircular or bracketlike, often
overlapping; zonate and multicolored in shades
of brown and rust; densely hairy. **Fertile
undersurface** smooth, pale buff. **Flesh** thin,
tough, leathery. **Spores** cylindrical, smooth,
5.5–7.5 × 2–3μ. Deposit white. **Habitat** on logs
and stumps of hardwoods, especially oak.
Found throughout North America. **Season** all
year. **Not edible.**

Cerrena unicolor (Bull. ex Fr.) Murr. **Cap**
0.5–7.5cm (¼–2¾in) across, starts as a
spreading resupinate, then forms crowded,
often overlapping brackets; white to gray; thin;
with dense covering of stiff hairs, often zoned

and with covering of algae on upper surface.
Tubes 0.4–4mm (¹⁄₃₂–³⁄₁₆in) deep, mazelike,
becoming toothlike or more rarely porelike;
whitish. **Pores** 2–3 per mm, white to gray.
Flesh thin, tough; white. **Spores** ellipsoid,
smooth, 4.5–5.5 × 2.5–3.5μ. Deposit white.
Habitat on wood, mostly deciduous. Found
throughout northern North America into the
Southeast and Midwest. **Season** all year. **Not
edible.**

Cystostereum murraii (Berk. & Curt.) Pouzar
Fruit body grows firmly attached along a
vertical surface in crusty patches up to 20cm
(8in) across; upper surface has edges turned a
few mm outward and distinctly bounded;
upper surface whitish to grayish or ochre-
white, spotting slightly brownish when bruised,
lower surface brown; upper surface uneven,
warty, with concentric ribs and cracks when
dry. **Flesh** hard, tough, crustlike. **Odor**
agreeable, like coconut (when fresh), odorless
when dry. **Spores** ellipsoid, smooth, 4.5–5 ×
2–3μ. **Habitat** growing on dead trunks
and branches of hardwoods at higher
elevations. Found in New York and
presumably other states. **Season** all year.
Not edible.

Bjerkandera adusta (Willd. ex Fr.) Karsten syn.
Polyporus adustus Willd. ex Fr. **Bracket** 2–4cm
(¾–1½in) across, 1–2cm (½–¾in) wide,
1–6mm (¹⁄₁₆–¼in) thick, overlapping,
undulate; upper surface gray-brown with a
white margin when young, becoming darker
and blackening at the margin with age;
suedelike; leathery, drying hard. **Tubes**
1–2mm (¹⁄₁₆–³⁄₃₂in) deep; gray. **Pores** minute
(5–7 per mm), subcircular; smoke gray,
darkening with age. **Flesh** pale buff with a thin

layer of matted hair up to 6mm (¼in) thick.
Odor strongly fungusy. **Taste** sourish. **Spores**
ellipsoid to almost cylindrical, smooth,
5–6 × 2.5–3.5μ. Deposit white. Hyphal
structure monomitic; generative hyphae with
clamp connections. **Habitat** on dead logs and
slash of deciduous and coniferous wood; often
several fused together in tiers or in an
overlapping group which may be 20cm (8in)
across. Common. Found widely distributed
throughout North America. **Season**
July–November. **Not edible.**

Gloeophyllum protractum (Fr.) Imaz. **Fruit body**
annual or possibly perennial; no stem; broadly
attached along fallen logs. **Bracket** up to 10cm
(4in) across, 4cm (1½in) wide, 0.5–1.5
(¼–½in) thick at base, somewhat triangular in
section, with a rather sharp edge; yellowish
brown to brown when young, becoming
grayish to black in age; upper surface distinctly
zoned and grooved, smooth and somewhat
glossy at first becoming lined, uneven and a bit
cracked near the base; tough and leathery.
Tubes up to 10mm (½in) deep, with a light
bloom. **Pores** 1–2 per mm, up to 3 × 1mm
(⅛–¹⁄₁₆in), entire and angular, then somewhat
elongated; dingy yellowish brown to tawny,
bruising darker. **Flesh** up to 10mm (½in)
thick, leathery; tawny to deep brown with a
distinct black cuticle on top. **Spores**
cylindrical, smooth, 8.5–11 × 3–4μ. Hyphal
structure trimitic; clamps present. **Habitat** on
coniferous logs in open woods and forests.
Found widely distributed in North America.
Season June–October. **Not edible.**

Datronia mollis (Sommerf. ex Fr.) Donk **Fruit
body** annual; no stem. **Bracket** up to 7cm
(2¾in) across, 2.5cm (1in) wide, 0.6cm (¼in)

thick, undulating, shelflike, thin but tough, leathery brackets, often in tiers; upper surface umber-brown becoming darker brown to black; velvety then smooth. **Tubes** up to 3mm (⅛in) deep; pale buff. **Pores** 1–2 per mm, angular, irregularly elongated or slotlike; surface buff to umber-brown. **Flesh** up to 1mm (¹⁄₁₆in) thick; lower layer pale buff, upper layer dark brown, separated by thin black layer. **Spores** subcylindric, smooth, 10–7×3–4.5μ. Hyphal structure dimitic. **Habitat** on dead wood of numerous hardwood species. Found widely distributed in forest regions of northern North America. **Season** all year. **Not edible.**

Turkey Tail *Trametes versicolor* (L. ex Fr.) Pilát syn. *Coriolus versicolor* (L. ex Fr.) Quél. **Bracket** 4–10cm (1¼–4in) across, 3–5cm (1¼–2in) wide, leathery, scaly, forming large overlapping tiered groups; color variable, concentrically zoned black-green, gray-blue, gray-brown, or ochraceous rust, with a white to cream margin; upper surface velvety becoming smooth with age. **Tubes** 0.5–1mm (¹⁄₃₂–¹⁄₁₆in) deep; white drying yellowish. **Pores** 3–5 per mm, circular or irregularly angular; white, yellowish, or light brown. **Stem** none or rudimentary. **Flesh** 1–3mm (¹⁄₁₆–⅛in) thick, tough, and leathery; white. **Odor** not distinctive. **Taste** not distinctive. **Spores** ellipsoid, smooth, 5–6×1.5–2.2μ. Deposit white. **Habitat** in groups, rows, or overlapping clusters on decaying logs and stumps of hardwoods or in wounds of living trees. Very common. Found widely distributed throughout North America. **Season** generally fruits May–December (November–May in the West), but persists all year. **Not edible.**

Trametes pubescens (Schum. ex Fr.) Pilát **Bracket** 1.5–5cm (½–2in) across, 2.5–8cm (1–3in) wide, 0.3–1cm (⅛–½in) thick, in circular clusters, often overlapping; white or grayish yellow when fresh, grayish or yellow when dry; downy to velvety to almost smooth, often radially lined toward the margin. **Tubes** 1–4mm (¹⁄₁₆–³⁄₁₆in) deep. **Pores** 3–4 per mm, angular; white when fresh, drying yellowish or umber at times. **Stem** none or rudimentary. **Flesh** tough and watery, drying rigid, reviving; white. **Spores** cylindrical, smooth, 5–8×2–2.5μ. Deposit white. **Habitat** in clusters on dead wood of deciduous trees. Found widely distributed in North America. **Season** June–October. **Not edible.**

Cerrena unicolor ⅓ life-size

Cystostereum murraii ⅓ life-size

Bjerkandera adusta ⅓ life-size

Gloeophyllum protractum ⅔ life-size

Datronia mollis ⅓ life-size

Turkey Tail *Trametes versicolor* ⅓ life-size

Trametes pubescens ¼ life-size

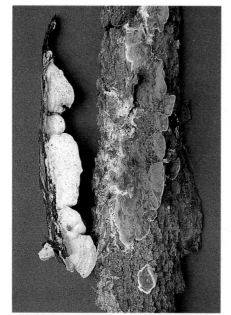

Gloeoporus dichrous ⅓ life-size

Phlebia radiata ¼ life-size

Gloeoporus dichrous (Fr.) Bres. **Fruit body** annual. Resupinate and overlapping with the edges often curving up to form small, elongated, narrow shelves up to 10cm (4in) across, 4cm (1½in) wide, 0.5cm (¼in) thick at the base, with a sharp, undulating margin; upper surface white to cream, with concentric bands of color; finely felty becoming rough and tufted or smoother, depending on weather conditions during growth. **Tubes** rubbery when fresh, resinous to horny when dry and old, gelatinous. **Pores** 4–6 per mm, round to angular; surface light reddish, becoming dark purplish then browner when old, often covered with a bloom. **Flesh** up to 4mm (³⁄₁₆in) thick, thicker than the tubes, cottony to loose; pure white. **Spores** cylindrical, smooth, 3.5–5.5 × 0.7–1.5μ. Hyphal structure monomitic; clamps present. **Habitat** on the dead wood of numerous species of hardwood and sometimes on conifers and dead polypores. Found widely distributed throughout North America. **Season** September–October. **Not edible.**

Phlebia radiata Fr. syn. *Phlebia merismoides* Fr. **Fruit body** annual; up to 25cm (10in) across,

Sarcodontia setosa ⅓ life-size

Pseudomerulius aureus ¾ life-size

Stereum striatum ¾ life-size

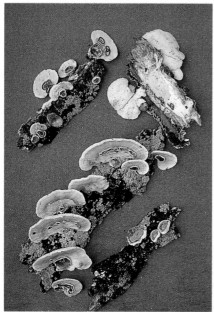

Poronidulus conchifer ⅔ life-size

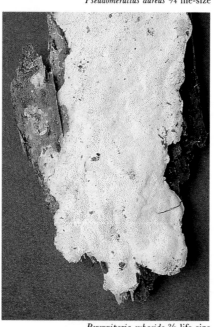

Perenniporia subacida ⅔ life-size

Stereum complicatum ⅓ life-size

2–3mm (³⁄₃₂–¹⁄₈in) thick, resupinate forming round, oval, or irregular patches with a fringed margin; color varies from dull flesh-color or purplish to bright fluorescent orange, especially at the margin; surface very wrinkled with radiating ridges. **No stem. Flesh** thin, softish or slightly gelatinous when fresh, tough in age. **Spores** sausage-shaped, smooth, 3.5–7 × 1.3μ. Deposit whitish. **Habitat** on dead or decaying wood of deciduous and coniferous trees. Found widely distributed in North America. **Season** September–December. **Not edible.**

Sarcodontia setosa (Pers.) Donk **Fruit body** 5–20cm (2–8in) across, forming crustlike spreading patches on the surface of logs; often stained with wine-red areas. **Fertile surface** bright yellow, formed of downward pointed teeth or spines 5–10mm (¼–½in) long. **Odor** strong, very sweet-fruity to unpleasant. **Taste** mild. **Spores** teardrop shape, smooth, 5–6 × 3.5–4μ. **Habitat** on logs or standing wood of fruit trees, especially apples. Rather uncommon. Found widely distributed in North America. **Season** July–October. **Not edible.**

Pseudomerulius aureus (Fr.) Jul. **Fruit body** 5–20cm (2–8in) across, about 2mm (³⁄₃₂in) thick, a spreading, wrinkled crust on the underside of logs; bright orange to golden brown, paler yellow-gold at margin; surface wrinkled and folded. **Spores** cylindrical, smooth, 3.5–4.5 × 1.3–1.8μ. Deposit yellowish. **Habitat** on undersides of dead, barkless conifer logs. Found widely distributed throughout North America. **Season** August–October. **Not edible.**

Stereum striatum (Fr.) Fr. **Bracket** 0.5–3cm (¼–1¼in) across, flat and round, fan-shaped, or sometimes conical if small; whitish, silvery, pale gray, or buff; dry with long silky radiating hairs. **Fertile surface** buff to pale brown or yellowish buff to brown, fading to whitish in age; smooth, sometimes concentrically zoned. **Stem** none or present as small umbo on cap. **Flesh** very thin, tough and pliant when fresh. **Spores** cylindrical, smooth, 6–8.5 × 2–3.5μ. Deposit white. **Habitat** often fused laterally to form lines up to 10cm (4in) long, in groups or masses on twigs and dead branches of hardwoods, especially hornbeam. Found widely distributed in central and northeastern North America. **Season** all year. **Not edible.**

Stereum complicatum (Fr.) Fr. **Bracket** 0.3–1.5cm (⅛–½in) across, fan-shaped or semicircular, with a wavy margin; pinkish or orangish to cinnamon fawn or grayish, with a paler whitish margin; silky-hairy, concentric zones and furrows, smooth and shiny near margin. **Fertile surface** orange, fading to cream-fawn or cinnamon fawn; smooth or slightly ridged where caps meet. **Spores** cylindrical to slightly curved, smooth, 5–6.5 × 2–2.5μ. Deposit white. **Habitat** in overlapping or laterally fused groups on dead twigs and stumps of hardwoods, especially oak. Found throughout North America except the Rocky Mountains. **Season** July–January, fruit body overwinters. **Not edible.**

Poronidulus conchifer (Schw.) Murr. **Fruit body** small cups 0.5–2cm (¼–¾in) wide; white, zoned grayish on inner surface. The cups first appear on the wood and then develop into shelflike caps 1–5cm (½–2in) across, semicircular or kidney-shaped, sometimes with the small cups still adhering to them. Upper surface of cap whitish to grayish-white or yellowish, with concentric zones, particularly on the margin; smooth to minutely hairy or wrinkled. **Tubes** 1–2mm (¹⁄₁₆–³⁄₃₂in) deep.

Pores 2–4 per mm, angular; white to yellowish. **Stem** lateral, knoblike, or none. **Flesh** 0.5–1mm (¹⁄₃₂–¹⁄₁₆in) thick; white. **Spores** cylindrical, smooth, 5–7 × 1.5–2.5μ. Deposit white. Cup-shaped structure sterile. **Habitat** on dead elms and probably other deciduous wood. Found in central and eastern North America. **Season** June–November, fruit body overwinters. **Not edible. Comment** In the first stage of development this fungus can easily be misidentified as a cup fungus.

Perenniporia subacida (Pk.) Donk **Fruit body** a simple layer of tubes spreading over the surface of logs, spongelike or crustlike; cream to pale yellowish. **Tubes** 1–5mm (¹⁄₁₆–¼in) deep; often perennial, and then with more than one tube layer. **Pores** minute (4–6 per mm). **Spores** 5–8 × 3–5μ. Deposit white. **Habitat** on both hardwoods and conifers. Often abundant. Found in the Pacific Northwest. **Season** July–November. **Not edible.**

Sweet Tooth *Hydnum repandum* ⅔ life-size

TOOTHED FUNGI pp. 273–279. *These fungi have needlelike teeth in place of gills or pores.*

Sweet Tooth *Hydnum repandum* L. ex Fr. syn. *Dentinum repandum* (Fr.) S. F. Gray **Cap** 3–15cm (1¼–6in) across, convex then slightly depressed, often irregular in outline; thick-fleshed; pale tan to rusty cinnamon or pale orange; dry, smooth to slightly roughened. **Spines** on undersurface 5–6mm (¼in) long, decurrent on stem; pale cream-orange. **Stem** 30–80 × 10–35mm (1¼–3 × ½–1¼in), tapered at base; colored as cap. **Flesh** thick, firm; cream staining orange. **Odor** pleasant. **Taste** slightly bitter. **Spores** globose, smooth, 6.5–8.5 × 6.5–8.5μ. Deposit white. **Habitat** in mixed woods. Common. Found throughout North America. **Season** July–September. **Edible** — good. **Comment** A pure white form, var. *album*, is frequently found.

Hydnellum aurantiacum ⅔ life-size

Hydnellum regium ⅓ life-size

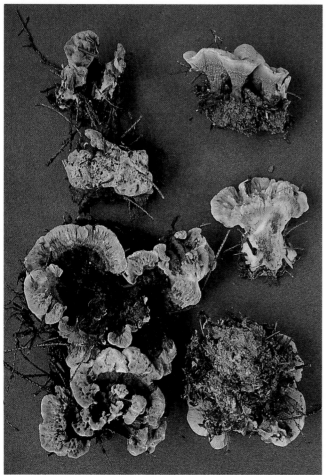

Hydnellum geogenium ½ life-size

Hydnellum caeruleum ½ life-size

Hydnellum aurantiacum (Fr.) Karsten **Fruit body** often fused together. **Cap** 3–15cm (1¼–6in) across, flattened-depressed; orange-brown to rusty cinnamon; tomentose-velvety, often with coarse lumps and protrusions at center when mature. **Spines** on undersurface white then brownish with white tips. **Stem** 30–60×10–20mm (1¼–2¼×½–¾in); orange to dark brown. **Flesh** distinctly zoned; orange to cinnamon. **Odor** fragrant, persistent. **Taste** not distinctive. **Spores** strongly tuberculate, 5.5–7.5×5–6µ. Deposit buff. **Habitat** under conifers, often in masses. Common. Found throughout North America. **Season** July–August. **Not edible.**

Hydnellum regium Harrison **Fruit body** usually in fused masses. **Cap** 5–10cm (2–4in) across, fused in groups up to 25cm (10in) across and 15cm (6in) high from a central stem; violaceous black; surface rough, fibrous, radially striate at margin. **Spines** on undersurface vinaceous brown. **Stem** 50–150×25–50mm (2–6×1–2in), compound, tapered at base; pinkish cinnamon below. **Flesh** violaceous at apex, dull tan at base. **Odor** fragrant, heavy. **Taste** slightly acrid. **Spores** subglobose, tuberculate, 4.5–6×3.5–4.5µ. Deposit brown. **Habitat** gregarious under conifers. Found in western North America. **Season** September–December. **Edibility not known.**

Hydnellum geogenium (Fr.) Banker syn. *Hydnum geogenium* Fr. **Fruit body** single or often fused with others. **Cap** 5–25cm (2–10in) across, irregular in outline, flattened; thin, pliant, rather zoned; olive-brown, deep brown to buff, and often with bright yellow tomentum, especially when young; surface fibrous, knobby, and pitted at center. **Spines** on undersurface decurrent for entire length of stem; olive-yellow to olive-brown, but usually brighter yellow at margin. **Stem** very short or almost absent, concolorous with cap, with numerous strands of yellow mycelium in soil around base. **Flesh** thin; olive. **Odor** not distinctive. **Taste** not distinctive. **Spores** angular to subglobose, tuberculate, 3–4.5×3–4µ. Deposit brown. **Habitat** in coniferous woods. Common. Found on Atlantic seaboard from Quebec to North Carolina. **Season** August–October. **Not edible.**

Hydnum umbilicatum Pk. **Cap** 3–5cm (1¼–2in) across, slightly irregular in shape (less so than in *Hydnum repandum*), soon deeply umbilicate, often with the depression continuing into the stem; deep reddish buff to orange-brown; smooth to slightly felty. **Spines** on undersurface up to 7mm (¼in) long; pale buff. **Stem** 25–80×5–10mm (1–3×¼–½in), usually central; slightly paler than the cap, bruising darker. **Flesh** thin; whitish buff. **Odor** none. **Taste** mild. **Spores** subglobose, tuberculate, smooth, 7.5–9×6–7.5µ. Deposit white. **Habitat** usually in some numbers, in wet boggy woods under spruce, balsam, or cedar, occasionally deciduous woods. Common. Found in eastern North America and the Great Lakes area. **Season** September–October. **Edible** — good.

Hydnellum scrobiculatum (Fr. ex Secr.) Karst var. *Hydnellum scrobiculatum* var. *zonatum* *zonatum* (Batsch ex Fr.) Harrison **Fruit body** single or fused together. **Cap** 3–6cm (1¼–2¼in) across, funnel-shaped to flattened, thin at margin, concentrically zonate; pinkish at margin, darker rusty cinnamon elsewhere. **Spines** on undersurface very short, crowded; dark vinaceous cinnamon. **Stem** 5–30×5–25mm (¼–1¼×¼–1in), single or compound; concolorous with cap. **Flesh** fibrous, tough, zonate; light reddish brown. **Odor** slight. **Taste** mild. **Spores** subglobose, tuberculate, 4.5–5.5×4–4.5µ. Deposit dull brown. **Habitat** in mixed woods. Common. Found widespread throughout North America. **Season** August–September. **Edibility poor.**

Hydnellum suaveolens (Scop. ex Fr.) Karsten syn. *Hydnum suaveolens* Scop. **Fruit body** usually solitary. **Cap** 4–15cm (1½–6in) across, convex then expanded and flattened; dull white when young, then turning to deep brown from center outward, margin remaining white; surface tomentose, soft. **Spines** on undersurface 2–5mm (³⁄₃₂–¼in) long, decurrent; pallid, buff to deep brown. **Stem** 30–50×10–25mm (1¼–2×½–1in), short, hard, woody; bright blue-violet; surface tomentose. **Flesh** fibrous; zoned with violet to violet-black hues, especially in stem. **Odor** strong, sweet or sickly. **Taste** mild to slightly cinnamon-like. **Spores** oblong, tuberculate, angular, 4.5–6×3–4µ. Deposit brown. **Habitat** under conifers, especially in mountain or northern areas, from eastern to western North America. **Season** August–October. **Not edible. Comment** The closely related *Hydnellum caeruleum* (Hornem. ex Pers.) Karsten is distinguished by the lack of violet on its stem, by its spines being shaded blue at the tips, and by the young caps being pale blue.

Hydnellum pineticola Harrison **Fruit body** single to joined in masses. **Cap** 5–15cm (2–6in) across; purplish cinnamon to vinaceous, then dark brown, often with pink droplets in wet weather, bruising blackish. **Spines** on undersurface dull brown, tips paler. **Stem** 25–50×10–25mm (1–2×½–1in), bulbous; paler than cap; woolly-felted. **Flesh** firm, tough; dark cinnamon in cap. **Odor** farinaceous. **Spores** subglobose, tuberculate, 4.5–6×4–5µ. Deposit dull brown. **Habitat** under pine on sandy soils. Found in Canada and Michigan. **Season** August–October. **Edibility not known.**

Hydnellum scrobiculatum var. *zonatum* ⅓ life-size

Hydnellum suaveolens ¼ life-size

Hydnum umbilicatum ⅓ life-size

Hydnellum pineticola ⅓ life-size

Hydnellum mirabile ⅓ life-size

Irpex lacteus ⅓ life-size

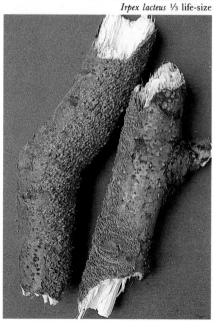

Hydnochaete olivacea ⅓ life-size

Sarcodon fennicum ⅓ life-size

Hydnellum mirabile (Fr.) Karsten **Fruit body** 5–15cm (2–6in) across. **Cap** flattened; dull brownish white to dark brown, margin dirty yellow, darker where bruised; weeping a dark brown fluid when young; rough or roughly hairy. **Spines** on undersurface light brown with paler tips. **Stem** 10–50×10–25mm (½–2×½–1in), tapered below; pale brown with a darker brown tomentum at base. **Flesh** pallid to brownish. **Odor** mealy. **Spores** subglobose, tuberculate, 5–7×4.5–6μ. Deposit buff-brown. **Habitat** solitary to clustered in mixed woods. Found in Canada across to the Pacific Northwest. **Season** August–November. **Edibility not known.**

Irpex lacteus (Fr. ex Fr.) Fr. **Fruit body** annual; no stem; resupinate and overlapping or laterally fused, with the edges curling up from the elongated, spreading, crustlike mass; up to 7cm (2¾in) across, 1cm (½in) wide, 0.5cm (¼in) thick; upper surface white, cream, or pale buff; densely woolly or hairy. **Tubes** up to 3mm (⅛in) deep, continuous with the flesh; white to pale tan. **Pores** 2–3 per mm near the margin, angular; surface white to cream. **Flesh** up to 2mm (³⁄₃₂in) thick, soft-fibrous; white to pale tan. **Spores** ellipsoid to cylindrical, smooth, 5–7×2–3μ. Deposit white. Cystidia conspicuous, long, encrusted; hyphal structure dimitic. **Habitat** on logs, stumps, branches, or trunks of dead hardwoods and occasionally on dead conifers. Common. Found widely distributed throughout North America except the Southwest. **Season** all year. **Not edible.**

Hydnochaete olivacea (Schw.) Banker **Fruit body** annual or persisting; no stem; broadly spreading crust with jagged surface, up to 30cm (12in) long, 10cm (4in) wide; dull yellowish or olive-brown; nearly porelike at the margin but strongly toothlike elsewhere. **Fertile surface** teeth up to 6mm (¼in) long, flattened to cylindrical, blunt and ragged near tip. Bristles numerous over the entire surface of the teeth; sharp, tapered at both ends, thick-walled; dark reddish brown; 55–90×9–15μ. **Flesh** 1–3mm (¹⁄₁₆–⅛in) thick; brown; leathery. **Spores** cylindrical, smooth, 5.5–7×1–1.5μ. Deposit white. **Habitat** on the underside of dead branches of hardwoods,

especially oak. Common. Found widely distributed in eastern North America. **Season** June–October, may persist for years. **Not edible.**

Sarcodon fennicum Karsten **Cap** 5–15cm (2–6in) across, convex then flattened, often irregular in outline; dull red-brown, paler at margin; smooth then slightly scaly, especially at center. **Spines** on undersurface 3–5mm (⅛–¼in) long, decurrent on stem, crowded; pale buff with darker brown tips. **Stem** 20–50× 10–25mm (¾–2×½–1in); reddish brown with darker blue-green to blackish-olive base. **Flesh** firm; pale buff, blue-green in stem base. **Odor** pleasant. **Taste** acrid and unpleasant. **Spores** subglobose, tuberculate, 5.5–6.5×6.8–7.5μ. Deposit brown. **Habitat** in groups under mixed woods, especially beech. Found in eastern North America. **Season** September–October. **Not edible. Comment** KOH on flesh does not give a green reaction as in some similar species, such as *Sarcodon scabrosus* (Fr.) Karsten.

Sarcodon atroviridis (Morgan) Banker syn. *Hydnum atroviride* Morgan **Cap** 2–8cm (¾–3in) across, convex to concave with age; pale tan to fawn when young, then becoming black where bruised, entirely black when old, with greenish hues on drying; dry, felty, smooth or with a few pits and ridges. **Spines** on undersurface 5–15mm (¼–½in) long; gray buff to deep brown with paler tips; brittle. **Stem** 50–80×5–10mm (2–3×¼–½in), equal; buff, soon blackish like cap; smooth. **Flesh** firm; pale buff. **Odor** none. **Taste** bitter. **Spores** subglobose, 6–8.5×6–8.5μ. Deposit buffy brown. **Habitat** in mixed woods. Found in eastern and southeastern North America from Massachusetts to Florida. **Season** July–September. **Edibility poor.**

Sarcodon imbricatus (L. ex Fr.) Karsten syn. *Hydnum imbricatum* L. ex Fr. **Cap** 5–20cm (2–8in) across, flattened-convex then depressed; dark reddish to purplish brown against paler pinkish flesh; velvety then cracked into coarse overlapping scales. **Spines** on undersurface 1–10mm (¹⁄₁₆–½in) long; white to purplish brown. **Stem** 50–80×20–50mm (2–3×¾–2in), tapered or swollen at base; whitish to purplish brown. **Flesh** firm; white. **Odor** not farinaceous, not distinctive. **Taste** soon bitter. **Spores** ellipsoid, tuberculate, 7–8×5–5.5μ. Deposit brown. **Habitat** coniferous woods. Found in northern and eastern North America. **Season** July–September. **Edible** but poor.

Sarcodon joeides (Pass.) Bat. **Cap** 5–15cm (2–6in) across, irregular in outline, convex cushion-shaped; pale pinkish buff to vinaceous brown; dry, smooth to slightly scaly at center. **Spines** on undersurface up to 1cm (½in) long, decurrent down stem; pale lilac brown. **Stem** 20–50×10–30mm (¾–2×½–1¼in), tapered at base; concolorous with cap. **Flesh** thick, firm; violet-purple, deep green with KOH. **Odor** pleasant. **Taste** acrid or bitter. **Spores** subglobose, tuberculate, 5.4–5.8×3.6–4.2μ. Deposit brown. **Habitat** in deciduous woods, especially beech. Common but hard to see. Found in eastern North America. **Season** September–October. **Not edible.**

Sarcodon underwoodii Banker **Cap** 4–8cm (1½–3in) across, convex to flattened, depressed at center; vinaceous brown to fawn; covered with small adpressed scales or even smooth when young. **Spines** on undersurface short, 1–3mm (¹⁄₁₆–⅛in); white, then brown with grayish tips. **Stem** 40–80×6–10mm (1½–3×¼–½in), usually strongly curved at

(*continued on page 278*)

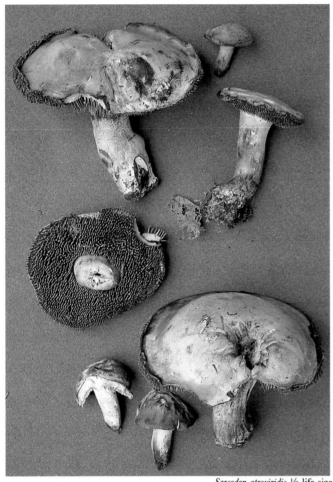

Sarcodon atroviridis ½ life-size

Sarcodon imbricatus ⅓ life-size

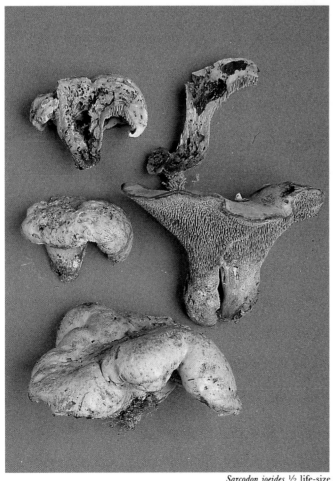

Sarcodon joeides ½ life-size

Sarcodon underwoodii ⅓ life-size

Climacodon septentrionalis ⅓ life-size

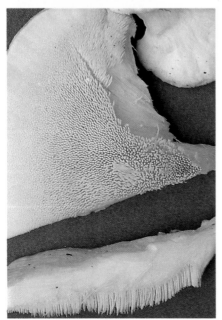

Climacodon septentrionalis nearly life-size

brownish yellow when dry, with very faint zones; densely hairy and roughened. **Spines** on undersurface 0.5–2cm (¼–¾in) long, narrow, with lacerated tips, crowded, pliant; dull white drying yellowish. **Flesh** up to 4cm (1½in) thick, fibrous, tough, elastic; white, zoned. **Odor** none or mild when fresh, of ham when dry. **Taste** none or mild when fresh, bitter when old. **Spores** ellipsoid, smooth, 2.5–3× 4–5.5µ. Deposit white. Cystidia thick-walled with encrusted tip. **Habitat** high up or in the wounds of living deciduous trees such as maple, beech, and birch. Found widely distributed in northeastern North America as far south as Tennessee. **Season** July–October. **Not edible.**

Hericium ramosum (Bull. ex Mérat) Let. **Fruit body** a mass of white, long, multiple branched stems covered with large numbers of very small spines. Whole fungus 10–25cm (4–10in) across, 10–15cm (4–6in) high; stems thin, branched, covered on both sides with spines 0.5–1cm (¼–½in) long. **Flesh** soft, brittle; white. **Odor** pleasant. **Taste** pleasant. **Spores** ellipsoid, smooth to finely roughened, 3–5×3–4µ. Deposit white. **Habitat** on fallen logs of beech and maple. Found throughout North America. **Season** August–October. **Edible** — good. **Comment** Easily distinguished from the rather similar *Hericium coralloides* (below) by its much shorter spines growing along the *whole* length of the thin branches. A pink form of *Hericium ramosum* turned up at a foray in Oneonta, New York; it was referred to as *Hericium ramosum* var. *rosea*, of which I can find no trace in the literature.

Hericium coralloides (Fr.) S. F. Gray **Fruit body** a mass of tufted stems, each with many pendant spines. Whole fungus may be 15–30cm (6–12in) across and 20–40cm (8–16in) high; branches are white and stout and branch repeatedly from a central, basal point; tip of each branch has white spines 0.5–2cm (¼–¾in) long, in clusters like hands. **Flesh** firm; white. **Odor** pleasant. **Taste** pleasant. **Spores** ellipsoid, smooth or very finely roughened, 5–7×4.5–6µ. Deposit white. **Habitat** on both fallen timber and living trees, especially beech and maple. Found in northeastern North America, south to North Carolina. **Season** August–October. **Edible** — good. **Comment** The true *Hericium coralloides*, according to some mycologists, is strictly European, and the correct name for this fungus may eventually be *Hericium americanum*.

Indian Paint Fungus *Echinodontium tinctorium* (Ellis & Ev.) Ellis & Ev. **Fruit body** perennial; no stem. **Bracket** up to 40cm (16in) across, 30cm (12in) wide, 20cm (8in) thick, hoof-shaped; upper surface dark dull brown to olive-black, bristly and hairy becoming hard, brittle, and cracking concentrically into blocks. **Teeth** up to 3cm (1¼in) long, flattened, thin and brittle at first, becoming thick and stout with blunt ends and crowded in age; pale to pinkish, grayish buff. **Flesh** up to 5cm (2in) thick, with blackish brittle upper layer 2mm (³⁄₃₂in) thick, hard, woody, zoned; brick red. **Spores** ellipsoid, smooth to minutely spiny, amyloid, 6–8×4.5–6µ. Deposit white. Hyphal structure dimitic. **Habitat** under branch stubs on coniferous wood, usually western grand fir and western hemlock. Found in western North America from Alaska to Mexico. **Season** all year. **Not edible.**

(continued from page 276)
base; deep, dirty brown, white at base. **Flesh** tough; brownish. **Odor** not distinctive. **Taste** bitter. **Spores** subglobose, strongly tuberculate, 5.5–6.5×6–7µ. Deposit buff-brown. **Habitat** in deciduous woods. Rare. Found from Massachusetts to North Carolina and Tennessee; also in Colorado and Iowa. **Season** August–September. **Edibility not known.**

Climacodon septentrionalis (Fr.) Karsten **Fruit body** huge, consisting of overlapping fan-shaped caps growing in horizontal clusters 15–30cm (6–12in) high, arising from a solid base which narrows to an attachment about 2cm (¾in) wide where it enters the wood. **Cap** 10–15cm (4–6in) across, 2–5cm (¾–2in) thick near the base, shelflike, thinning toward the margin; whitish to yellowish or buff, turning

Hericium ramosum var. *rosea* ¾ life-size

Hericium coralloides ½ life-size

Indian Paint Fungus *Echinodontium tinctorium* life-size

Hericium ramosum ¾ life-size

Longula texensis ⅓ life-size

PUFFBALLS pp. 280–291. *This section includes stinkhorns, earthstars, and bird's-nest fungi as well as the puffballs. All are part of a large group of fungi known as Gasteromycetes, or stomach fungi. The spores form within the fungus; when they are mature, the surface of the fungus cracks or breaks up, allowing the spores to disperse.*

Longula texensis (Berk. & Curt.) Zeller **Cap** 3–9cm (1¼–3½in) across, oval to round or broadly convex; white to buff, becoming more brownish in age; smooth or breaking up into scales or warts which may wear away; becoming fragile and often splitting as it dries out in maturity. **Spore mass** comprising crowded, convoluted, folded plates and cavities; brownish becoming chocolate brown or blackish in maturity. **Stem** 20–100 × 15–35mm (¾–4 × ½–1¼in), slightly thicker toward the base; white or similar to cap; smooth or longitudinally lined, becoming tough and woody as it dries. **Veil** two-layered, at first over cap margin and stem, then separating

from cap and forming a ring on the upper part of the stem. **Flesh** firm, solid; white bruising yellowish or pinkish in the stem. **Spores** nearly globose, smooth, 6–7.5 × 5–6.5μ. **Habitat** singly, scattered, or in groups in poor soil, waste ground, fields, lawns, and arid areas. Found in southwestern North America from Texas west and as far north as Oregon. **Season** July–August. **Edibility not known.**

Macowanites luteolus Smith and Trappe **Fruit body** 2–4cm (¾–1½in) across, uneven, lumpy button mushroom–shaped with a smooth surface; white to creamy buff, drying yellowish; slightly viscid at first. **Spore mass** creamy spongelike mass of half-formed gills. **Stem**

10–20 × 5–10mm (½–1 × ¼–½in), short and stubby; whitish or creamy like the cap. **Flesh** pallid. **Odor** slight. **Taste** like raw potatoes. **Spores** globose to subglobose, with warts or spines like a russula, 7.9–9.4 × 7–8.5μ; warts up to 1mm (¹⁄₁₆in) long, amyloid when viewed under a microscope. **Habitat** just under or on the surface of the ground under conifers. Found in Oregon and on the Washington coast. **Season** September–October. **Edibility not known. Comment** Like an underground russula that does not properly form its gills. One form, which has no stem, was collected in Washington with the normal type. The spores of this collection are similar but a little larger, 9–14.5 × 8.4–11.5μ.

Endoptychum agaricoides Czerniaiev **Cap** 1–7cm (½–2¾in) across, 2–10cm (¾–4in) high, oval to rounded and generally wider at the base; white becoming dingy to tan; smooth with minute hairs, sometimes becoming scaly. **Spore mass** like contorted gills, chambered; whitish becoming pale brown in maturity; sometimes slightly powdery. **Stem** barely exposed, extending up into gleba, attached by a cord to the ground; whitish becoming yellowish. **Spores** ellipsoid, smooth, brownish, 6.5–8 × 5.5–7μ. **Habitat** scattered, in groups, or even in dense clusters on lawns, flower beds, pastures, and cultivated or wasteland. Sometimes abundant. Found widely distributed throughout North America. **Season** May–October. **Edibility not known.**

Thaxterogaster pingue (Zeller) Smith & Singer **Fruit body** 1–4cm (½–1½in) across, club-shaped to flattened; ochre to olive yellowish in color; greasy to the touch. **Spore mass** spongelike mass of yellowish-brown unformed gills, which darken to rusty brown as the spores mature. **Stem** 1–4cm (½–1½in) long, rudimentary; color variable from whitish to ochre or sometimes with violaceous tints. **Odor** yeasty. **Taste** nasty. **Spores** ellipsoid, slightly roughened, 14–17.5 × 7.5–9.5μ, brownish in 3 percent KOH. **Habitat** under conifers, especially spruce. Found in the Pacific Northwest and the Rockies. **Season** September–November. **Not edible,** it is thought. **Comment** Closely related to the genus *Cortinarius*, probably an underground form that can develop in drier conditions than an ordinary, above-ground mushroom.

Leucogaster rubescens Zeller & Dodge **Fruit body** globose to irregular, partially or entirely underground; surface whitish when young, darkening to pink then reddish brown; viscid when wet. **Spore mass** cream-white with tiny cells 2mm (³⁄₃₂in) across filled with clear gel, exuding a latex when cut. **Odor** not distinctive. **Taste** not distinctive. **Spores** globose, nonamyloid, surface ornamented, alveolate-reticulate, white, 10–15μ. **Habitat** in soil and leaf litter. Found in the Pacific Northwest. **Season** April–June. **Edible.**

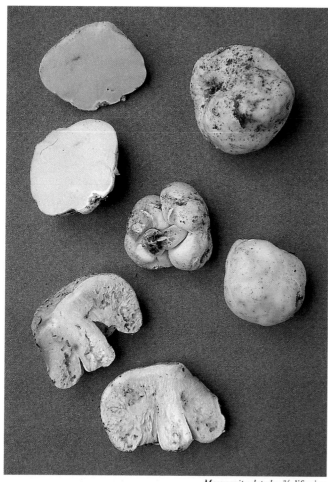

Macowanites luteolus ¾ life-size

Endoptychum agaricoides ½ life-size

Thaxterogaster pingue life-size

Leucogaster rubescens ½ larger than life-size

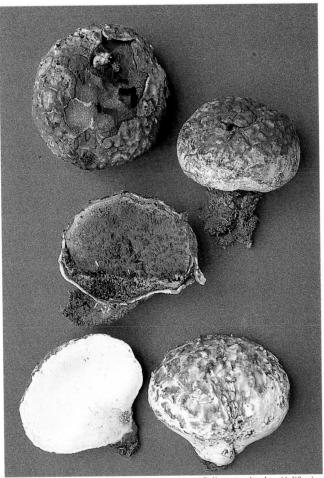

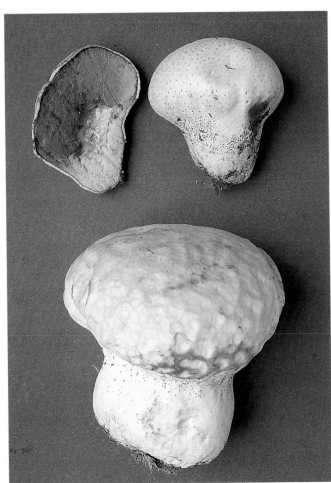

Calbovista subsculpta ⅓ life-size

Purple-spored Puffball *Calvatia cyathiformis* ⅓ life-size

Calbovista subsculpta Morse **Fruit body** 8–15cm
(3–6in) across, 6–9cm (2¼–3½in) high, nearly
round or sometimes a bit broader; whitish to
dingy; covered with flattened warty scales with
grayish tips and brownish hairs at the center.
Spore mass white becoming brownish. Sterile
base one-quarter to one-third of mushroom;
dull, white, firm. **Spores** globose, almost
smooth, 3–5×3–5μ. **Habitat** singly or
scattered or in small groups in open areas
along roadsides and wood edges in subalpine
places. Sometimes abundant. Found in the
Rocky Mountains and the Pacific coastal
ranges. **Season** April–August. **Edible** but only
when the spore mass is white; excellent.

Purple-spored Puffball *Calvatia cyathiformis*
(Bosc) Morgan **Fruit body** 5–16cm (2–6¼in)
across, 5–11 (15)cm (2–4½[6]in) high,
hemispherical with a rudimentary stem, forcing
up through turf; pallid to dirty tan or brownish
with age; smooth at first, then soon cracking
into smaller areas to give a tesselated effect;
breaking up to reveal spore mass (gleba) in a
persistent cuplike base. **Spore mass** of cottony
consistency; white at first, then soon deep lilac
to dark purple-brown. **Spores** globose, with
distinct spines, 3.5–7.5×3.5–7.5μ. **Habitat** in
open fields and lawns. Common. Found
widespread in eastern North America and the
Great Plains. **Season** July–November. **Edible**
when young; good. **Comment** The large violet-
brown cuplike base left after the spores are
dispersed can often be found through the
winter.

Lycoperdon perlatum Pers. syn. *Lycoperdon
gemmatum* Batsch **Fruit body** 2.5–6cm
(1–2¼in) across, 2–9cm (¾–3½in) high,
subglobose with a distinct rudimentary stem;
white at first, becoming yellowish brown; outer
layer of short pyramidal warts, especially dense
at the head, rubbing off to leave an indistinct
meshlike pattern beneath, which opens by a
pore. **Spore mass** white, then olive-brown at
maturity. Sterile base spongy, occupying the
stem. **Spores** globose, minutely warted, olive-
brownish, 3.5–4.5×3.5–4.5μ. **Habitat** singly,
scattered, or in clusters in waste areas and
open woods and along wood edges. Found
widely distributed in North America. **Season**
July–October. **Edible** when flesh is completely
white; excellent.

Lycoperdon echinatum Pers. **Fruit body** 2–5cm
(¾–2in) across, subglobose tapering into a
short stem; white becoming brown; outer layer
forming pointed spines 3–5mm (⅛–¼in) long,
convergent at the tips in groups of 3 or 4; a
netlike pattern is left when the spines have
become detached; opening by a central pore.
Spore mass white, becoming purple-brown in
maturity. Sterile base occupying stem, often
small. **Spores** globose, warted, brown,
4–6×4–6μ. **Habitat** among leaves and on
humus in woods. Sometimes abundant. Found
in northeastern North America. **Season**
August–November. **Edible** when skinned and
white.

Lycoperdon molle Pers. **Fruit body** 1–4cm

(½–1½in) across, 6cm (2¼in) high, usually
pear-shaped; grayish brown to milky-coffee-
colored; minutely spiny or granular. **Spore
mass** white, dark brown in maturity; inner
spore case opening by a wide, irregular pore.
Sterile base has large chambers; sometimes
wide and sometimes narrowed to a distinct
stalk. **Spores** globose, 3.5–5×3.5–5μ. **Habitat**
on soil or humus in deciduous or coniferous
forests. Frequent but not abundant. Found
widely distributed in North America. **Season**
August–October. **Edible** when flesh completely
white.

Lycoperdon marginatum Vitt. **Fruit body** 1–5cm
(½–2in) high, globose, sometimes with a small
rooting base; white becoming brownish; with
an outer covering of pointed warts or spines,
which fall away in irregular sheets, exposing
the olive-brown inner wall, or endoperidium.
Spore mass olive to gray-brown. Sterile base
well developed, chambers about 1mm (1⁄16in)
across, capillitial threads 3–6μ wide. **Spores**
globose, minutely ornamented, olive-brown,
3.5–4.2×3.5–4.2μ. **Habitat** on sandy soil.
Very common. Found widespread throughout
North America. **Season** June–October. **Edible**.

Pear-shaped Puffball *Lycoperdon pyriforme*
Schaeff. ex Pers. **Fruit body** 1.5–4cm
(½–1½in) across, 3.5 cm (1¼in) high,
subglobose to club- or pear-shaped, attached to
the substrate by prominent mycelial strands.
White when young, then soon pale yellowish
buff or gray-brown; outer layer of scurfy spines,

warts, or granules, inner wall becoming smooth and papery, opening by an apical pore. **Spore mass** olive-brown. Sterile base occupying the stem is spongy with rather small cells. Capillitium distinctive in being formed of brownish branched threads all lacking in any trace of tiny pores found in other members of the genus. **Spores** globose, olive-brown, 3–4 × 3–4μ. **Habitat** in large clusters on decaying wood, logs, etc. Common. Found widespread throughout North America. **Season** July–November. **Edible** — good when young.

Vascellum pratense (Pers.) Kreisel syn. *Vascellum depressum* (Bonord.) Smarda **Fruit body** 2–4cm (¾–1½in) across, subglobose narrowed into a short, squat stem; white at first, then yellowish flesh-colored, finally light brown; outer layer scurfy with some small white spines; inner wall smooth and shining, opening by a small pore, but eventually the upper part breaking away, totally leaving the fruit body bowl-shaped. **Spore mass** firm; white, becoming olive-brown then powdery. Sterile base well-developed, separated from the spore mass by a distinct membrane. **Spores** globose, finely warted, olive-brown, 3–5.5 × 3–5.5μ. **Habitat** on lawns, golf courses, or pastures. Common. Found widely distributed in North America. **Season** September–November. **Edible** when white inside.

Morganella subincarnata (Pk.) Kreisel & Dring **Fruit body** 1–3cm (½–1¼in) across, roundish to pear-shaped; surface with cinnamon-buff to brownish spines and warts in groups when young; conspicuously pitted in maturity when spines have worn off. **Spore mass** grayish or brownish, purple-tinged in maturity. Sterile base well developed or almost none. **Spores** globose, ornamented, 3.5–4 × 3.5–4μ. **Habitat** scattered or in groups on damp, mossy logs in hardwood forests, or on beech, birch, or maple slash. Sometimes common. Found in eastern North America. **Season** August–October. **Edibility not known**.

Lycoperdon perlatum ⅓ life-size

Lycoperdon echinatum ⅓ life-size

Lycoperdon molle ½ life-size

Lycoperdon marginatum ⅓ life-size

Morganella subincarnata ⅔ life-size

Vascellum pratense ⅔ life-size

Pear-shaped Puffball *Lycoperdon pyriforme* ⅓ life-size

Scleroderma cepa (Vaill.) Pers. **Fruit body** 1.5–9cm (½–3½in) across, subglobose, flattened, or lobed; no stem or almost none, attached by a thick mass of tough, hairy mycelium. Peridium (outer skin) 1–3mm (¹⁄₁₆–¹⁄₈in) thick; when fresh, hard, quite tough; white in cross-section, becoming reddish or pinkish brown when cut. Surface whitish when young, becoming straw-colored to yellowish brown or leather brown, turning deep pinky-brown if rubbed; smooth becoming very finely cracked and scaly, especially on the top where exposed to light. **Spore mass** white and firm when young, soon becoming black or purple-black, then paler or browner and powdery. **Odor** none. **Spores** globose, spiny but not reticulate, 7–10×7–10μ. **Habitat** singly, scattered, or in groups under deciduous and coniferous trees in woods, in gardens, and along roadsides. Common. Found widely distributed in North America. **Season** July–October. **Poisonous**.

Scleroderma areolatum Ehrenberg **Fruit body** 3–10cm (1¼–2in) across, a globose, leathery ball; yellow-brown to tawny-gold; dotted with tiny darker brown scales, each surrounded with a clear area (areola). The outer skin (peridium) is rather thin when mature. **Spore mass** starts pallid but soon turns purplish and then olive-brown and powdery. **Odor** strong when cut open, unpleasant. **Spores** globose, 10–18×10–18μ, with abundant spines up to 1.5μ high. **Habitat** in woods and gardens. Common. Found in eastern North America. **Season** June–October. **Poisonous**.

Tulostoma brumale Pers. **Fruit body** consists of a globose, somewhat collapsed head 1–2cm (½–¾in) across, rusty brown; attached to a slender, fibrous stem 20–50×3–4mm (¾–2×¹⁄₈–³⁄₁₆in). Head opening by a circular pore surmounting a pale ochre to whitish cylindrical mouth. **Spore mass** rusty salmon and powdery at maturity. **Spores** globose, finely warted, 3–5×3–5μ. **Habitat** scattered or in groups on sandy soil, in gravel, or in waste places. Frequent. Found widely distributed in North America. **Season** September–December, or almost anytime for weathered specimens. **Not edible**.

Astraeus hygrometricus (Pers.) Morgan **Fruit body** a roundish spore sac encircled by starlike rays; stalkless, arising from black, hairlike basal rhizomorphs. **Spore sac** 1–3cm (½–1¼in) across, nearly round, with an irregular slit or tear at the top; whitish becoming grayish or brown; felty. **Spore mass** pure white, becoming brown in age. **Rays** 2.5–5cm (1–2in) long, formed from the outer gray-brown wall of the spore sac splitting into 7–15 pointed rays which bend backward when wet and curve inward when dry. **Spores** globose, warty, reddish brown, 7.4–10.8×7.4–10.8μ. **Habitat** scattered or in groups, it is found in almost any habitat ranging from sandy dunes to high mountains. It develops just under the soil and becomes exposed when mature. Common. Found widely distributed throughout North America. **Season** September–November, but sometimes keeps its shape and condition for a year or more. **Not edible**. **Comment** This earthstar is hard and leathery when dry but quickly revives when soaked in water.

Earthstar Puffball *Scleroderma geaster* Fr. syn. *Scleroderma polyrhizon* Pers. **Fruit body** 5–10cm (2–4in) across, subglobose, flattened on top, tapering below into a stemlike base with a large basal mycelial mass binding together the soil into a large mass. Surface of the very thick cuticle a dirty tan-brown to

ochre; roughened, granular, splitting into irregular starlike segments, soon peeling back to a varying extent, exposing the blackish cracked inner surface. **Spore mass** deep purple-brown; powdery. **Spores** globose, warted, 5–10×5–10μ. **Habitat** at first almost completely below ground, pushing up through sandy soils. Found widely distributed in North America. **Season** August–November. **Not edible**.

Scleroderma citrinum Pers. syn. *Scleroderma aurantium* (Vaill.) Pers. **Fruit body** 3–15cm (1¼–6in) across, a leathery ball; dull yellow-brown to tawny or even whitish yellow; surface coarsely scaly or warty. The peridium is thick and splits open at the apex when mature. **Spore mass** starts pallid, then turns purplish before becoming blackish brown when mature. **Odor** strong and pungent when cut open. **Spores** globose, with a distinct reticulum, 9–13×9–13μ. **Habitat** in mixed woods and gardens. Common and often abundant. Found throughout North America. **Season** July–October. **Poisonous**.

Dye-maker's False Puffball *Pisolithus tinctorius* (Pers.) Coker & Couch syn. *Pisolithus arhizus* (Pers.) Rausch. syn. *Pisolithus arenarius* A. & S. **Fruit body** 6–12cm (2¼–4¾in) across, 5–25cm (2–10in) high, forming a large irregular club with the base narrowing below into a thick stemlike base submerged in the ground; ochraceous to dirty olive-brown and resembling balls of horse dung lying on the ground, with chrome yellow powdery markings on the submerged part. Outer wall thin and soon becoming brittle, readily breaking open to expose the dark brown, stony, gravel-like peridioles (pea-shaped chambers which contain the spores). **Spores** globose, warted, 7–11.5(12)×7–11.5(12)μ. **Habitat** in sandy or well-drained soils, old lawns, roadsides, and pine woods. Found widely distributed in North America. **Season** July–October. **Not edible**.

Stalked Puffball-in-aspic *Calostoma cinnabarina* Desv. **Fruit body** a round, bright red spore sac, covered with a thick clear jelly and raised up on a spongy red stem. **Spore sac** has an outer wall which splits apart, leaving bright red fragments in the jelly at the base of

(continued on page 286)

Scleroderma cepa ⅔ life-size

Scleroderma areolatum ½ life-size

Tulostoma brumale ½ life-size

Astraeus hygrometricus ½ life-size

Earthstar Puffball *Scleroderma geaster* ⅓ life-size

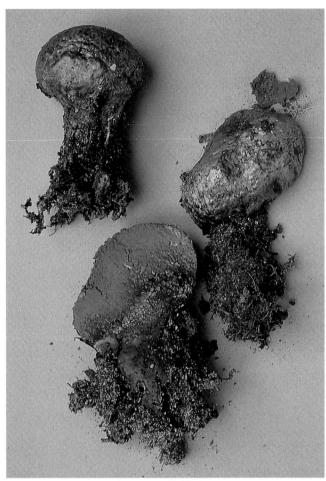

Dye-maker's False Puffball *Pisolithus tinctorius* ½ life-size

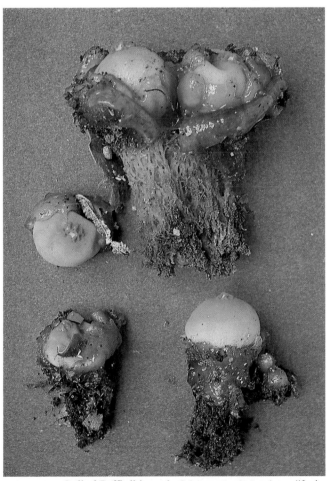

Stalked Puffball-in-aspic *Calostoma cinnabarina* almost life-size

Scleroderma citrinum ⅔ life-size

Geastrum saccatum ¾ life-size

Geastrum triplex ⅔ life-size

(*continued from page 284*)
the fungus; the inner spore sac is pale yellow dusted with a bright red powdery coating, 1–2cm (½–¾in) across with a bright cinnabar-red apical opening in the shape of a cross. Protruding from the ground at first, it will slowly be raised up on a spongelike short stem, 15–30×10–15mm (½–1¼×½in), pale cinnabar red. **Spores** elliptical, pitted, cream-yellow, 14–20×6–9μ. **Habitat** on soil, especially stream banks, pathsides. Found throughout eastern and southeastern North America. **Season** April–May, September–October. **Edibility not known. Comment** This remarkable species is very easily distinguished, looking rather like a small red tomato surrounded by jelly and red pips.

Geastrum saccatum Fr. **Fruit body** a round, bulblike sac whose outer wall splits, unfolds, and bends back into 4–8 starlike rays. **Spore sac** 0.5–2cm (¼–¾in) across, round to flattened with a disclike depression or mouth area set off from the rest of the spore sac by a distinct ring or shallow groove; buff, dull gray, or brownish, paler at the mouth area; smooth. **Spore mass** firm; white becoming brownish and powdery. **Rays** 2–4cm (¾–1½in) long; upper surface pallid to tan or ochre-brown, undersurface buff to pale tan; rubbery when fresh, sometimes cracking. **Spores** globose, warty, 3.5–4.5×3.5–4.5μ. **Habitat** singly or in groups around decaying stumps or in leaf litter in hardwood forests or under juniper and

conifers. Quite common. Found widely distributed in North America. **Season** July–October, but often persisting for months. **Not edible**.

Geastrum triplex Jung. **Fruit body** when unopened 3–5cm (1¼–2in) across, bulb-shaped; opening to 5–10cm (2–4in) across, with the outer wall splitting into 4–8 pointed rays and covered in a thick, pinkish-brown, fleshy layer which cracks as the rays bend back under the fruit body, leaving the spore sac sitting in a saucerlike base. **Spore sac** no stem; pale gray-brown with a paler ring around the slightly raised mouth. **Spore mass** firm; white becoming dark brown and powdery. **Rays** 2–4cm (¾–1½in) long, up to 4mm (³⁄₁₆in) thick; creamy-colored; rubbery. **Spores** globose, warted, brown, 3.5–4.5×3.5–4.5μ. **Habitat** singly, scattered, or in groups in leaf litter in deciduous woods or occasionally in the open. Quite common. Found widely distributed in North America. **Season** August–October, but often persists, without decaying, for months. **Not edible**.

Geastrum coronatum Pers. **Fruit body** an oval or egg-shaped sac whose outer wall splits into 5–8 starlike rays. **Spore sac** 0.5–2cm (¼–¾in) across, raised on a short stalk; dark purple-brown with a paler mouth area outlined by a distinct groove and lined with fine hairs. **Spore mass** dark chocolate brown at maturity. **Rays** 1–2cm (½–¾in) long when expanded; recurve to support the fruiting body on their tips; outer skin light yellow-brown on either side. **Spores** subglobose to globose, warty, 3.5–5×3.5–5μ.

Habitat in small groups in conifer woods. Found widely distributed in North America. **Season** July–October. **Not edible**.

Geastrum floriforme Vitt. **Fruit body** 2–2.5cm (¾–1in) across when expanded; round spore case with an outer wall consisting of pointed lobes that are bent inward when dry and expanded when wet. **Spore sac** covered in soil and debris at first, then smooth and pale brownish; indistinct mouth opening. **Spore mass** brown at maturity. **Spores** globose, warty, 2.5–3.5×2.5–3.5μ. Thick-walled capillitial threads approximately 10μ in diameter. **Habitat** singly or in small groups on sandy soil. Found widely distributed in North America. **Season** July–October. **Not edible**.

Geastrum fornicatum (Huds.) Fr. **Fruit body** 1–2.5cm (½–1in) across; comprises a brownish central globose spore sac with a fringed apical pore, borne on a short stem, and carried clear of the surrounding leaf litter on 4 tall, narrow, downward-pointing rays standing on the tips of 4 similar but shorter, broader, upward-pointing rays which form a mycelial mat or cuplike structure at the base of the fungus. **Spore mass** firm; white becoming blackish brown and powdery. **Spores** globose, warted, dark brown, 3.5–4.5×3.5–4.5μ. **Habitat** scattered or in dense groups in leaf litter, humus, or organic debris in woods, near stables, or on rubbish or waste ground. Quite common in the Southwest but rare elsewhere. Found widely distributed in North America. **Season** October–March. **Not edible**.

Geastrum coronatum ½ life-size

Rhizopogon rubescens ¼ life-size

Rhizopogon rubescens Tul. **Fruit body** 1.5–6cm (½–2¼in) across, generally a lumpy potato shape; at first white, but the surfaces above ground take on a greenish-olive color in age, distinctly reddening when handled. **Spore mass** white at first, then olive-buff as it matures, staining reddish when cut; finely granular. **Odor** slight. **Taste** slight. **Spores** fusoid to subfusoid, 6–10×2.5–4.5μ. **Habitat** under conifers, especially pine. Frequent. Found throughout North America. **Season** July–November. **Not edible**.

Rhizopogon parksii Smith **Fruit body** 1–3cm (½–1¼in) across, roundish, slightly flattened; pallid to brownish or pinkish, staining dingy purple or bluish black on the surface when bruised or with age; minutely hairy. **Spore mass** olive. **Spores** subellipsoid, 5.5–6.5×2.3–3μ. **Habitat** under Sitka spruce and Douglas fir. Very common and abundant in the Sitka spruce zone. Found along the West Coast. **Season** August–October. **Not edible**.

Rhizopogon ellenae Smith **Fruit body** 1–9cm (½–3½in) across or long, oval, elongated, or roundish and sometimes flattened; outer surface white at first, then developing pinky-gray or purplish smoky-gray tones in age or with handling; smooth or minutely hairy with a few strands of mycelium. **Spore mass** minutely chambered; white when young. **Spores** ellipsoid to oblong, smooth, weakly amyloid, 7–9×3–4μ. **Habitat** scattered or in groups or clusters in soil or needle duff under conifers, particularly pine. Common and sometimes abundant. Found in western North America. **Season** September–November. **Edibility not known**.

Rhizopogon occidentalis Zeller & Dodge, **Fruit body** 1–5cm (½–2in) across, irregularly potato-shaped or pear-shaped; whitish at first, then lemon yellow in age, bruising orange or reddish brown; minutely hairy with loose, cordlike brownish rhizomorphs. **Spore mass** minutely chambered; pale yellow-orange, drying cinnamon buff. **Odor** of sourdough. **Spores** ellipsoid, smooth, 5.5–7×2.3–2.6μ. **Habitat** singly or scattered in sandy soil under mixed conifers. Found in the Pacific Northwest, Idaho, and California. **Season** May–September. **Not edible**.

Geastrum floriforme life-size

Rhbizopogon parksii ½ life-size

Geastrum fornicatum ½ life-size

Rhizopogon ellenae ⅓ life-size

Rhizopogon occidentalis ⅓ life-size

Crucibulum laeve ⅔ life-size

Nidula candida ½ larger than life-size

Cyathus stercoreus almost life-size

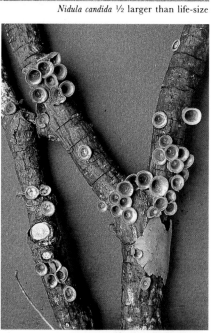

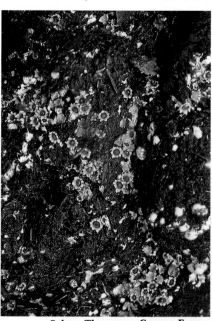

Cyathus striatus ¾ life-size

Cyathus olla ⅔ life-size

Sphere Thrower or **Cannon Fungus**
Sphaerobolus stellatus life-size

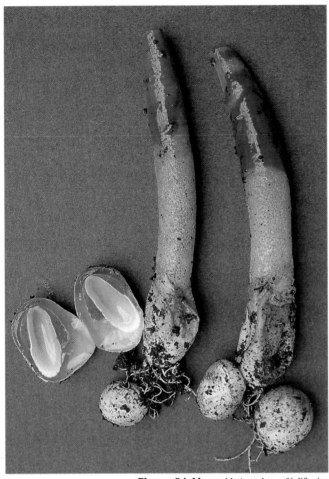

Elegant Stinkhorn *Mutinus elegans* ¾ life-size

Mutinus elegans (pink-stemmed form) ¾ life-size

Nidula candida (Pk.) White **Fruit body** a small cup containing tiny "eggs." **Cup** 5–15mm (¼–½in) high, flared at the apex, outer surface scurfy; gray-brown to ochre; inner surface whitish, with a layer of mucilage in which the eggs (peridioles) lie. **Eggs** 1.5–2mm (1/16–3/32in) across; light brown; attached to the walls by a fine cord. **Spores** ellipsoid, 8–10 × 4–6μ. **Habitat** in groups on old berry canes, rotten wood, or rich soil. Found in the Pacific Northwest, north to Alaska. **Season** June–October. **Not edible.**

Crucibulum laeve (Huds.) Kamb. **Fruit body** a small cup or goblet containing a number of small flattened "eggs." **Cup** 0.5–1cm (¼–½in) high, 1cm (½in) across, tapered downward; yellow-ochre to tawny brown; outer surface velvety, inner surface pallid, smooth, and shiny; mouth at first covered with a densely hairy lid. **Eggs** 1.5mm (1/16in) across; white; attached to cup by long thin cord. **Spores** ellipsoid, smooth, 4–10 × 4–6μ. **Habitat** on decaying logs and twigs. Common. Found throughout most of North America. **Season** July–October. **Not edible.**

Cyathus stercoreus (Schw.) de Toni **Fruit body** 4–8mm (3/16–5/16in) across, 5–15mm (¼–½in) high, shaped like a goblet or narrow inverted cone; outer surface brownish orange to grayish or yellowish brown, covered with shaggy hairs that wear off in maturity leaving a smooth surface; inner surface lead gray to blackish, smooth. Goblet contains flattened dark grayish-brown "eggs" initially concealed by a thin whitish membrane across the mouth of the cup, which withers in maturity. **Eggs** usually attached by a slender, fragile cord, but this is sometimes not evident. **Spores** subglobose to

ovoid, thick-walled, 22–35 × 18–30μ. **Habitat** scattered or in dense clusters on dung, manured soil, and sawdust piles. Common. Found widely distributed throughout North America. **Season** July–October. **Not edible**.

Cyathus striatus Huds. ex Pers. **Fruit body** variable but generally 6–8mm (¼–5/16in) across, 7–10mm (¼–½in) high, shaped like an obtuse inverted cone; outer surface shiny, light to dark gray, and fluted; containing several gray "eggs" initially concealed by a whitish membrane across the mouth of the cup, which withers at maturity. **Eggs** flattened and attached by a strong, elastic cord. **Spores** ellipsoid, notched at one end, smooth, 15–22 × 8–12μ. **Habitat** scattered or in groups on dead wood or vegetable debris in open woods. Common. Found widely distributed throughout North America. **Season** July–October. **Not edible**.

Cyathus olla Batsch ex Pers. **Fruit body** 8–12mm (5/16–½in) across, 8–15mm (5/16–½in) high, in the form of a trumpet; with a velvety yellowish-gray outer surface and a smooth silver-gray inner surface; contains several "eggs," actually the spore sacs. **Eggs** up to 3mm (⅛in), disc-shaped, pallid. **Spores** ovoid, 10–14 × 6–8μ. **Habitat** on soil, twigs, and other organic debris. Found widely distributed in North America. **Season** July–October. **Not edible**.

Sphere Thrower or **Cannon Fungus**
Sphaerobolus stellatus Tode ex Pers. **Fruit body** 1–3mm (1/16–⅛in) across; initially globose and whitish, becoming more ochraceous and splitting above into 4–9 minute orange-colored rays, exposing the peridiole (the "egg" containing the spores) as a brownish ball. By

the sudden reversal of the receptacle, which then appears as a translucent white sphere sitting on the star-shaped outer wall, the egg is projected over a range of up to 14 feet to disperse the spores. **Spores** oblong, smooth, 7–10 × 3.5–5μ. **Habitat** in groups on rotting wood, sawdust, dung, and often organic debris. Frequent but easily overlooked. Found widely distributed in eastern North America west to Michigan and Texas. **Season** May–September. **Not edible**.

Elegant Stinkhorn *Mutinus elegans* (Montagne) Fisch. syn. *Mutinus curtisii* (Berk.) Fisch. **Fruit body** initially a semisubmerged "egg" 1–2cm (½–¾in) across, white to rose-pink; this ruptures and the spongy, spore-bearing stem emerges. **Stem** 100–180 × 10–20mm (4–7 × ½–¾in), tapering at apex to an acute point, with small opening at tip; bright reddish orange, fading at base; composed of uniform, undifferentiated cells visible to the naked eye (0.25–0.5mm [1/64–1/32in]), surface of cells sealed, not open like sponge; most of upper half covered in dark olive to blackish spore mass, which soon liquefies. **Odor** strong, not especially unpleasant, sickly sweet or metallic. **Spores** elliptical, smooth, 4–7 × 2–3μ. **Habitat** in leaf litter, woody debris, and rich soil. Common. Found from Quebec to Florida and west to Great Lakes. **Season** July–September. **Edible** in egg stage but not recommended. **Comment** Also illustrated is a pink-stemmed form, in which the dark spore mass is limited to the upper position and the stem itself narrows beneath that division. There has been much confusion over this stinkhorn, but I think it will turn out to be *Mutinus ravenelii* (Berk. & Curt.). The two stinkhorns illustrated need further study.

Sparassis herbstii ¾ life-size

Pseudocolus fusiformis ¾ life-size

Sparassis herbstii Pk. **Fruit body** a large mass of flattened, wavy, or lobed branches, the whole fungus resembling a cabbage or cauliflower-like mass, 15–30cm (6–12in) across, 15–20cm (6–8in) high. Individual branches are a pale creamy yellow and are variously lobed, curled, and twisted, uniting at the base into a rootlike structure. **Flesh** tough; white. **Odor** pleasant. **Taste** pleasant. **Spores** ovate, smooth, 4–7 × 3–4μ. Deposit white. **Habitat** at base of trees, often pine or oak. Found in eastern North America. **Season** July–October. **Edible** — good. **Comment** This fungus is often called *Sparassis crispa* Wulf. ex Fr., but the true *crispa* has much tighter, more densely packed, smaller lobes.

Pseudocolus fusiformis (Fisch.) Lloyd syn. *Pseudocolus schellenbergiae* Sumstine **Fruit body** a stem with usually 3 apical arms united at their tips, 5–10cm (2–4in) tall. **Stem** white to flushed pale orange at apex; arms are transversely wrinkled, bright orange, inner surface covered with a fetid, deep olive-green spore mass (gleba). Fruit body emerges from a small white "egg," which is left at base of stem as a volva. **Spores** ellipsoid, 4.5–5.5 × 2–2.5μ. **Habitat** on wood mulch or soil enriched with wood chips, in gardens and wood borders. Found in eastern North America and spreading. **Season** July–September. **Not edible. Comment** This was probably an introduction from more tropical climes but has adapted well in North America and is reported quite frequently now.

Netted Stinkhorn *Dictyophora duplicata* (Bosc) Fisch. syn. *Phallus duplicatus* Bosc **Fruit body** starting as a large white "egg" 4–6cm (1½–2in) across, then rupturing to release the spongy stem and head. **Head** bell-shaped, deeply pitted-reticulate, 4–5cm (1½–2in) deep, attached to stem at center by a white circlet surrounding the open pore at top of stem; lower margin of head is free with a prominent netlike indusium, or veil, which is 3–6cm (1¼–2¼in) deep. **Stem** 100–150 × 30–45mm (4–6 × 1¼–1¾in), hollow, of a cellular, spongelike structure; white. **Spore mass** on outer surface of head is deep olive; solid then soon liquid, and with a very fetid odor. **Spores** ellipsoid, smooth, 3.5–4.5 × 1–2μ. **Habitat** frequent in soil around deciduous trees and stumps. Found in eastern North America. **Season** June–October. **Edible** in egg stage but not recommended. **Comment** The common

Netted Stinkhorn *Dictyophora duplicata* ¼ life-size

stinkhorn of Europe, *Phallus impudicus* Pers., is sometimes reported in North America, but many or all of these records actually refer to the closely related species *Phallus hadriani* Vent. ex Pers. (syn. *Phallus imperialis* Schulzer), which differs in its pink rather than white egg, broader apical disc on head, and preference for warmer regions.

CLUB AND CORAL FUNGI pp. 291–299.
*This section includes the simple club-shaped fungi and the ground fans (*Thelephora*) as well as the more branched corals.*

Pestle-shaped Coral *Clavariadelphus pistillaris* (Linn.: Fr.) Donk syn. *Clavaria pistillaris* Fr. **Fruit body** 7–30cm (2¾–12in) high, 2–6cm (¾–2¼in) wide, forming a large club swollen at its apex; light yellow to deep ochre, with a purplish-lavender bloom over the stem; dry, smooth, with longitudinal wrinkles on the lower part of the club; never branching. **Flesh** firm then soft and spongy; white, bruising brownish. **Odor** sickly, mushroomy. **Taste** mild to bitter. **Spores** ellipsoid, smooth, (7.5) 9–13.8(16) × 4.5–7.5 Deposit white to yellowish. **Habitat** gregarious to solitary in leaf litter, especially beech. Found widespread in North America. **Season** July–October. **Edible**.

Clavariadelphus truncatus (Quél.) Donk **Fruit body** 5–15cm (2–6in) high, 3–8cm (1¼–3in) wide at the top, club-shaped, often broad and flattened at sterile top, narrowing down to a bulbous base; yellowish ochre to dark apricot orange; wrinkled. **Stem** indistinct; white-hairy at base. **Flesh** firm to spongy; whitish to ochre, darker on bruising. **Odor** none. **Taste** sweet. **Spores** ellipsoid, smooth, 9–12 × 5–8µ. Deposit pale ochre. **Habitat** scattered or in groups or clumps on the ground in coniferous woods. Widely distributed throughout North America. **Season** August–October. **Edible** —good.

Clavariadelphus ligula (Fr.) Donk **Fruit body** 3–10cm (1¼–4in) high, 5–15mm (¼–½in) wide, elongated or club-shaped, often flattened and spoonlike; yellowish to ochre to reddish or dark apricot. **Stem** indistinct; white-hairy at base. **Flesh** rather spongy; white. **Spores** narrowly ellipsoid, smooth, 8–15 × 3–6µ. Deposit pale yellowish or white. **Habitat** scattered or in groups on humus or pine needles under conifers. Widely distributed in northern North America and California. **Season** July–November. **Not edible**.

Macrotyphula juncea (Fr.) Berthier syn. *Clavaria juncea* Fr. **Fruit body** 3–10cm (1¼–4in) high, 0.5–2mm (¹⁄₃₂–³⁄₃₂in) wide, acute becoming blunt; pale brownish yellow; rather stiff and rigid, becoming flaccid in age. **Stem** 1–5cm (½–2in), distinct, slightly narrower than fertile club. **Flesh** firm, not brittle, juicy. **Odor** sour. **Taste** acrid. **Spores** almond-shaped, smooth, white, 6–12 × 3.5–5.5µ. **Habitat** leaf litter and debris in hardwoods and mixed woods. Found in New York. **Season** September. **Not edible**.

Golden Spindles *Clavulinopsis fusiformis* (Sow.: Fr.) Corner syn. *Clavaria fusiformis* Fr. **Fruit body** 5–15cm (2–6in) high, simple, thin, clubs, tips narrow and pointed, stems often flattened; bright yellow; found in dense clumps. **Flesh** firm, brittle, becoming hollow in interior; yellow. **Odor** none. **Taste** bitter. **Spores** oval, smooth, 5–9 × 4.5–9µ. Deposit white. **Habitat** in grass in woods and fields. Common. Found widely distributed throughout North America. **Season** July–October. **Edible**.

White Spindles *Clavaria vermicularis* Micheli: Fr. **Fruit body** 6–15cm (2¼–6in) high, 3–5mm (⅛–¼in) wide, simple, cylindrical,
(continued on next page)

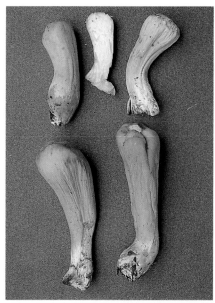

Pestle-shaped Coral *Clavariadelphus pistillaris* ⅓ life-size

Clavariadelphus truncatus ½ life-size

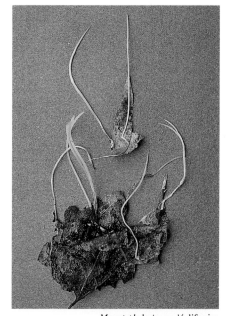

Clavariadelphus ligula ⅓ life-size

Macrotyphula juncea ½ life-size

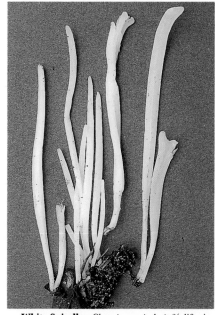

Golden Spindles *Clavulinopsis fusiformis* ⅓ life-size

White Spindles *Clavaria vermicularis* ⅔ life-size

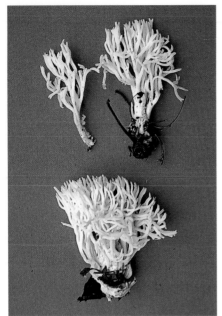

Ramariopsis kunzei ⅓ life-size

Lentaria micheneri ½ life-size

Lentaria micheneri (Berk. & Curt.) Corner **Fruit body** up to 4cm (1½in) high, with a short stem and many branches that split dichotomously and have numerous acute tips; pale orangish pink or salmon, drying dingy gray or dull yellow; fertile surface in patches on the stem. **Flesh** tough. **Taste** bitter. **Spores** long ellipsoid, smooth, white, 7–9 × 2.3–4.5μ. **Habitat** singly or in groups, arising from white mycelial patches on leaves, particularly oak and beech. Found in central and northeastern North America. **Season** July–September. **Not edible**.

Clavulina cristata (Fr.) Schroet. syn. *Clavaria cristata* Fr. **Fruit body** 2.5–8cm (1–3in) high, 2–4cm (¾–1½in) wide overall, densely branched tufts with jagged, fringed, or cristate (crested) tips; whitish, often becoming yellow-tinged or ochre, rarely pinkish white. **Stem** up to 3cm (1¼in) high, sometimes none, branching above and below; whitish or grayish. **Spores** subglobose, smooth, 7–11 × 6.5–10μ. Deposit white. Basidia 2-spored. **Habitat** singly or scattered on the ground in fields or under hardwoods and conifers. Common. Found widely distributed throughout North America. **Season** June–September. **Edible**.

Clavulina cinerea (Fr.) Schroet. syn. *Clavaria cinerea* Fr. **Fruit body** 2–10cm (¾–4in) high, numerously and compactly branched, upper branches dichotomous with blunt tips, sometimes flattened and toothed; gray or ash-colored, sometimes tinged purplish, often brownish in age. **Stem** up to 1cm (½in) thick, sometimes absent. **Flesh** firm; grayish white. **Spores** subglobose or broadly ellipsoid, smooth, 6.5–11 × 6–10μ. Deposit whitish or yellowish. **Habitat** scattered or in groups or clumps in moss on the ground or pine needles in coniferous or mixed woods. Common. Found widely distributed throughout North America. **Season** August-October. **Edible**.

Clavaria fumosa Fr. **Fruit body** 2–15cm (¾–6in) high, 2–7mm (³⁄₃₂–¼in) wide individually; numerous spindle-shaped, often twisted and compressed fruit bodies in dense tufts; cream, whitish, pale mouse gray, or very pale flesh. **Stem** indistinct or none. **Flesh** brittle; whitish. **Spores** ellipsoid, white, 5–8 × 3–4μ. **Habitat** in dense clusters on humus under hardwoods. Rare. Found in northeastern North America. **Season** July–September. **Not edible**.

Clavaria purpurea Fr. **Fruit body** 3–12cm (1¼–4¾in) high, 2–6mm (³⁄₃₂–¼in) wide individually; numerous slender cylindrical to spindle-shaped, compressed fruit bodies in a tuft; purple, lavender, amethyst, or pale brownish or smoky purple fading to pinky-buff. **Stem** indistinct, white-hairy at base. **Flesh** brittle; white or similar but paler than fruit body. **Odor** not distinctive. **Spores** ellipsoid to oblong, smooth, 5.5–9 × 3–5μ. Deposit white. **Habitat** in groups or clusters on wet soil near conifers in mountainous areas. Common in the Rocky Mountains and Pacific Northwest but rare in other parts of northern North America. **Season** July–October. **Edible**.

Clavaria zollingeri Lév. syn. *Clavaria lavendula* Pk. **Fruit body** 1.5–8cm (½–3in) high, sparingly or numerously branched, often from base, branches divided one to four times, round to somewhat flattened; deep amethyst or violet. **Stem** up to 3cm (1¼in) long, variable; paler, often grayish then yellowish with age. **Flesh** brittle. **Spores** broadly ellipsoid or subglobose, smooth, 4–7 × 3–5μ. Deposit white. Basidia mostly 4-spored. **Habitat** singly or more usually in groups or dense clusters on the

Clavulina cristata ½ life-size

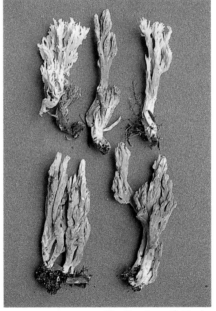

Clavulina cinerea ⅓ life-size

Clavaria fumosa ⅓ life-size

(continued from previous page) becoming flattened, grooved; white, yellowing with age. **Stem** indistinct, cluster branched only at base. **Flesh** very fragile, brittle; white. **Spores** ellipsoid, smooth, white, 5–7 × 3–4μ. Deposit white. **Habitat** in tufts or clusters in moist soil, grass, fields, and woods. Sometimes common. Found widely distributed in many parts of North America. **Season** July–September. **Edible** but not worthwhile.

Ramariopsis kunzei (Fr.) Donk **Fruit body** 2–10cm (¾–4in) high, usually numerously branched, compound; snow-white at first, tinged with pink in maturity; scurfy. **Stem** up to 1cm (½in) or none, scruffy-hairy. **Flesh** pliant to fragile; white. **Spores** broadly ellipsoid to subglobose, with minute spines, 3–5.5 × 2.5–4.5μ. Deposit white. **Habitat** in groups or clusters on humus in the woods. Sometimes common. Found widely distributed in northeastern North America and from the Pacific Northwest to southern California. **Season** July–October. **Edible**.

ground in woods or under trees in the open. Not common. Found in northeastern North America. **Season** August. **Not edible.** **Comment** A cosmopolitan and very variable fungus sometimes confused with *Clavulina amethystina* (Fr.) Donk, which has numerous branches, larger spores, and 2-spored basidia.

Clavicorona pyxidata (Pers.: Fr.) Doty **Fruit body** 5–12 cm (2–4¾in) high, 2–8cm (¾–3in) wide overall; numerously branched with cup-shaped, crownlike tips; pallid to pale yellow when young, becoming dull ochre, tan, or pinkish. **Stem** 1–3mm (¹⁄₁₆–⅛in), very short; whitish or brownish pink; smooth, densely felty. **Flesh** pliable, tough; whitish. **Odor** faintly of newly dug potatoes. **Taste** slowly rather peppery. **Spores** ellipsoid, nearly smooth, amyloid, 4–5×2–3µ. Deposit white. **Habitat** scattered, in groups, or in dense clusters on rotting logs, particularly of aspen, willow, or poplar. Common. Found widely distributed east of the Rocky Mountains. **Season** June–September. **Edible.**

Ramaria concolor (Corner) Petersen **Fruit body** up to 14cm (5½in) high, 10cm (4in) wide, several major branches dividing into numerous erect smaller branches with longish dichotomous tips; branches pale ochraceous salmon to tan, becoming darker in age, tips similar or some shade of pale tan or ochre tan. **Base** a white felty basal mat of mycelium and the tangle of stout rhizomorphic strands penetrating into the earth, from which a variable stem up to 1.5cm (½in) long branches almost immediately; stem cinnamon- or clay-colored, becoming more reddish brown or deep chocolate. **Odor** strongly of anise or aromatic. **Taste** mildly bitter to mildly acrid. **Spores** ellipsoid, ornamented with very obscure low ridges or warts, 7.8–10×3.7–4.8µ. Clamps present. **Habitat** singly or in dense clusters on rotting wood, deciduous or coniferous. Common in eastern North America, occasional from Montana westward to the Pacific. **Season** July–September. **Edibility not known.**

Ramaria conjunctipes (Coker) Corner var. *tsugensis* Marr & Stuntz **Fruit body** 4.5–18cm (1¾–7in) high, 3–7cm (1¼–2¾in) wide; branches of individual fruit bodies slender, hollow, branching, compact, and almost parallel, divided near the tips; salmon- or peach-colored with a waxy, translucent quality, light yellow tips, faint mauve areas where bruised. **Base** generally a close cluster of up to 10 steeply tapered to slightly bulbous stems; underground portion white, covered with white matted hairs; nonamyloid. **Flesh** fleshy-pliable, rubbery, drying brittle and looking like translucent plastic; same color as fruit body. **Odor** not distinctive. **Taste** not distinctive. **Spores** ovoid or ellipsoid, finely ornamented with linearly lobed warts, 6–10×4–6.5µ. Deposit golden yellow. No clamps present. **Habitat** on the ground under western hemlock. Found in the Pacific Northwest. **Season** September–October. **Edibility not known.**

Clavulinopsis corniculata (Schaeff.: Fr.) Corner syn. *Clavaria corniculata* Fr. **Fruit body** 2–8cm (¾–3in) high, generally dichotomously branched (occasionally simple) with incurved crescentlike tips; egg yellow to ochraceous yellow, with white down near the base. **Stem** up to 6cm (2¼in) long, sometimes none, gradually enlarged upward. **Flesh** tough and firm. **Odor** mealy. **Taste** mealy. **Spores** nearly globose, smooth, 4.5–7×4.5–7µ. Deposit white. **Habitat** singly, scattered, or in groups on humus in woods. Common. Found widely distributed in northern and eastern North America. **Season** June–October. **Edible.**

Clavaria purpurea ⅔ life-size

Clavaria zollingeri life-size

Clavicorona pyxidata ½ life-size

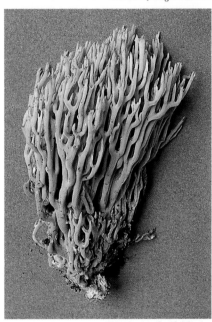

Ramaria concolor ¾ life-size

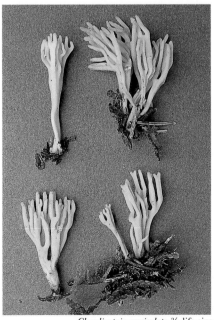

Clavulinopsis corniculata ⅔ life-size

Ramaria conjunctipes ⅓ life-size

Ramaria araiospora var. *araiospora* Marr & Stuntz
Fruit body 5–13cm (2–5in) high, 2–8cm
(¾–3in) wide; up to 6 branches coming from
the base then branching repeatedly, most quite
slender and forking at the tips; red fading to
light red with red tips, becoming yellow or
orange. **Base** 20–30×15mm (¾–1¼×½in),
single, slightly bulbous; white or yellowish
white discoloring brownish white; base covered
with thin white matted hairs; nonamyloid.
Flesh fleshy-fibrous becoming brittle; same
color as fruit body. **Odor** not distinctive. **Taste**
not distinctive. **Spores** subcylindrical,
ornamented with linearly lobed warts,
8–13×3–4.5μ. Deposit yellowish. No clamps
present. **Habitat** on the ground under western
hemlock. Found in the Pacific Northwest and
California. **Season** September–November.
Edible — good. **Comment** *Ramaria araiospora*
var. *rubella* Marr & Stuntz has branches that
are magenta-red with red or slightly paler tips
which do not turn yellow. In other respects, it
is very similar to var. *araiospora*.

Ramaria primulina Petersen **Fruit body** up to
15cm (6in) high, 8cm (3in) wide, up to 5

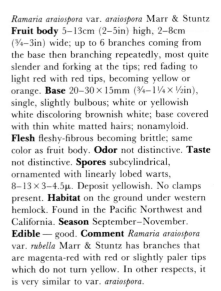

Ramaria araiospora var. *rubella* ¾ life-size

Ramaria primulina ¼ life-size

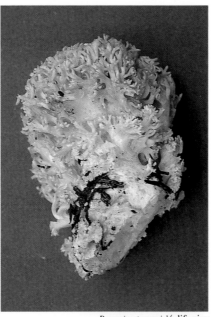

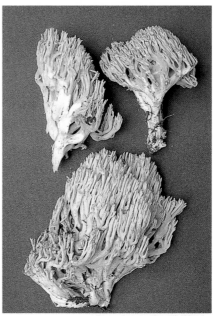

Ramaria araiospora var. *araiospora* ⅓ life-size

Ramaria strasseri ½ life-size

Ramaria acrisiccescens ⅓ life-size

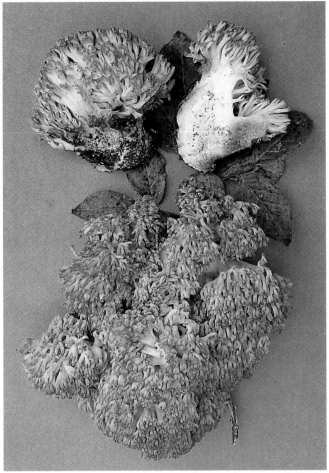

Ramaria rubrievanescens ⅓ life-size

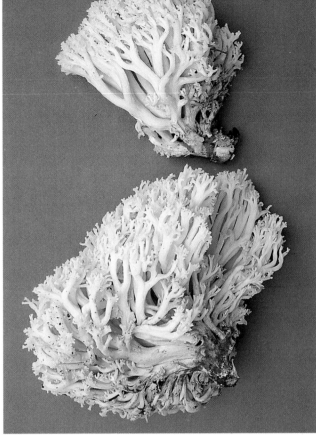

Ramaria velocimutans ¼ life-size

major ascending branches dividing into 3–6 ranks of smaller branches, terminating in stiff, thin, crowded tips divided into "fingers"; branches yellow, sometimes with a greenish tinge, tips bright, light yellow. **Base** up to 25 × 15mm (1 × ½in), single, irregularly shaped, knobby; with numerous abortive branchlets; off-white; smooth becoming slippery in age. **Flesh** solid, mottled, gelatinous, becoming loosely gelatinous in age, translucent; off-white to pale yellow. **Odor** mealy, beanlike, or slightly aromatic. **Taste** mealy, faintly bitter. **Spores** ellipsoid, ornamented with scattered isolated warts, 9–12.2 × 4–4.7μ. Clamps present. **Habitat** on the ground in coniferous or mixed forests. Found in eastern North America. **Season** August–September. **Edibility not known.**

Ramaria strasseri (Bres.) Corner **Fruit body** 8–23cm (3–9in) high, 6–23cm (2¼–9in) wide; a dense cluster of terminal branches, often incurving, growing on larger branches; light yellow or orange, slightly more pinky-brown at the tips, turning grayish orange to dark brown with age. **Base** 25–70 × 15–70mm (1–2¾ × ½–2¾in), short, thick, tapered, with 2–3 primary branches up to 4cm (1½in) in diameter; white when fresh, staining light yellow to grayish orange; strongly amyloid. **Flesh** white. **Odor** sweet. **Taste** not distinctive. **Spores** subcylindrical, striate-ornamented, 11–20 × 3.5–6μ. Deposit grayish yellow to orange. Clamps present. **Habitat** on the ground under western hemlock. Found in the Pacific Northwest. **Season** September–November. **Edibility not known.**

Ramaria acrisiccescens Marr & Stuntz **Fruit body** 5–29cm (2–11½in) high, 1.5–18cm (½–7in) wide; many slender, elongated, almost parallel, numerously branching dichotomous branches with rounded tips; pale buff-yellow or buff-orange, browner toward the base, paler toward the tips, which have a faint pinkish tinge. **Base** 15–90 × 10–30mm (½–3½ × ½–1¼in), often deeply buried, slender, tapered, single or in a close cluster; white when fresh, becoming darker; nonamyloid. **Flesh** fleshy-fibrous when fresh, becoming brittle and crumbly; brownish white. **Odor** faintly musty-sweet to beanlike. **Taste** not distinctive or slightly acid when fresh, developing a distinctly bitterish acid taste when dry. **Spores** subcylindric to ellipsoid, with a prominent lateral apiculus, ornamented with distinct lobed warts, 8–14 × 4–6μ. Deposit grayish yellow. No clamps present. **Habitat** on the ground under western hemlock. Found in the Pacific Northwest and California. **Season** September–November. **Edibility not known.**

Ramaria rubrievanescens Marr & Stuntz **Fruit body** 7–15cm (2¾–6in) high, 6.5–10cm (2½–4in) wide; numerous branches arising from a massive stem and curving inwards, branching up to five times, lower branches very short and broad, tips of branches rounded or blunt; yellowish white with tips flushed shell-pink only when young, then becoming white or creamy in age. **Base** 35–90 × 25–50mm (1¼–3½ × 1–2in), single, cylindrical or cone-shaped; milk-white discoloring yellowish, bruising brownish violet. **Flesh** firm when fresh, becoming hard;

whitish. **Odor** faintly sweet. **Taste** faintly nutlike. **Spores** mummy-shaped, ornamented with lines, 11–13 × 4–5.5μ. Deposit light yellow. **Habitat** under conifers. Occasional. Found in the Pacific Northwest. **Season** September–November. **Edibility not known.** **Comment** One of the most striking ramarias to find.

Ramaria velocimutans Marr & Stuntz **Fruit body** 7–30cm (2¾–12in) high, 3.5–26cm (1¼–10¼in) wide; a cauliflower-like mass of short branches, branching up to eight times from the stem, dividing into many "fingers" near the narrowly rounded tips; branches yellowish or brownish white or light yellow, tips similar or paler. **Base** 20–90 × 10–45mm (¾–3½ × ½–1¾in), single or occasionally compound, underground, tapering or cylindrical; white to pale yellow with extensive dark brown areas; nonamyloid. **Flesh** fleshy-fibrous, drying hard and brittle; white. **Odor** sweet, sometimes slightly unpleasant. **Taste** not distinctive. **Spores** subcylindrical, ornamented with fine lobed warts, 8–12 × 3.5–5μ. Deposit light yellow to grayish orange. Clamps present. **Habitat** on the ground under western hemlocks or western yews. Found in the Pacific Northwest and California. **Season** September–November. **Edibility not known.**

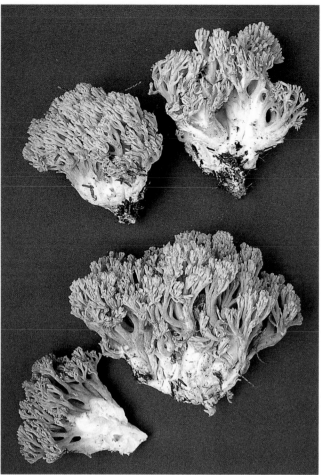

Ramaria testaceo-flava ⅓ life-size

Ramaria gelatinosa var. *oregonensis* ⅓ life-size

Ramaria testaceo-flava (Bres.) Corner **Fruit body**
up to 14cm (5½in) high, 9cm (3½in) wide;
several major branches ascending or laterally
compressed in luxuriant material, subdividing
into 3–5 ranks of smaller branches with cusped
tips that lengthen into "fingers" in age;
branches light chocolate brown bruising
darker, tips golden yellow to buff-yellow. **Base**
up to 50 × 30mm (2–1¼in), single or falsely
clustered, especially in deep moss; white
bruising chocolate brown; smooth, hoary or
covered with mycelium. **Flesh** solid, dry;
white, slowly bruising brown when cut. **Odor**
none or mildly pleasant or earthy. **Taste**
mildly or moderately bitter, sometimes
astringent. **Spores** ellipsoid, ornamented with
scattered low warts, 9.5–14 × 4.2–6.4μ. Deposit
yellow-ochre. Clamps present. **Habitat** on the
ground under hemlock or mixed conifers.
Found in western and eastern North America.
Season August–September. **Edibility not
known.**

Ramaria gelatinosa (Coker) Corner var.
oregonensis Marr & Stuntz **Fruit body** 8–15cm
(3–6in) high, 5–12cm (2–4¾in) wide;
numerous branch systems arising from the
compound basal mass and branching up to
seven times, almost parallel, sometimes
somewhat flattened, finely divided near
the narrowly rounded tips; branches light
orange, then developing darker shades such as
grayish orange or agate brown, sometimes
with a definite violet gray, tips the same or
distinctly paler. **Base** 40–70 × 30–80mm
(1½–2¾ × 1¼–3in), a broad, compound,
wrinkled gelatinous mass of fused axes; areas of
white, yellow, or light orange; base covered

with thin white matted hairs; nonamyloid.
Flesh gelatinous and translucent, drying hard
and brittle; pale grayish orange. **Odor** musty,
sweet. **Taste** not distinctive. **Spores** broadly
ovoid to cylindrical, with a prominent lateral
apiculus, coarsely ornamented with lobed
warts, 7–10 × 4.5–6μ. Deposit golden yellow.
Clamps present. **Habitat** on the ground under
western hemlock. Found in the Pacific
Northwest. **Season** September–November.
Edibility suspect.

Ramaria cystidiophora (Kauffman) Corner var.
fabiolens Marr & Stuntz **Fruit body** 9–18cm
(3½–7in) high, 7–15cm (2¾–6in) wide;
branching up to ten times from the base,
dichotomous nodes, slender diverging branches
that fork or divide into many "fingers" near
the tips; light yellow with bright, sunflower
yellow tips, although this fades somewhat in
older specimens. **Base** 30–70 × 15–40mm
(1¼–2¾ × ½–1½in), single or in a close
cluster with several stems growing from a
united basal clump; underground white or
yellowish white, covered with white, matted
hairs; nonamyloid. **Flesh** branches cartilagino-
gelatinous, base firmly gelatinous becoming
hard and brittle; light yellow. **Odor** beanlike. **Taste** slightly bitter or not distinctive.
Spores subcylindric to elongate ellipsoid,
ornamented with small irregularly shaped
warts, 8–11 × 3.5–5μ. Deposit golden yellow.
Clamps present. **Habitat** on the ground under
western red cedar, western yew, and western
hemlock. Found in the Pacific Northwest.
Season September–November. **Edibility not
known.**

Ramaria xanthosperma (Pk.) Corner **Fruit body**
up to 14cm (5½in) high, 8cm (3in) wide;
several major ascending branches dividing into
4–6 ranks of smaller dividing branches with
crowded, doubly divided tips that broaden out
to minute "fingers" in age; branches dull
creamy yellow to ochre, tips clear yellow,
purplish brown at the ends when bruised. **Base**
up to 20 × 20mm (¾ × ¾in), single stem
tapering sharply to a bluntly pointed base;
white underground, yellowish above, staining
maroon on the base and lower branches, slowly
turning brown; minutely felty underground,
smooth above. **Flesh** cartilaginous, fibrous;
white. **Odor** mildly aromatic. **Taste** musty,
bitter, or beanlike. **Spores** cylindrical, finely
and sparsely ornamented, 11.2–14.4 × 3.6–5μ.
No clamps present. **Habitat** on the ground
under hemlock or mixed conifers. Found in
northeastern North America. **Season**
August–September. **Edibility not known.**

Ramaria flavo-saponaria Petersen **Fruit body** up
to 12cm (4¾in) high, 8cm (3in) wide;
numerous short branches that become very
long in mature specimens and divide several
times to the crowded tips; light, bright yellow
branches with cadmium yellow tips, all parts
staining weakly vinescent around soil particles.
Base up to 30 × 15mm (1¼–½in), single,
knobby and irregularly shaped; white
underground, yellow-buff above. **Flesh** slippery
to soapy but not gelatinous; white, mottled.
Odor distinctly beanlike when fresh, of
fenugreek when dry. **Taste** moderately
beanlike. **Spores** narrowly ovoid to
subcylindrical, ornamented with complex,
meandering ridges forming a reticulum,

7.2–11.2 × 3.6–5.4μ. No clamps present.
Habitat on the ground under hemlock and
mixed conifers. Found in northeastern North
America. **Season** August–September. **Edibility
not known.**

Ramaria magnipes Marr & Stuntz **Fruit body**
9–25cm (3½–10in) high, 14–25cm (5½–9½in)
wide; several thick primary branches dividing
into numerous compact, cauliflower-like branch
systems that end in crowned molar-like tips;
young branches and tips butter yellow
becoming light yellow, then aging to brownish
pale orange, often changing to brick red when
bruised or exposed to frost. **Base** 7–14cm
(2¾–5½in), single, large, tapering steeply or
broadly conical in shape, rooting; off-white to
brownish; weakly amyloid. **Flesh** fleshy-fibrous
becoming hard or rather chalky-friable; white.
Odor mild or rather unpleasant. **Taste** mild
when fresh, becoming slightly bitter with
cooking. **Spores** cylindrical, no ornamentation
or very obscure warts, 10–14 × 3–4.5μ. Clamps
present. **Habitat** on the ground under vine
maple or in mixed coniferous forests. Found in
Idaho and westward to the Pacific. **Season**
May–August. **Edibility not known.**

Ramaria longispora Marr & Stuntz **Fruit body**
4–18cm (1½–7in) high, 2–9 (¾–3½in) wide;
up to 6 slender dividing branches arising from
the primary axes, sometimes hollow and
slightly divergent, finely divided near the tips;
branches light to deep orange, tips chrome
yellow when young, becoming orange in
maturity. **Base** 30 × 15mm (1¼–½in), single,
slightly bulbous or subcompound, consisting of
up to 6 axes arising from a rootlike structure;
underground section white, yellow above,
nonamyloid. **Flesh** fleshy-fibrous, becoming
brittle when dry; same color as branches or
paler. **Odor** not distinctive. **Taste** not
distinctive. **Spores** subcylindrical, ornamented
with numerous distinct warts, 10–18 × 4–6μ.
Deposit apricot yellow. No clamps present.
Habitat on the ground under western hemlock.
Found in the Pacific Northwest. **Season**
September–October. **Edibility not known.**

Ramaria formosa var. *concolor* McAfee & Grund
Fruit body 5–10cm (2–4in) high, 3.5–9cm
(1¼–3½in) wide; several main ascending
branches, dividing in several ranks of smaller
branches, which divide into "fingers" at the
tips; pale buffy salmon branches with similar
color tips that brown when bruised or in age;
branch texture often wrinkled. **Base** single or
falsely clustered stems; white browning on
handling; smooth. **Flesh** solid drying soft, not
gelatinous; white. **Odor** virtually none. **Taste**
mild when fresh, becoming acidic. **Spores**
cylindrical, 10.4–13 × 4.7–5.4μ, ornamented
with short wide ridges up to 0.2μ high. Clamps
present. **Habitat** on the ground under hemlock
and mixed conifers. Found in northeastern
North America. **Season** August–September.
Edibility not known.

Ramaria cystidiophora var. *fabiolens* ⅓ life-size

Ramaria xanthosperma ¼ life-size

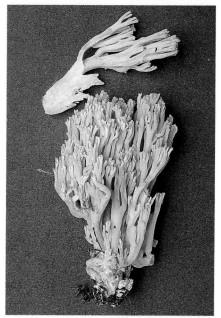
Ramaria flavo-saponaria ⅓ life-size

Ramaria magnipes ⅔ life-size

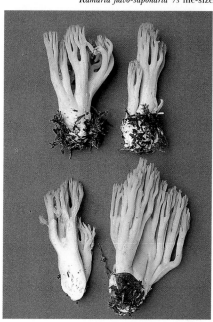
Ramaria longispora ⅓ life-size

Ramaria formosa var. *concolor* ⅓ life-size

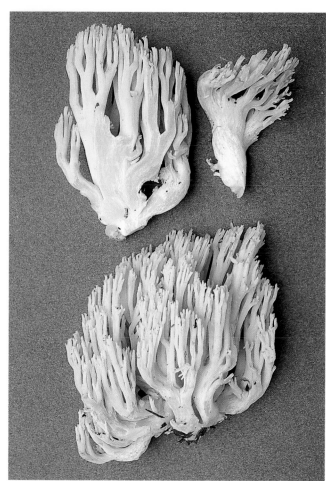

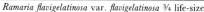

Ramaria flavigelatinosa var. *flavigelatinosa* ¾ life-size

Ramaria flavigelatinosa var. *megalospora* ⅔ life-size

Ramaria flavigelatinosa var. *flavigelatinosa* Marr & Stuntz **Fruit body** 5–14cm (2–5½in) high, 3–24cm (1¼–9½in) wide; either several or numerous connate branches growing from the base becoming free, divided, and divergent higher up and forked or divided into many "fingers" near the narrowly rounded tips; branches light yellow to maize yellow with the tips similar or slightly brighter, sometimes small areas bruising mauvish. **Base** 15–55 × 10–60mm (½–2 × ½–2¼in), a compound, conical mass of connate axes; white, becoming light yellow higher up; nonamyloid. **Flesh** firmly gelatinous drying hard; base translucent white, branches yellow. **Odor** not distinctive. **Taste** not distinctive. **Spores** subcylindric, ornamented with irregularly shaped warts, 8–11 × 3.5–4.5μ. Deposit maize to apricot yellow. No clamps present. **Habitat** on the ground under western hemlock. Found in the Pacific Northwest and California. **Season** September–November. **Edibility not known.**

Ramaria flavigelatinosa var. *megalospora* Marr & Stuntz **Fruit body** up to 10cm (4in) high, 6cm (2¼in) wide; 2 or 3 major branches growing nearly erect from the base and dividing into numerous ranks of erect smaller branches with sharply finger-shaped tips that become more rounded in age; branches yellow lower down, above pallid salmon color, becoming more orange-salmon, with bright yellow tips when young, fading to buff-yellow. **Base** 15 × 10mm (½ × ½in), expanding upward into a dense cluster of small irregularly shaped stalks; white

underground, light yellow higher up where exposed. **Flesh** solid, firmly gelatinous to soapy, drying hard and cartilaginous; white with small somewhat translucent spots. **Odor** mildly beanlike. **Taste** none or faintly beanlike. **Spores** ellipsoid, ornamented with coarse warts and some complex ridges, 9.7–12.6 × 4.3–5.4μ. Deposit clay color. No clamps present. **Habitat** coniferous woods in the Pacific Northwest and in leafy woods of eastern North America. **Season** August–September. **Edibility not known.**

Ramaria stricta (Pers.: Fr.) Quél. **Fruit body** 5–10cm (2–4in) high, 2–7cm (¾–2¾in) wide; many slender, straight, almost parallel dichotomous branches; lower section grayish orange, pale yellow above, including tips, all parts bruising purplish brown. **Base** 2–20 × 3–15mm (³⁄₃₂–¾ × ⅛–¾in) when single and distinct, sometimes indistinct and broader, branching from the substratum; base thickly matted with white hairs, velvety. **Flesh** leathery when fresh; brownish white, darkening when cut. **Odor** slightly resembling anise. **Taste** bitter. **Spores** ellipsoid, ornamented with minute shallow warts, 7–10 × 3.5–5.5μ. Deposit golden yellow. Clamps present. **Habitat** mostly on coniferous wood, particularly hemlock. Found throughout North America. **Season** July–October. **Not edible.**

Thelephora terrestris Fr. **Fruit body** 2–6cm (¾–2¼in) across, fan-shaped or vaselike, vertical to horizontal, forming large clustered

groups; reddish brown to chocolate brown, darkening to almost black in age; covered in radiating fibers and becoming tinged on the paler-colored margin. **Fertile undersurface** clay-brown to pallid; smooth to irregularly wrinkled. **Stem** none or very short. **Spores** angularly ellipsoid, sparsely spiny, 8–12 × 6–9μ. Deposit purplish to purplish brown. **Habitat** occasionally singly but more generally in groups or clusters on the ground in sandy soil or on roots, stumps, and tree seedlings. Common. Found widely distributed in northern North America as far south as central California and Georgia. **Season** July–September. **Not edible.**

Thelephora palmata Scop.: Fr. **Fruit body** 2–10cm (¾–4in) high, 1–5cm (½–2in) wide, comprising several erect, flattened, palmate branches arising from a common stem; white when young, then purple-brown. **Stem** 10–20 × 2–5mm (½–¾ × ³⁄₃₂–¼in), a short "trunk" below the branches. **Flesh** leathery. **Odor** fetid or strongly of garlic. **Spores** ellipsoid-angular, spiny, 8–11 × 7–8μ. Deposit dark reddish brown. **Habitat** singly or in groups on moist ground in woods or wood edges. Not common. Found widely distributed in North America. **Season** August–November (later in California). **Edibility not known.**

Thelephora vialis Schw. **Fruit body** 2.5–10cm (1–4in) high, 2.5–15cm (1–6in) wide, spoon- or fan-shaped folds arising from a common base to form a vaselike or cuplike mushroom;

Ramaria stricta ⅓ life-size

Thelephora terrestris ½ life-size

dirty white or yellowish to brownish gray or mauvish gray; radially lined, slightly scaly-hairy toward the base, becoming smooth. **Fertile surface** pale yellow to grayish brown; smooth becoming slightly wrinkled with tiny projections. **Stem** 5–50 × 5–50mm (¼–2 × ¼–2in); whitish to gray; slightly hairy. **Flesh** thick, leathery. **Odor** sharp to fetid when drying. **Spores** angular and lobed, slightly spiny, 4.5–8 × 4.5–6.5μ. Deposit buff. **Habitat** on the ground in hardwood forests, particularly under oak. Found in eastern North America from Vermont to South Carolina and west to Illinois. **Season** August–October. **Not edible**.

Laxitextum bicolor (Pers.: Fr.) Lentz syn. *Stereum fuscum* (Schrad.) Quél. **Fruit body** 2–5cm (¾–2in) across, or spreading in sheets up to 7–15cm (2¾–6in); shell-like, overlapping with reflexed edges, upper surface pale brown to nut brown; suedelike to wrinkled sometimes with faint lines and zones of color; lower surface white. **Fertile surface** white drying to pale buff or pale pinky-buff. **Flesh** spongy, pliant. **Spores** oblong ellipsoid, smooth, amyloid, 3.5–4.5 × 2–3μ. Deposit whitish. Gloeocystidia 80–190 × 8–10μ. **Habitat** on rotting deciduous wood, particularly alder. Found in eastern North America, west to Indiana and New Mexico, also in the Pacific Northwest. **Season** August–December, sometimes overwinters. **Not edible**.

JELLY FUNGI pp. 299–301. *The members of this group (Tremellales) have gelatinous fruit bodies and nearly all are found on twigs or logs.*

Tree Ear *Auricularia auricula* (Hooker) Underwood **Fruit body** 2–15cm (¾–6in) across, centrally or laterally cuplike or ear-shaped; outer surface tan-brown with minute grayish downy hairs, inner surface gray-brown, smooth or often wrinkled, and earlike. **Flesh** gelatinous when fresh, drying hard and horny. **Spores** sausage-shaped, smooth, 16–18 × 6–8μ. Deposit white. Basidia cylindrical with 3 transverse septa. **Habitat** severally or numerously on hardwood and coniferous trees. Common in the mountainous West. Found widely distributed throughout North America. **Season** May–November. **Edible**.

Laxitextum bicolor ½ life-size

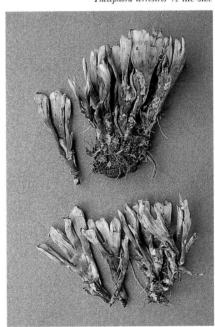

Thelephora palmata ¾ life-size

Tree Ear *Auricularia auricula* ½ life-size

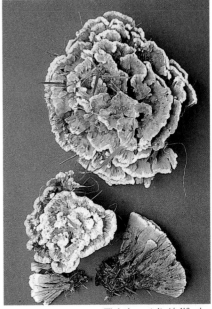

Thelephora vialis ¼ life-size

Pseudohydnum gelatinosum (Scop. ex Fr.) Karsten **Fruit body** 3–6cm (1¼–2¼in) across, tonguelike, laterally attached to wood; whole body is translucent, jellylike, upper surface smooth or dull, whitish to gray; underside with numerous small, toothlike gelatinous spines. **Spores** globose, 5–7 × 5–7μ. Deposit white. **Habitat** on fallen logs and stumps of conifers. Found throughout North America. **Season** July–November. **Edible**.

Candied Red Jelly Fungus *Phlogiotis helvelloides* (Fr.) Mar. syn. *Tremiscus helvelloides* (DC ex Pers.) Donk. **Fruit body** 2.5–10cm (1–4in) high, 4–6cm (1½–2¼in) wide, funnel-shaped or spoon-shaped with a lobed margin; translucent pinkish white to deep rose or apricot; smooth or slightly wrinkled. **Stem** short, off-center. **Flesh** firm-gelatinous. **Spores** oblong to ellipsoid, 10–12 × 4–5μ. Deposit white. **Habitat** on rotting wood or on the ground under conifers. Common in the Pacific Northwest. Found widely distributed in North America. **Season** May–July, August–October. **Edible**.

Guepiniopsis alpina (Tracy & Earle) Bres. **Fruit body** 0.2–1cm (³⁄₃₂–½in) high, 0.3–2cm (⅛–¾in) wide, cone-shaped or cup-shaped; bright orange to pale amber when moist, deep orange-red when dry. **Fertile surface** concave, more or less smooth. **Flesh** gelatinous. **Spores** sausage-shaped, smooth, becoming 4–5-celled, 15–17.5 × 5–6μ. Deposit yellowish. Basidia shaped like a tuning fork. **Habitat** on coniferous logs, stumps, and debris, also on living conifer twigs. Frequent and often abundant. Found in the Rocky Mountains, west to Washington and south to Colorado. **Season** May–June. **Not edible**.

Tremellodendron pallidum (Schw.) Burt syn. *Tremellodendron schweinitzii* (Pk.) Atkinson **Fruit body** 3–10cm (1¼–4in) high, 5–15cm (2–6in) wide, coral-like with several or many upright stems arising from a tough mycelial base; branches flattened and fused in parts; whitish to warm buff, frequently green in age from algae living in the moist tissue. **Flesh** tough, quite gelatinous. **Spores** subglobose to sausage-shaped, smooth, 7.5–10 × 4–6μ. Deposit white. **Habitat** on humus or on the ground in mixed or deciduous woods. Frequent. Found in eastern North America, south to South Carolina and west to Minnesota, also in New Mexico. **Season** July–November. **Edible** but not worthwhile.

Tremella reticulata (Berk.) Farlow **Fruit body** 3–10cm (1¼–4in) high, 6–15cm (2¼–6in) wide, a rosette of erect hollow lobes or branches, tufted when young, blunt when mature; white becoming pale dingy yellow in age. **Flesh** gelatinous-cartilaginous. **Spores** broadly ovoid, depressed on one side, 9–11 × 5–6μ. **Habitat** on the ground or on rotting wood, under hardwoods. Occasional. Found in northeastern North America, west to the Great Plains. **Season** July–October. **Edibility not known**.

Dacrymyces palmatus (Schw.) Bres. **Fruit body** 1–2.5cm (½–1in) high, 1–6cm (½–2¼in) wide, a large, lobed, brainlike mass; bright orange to deep orange-red, with a white rooting attachment at the base; horny when dry. **Flesh** tough, gelatinous, becoming softer and finally melting away. **Spores** cylindrical to sausage-shaped, smooth, becoming 8–10-celled, 17–25 × 6–8μ. Deposit yellowish. **Habitat** on coniferous wood. Found throughout North America. **Season** May–November. **Not edible**.

Pseudohydnum gelatinosum ½ life-size

Candied Red Jelly Fungus *Phlogiotis helvelloides* ½ life-size

Guepiniopsis alpina ⅓ life-size

Tremellodendron pallidum ⅓ life-size

Tremella reticulata ½ life-size

Dacrymyces palmatus ⅔ life-size

Black Morel *Morchella elata* ⅓ life-size

Yellow Morel *Morchella esculenta* ⅓ life-size

Ascomycetes

are the spore shooters. Instead of the spores forming on the surface, as in Basidiomycetes (everything previous to this page is a Basidiomycete, or spore dropper), they form in sacs (Asci) within the surface of the fungus and on maturity they are expelled.

Black Morel *Morchella elata* Fr. **Fruit body** 5–10cm (2–4in) tall, the head ovoid to conical, tending to be completely conical as it dries; the fertile head is black or dark brown with vertical ribs, the pits are rather regular. **Stem** 40–100 × 15–50mm (1½–4 × ½–2in), base often swollen; white, staining a little brownish in age; coarsely granular. **Flesh** hollow. **Odor** pleasant. **Taste** pleasant. **Spores** elliptical, smooth, 20–28 × 12–15μ. Deposit cream. **Habitat** with conifers, especially pine, and also with poplar. Found throughout North America, but especially Washington and Michigan. **Season** April–May. **Edible** — good, but only to be eaten cooked. Some people seem to be allergic to all morels. **Comment** This species is part of a complex that some mycologists split up into various species: *Morchella conica* Pers. has a more conical shape; *Morchella angusticeps* Pk. has a narrow head with fewer pits.

Yellow Morel *Morchella esculenta* Fr. and related forms. **Fruit body** 5–15cm (2–6in) tall, a rounded to elongated spongelike structure; the pits of the sponge are very irregular without sign of any vertical arrangement. Ridges are pallid at first, with pits dark smoky brown; as fruit body matures the cap flushes yellow-ochre to orange as a result of the mature spores. **Stem** 40–80 × 20–50mm (1½–3 × ¾–2in), often enlarged-swollen at base, hollow; white to pale cream. **Flesh** crisp; white. **Odor** pleasant. **Taste** pleasant. **Spores**

Morchella esculenta (under tulip-poplars) ¼ life-size

Morchella esculenta (under hickory trees) ¼ life-size

ellipsoid, smooth, 20–25 × 12–16μ. Deposit deepest ochre-orange (see Comment below). **Habitat** various, in old abandoned apple orchards, under dying or dead elms, gardens, roadsides, or chalk quarries. Found throughout North America. **Season** April–May. **Edible** — highly prized! **Comment** This is one of the most confused and uncertain species in the literature. The distinctive features of the fungus as originally described and as accepted by most authors are the *deep orange spores*, yellow-orange mature cap, and a preference for

dying or dead trees of apple and elm. Other distinctive species (unnamed) are shown here, such as the "tulip morel," which is small, with far fewer pits, and grows under tulip-poplars (*Liriodendron*). This species has very pale, cream-yellow spores and is abundant in eastern North America. Also recorded is a small, dark-capped, species with a bulbous stem; this has yellow spores that are distinctly bacillus-like in shape. It occurs under mixed hardwoods, especially under hickory (*Carya*) in eastern North America.

False Morel *Gyromitra esculenta* ⅓ life-size

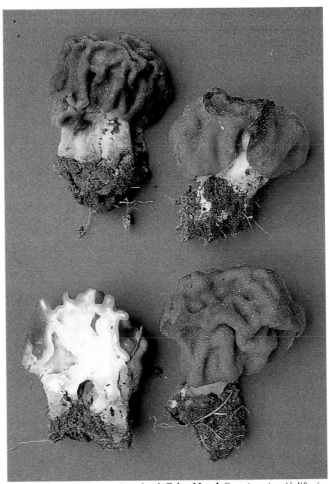

Snowbank False Morel *Gyromitra gigas* ½ life-size

White Helvella *Helvella crispa* nearly life-size

Verpa bohemica ½ life-size

False Morel *Gyromitra esculenta* (Pers.) Fr. **Cap** 3–11cm (1¼–4½in) across, brainlike, irregularly rounded, and somewhat flattened; reddish brown or darker, yellowish brown in some forms; sometimes almost smooth, but generally intricately wrinkled and folded but *not* pitted. **Stem** 20–50×15–40mm (¾–2×½–1½in), stuffed becoming hollow in chambers, equal or expanded at either end; pale flesh-colored; smooth or faintly grooved. **Flesh** thin and brittle. **Spores** ellipsoid, smooth, containing 2 or more yellowish oil droplets, 18–22×9–13µ. **Habitat** singly or in groups in coniferous and deciduous woods. Common in northern mountain forests. Found widely distributed in North America. **Season** March–May. **Deadly poisonous** when eaten raw. Prolonged cooking or drying may reduce the toxicity of this species so that a single meal of it may not be so poisonous. However, repeated doses at even the reduced toxicity may accumulate in the system to a fatal level.

Snowbank False Morel *Gyromitra gigas* (Krombh.) Quél. **Cap** 4–15cm (1½–6in) across and high, brainlike, round to ellipsoid, with an irregularly lobed margin that is sometimes fused to the stalk; yellow-brown to tan, more reddish brown in age; deeply convoluted and wrinkled, interior chambered; undersurface whitish. **Stem** 5–10cm (2–4in) long and wide; thick and short; interior multichanneled or folded in cross-section; surface whitish; ribbed, wrinkled, or grooved. **Flesh** brittle. **Spores** ellipsoid, smooth or finely warted, 24–36×10–15µ. **Habitat** singly or in groups on soil or humus under conifers around melting snowbanks. Often common. Found in mountainous forest areas from the Rocky Mountains westward to the Pacific. **Season** May–June. **Not edible** — contains poisonous hydrazines.

White Helvella *Helvella crispa* Fr. **Cap** 1.5–6cm (½–2¼in) across, saddle-shaped and deeply lobed, convoluted in the center, margin inrolled then expanded; whitish with pale buff or tan underside; hairy beneath. **Stem** 20–90×5–25mm (¾–3½×¼–1in), hollow with interior chambers, tapering toward the cap; whitish; fluted and ribbed, minutely hairy. **Asci** 300×18µ. **Spores** ellipsoid, containing 1 large central oil drop, 18–21×10–13µ. **Habitat** on the ground in deciduous or coniferous woods and open grassy spaces. Found widely distributed throughout North America. **Season** July–October (December–April in California). **Edible** but causes gastric upset in some people.

Verpa bohemica (Krombh.) Schroet. **Cap** 1–2.5cm (½–1in) high, thimble- to bell-shaped; dull yellow-brown to sepia; surface deeply wrinkled and convoluted; margin completely free from the stem, cap attached only at top. **Stem** 50–100×10–25mm (2–4×½–1in), slightly clavate; white; smooth to scurfy-granular, often ridged. **Flesh** white. **Odor** pleasant. **Taste** pleasant. **Asci** each ascus holds only 2 spores. **Spores** huge, ellipsoid, smooth, 60–80×15–18µ. Deposit yellow. **Habitat** in damp woods along stream banks, pathsides. Found throughout most of North America although rare in the East. **Season** March–early May. **Edible** with caution. Although widely eaten, it has caused adverse symptoms in some people.

Verpa conica Swartz ex Pers. **Cap** 1–2cm (⅜–¾in) high, thimblelike; dull brown; smooth to slightly wrinkled; margin free of stem, cap attached only at top. **Stem** 40–60×5–15mm (1½–2¼×¼–½in), hollow,

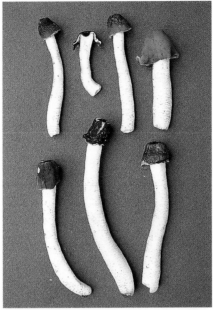

Verpa conica ⅓ life-size

Morchella semilibera ¼ life-size

equal; white to slightly brown; rough, scurfy. **Flesh** white. **Odor** pleasant. **Taste** pleasant. **Asci** with 8 spores. **Spores** ellipsoid, smooth, 22–26×12–16µ. Deposit yellow. **Habitat** in deciduous woods and orchards. Found throughout most of North America. **Season** April–May. **Edible**.

Morchella semilibera Fr. syn. *Mitrophora semilibera* (DC ex Fr.) Lév. **Cap** 1.5–4cm (½–1½in) high, bluntly to sharply conical, with conspicuous vertical ridges and pits; ridges are blackish brown while pits are yellowish brown; margin of cap is free of stem for nearly half of cap radius. **Stem** 50–100×5–40mm (2–4×¼–1½in), equal to clavate, hollow, fragile; white to dull ivory; strongly scurfy. **Taste** pleasant. **Odor** pleasant. **Spores** subglobose, 24–34×15–21µ. Deposit cream. **Habitat** in mixed hardwoods. Found throughout North America. **Season** early April–May. **Edible** — good.

Gyromitra ambigua (Karsten) Harmaja **Cap** 2–7cm (¾–2¾in) across, saddle-shaped to 3-lobed, with an incurved margin; dark reddish brown with a distinctly violet tinge, some-

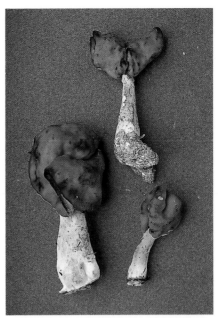

Gyromitra ambigua ½ life-size

Gyromitra infula ⅓ life-size

times almost black when dry; hymenium smooth when young, then becoming irregularly wrinkled. **Stem** 10–40×5–10mm (½–1½×¼–½in), sometimes irregular, thicker toward the base, hollow; whitish to buff with violet tints. **Odor** not distinctive. **Taste** mild. **Spores** subfusiform to broadly fusiform, with 2 large oil drops, 22–33×7.5–12µ. Deposit not known. **Habitat** singly or in groups on barren sandy soil along roads and paths near pine. In New Jersey and possibly other areas. **Season** September–October. **Edibility not known.**

Gyromitra infula (Schaeff. ex Fr.) Quél. **Cap** 3–10cm (1¼–4in) across, saddle-shaped or 3-lobed, with an incurved margin; reddish brown to dark brown; wrinkled to convoluted. **Stem** 10–60×20mm (½–2¼×¾in), hollow, sometimes irregular; whitish to buff. **Flesh** brittle. **Spores** ellipsoid, smooth, containing 2 large oil drops, 19–23×7–8µ. **Habitat** singly or scattered on humus and rotting wood or debris. Common. Found widely distributed throughout North America. **Season** July–October (November–April in the West). **Poisonous.**

Black Helvella *Helvella Lacunosa* ½ life-size

Tarzetta cupularis ⅔ life-size

Discina perlata ⅓ life-size

Black Helvella *Helvella lacunosa* Afz. ex Fr. **Cap** 1.5–5cm (½–2in) across, saddle-shaped with convoluted lobes, one lobe often pointing upward and recurved, margin attached to stem in places; gray to blackish with a paler underside; surface even or wrinkled toward the center. **Stem** 40–100×15–20mm (1½–4×½–¾in), hollow with interior chambers; pale gray; deeply furrowed with sharp ribs which are sometimes double-edged. **Asci** 350–18μ. **Spores** ellipsoid, smooth, containing 1 large central oil drop, 15–20×9–12μ. **Habitat** on the ground or on decaying wood in both coniferous and deciduous forests. Quite common. Found widely distributed in North America, but especially in the Pacific Northwest and California. **Season** August–October (January–February in the Southwest). **Edible** but not worthwhile.

Tarzetta cupularis (L. ex Fr.) Lamb. **Cup** 1–2cm (½–¾in) across, tightly cup-shaped even when mature, with an irregularly and finely toothed margin; inner surface pale gray-buff, outer surface appears lighter due to its covering of fine, pale down. **Stem** very slight or none. **Asci** 300×15μ, not blued by iodine. **Spores** broadly ellipsoid, smooth, containing 2 large oil drops, 19–21×13–15μ. **Habitat** on burned ground, damp soil, and moss in coniferous woods. Found in central and northern North America. **Season** June–September. **Not edible.**

Discina perlata (Fr.) Fr. **Cup** 4–10cm (1½–4in) across, disc-shaped, with an incurved, wavy margin; inner surface bay brown to tan, often wrinkled or convoluted, outer surface pale yellowy brown or flesh-colored, often veined. **Stem** 5–10mm (¼–½in) long, very short, stout, solid; brownish tan; furrowed. **Flesh** brittle. **Asci** 450–12μ. **Spores** fusoid, with a short extension at each end, minutely warted, containing 3 oil drops, 30–35×12–13μ. **Habitat** singly or in groups on humus or rotting wood or near melting snowbanks in coniferous areas. Found widely distributed in northeast and northwest North America and in California. **Season** April–July. **Not edible.**

Helvella villosa (Hedw. ex O. Kuntze) Dissing & Nannfeldt syn. *Cyathipodia villosa* (Hedw. ex O. Kuntze) Boud. **Cup** 1–3cm (½–1¼in) across, compressed cup-shaped when young, becoming saucer-shaped and splitting into several irregular lobes; inner surface pale brownish gray or steel gray; outer surface gray covered with soft hairs. **Stem** 10–30×3–5mm (½–1¼×⅛–¼in); brownish gray, pale yellowish at base. **Spores** ovoid, smooth, 17–21×9–12.5μ. **Habitat** on rich humus. Found in central and eastern North America. **Season** June–October. **Edibility not known.**

Helvella macropus (Pers. ex Fr.) Karsten syn. *Macroscyphus macropus* Pers. ex S. F. Gray **Cup** 2–4cm (¾–1½in) across, cup-shaped; inner surface brownish gray, outer surface appearing pale gray due to the dense covering of tufted downy hairs. **Stalk** 25–40×3–5mm (1–1½×⅛–¼in), slightly enlarged toward the base; gray, paler below; covered with dense tufts of downy hair. **Flesh** thin; white. **Asci** 350×20μ. **Spores** ellipsoid to subfusoid, containing 1 large central oil drop and 1 smaller drop at each end, 20–30×10–12μ. **Habitat** on the ground in rich soil or on decaying logs in coniferous or deciduous woods. Common. Found widely distributed in North America. **Season** June–November (December–January on the West Coast). **Not edible.**

Peziza badia Pers. ex Mérat **Cup** 3–8cm (1¼–3in) across, cup-shaped, irregularly wavy

with age; inner surface olive-brown, scurfy. **No stem.** **Flesh** thin; reddish brown, yielding watery juice. **Asci** 330×15μ, blued at the tip by iodine. **Spores** ellipsoid, containing 2 oil drops, irregular reticulations, 17–20×9–12μ. **Habitat** singly and in large numbers on the ground in woods, often on sandy soil. Uncommon. Found widely distributed in northern North America, Alabama, and California. **Season** August–November. **Edible.**

Peziza badioconfusa Korf **Cup** 3–10cm (1¼–4in) across, deeply cup-shaped, becoming irregular; inner surface dull reddish brown, outer surface similar but roughened or unpolished. **No stem.** **Flesh** brittle. **Asci** about 140×12μ. **Spores** ellipsoid, finely warted, containing 2 oil drops, 17–21×8–10μ. Spore sacs amyloid. **Habitat** singly or in groups on humus or beside old logs. Throughout North America and common in the East. **Season** May–June. **Edible.**

Peziza atrovinosa Cke. & Gerard **Cup** 2–5cm (¾–2in) across, quite deeply cup-shaped, sometimes becoming contorted in age or from pressure of overlapping cups; inner surface brown becoming almost black with an olivaceous tinge from a dusting of spores, outer surface cinnamon brown to smoky. **No stem.** **Asci** 275×14μ. **Spores** ellipsoid, smooth, containing 1 or 2 oil drops, 10×15–17μ. **Habitat** in groups or sometimes dense clusters on the ground or among moss in woods. Rather common. Found in eastern North America and Montana. **Season** June–September. **Not edible.**

Peziza succosa Berk. **Cup** 2–5cm (¾–2in) across, cup-shaped expanding to shallowly cup-shaped or even somewhat dislike, sometimes with a somewhat lobed margin; inner surface gray-brown with a slight olivaceous tint, outer surface whitish, grayish, or yellowish. **No stem.** **Flesh** thin; whitish yielding yellow juice that makes yellow stains on the edges of the cup and flesh. **Asci** 350×18μ. **Spores** ellipsoid, warty and ridged, containing 2 large oil drops, 17–22×9–12μ. **Habitat** on damp soil in woods. Quite frequent. Found in northeastern North America, west to Wisconsin. **Season** June–August. **Not edible.**

Peziza domiciliana Cke. **Fruit body** 2–10cm (¾–4in) across, at first gobletlike with a distinct stem, then soon expanding, flattened with an inrolled margin; white then soon pale yellow-ochre to brownish buff. **Flesh** fragile, thin; buff then golden yellow when broken. **Odor** not distinctive. **Taste** not distinctive. **Asci** 200–250×11–12μ. **Spores** smooth, ellipsoid, with 2 oil drops inside, 13–15×8–10μ. Paraphyses slender, septate, slightly swollen above. **Habitat** usually associated with buildings, often on damp sand, plaster, in cellars, on rotten wood in houses. Common. Found throughout North America. **Season** all year. **Not edible.**

Paxina acetabulum (L. ex St. Amans) O. Kuntze **Cup** 4–7cm (1½–2¾in) across, deeply cup-shaped; inner surface deep chestnut brown, outer surface paler, minutely downy, sometimes with prominently forked veins. **Stem** 10–30×5–25mm (½–1¼×¼–1in), more or less hollow; whitish; deeply furrowed and strongly ribbed with chambers fused into underside of cup. **Flesh** hollowed or chambered; white. **Asci** 400×20μ. **Spores** broadly ellipsoid, smooth, 18–22×12–14μ. **Habitat** singly or scattered on the ground in woods and open spaces. Common. Found widely distributed throughout North America. **Season** May–June (December–April in the Southwest). **Poisonous.**

(*text continued on page 306*)

Helvella villosa ⅔ life-size

Peziza badia ⅔ life-size

Peziza badioconfusa ⅓ life-size

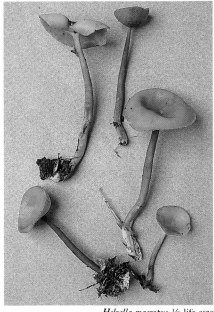

Helvella macropus ½ life-size

Peziza atrovinosa ⅓ life-size

Peziza succosa ⅓ life-size

Paxina acetabulum ½ life-size

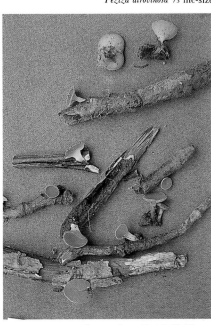

Sarcoscypha occidentalis ½ life-size

Peziza domiciliana ⅔ life-size

Caloscypha fulgens ⅔ life-size

Orange Peel Fungus *Aleuria aurantia* ¾ life-size

(*continued from page 304*)
Sarcoscypha occidentalis (Schw.) Sacc. **Cup**
0.5–1.5cm (¼–½in) across, shallow cup-
shaped; inner surface scarlet, outer surf.ce
whitish and smooth, though color on inner
surface may show through. **Stem** 10–30 × 2mm
(½–1¼ × ³⁄₃₂in), cylindrical; white. **Spores**
ellipsoid, usually containing 1 oil drop at each
end, often surrounded by smaller drops,
20–22 × 10–12μ. **Habitat** on wet sticks and
fallen branches in damp hardwood forests.
Common. Found in eastern North America.
Season May–June or later in cool years.
Edibility not known.

Caloscypha fulgens (Pers. ex Fr.) Boud. **Cup**
1–5cm (½–2in) wide, irregularly cup-shaped;
inner surface deep yellow staining blue-green
and drying orange, outer surface blue to
greenish blue. **No stem. Asci** 8-spored,
150 × 10μ. **Spores** globose, smooth, 5–7 ×
5–7μ. **Habitat** singly to clustered in wet,
boggy places in mountainous coniferous areas.
Sometimes common. Found in northern North
America and California. **Season** April–July.
Not edible.

Orange Peel Fungus *Aleuria aurantia* (Fr.)
Fuckel **Cap** 1–10cm (½–4in) across, saucer-
shaped to shallowly cup-shaped, becoming
wavy and flattened; inner surface bright
yellowish orange, outer surface whitish covered
in fine white down. **No stem. Flesh** thin,
brittle; whitish. **Asci** 220 × 13μ, not blued by
iodine. **Spores** ellipsoid, coarsely reticulate,
containing 2 oil drops, 17–24 × 9–11μ. **Habitat**
in groups or clusters on hard or disturbed soil
in gardens, in grass, or along roadsides.
Common. Found widely distributed throughout
North America. **Season** July–October. **Edible.**

Sarcosphaera crassa (Santi ex Steudl) Pouz. **Cup**
3–15cm (1¼–6in) across, starts under the soil
as smooth, hollow, and globelike, then splits
open to become deeply cup-shaped with
starlike rays; inner surface violet or grayish
lilac, outer surface white to creamy and
minutely felty; fleshy, thick-walled. **No
stem. Flesh** brittle, fragile; white. **Asci**
300–360 × 12–13μ, stained blue at tip by
iodine. **Spores** ellipsoid, with blunt ends,
smooth, containing 2 oil drops, 15–18 × 8–9μ.
Habitat singly or in clusters under coniferous
or decidous trees. Sometimes common. Found
widely distributed in northwestern North
America and also reported in the Northeast.
Season June–August. **Not edible.**

Sarcosoma mexicana (Ellis & Holway) Pad. &
Tyl. **Fruit body** 3–10cm (1¼–4in) high,
4–10cm (1½–4in) across; top-shaped becoming
disc-shaped or sometimes cup-shaped, with a
wavy margin; upper surface black or dark
brown. Outside undersurface dark gray to
black; velvety, often roughened with dark
granules; tapering down into a thick, short
base "stalk," which is generally deeply ribbed
or wrinkled or has large pockets. **Flesh**
watery-gelatinous; clear gray to black or
brownish. **Spores** ellipsoid to semisausage-
shaped, smooth, with 1 to 3 oil droplets,
23–34 × 10–14μ. **Habitat** singly, in groups, or
in clusters on rotting wood or needle duff
under conifers. Generally rare, but occasionally
common in mountain areas. Found in western
North America. **Season** February–September.
Edibility not known.

Peziza vesiculosa Bull. ex St. Amans **Cup** 6–8cm
(2¼–3in) across, deeply cup-shaped, often with
a strongly incurved margin; inner surface light

yellowish brown, outer surface yellowy buff or
fawn, scurfy. **No stem. Flesh** very brittle;
fawn; forms blisters in center of cup. **Asci** up
to 380 × 25μ. **Spores** ellipsoid, smooth,
20–24 × 11–14μ. Spore sacs amyloid. **Habitat**
in clusters on manure heaps, mushroom beds,
and richly manured soil. Common. Found
widely distributed throughout North America.
Season June–October (November–February
on West Coast). **Poisonous.**

Peziza bovina Phillips **Cup** up to 1.5cm (½in)
across, cup-shaped with an upturned margin;
inner surface cinnamon brown, outer surface a
little paler and scurfy. **No stem. Flesh** brittle.
Spores ellipsoid, smooth, 19–22 × 9–10μ.
Habitat growing on dung. Found in the Pacific
Northwest. **Season** April–June. **Not edible.**

Otidea grandis (Pers.) Rehm. **Cup** 1–2cm
(½–¾in) high, 1–4cm (½–1½in) across, cup-
shaped when young, expanding irregularly,
with lobed and wavy edges; inner surface tan
or dark apricot, outer surface orange-brown or
paler. **Stem** up to 10 × 5mm (½ × ¼in);
yellowish. **Asci** about 165 × 10μ. **Spores**
ellipsoid to subfusoid, 14–17 × 6–7μ. **Habitat**
singly or in clusters on the ground in mixed or
coniferous woods. Rare. Found in northeastern
North America. **Season** July–September.
Edibility not known.

Otidea alutacea (Pers.) Mass. syn. *Peziza alutacea*
Pers. **Cup** 2–5cm (¾–2in) high, 2–4cm
(¾–1½in) across, lopsided and irregularly
wavy with a split at the shorter side; inner
surface clay-buff, outer surface pale fawn buff
and slightly scurfy. **Stem** very slight to almost
none. **Flesh** thick; yellowish. **Asci** about
250 × 15μ, not blued by iodine. **Spores**
ellipsoid, containing 2 oil drops, 13–15 × 7–8μ.

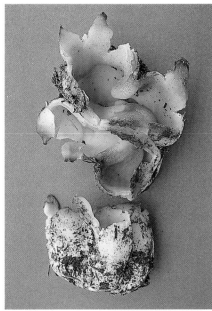

Sarcosphaera crassa ⅓ life-size

Sarcosoma mexicana ⅓ life-size

Peziza vesiculosa ⅓ life-size

Otidea grandis ⅓ life-size

Otidea alutacea ½ life-size

Peziza bovina almost life-size

Habitat densely clustered on the ground in woods. Found in the Pacific Northwest and probably elsewhere. **Season** August–October. **Not edible**.

Melastiza chateri (W. G. Smith) Boud. **Cup** 0.5–2cm (¼–¾in) across, saucer-shaped; inner surface vermilion-orange, smooth, outer surface paler, covered in minute, downy brown hairs, especially near the thick margin. **No stem**. **Asci** 300×15μ. **Spores** ellipsoid, coarsely reticulate, 17–20×10–13μ. **Habitat** scattered or in clustered groups on bare or mossy soil. Found in northern North America. **Season** May–July. **Edibility not known**.

Humaria hemisphaerica (Wigg. ex Fr.) Fuckel **Cup** 0.5–3cm (¼–1¼in) across, remaining deeply cup-shaped; inner surface whitish, outer surface light brown; densely covered in stiff, thick-walled brownish hairs. **No stem**. **Asci** 350×20μ, not blued by iodine. **Spores** broadly ellipsoid, coarsely warty, containing 2 oil drops, 20–40×10–12μ. **Habitat** scattered or in clusters on rotten wood or rich humus. Widely distributed throughout many parts of North America. **Season** July–August. **Not edible**.

Melastiza chateri ⅔ life-size

Humaria hemisphaerica life-size

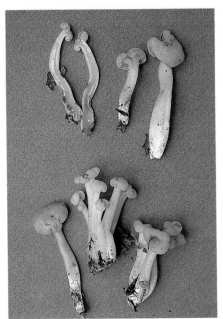

Jelly Babies *Leotia lubrica* ⅓ life-size

Leotia viscosa ½ life-size

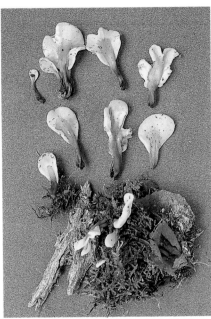

Velvety Fairy Fan *Spathularia velutipes* ½ life-size

Jelly Babies *Leotia lubrica* Fr. **Fruit body** a small stalked club with convoluted head. **Head** 1–4cm (½–1½in) across, convex and rather convoluted, margin inrolled; ochre, cinnamon to pale buff, often with olive tint; smooth, gelatinous. **Stem** 20–50×5–10mm (¾–2×¼–½in); pale ochre-yellow; minutely scaly-squamulose. **Spores** spindle-shaped, with rounded ends, often curved, 20–25×5–6µ; becoming 6–8-celled within. **Habitat** often gregarious on soil in mixed woods. Common. Found throughout North America. **Season** July–October. **Edibility not known.**

Leotia viscosa Fr. **Fruit body** 3–9cm (1¼–3½in) high, a club-shaped head with a long stalk. **Head** 2–3cm (¾–1¼in) high, 0.5–1cm (¼–½in) wide, convex with a convoluted, inrolled margin; olive-green to dark green. **Stem** 20–40×5–10mm (¾–1½×¼–½in); white, yellow, or orange with green dots or particles. **Spores** spindle-shaped with rounded ends, often slightly curved, 17–26×4–6µ. **Habitat** scattered or in groups or clusters on soil or rotten wood. Found throughout North America. **Season** July–October. **Edibility not known.**

Velvety Fairy Fan *Spathularia velutipes* Cke. & Farlow **Fruit body** 1.5–5cm (½–2in) high, forming a laterally compressed fan- to spoonlike head on a narrow stem. **Cap** pale yellow or buff, often irregular and lobed, rather soft-fleshed. **Stem** 3–6mm (⅛–¼in) long, brownish and minutely velvety. **Spores** needlelike, smooth, 33–43 ×2–3µ; divided into many cells. Deposit white. **Habitat** on decaying logs, in pine woods. Found from North Carolina to northeastern North America and west to Minnesota. **Season** August–September. **Not edible. Comment** The similar *Spathularia flavida* Pers. has a paler, whitish and nonvelvety stem; in California is found *Spathularia spathulata* (Imai) Mains, with a browner or reddish-brown head.

Scutellinia umbrorum (Fr.) Lamb. **Cup** 0.5–1cm (¼–½in) across, closed at first, becoming disc-shaped; inner surface bright red, outer surface appearing dark brown because of the covering of dark hairs that project from the margin, giving a fringelike appearance. **No stem. Asci**

up to 350×27µ. **Spores** ellipsoid, smooth at first, but becoming roughened with warts in maturity, usually containing 1 large oil drop, 12–14×23– 24µ. **Habitat** in large, dense groups on very damp soil or rotten wood. Found widely distributed in many parts of North America. **Season** July–September. **Not edible.**

Eyelash Fungus *Scutellinia scutellata* (L. ex Fr.) Lamb. **Cup** 0.5–2cm (¼–¾in) across, closed at first, then opening into a shallow disc shape; inner surface bright orange-red, outer surface pale brown covered in stiff dark brown or black hairs which look distinctly like eyelashes rimming the margin; the hairs have forked bases rooting in the flesh. **No stem. Asci** 300×25µ. **Spores** ellipsoid, minutely warty, containing several small oil droplets, 18–19× 10–12µ. **Habitat** usually growing in dense groups on damp soil or wet, rotten wood. Very common. Found widely distributed throughout North America. **Season** June–November. **Not edible.**

Trichoglossum octopartitum Mains **Fruit body** 1.5–4cm (½–1½in) high, club-shaped or compressed spoon-shaped; black; velvety. **Spores** fusoid or subfusoid, 100–120×4.5–5µ. Paraphyses cylindrical, enlarged at the apical cell. **Habitat** on soil. Not common. Found in Ohio, Tennessee, West Virginia, and New Jersey. **Season** June–August. **Not edible.**

Trichoglossum farlowii (Cke.) Durand **Fruit body** 3–8cm (1¼–3in) high, club-shaped to spear-shaped with a distinct head; black; velvety. **Spores** 0–5 septate, most commonly 3 septate (divided by 3 walls), 57–75×6–7µ. Paraphyses cylindrical, enlarged at apex like a comma. **Habitat** on moist or wet soil, in moss, or on rotten logs. Found in eastern North America. **Season** July–October. **Not edible.**

Neolecta irregularis (Pk.) Korf & Rog. **Fruit body** up to 7cm (2¾in) high, irregularly contorted and compressed club-shaped; yellow at the head with a white stem, satiny with a dusting of fine powder. **Spores** ellipsoid, smooth, 6–10×4–5µ. **Habitat** scattered or in groups on the ground or on moss among conifers. Found widely distributed throughout

North America in coniferous areas. **Season** June–October. **Not edible.**

Dasyscyphus virgineus S. F. Gray syn. *Lachnella virginea* (Batsch) Phillips **Cup** 0.1cm (¹⁄₁₆in) wide, round, closed, expanding into a shallow cup shape; cream; covered with white hairs. **Stem** 1.5–3mm (¹⁄₁₆–⅛in) long; densely covered in long white hairs. **Spores** spindle-shaped, 6–10×1.5–2.5µ. **Habitat** scattered or in large groups on dead wood, twigs, and plant stems. Found widely distributed in North America. **Season** April–June, September–October. **Not edible.**

Geoglossum difforme (Fr.) Durand **Fruit body** 3–12cm (1¼–4¾in) high, club-shaped, strongly compressed; black; smooth, sticky, particularly on the tapering stem. **Asci** up to 240×275µ. **Spores** club-shaped to cylindrical, smooth, 6–7×95–125µ. **Habitat** growing singly or in groups on soil, humus, rotting wood, or among pine needles. Found in eastern North America. **Season** July–September. **Not edible.**

Dasyscyphus bicolor Fuckel syn. *Lachnella bicolor* (Bull.) Phillips **Cup** 0.1–0.2cm (¹⁄₁₆–³⁄₃₂in) across, closed becoming expanded to cup-shaped; whitish; densely covered in white hairs. **Stem** tiny, densely covered in white hairs. **Spores** fusiform, 6–12×1.5–2µ. **Habitat** on small twigs. Found widely distributed in North America. **Season** July–September. **Not edible.**

Microglossum rufum (Schw.) Underwood **Fruit body** a spoonlike club. **Head** 1–4cm (½–1½in) high, 0.3–1.5cm (⅛–½in) wide; orange to yellow; compressed. **Stem** 10–40×1–3mm (½–1½×¹⁄₁₆–⅛in), cylindrical; orange to yellow, slightly lighter than the head. **Spores** sausage-shaped, smooth, 20–36×4–6µ. **Habitat** scattered or in clusters on sphagnum moss, decaying wood, and leaf debris. Common. Found widely distributed in North America. **Season** June–September. **Not edible.**

Scutellinia umbrorum ¾ life-size

Eyelash Fungus *Scutellinia scutellata* ¾ life-size

Trichoglossum octopartitum ½ life-size

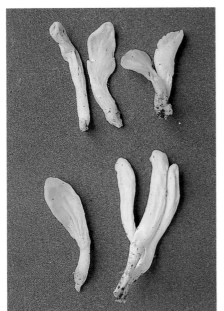

Neolecta irregularis ½ life-size

Dasyscyphus virgineus just over life-size

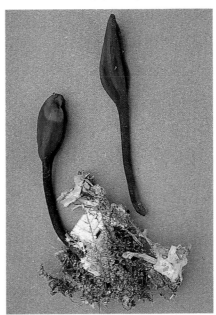

Trichoglossum farlowii ½ life-size

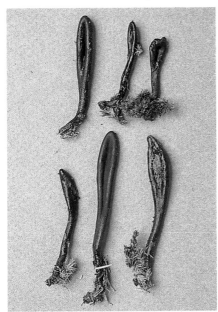

Geoglossum difforme ½ life-size

Dasyscyphus bicolor just over life-size

Microglossum rufum ⅔ life-size

Pachyella clypeata ⅓ life-size

Bulgaria rufa ¼ life-size

Ascocoryne cylichnium ¾ life-size

Pachyella clypeata (Schw.) Le Gal **Cup** 1–4cm
(½–1½in) across, disc-shaped, closely attached
to the substrate; upper surface blackish and
shining, undersurface pallid with a faint
greenish tinge. **No stem. Spores** ellipsoid,
25–27 × 12–14µ. **Habitat** on very wet decayed
wood of deciduous trees. Fairly common.
Found widely distributed in eastern North
America. **Season** August–October. **Edibility
not known.**

Bulgaria rufa Schw. **Fruit body** 2–7cm
(¾–2¾in) across, closed at first, then opening
to become shallowly cup-shaped with an
incurved margin; inner surface pale reddish or
reddish brown, with a gelatinous layer giving a
rubbery consistency; outer surface blackish
brown with clusters of hairs. **Stem** up to
10 × 5mm (½ × ¼in), attached below by dense
mass of black mycelium. **Asci** narrow, up to
275–300µ. **Spores** ellipsoid, with ends strongly
narrowed, 10 × 20µ. **Habitat** in groups or
dense clusters on buried sticks under leaf mold
or soil. Often common. Found in eastern North
America. **Season** May–June. **Not edible.**

Ascocoryne cylichnium (Tul.) Korf **Fruit body**
up to 2cm (¾in) across, lobed or irregularly
cup-shaped with a flattened or concave disc;
dark reddish brown or purplish brown; moist,
slimy, minutely scurfy. **Stem** none or very
short. **Asci** 200–220 × 10–12µ. **Spores** fusoid,
18–30 × 4–6µ, multiseptate and bud off secondary spores while still in the ascus. **Habitat**
clustered on stumps and fallen logs. Common.
Found in the Pacific Northwest and probably
other areas. **Season** August–September. **Edibility not known.**

Xylaria polymorpha (Pers. ex Mérat) Grev. **Fruit
body** 2–8cm (¾–3in) high, 1–3cm (½–1¼in)
wide, irregularly club-shaped passing into a
short cylindrical stalk below; black with a
finely wrinkled or roughened surface. **Asci**
200 × 10µ. **Flesh** tough; white. **Spores**
fusiform, 20–32 × 5–9µ. Deposit dark brown to
black. **Habitat** in groups or clusters on rotting
wood and stumps, often beech and maple.
Sometimes abundant. Found throughout North
America. **Season** June–October and persisting
all year. **Not edible.**

Xylaria hypoxylon (L. ex Hooker) Grev. **Fruit
body** 2–8cm (¾–3in) high, subcylindrical at
first, becoming flattened and branched into an

antler shape; the upper branches powdered
white, finally tipped black when mature; lower
part black and hairy. **Flesh** woody, tough;
white. **Asci** 100 × 8µ. **Spores** bean-shaped,
black, 11–14 × 5–6µ. Deposit blackish. **Habitat**
on dead wood. Common. Found throughout
North America. **Season** June–August in the
East; October in the Pacific Northwest;
December–February in California. **Not edible.**

Rutstroemia luteovirescens (Roberge) White **Cup**
up to 0.4cm (³⁄₁₆in) across, disc concave at
first, then flattened; inner surface olivaceous
yellow-buff, outer surface similar, smooth and
slightly striate. **Stem** up to 12mm (½in) long;
similar color to cap. **Asci** 150 × 12µ. **Spores**
ellipsoid, slightly pointed at each end, often
containing 2 oil drops, 12–15 × 5–7µ. **Habitat**
growing from the blackened patches on
stems of fallen sycamore leaves. Found in
Washington. **Season** September–October. **Not
edible. Comment** This would seem to be a
first record in America for this species.

Bisporella citrina (Batsch ex Fr.) Korf &
Carpenter syn. *Calycella citrina* ([Hedw.] Fr.)
Boud. **Fruit body** 0.5–3mm (¹⁄₃₂–⅛in) across,
saucer-shaped, tapered below to a tiny stem;
bright golden yellow to orange with age;
exterior surface smooth. **Asci** 135 × 10µ.
Spores ellipsoid, with 2 oil drops at each end,
9–14 × 3–5µ; often becoming 1-septate.
Habitat in large numbers on dead wood of
deciduous trees. Found throughout North
America. **Season** July–November.

Picoa carthusiana Tul. **Fruit body**
0.5–8cm (¼–3in) across, subglobose to
globose; black to sooty violet-gray; minutely
warted. Interior solid, comprising fertile tissue
streaked with paler sterile veins. Fertile tissue
is whitish to buff becoming grayish green or
greenish blue; sometimes exudes a clear milk
which slowly stains white paper light mauve.
Asci 8-spored, buried in the fertile tissue.
Spores lemon- or spindle-shaped, smooth
when mature, 74–84 × 20–35µ. **Habitat** singly,
scattered, or in small groups in soil or humus
in woods. Found in western North America.
Season August–October. **Edible — good.**

Chlorociboria aeruginascens (Nyl.) Karsten ex
Ram syn. *Chlorosplenium aeruginascens* (Nyl.)
Karsten **Fruit body** 0.1–0.5cm (¹⁄₁₆–¼in)
across, cup-shaped then flattened with a wavy

irregular margin, attached to the substrate by
a short stalk; bright blue-green throughout,
sometimes with yellowish tints. **Asci** 70 × 5µ.
Spores fusiform, containing 2 small oil drops
situated at opposite ends of the spore, 6–10 ×
1.5–2µ. **Habitat** often several growing from a
common base on rotting or barkless wood,
especially oak. Quite common. Found widely
distributed throughout North America. **Season**
June–November. **Not edible. Comment** The
mycelium growing through the wood stains it
conspicuously blue-green. The stained wood is
often seen but the fruit bodies are less frequent.

Cudoniella clavus (A. & S. ex Fr.) Dennis. **Cap**
up to 1cm (½in) across, concave at first,
becoming strongly convex with a reflexed
margin; whitish to grayish or ochraceous buff,
usually flushed with violaceous tints. **Stem**
short; white, but often brown or black at base;
smooth. **Asci** up to 115 × 10µ, cylindrical-
clavate; varying greatly in length; 8-spored.
Spores oblong-fusiform, often narrower at one
end, 10–17 × 3–5µ. **Habitat** on wet, rotting
twigs and leaves, in ditches and swamps, also
on dead stems. Found in the Pacific Northwest.
Season August–October. **Not edible.**

Ustulina deusta (Fr.) Petrak **Fruit body** 4–10cm
(1½–4in) wide, sometimes forming sheets up
to 50cm (20in) long; forms irregular wavy
cushions or thick soft crusts on the substrate;
grayish white in the early stages, finally black
and very brittle, resembling charred wood.
Asci 300 × 15µ. **Spores** fusiform, with one side
more or less flattened, 28–34 × 7–10µ. Deposit
black. **Habitat** on old dead roots or stumps of
deciduous trees, especially beech, maple, or
ash. Common. Found widely distributed in
North America. **Season** July–November,
overwinters. **Not edible.**

Taphrina amentorum (Sadeback) Rostrup **Fruit
body** consists of asci, about 30–50 × 10–20µ,
which cover both surfaces of swollen catkin
scales and malform the infected scales so that
they become curved, tonguelike outgrowths;
light green when fresh, and stained or streaked
with red. **Ascospores** often budding in the
ascis, 3–5 × 4–5µ. **Habitat** densely on female
catkins of alders. Frequent. Found in
southwestern North America. **Season**
June–September. **Not edible.**

Xylaria polymorpha ⅓ life-size

Xylaria hypoxylon ½ life-size

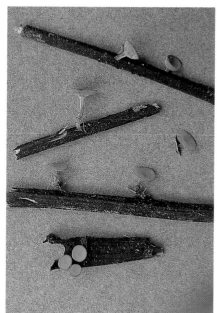

Rutstroemia luteovirescens just over life-size

Bisporella citrina ⅔ life-size

Picoa carthusiana almost life-size

Chlorociboria aeruginascens ½ life-size

Cudoniella clavus ¾ life-size

Ustulina deusta ⅓ life-size

Taphrina amentorum ½ life-size

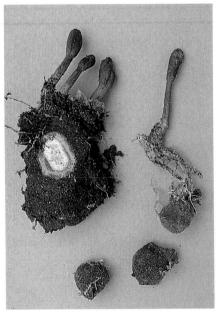

Cordyceps ophioglossoides ½ life-size

Cordyceps militaris life-size

Cordyceps capitata ⅔ life-size

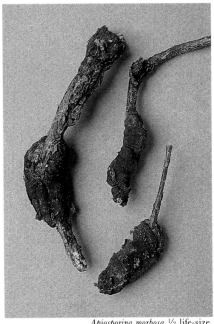

Apiosporina morbosa ½ life-size

Cordyceps ophioglossoides (Ehrhart ex Fr.) Link **Fruit body** 2–10cm (¾–4in) high, 0.3–0.8cm (⅛–⁵⁄₁₆in) thick; the fertile club-shaped to oval head is initially yellow becoming blackish; smooth then finely roughened; head tapering into a smooth yellow stem which is attached by basal threads to the underground truffle. **Spores** breaking into ellipsoid, cylindrical part-spores, 2.5–5 × 2μ. **Habitat** singly or severally, parasitically on species of underground false truffles (*Elaphomyces*). Found in eastern North America. **Season** August–October. **Not edible**.

Cordyceps militaris (L. ex St. Amans) Link **Fruit body** up to 7cm (2¾in) high, cylindrical or club-shaped; bright orange-red; the slightly swollen fertile head has a finely roughened surface and tapers into a smooth, paler, wavy stem. **Asci** very long, about 4μ wide. **Spores** threadlike, breaking into barrel-shaped part-spores, 3.5–6 × 1–1.5μ. **Habitat** singly or

numerously on larvae and pupae of butterflies and moths. Quite common. Found throughout North America. **Season** September–November. **Not edible**.

Cordyceps capitata (Holmsk. ex Fr.) Link **Fruit body** up to 10cm (4in) high with a round head; brown to olive-black; surface finely roughened; arising from a stout yellow stem, smooth or furrowed, which tapers upward. **Asci** about 15μ wide, cylindrical. **Spores** threadlike, breaking into cylindrical, smooth part-spores, 14–20 × 2–3μ. **Habitat** singly or in small groups parasitically on underground species of elaphomyces. Found widely distributed throughout North America. **Season** September–November. **Not edible**.

Apiosporina morbosa (Schw.) v. Arx **Fruit body** 3–15cm (1¼–6in) long, spindle-shaped, knotlike growths; black. **Spores** narrowly ellipsoid, smooth, 2-celled, 16–22 × 5–6.5μ. **Habitat** severally or numerously on the twigs of living cherry trees, but the twigs affected are usually dead. Common. Found in northeastern North America and possibly elsewhere. **Season** June–September. **Not edible**.

Lobster Mushroom *Hypomyces lactifluorum* (Schw. ex Fr.) Tul. This is a moldlike parasitic fungus found only on species of white russula and lactarius mushrooms, e.g., *Russula brevipes* and *Lactarius piperatus*. The fungus completely covers its host mushroom with a vivid orange to cinnabar-red coating which has minute pustules when viewed under a lens. Each bump is a flasklike vessel in which the spores are produced (perithecium). The gills of the host mushroom are almost completely aborted by the parasite. **Spores** spindle-shaped, strongly warted, clear, 35–50 × 4–5μ; equally divided into 2 cells. **Habitat** in woods on white lactarius and russula mushrooms. Found throughout North America. **Season** July–September. **Edible**. This can be a delicious fungus, greatly improving the flavor of the host mushroom, but caution should be observed since accurate identification of the host is sometimes impossible; it just might parasitize a poisonous species.

Hypomyces hyalinus (Schw. ex Fr.) Tul. **Fruit body** a whitish to pink-tinged mold that

produces reddish or amber perithecia. It forms a dense, thick covering on its amanita hosts, distorting them into thick, club-shaped forms that are unrecognizable. **Spores** spindle-shaped, strongly warted, transparent, unequally 2-celled; 13–22 × 4.5–6.5μ. **Habitat** on *Amanita rubescens, Amanita flavorubescens, Amanita frostiana,* and possibly *Amanita bisporigera.* Found in northeastern North America west to Colorado and south to North Carolina. **Season** June–October. **Not edible**.

Hypomyces chrysospermus Tul. **Fruit body** a 3-stage mold that grows on boletes. At first it is white and moldy, then it becomes yellow and powdery, and finally it becomes reddish brown and pimpled (this stage is rarely seen). **Spores** in white stage (asexual), ellipsoid, smooth, 10–30 × 5–12μ; in yellow stage (asexual), globose, warty, thick-walled, 10–25 × 10–25μ; in final stage (sexual), spindle-shaped, transparent, 25–30 × 5–6μ. **Habitat** singly or in groups on boletes; also reported on paxillus and rhizopogon mushrooms. Common. Found widely distributed in North America. **Season** June–September. **Not edible** — possibly poisonous.

Hypomyces luteovirens (Fr.) Tul. **Fruit body** a white then olive-green to dark green mold which produces dark green perithecia that project as small pimples. The mold covers the gills and stalk of lactarius and russula mushrooms. **Spores** spindle-shaped, finely warted, transparent, 1-celled, 28–35 × 4.5–5.5μ. **Habitat** in woods on various species of russula and lactarius mushrooms. Common in wet weather wherever the host species are found, throughout many parts of North America. **Season** July–November. **Not edible**.

Lobster Mushroom *Hypomyces lactifluorum* ⅓ life-size

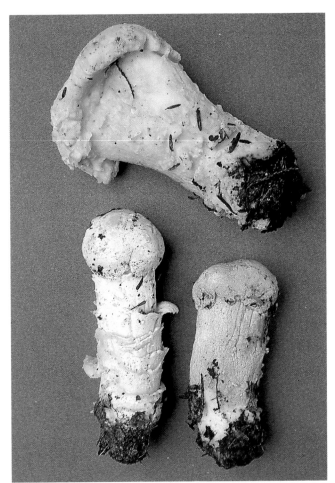

Hypomyces hyalinus ½ life-size

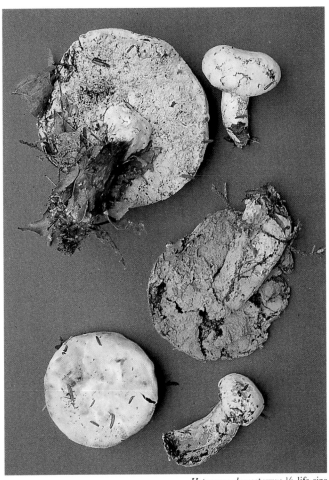

Hypomyces chrysospermus ½ life-size

Hypomyces luteovirens ½ life-size

Bibliography

These are the main works consulted in the preparation of this book. In addition I have consulted hundreds of papers in the scientific journals.

GENERAL BOOKS

Arora, David. *Mushrooms Demystified*. Berkeley, California: Ten Speed Press, 1986.

Lincoff, Gary H. *The Audubon Society Field Guide to North American Mushrooms*. New York: Alfred A. Knopf, 1981.

McKnight, Kent H., and McKnight, Vera B. *Mushrooms*. Boston: Houghton Mifflin Company, 1987.

Miller, Orson K., Jr. *Mushrooms of North America*. New York: Dutton, 1978.

Moser, Meinhard. *Keys to Agarics and Boleti*. London: Roger Phillips, 1983.

Smith, Helen V., and Smith, Alexander H. *How to Know the Non-Gilled Fleshy Funghi*. Dubuque, Iowa: Wm. C. Brown, 1973.

Smith, Helen V., Smith, Alexander H., and Weber, Nancy S. *How to Know the Gilled Mushrooms*. Dubuque, Iowa: Wm. C. Brown, 1979.

MONOGRAPHS AND MORE SPECIALIZED VOLUMES

Alessio, C. L. *Bresadola Iconographia Mycologica Inocybe*. Trento, Italy: Tridenti, 1980.

Bigelow, Howard E. "The Cantharelloid Fungi of New England and Adjacent Areas." *Mycologia*, LXX, 1978.

————. *North American Species of Clitocybe, Part I*. Vaduz, West Germany: J. Cramer, 1982.

————. *North American Species of Clitocybe, Part II*. Vaduz, West Germany: J. Cramer, 1985.

Brodie, Harold J. *The Bird's Nest Fungi*. Toronto: University of Toronto Press, 1975.

Coker, William C., and Beers, Alma H. *The Stipitate Hydnums*. Vaduz, West Germany: J. Cramer, 1970.

Coker, W. C., and Couch, J. N. *The Gasteromycetes of the Eastern United States and Canada*. Vaduz, West Germany: J. Cramer, 1969.

Corner, E. J. H. *A Monograph of Clavaria and Allied Genera*. London: Dawsons, 1967.

————. *A Monograph of Thelephora (Basidiomycetes)*. Vaduz, West Germany: J. Cramer, 1968.

————. *Supplement to "A Monograph of Clavaria and Allied Genera."* Vaduz, West Germany: J. Cramer, 1970.

Dennis, R. W. G. *British Ascomycetes*. Vaduz, West Germany: J. Cramer, 1978.

Dring, C. M. "Clathraceae." *Kew Bulletin* 35 (1), 1980.

Fries, Ella. *Icones selectae Hymenomycetum*, vols. I and II. Holmiae et Upsaliae, Sweden: P. A. Norstedt & Filii, 1877–84.

Gilbert, E. J. *Bresadola Iconographia Mycologica Amanitaceae*. Trento, Italy: Mediolani, 1940.

Gilbertson, R. L., and Ryvarden, L. *North American Polypores*, vols. I and II. Oslo, Norway: Fungiflora, 1986, 1987.

Groves, J. Walton, *Edible and Poisonous Mushrooms of Canada*. Ottawa: Research Branch Agriculture, 1979.

Grund, Darryl W., and Harrison, Kenneth A. *Nova Scotia Boletes*. W. Germany: J. Cramer, 1976.

Hesler, L. R. *Entoloma in Southeastern North America*. Vaduz, West Germany: J. Cramer, 1967.

Hesler, L. R. and Smith, Alexander H. *North American Species of Hygrophorus*. Knoxville: University of Tennessee Press, 1963.

————. *North American Species of Lactarius*. Ann Arbor: University of Michigan Press, 1979.

Hilber, Oswald. *Die Gattung Pleurotus*. Vaduz, West Germany: J. Cramer, 1982.

Jenkins, David T. *Amanita of North America*. Eureka, California: Mad River Press, 1986.

————. *A Taxonomic and Nomenclatural Study of the Genus Amanita section Amanita for North America*. Vaduz, West Germany: J. Cramer, 1977.

Kauffman, C. H. *The Agaricaceae of Michigan*. Lansing, Michigan: Johnson Reprint Corporation, 1976.

Kauffman, Murrill, et al. *North American Flora*, various vols. New York: Stechert-Hafner Service Agency Inc., 1917–1932.

Largent, David L. *The Genus Leptonia on the Pacific Coast of the United States*. Vaduz, West Germany: J. Cramer, 1977.

Lincoff, Gary, and Mitchel, D. H., (M.D.). *Toxic and Hallucinogenic Mushroom Poisoning*. New York: Van Nostrand Reinhold, 1977.

McIlvane, Charles, and Macadam, Robert K. *One Thousand American Fungi*. New York: Dover, 1973.

Marr, Currie D., and Stuntz, Daniel E. *Ramaria of Western Washington*. Vaduz, West Germany: J. Cramer, 1973.

Martin, G. W. *Revision of the North Central Tremellales*. Vaduz, West Germany: J. Cramer, 1969.

Miller, Orson K., Jr., and Farr, David F. *An Index to the Common Fungi of North America (Synonymy and Common Names)*. Vaduz, West Germany: J. Cramer, 1975.

Moser, Meinhard. *Die Gattung Phlegmaccium (Schleimkopfe)*. West Germany: J. Cramer, 1979.

Moser, Meinhard, and Julich, Walter. *Farbatlas der Basidiomyceten*. Stuttgart, West Germany: Gustav Fischer, 1987.

Pegler, David N. *The Genus Lentinus: A World Monograph*. London: Her Majesty's Stationery Office, 1983.

Petersen, Ronald H. *The Genera Gomphus and Gloeocantharellus in North America*. Vaduz, West Germany: J. Cramer, 1971.

————. *Ramaria subgenus Echinoramaria*. Vaduz, West Germany: J. Cramer, 1981.

————. *Ramaria subgenus Lentoramaria with Emphasis on North American Taxa*. Vaduz, West Germany: J. Cramer, 1975.

Phillips, Roger. *Mushrooms and Other Fungi of Great Britain and Europe*. London: Pan Books, 1981.

Pomerleau, Rene. *Flore des Champignons au Quebec*. Ottawa: Les Editions la Presse, 1980.

Rice, Miriam, and Beebee, Dorothy. *Mushrooms for Color*. Eureka, California: Mad River Press, 1980.

Romagnesi, Henri. *Les Russules d'Europe et d'Afrique du Nord*. Vaduz, West Germany: J. Cramer, 1985.

Seaver, Fred J. *The North American Cup-Fungi*, vols. I and II. New York: Lubrecht & Cramer, 1978.

Singer, Rolf. *The Boletineae of Florida*. Vaduz, West Germany: J. Cramer, 1977.

Singer, Rolf, and Harris, Bob. *Mushrooms and Truffles*. Koenigstein, West Germany: Koeltz Scientific Books, 1987.

Smith, Alexander H. *A Field Guide to Western Mushrooms*. Ann Arbor: University of Michigan Press, 1975.

————. *North American Species of Mycena*. Vaduz, West Germany: J. Cramer, 1971.

Smith, Alexander H., and Hesler, L. R. *The North American Species of Pholiota*. New York: Lubrecht & Cramer, 1968.

Smith, Alexander H., and Singer, Rolf. *A Monograph of the Genus Galerina Earle*. New York: Hafner, 1964.

Smith, Alexander H., and Thiers, Harry D. *The Boletes of Michigan*. Ann Arbor: University of Michigan Press, 1971.

————. *A Contribution Toward a Monograph of North American Species of Suillus*. Ann Arbor: University of Michigan Press, 1964.

Smith, Alexander H., and Weber, Nancy Smith, *A Field Guide to Southern Mushrooms*. Ann Arbor: University of Michigan Press, 1985.

————. *The Mushroom Hunter's Field Guide*. Ann Arbor: University of Michigan Press, 1980.

Snell, Walter H., and Dick, Esther A. *The Boleti of Northeastern North America*. Vaduz, West Germany: J. Cramer, 1970.

Thiers, Harry D. *Agaricales of California*. Eureka, California: Mad River Press, 1982.

————. *California Mushrooms: A Field Guide to the Boletes*. New York: Hafner, 1975.

Wolfe, Carl B., Jr. *Austroboletus and Tylopilus subg. Porphyrellus*. Vaduz, West Germany: J. Cramer, 1979.

Index

Each fungus is indexed to the main entry under its specific name, including any synonyms; the genera are indexed in bold to the first illustration entry. Common names are also indexed.